Problems and Solutions in Mechanical Engineering

Problems and Solutions in Mechanical Engineering

U.K. Singh
Assistant Professor
Department of Mechanical Engineering
Kamla Nehru Institute of Technology
Sultantpur

Manish Dwivedi
Assistant Professor
Department of Mechanical Engineering
College of Engineering, Science and
Technology, Gaura, Mohanlal Gunj, Lucknow

PUBLISHING FOR ONE WORLD
NEW AGE INTERNATIONAL (P) LIMITED, PUBLISHERS
New Delhi • Bangalore • Chennai • Cochin • Guwahati • Hyderabad
Jalandhar • Kolkata • Lucknow • Mumbai • Ranchi
Visit us at www.newagepublishers.com

Copyright © 2007, New Age International (P) Ltd., Publishers
Published by New Age International (P) Ltd., Publishers
First Edition: 2007
Reprint: 2008

All rights reserved.
No part of this book may be reproduced in any form, by photostat, microfilm, xerography, or any other means, or incorporated into any information retrieval system, electronic or mechanical, without the written permission of the copyright owner.

Branches:

- 36, Malikarjuna Temple Street, Opp. ICWA, Basavanagudi, **Bangalore**. © (080) 26677815
- 26, Damodaran Street, T. Nagar, **Chennai**. © (044) 24353401
- Hemsen Complex, Mohd. Shah Road, Paltan Bazar, Near Starline Hotel, **Guwahati**. © (0361) 2543669
- No. 105, 1st Floor, Madhiray Kaveri Tower, 3-2-19, Azam Jahi Road, Nimboliadda, **Hyderabad**. © (040) 24652456
- RDB Chambers (Formerly Lotus Cinema) 106A, Ist Floor, S.N. Banerjee Road, **Kolkata**. © (033) 22275247
- 18, Madan Mohan Malviya Marg, **Lucknow**. © (0522) 2209578
- 142C, Victor House, Ground Floor, N.M. Joshi Marg, Lower Parel, **Mumbai**. © (022) 24927869
- 22, Golden House, Daryaganj, **New Delhi**. © (011) 23262370, 23262368

ISBN: 978-81-224-2100-2

Rs. 250.00

C-08-05-2554

2 3 4 5 6 7 8 9 10

Printed in India at Print 'O' Pack, Delhi.
Typeset at River Edge, Delhi.

PUBLISHING FOR ONE WORLD
NEW AGE INTERNATIONAL (P) LIMITED, PUBLISHERS
4835/24, Ansari Road, Daryaganj, New Delhi-110002
Visit us at www.newagepublishers.com

Preface

Mechanical Engineering being core subject of engineering and Technology, is taught to almost all branches of engineering, throughout the world. The subject covers various topics as evident from the course content, needs a compact and lucid book covering all the topics in one volume. Keeping this in view the authors have written this book, **basically covering the cent percent syllabi of Mechanical Engineering (TME-102/TME-202) of U.P. Technical University, Lucknow (U.P.),** India.

From 2004–05 Session UPTU introduced the New Syllabus of Mechanical Engineering which covers Thermodynamics, Engineering Mechanics and Strength of Material. Weightage of thermodynamics is 40%, Engineering Mechanics 40% and Strength of Material 20%. Many topics of Thermodynamics and Strength of Material are deleted from the subject which were included in old syllabus *but books available in the market give these useless topics, which may confuse the students. Other books cover 100% syllabus of this subject but not covers many important topics which are important from examination point of view.* Keeping in mind this view this book covers 100% syllabus as well as 100% topics of respective chapters.

The examination contains both theoretical and numerical problems. So in this book the reader gets matter in the form of questions and answers with concept of the chapter as well as concept for numerical solution in stepwise so they don't refer any book for Concept and Theory.

This book is written in an objective and lucid manner, focusing to the prescribed syllabi. This book will definitely help the students and practicising engineers to have the thorough understanding of the subject.

In the present book most of the problems cover the Tutorial Question bank as well as Examination Questions of U.P. Technical University, AMIE, and other Universities have been included. Therefore, it is believed that, it will serve nicely, our nervous students with end semester examination. Critical suggestions and modifications by the students and professors will be appreciated and accorded

Dr. U.K. Singh
Manish Dwivedi

Feature of book
1. Cover 100% syllabus of TME 101/201.
2. Cover all the examination theory problems as well as numerical problems of thermodynamics, mechanics and strength of materials.
3. Theory in the form of questions – Answers.
4. Included problems from Question bank provided by UPTU.
5. Provided chapter-wise Tutorials sheets.
6. Included Mechanical Engineering Lab manual.
7. No need of any other book for concept point of view.

Syllabus TME-101/201-Mechanical Engineering

A. Thermodynamics (40 MARKS)

UNIT I. Fundamental Cocnepts and Definitions
Definition of thermodynamics, system, surrounding and universe, phase, concept of continuum, macroscopic and microscopic point of view, Density, specific volume, pressure, temperature. Thermodynamic equilibrium, property, state, path process, cyclic process, Energy and its form, work and heat, Enthalpy.

Law of Thermodynamics
ZEROTH LAW. Concepts of Temperature, zeroth law.
FIRST LAW. First law of thermodynamics. Concept of processes, flow processes and control volume. Flow work, steady flow energy equation, Mechanical work in a steady flow of process.
SECOND LAW. Essence of second law, Thermal reservoir, Heat engines. COP of heat pump and refrigerator. Statements of second law. Carnot cycle, Clausius inequality, Concept of Entropy.

UNIT II. Properties of Steam and Thermodynamics Cycles
Properties of steam, use of property diagram. Steam-Tables, processes involving steam in closed and open systems, Rankine cycle.
Introduction of I.C. Engines-two and four stroke S.I. and C.I. engines. Otto cycle, Diesel cycle.

B. Engineering Mechanics (40 MARKS)

UNIT III. Force System and Analysis
Basic Concept. Laws of motion. Transfer of force to parallel position. Resultant of planer force system. Free Body Diagrams, Equilibrium and its equation.
Friction. Introduction, Laws of Coulomb friction, Equilibrium of bodies involving dry friction-Belt Friction.

UNIT IV. Structure Analysis
Beams. Introduction, Shear force and Bending Moment, Shear force and Bending Moment Diagram for statically determinate beams.
Trusses. Introduction, Simple Trusses, Determination of Forces in simple trusses members, methods of joints and method of section.

C. Strength of Materials (20 MARKS)

UNIT V. STRESS AND STRAIN ANALYSIS
Simple Stress and Strain. Introduction, Normal shear stresses, stress-strain diagrams for ductile and brittle materials, Elastic constants, one dimensional loading of members of varying cross sections, strain energy.

Compound Stress and Strains. Introduction, state of plane stress, Principal stress and strain, Mohr's stress circle.
Pure Bending of Beams. Introduction, Simple Bending theory, Stress in Beams of different cross-sections.
Torsion. Introduction, Torsion of Shafts of circular section, Torque and Twist, Shear stress due to Torque.

IMPORTANT CONVERSION/FORMULA

1. Sine Rule

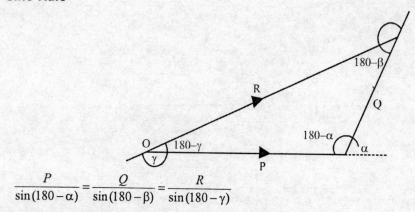

$$\frac{P}{\sin(180-\alpha)} = \frac{Q}{\sin(180-\beta)} = \frac{R}{\sin(180-\gamma)}$$

2. Important Conversion

$1\ N\ \ \ = 1\ kg \times 1\ m/sec^2$
$\ \ \ \ \ \ \ \ = 1000\ gm \times 100\ cm/sec^2$
$g\ \ \ \ \ \ = 9.81\ m/sec^2$
$1\ H.P.\ = 735.5\ KW$
$1\ Pascal(Pa) = 1N/m^2$
$1\ KPa\ = 10^3\ N/m^2$
$1\ MPa\ = 10^6\ N/m^2$
$1\ GPa\ = 10^9\ N/m^2$
$1\ bar\ \ = 10^5\ N/m^2$

3. Important Trigonometrical Formulas

1. $\sin(A+B) = \sin A.\cos B + \cos A.\sin B$
2. $\sin(A-B) = \sin A.\cos B - \cos A.\sin B$
3. $\cos(A+B) = \cos A.\cos B - \sin A.\sin B$
4. $\cos(A-B) = \cos A.\cos B + \sin A.\sin B$
5. $\tan(A+B) = (\tan A + \tan B)/(1 - \tan A.\tan B)$
6. $\tan(A-B) = (\tan A - \tan B)/(1 + \tan A.\tan B)$
7. $\sin 2A = 2\sin A.\cos A$
8. $\sin^2 A + \cos^2 A = 1$
9. $1 + \tan^2 A = \sec^2 A$
10. $1 + \cot^2 A = \csc^2 A$
11. $1 + \cos A = 2\cos^2 A/2$

12. $1 - \cos A = 2\sin^2 A/2$
13. $2\cos A.\sin B = \sin(A + B) - \sin(A - B)$
14. $\sin(-A) = -\sin A$
15. $\cos(-A) = \cos A$
16. $\tan(-A) = -\tan A$
17. $\sin(90° - A) = \cos A$
18. $\cos(90° - A) = \sin A$
19. $\tan(90° - A) = \cot A$
20. $\sin(90° + A) = \cos A$
21. $\cos(90° + A) = -\sin A$
22. $\tan(90° + A) = -\cot A$
23. $\sin(180° - A) = \sin A$
24. $\cos(180° - A) = -\cos A$
25. $\tan(180° - A) = -\tan A$
26. $\sin(180° + A) = -\sin A$
27. $\cos(180° + A) = -\cos A$
28. $\tan(180° + A) = \tan A$

4. Important Assumptions

In the part of mechanics we take
1. Upwards force as positive.
2. Downwards force as negative.
3. Towards Right hand force as positive.
4. Towards left hand force as negative
5. Clockwise moment as positive.
6. Anticlockwise moment as negative.

CONTENTS

Preface   *v*
Syllabus   *vi*
Important Conversion/Formula   *viii*

Part– A: Thermodynamics

1. Fundamental concepts, definitions and zeroth law   1
2. First law of thermodynamics   30
3. Second law of thermodynamics   50
4. Introduction of I.C. engines   65
5. Properties of steam and thermodynamics cycle   81

Part – B: Engineering Mechanics

6. Force : Concurrent Force system   104
7. Force : Non Concurrent force system   141
8. Force : Support Reaction   166
9. Friction   190
10. Application of Friction: Belt Friction   216
11. Law of Motion   242
12. Beam   265
13. Trusses   302

Part – C: Strength of Materials

14. Simple stress and strain   331
15. Compound stress and strains   393
16. Pure bending of beams   409
17. Torsion   432

Appendix

1. Tutorials Sheets   448
2. Lab Manual   474
3. Previous year question papers (New syllabus)   503

CONTENTS

Preface
Syllabus
Important (Cover seed quanta)

Part – A: Thermodynamics

1. Fundamental concepts, definitions and zeroth law 1
2. First law of thermodynamics 30
3. Second law of thermodynamics 50
4. Introduction of I.C. engine 70
5. Properties of steam and thermodynamics cycle 81

Part – B: Engineering Mechanics

6. Force: Concurrent Force system 101
7. Force: Non-Concurrent force system 141
8. Engg. Support Reaction 166
9. Friction 186
10. Application of Friction: Belt Friction 216
11. Law of Motion 243
12. Beam 266
13. Trusses 292

Part – C: Strength of Materials

14. Simple stress and strain 321
15. Compound stress and strain 362
16. Pure bending of beams 399
17. Torsion 432

Appendix

1. Formula Sheets 449
2. Lab Manual 474
3. Previous year question papers (New syllabus) 495–503

CHAPTER 1

FUNDAMENTAL CONCEPTS, DEFINITIONS AND ZEROTH LAW

Q. 1: Define thermodynamics. Justify that it is the science to compute energy, exergy and entropy.
(Dec–01, March, 2002, Jan–03)

Sol : Thermodynamics is the science that deals with the conversion of heat into mechanical energy. It is based upon observations of common experience, which have been formulated into thermodynamic laws. These laws govern the principles of energy conversion. The applications of the thermodynamic laws and principles are found in all fields of energy technology, notably in steam and nuclear power plants, internal combustion engines, gas turbines, air conditioning, refrigeration, gas dynamics, jet propulsion, compressors, chemical process plants, and direct energy conversion devices.

Generally thermodynamics contains four laws;

1. Zeroth law: deals with thermal equilibrium and establishes a concept of temperature.

2. The First law: throws light on concept of internal energy.

3. The Second law: indicates the limit of converting heat into work and introduces the principle of increase of entropy.

4. Third law: defines the absolute zero of entropy.

These laws are based on experimental observations and have no mathematical proof. Like all physical laws, these laws are based on logical reasoning.

Thermodynamics is the study of energy, energy and entropy.

The whole of heat energy cannot be converted into mechanical energy by a machine. Some portion of heat at low temperature has to be rejected to the environment.

The portion of heat energy, which is not available for conversion into work, is measured by entropy.

The part of heat, which is available for conversion into work, is called energy.

Thus, thermodynamics is the science, which computes energy, energy and entropy.

Q. 2: State the scope of thermodynamics in thermal engineering.

Sol: Thermal engineering is a very important associate branch of mechanical, chemical, metallurgical, aerospace, marine, automobile, environmental, textile engineering, energy technology, process engineering of pharmaceutical, refinery, fertilizer, organic and inorganic chemical plants. Wherever there is combustion, heating or cooling, exchange of heat for carrying out chemical reactions, conversion of heat into work for producing mechanical or electrical power; propulsion of rockets, railway engines, ships, etc., application of thermal engineering is required. Thermodynamics is the basic science of thermal engineering.

Q. 3: Discuss the applications of thermodynamics in the field of energy technology.

Sol: Thermodynamics has very wide applications as basis of thermal engineering. Almost all process and engineering industries, agriculture, transport, commercial and domestic activities use thermal engineering. But energy technology and power sector are fully dependent on the laws of thermodynamics.

For example:
  (i) Central thermal power plants, captive power plants based on coal.
  (ii) Nuclear power plants.
  (iii) Gas turbine power plants.
  (iv) Engines for automobiles, ships, airways, spacecrafts.
  (v) Direct energy conversion devices: Fuel cells, thermoionic, thermoelectric engines.
  (vi) Air conditioning, heating, cooling, ventilation plants.
  (vii) Domestic, commercial and industrial lighting.
  (viii) Agricultural, transport and industrial machines.

All the above engines and power consuming plants are designed using laws of thermodynamics.

Q. 4: Explain thermodynamic system, surrounding and universe. Differentiate among open system, closed system and an isolated system. Give two suitable examples of each system. *(Dec. 03)*

Or

Define and explain a thermodynamic system. Differentiate between various types of thermodynamic systems and give examples of each of them. *(Feb. 2001)*

Or

Define Thermodynamics system, surrounding and universe. *(May–03)*

Or

Define closed, open and isolated system, give one example of each. *(Dec–04)*

Sol: In thermodynamics the system is defined as the quantity of matter or region in space upon which the attention is concentrated for the sake of analysis. These systems are also referred to as **thermodynamics system.**

It is bounded by an arbitrary surface called **boundary.** The boundary may be real or imaginary, may be at rest or in motion and may change its size or shape.

Everything out side the arbitrary selected boundaries of the system is called **surrounding or environment.**

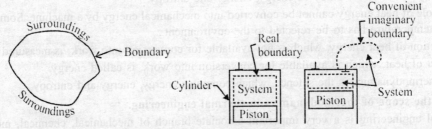

Fig. 1.1 *The system*   **Fig. 1.2** *The real and imaginary boundaries*

The union of the system and surrounding is termed as universe.
Universe = System + Surrounding

Types of system

The analysis of thermodynamic processes includes the study of the transfer of mass and energy across the boundaries of the system. On the basis the system may be classified mainly into three parts.

(1) Open system (2) Closed System (3) Isolated system

(1) Open system

The system which can exchange both the mass and energy (Heat and work) with its surrounding. The mass within the system may not be constant. The nature of the processes occurring in such system is flow type. For example

1. Water Pump: Water enters at low level and pumped to a higher level, pump being driven by an electric motor. The mass (water) and energy (electricity) cross the boundary of the system (pump and motor).

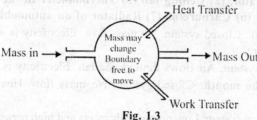

Fig. 1.3

2. Scooter engine: Air arid petrol enter and burnt gases leave the engine. The engine delivers mechanical energy to the wheels.

3. Boilers, turbines, heat exchangers. Fluid flow through them and heat or work is taken out or supplied to them.

Most of the engineering machines and equipment are open systems.

(2) Closed System

The system, which can exchange energy with their surrounding but not the mass. The quantity of matter thus remains fixed. And the system is described as control mass system.

The physical nature and chemical composition of the mass of the system may change.

Water may evaporate into steam or steam may condense into water. A chemical reaction may occur between two or more components of the closed system.

For example

1. Car battery, Electric supply takes place from and to the battery but there is no material transfer.

2. Tea kettle, Heat is supplied to the kettle but mass of water remains constant.

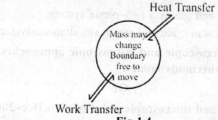

Fig 1.4

3. Water in a tank
4. Piston – cylinder assembly.

(3) Isolated System

In an Isolated system, neither energy nor masses are allowed to cross the boundary. The system has fixed mass and energy. No such system physically exists. Universe is the only example, which is perfectly isolated system.

Other Special System

1. Adiabatic System: A system with adiabatic walls can only exchange work and not heat with the surrounding. All adiabatic systems are thermally insulated from their surroundings.

Example is Thermos flask containing a liquid.

2. Homogeneous System: A system, which consists of a single phase, is termed as homogeneous system. For example, Mixture of air and water vapour, water plus nitric acid and octane plus heptanes.

3. Hetrogeneous System: A system, which consists of two or more phase, is termed as heterogeneous system. For example, Water plus steam, Ice plus water and water plus oil.

Q. 5: Classified each of the following systems into an open or closed systems.

(1) Kitchen refrigerator, (2) Ceiling fan (3) Thermometer in the mouth (4) Air compressor (5) Pressure Cooker (6) Carburetor (7) Radiator of an automobile.

(1) Kitchen refrigerator: Closed system. No mass flow. Electricity is supplied to compressor motor and heat is lost to atmosphere.

(2) Ceiling fan: Open system. Air flows through the fan. Electricity is supplied to the fan.

(3) Thermometer in the mouth: Closed system. No mass flow. Heat is supplied from mouth to thermometer bulb.

(4) Air compressor: Open system. Low pressure air enters and high pressure air leaves the compressor, electrical energy is supplied to drive the compressor motor.

(5) Pressure Cooker: Closed system. There is no mass exchange (neglecting small steam leakage). Heat is supplied to the cooker.

(6) Carburetor: Open system. Petrol and air enter and mixture of petrol and air leaves the carburetor. There is no change of energy.

(7) Radiator of an automobile: Open system. Hot water enters and cooled water leaves the radiator. Heat energy is extracted by air flowing over the outer surface of radiator tubes.

Q. 6: Define Phase.

Sol: A phase is a quantity of matter, which is homogeneous throughout in chemical composition and physical structure.

If the matter is all gas, all liquid or all solid, it has physical uniformity. Similarly, if chemical composition does not vary from one part of the system to another, it has chemical uniformity.

Examples of one phase system are a single gas, a single liquid, a mixture of gases or a solution of liquid contained in a vessel.

A system consisting of liquid and gas is a two–phase system.

Water at triple point exists as water, ice and steam simultaneously forms a three–phase system.

Q. 7: Differentiate between macroscopic and microscopic approaches. Which approach is used in the study of engineering thermodynamics. *(Sept. 01; Dec., 03, 04)*

Or

Explain the macroscopic and microscopic point of view. Dec–2002

Sol: Thermodynamic studies are undertaken by the following two different approaches.

1. Macroscopic approach–(Macro mean big or total)
2. Microscopic approach–(Micro means small)

The state or condition of the system can be completely described by measured values of pressure, temperature and volume which are called macroscopic or time–averaged variables. In the classical thermodynamics, macroscopic approach is followed. The results obtained are of sufficient accuracy and validity.

Statistical thermodynamics adopts microscopic approach. It is based on kinetic theory. The matter consists of a large number of molecules, which move, randomly in chaotic fashion. At a particular moment, each molecule has a definite position, velocity and energy. The characteristics change very frequently due to collision between molecules. The overall behaviour of the matter is predicted by statistically averaging the behaviour of individual molecules.

Microscopic view helps to gain deeper understanding of the laws of thermodynamics. However, it is rather complex, cumbersome and time consuming. Engineering thermodynamic analysis is macroscopic and most of the analysis is made by it.

These approaches are discussed (in a comparative way) below:

Macroscopic approach	*Microscopic approach*
1. In this approach a certain quantity of matter is considered without taking into account the events occurring at molecular level. In other words this approach to thermodynamics is concerned with gross or overall behaviour. This is known as classical thermodynamics.	1. The approach considers that the system is made up of a very large number of discrete particles known as molecules. These molecules have different velocities and energies. The values of these energies are constantly changing with time. This approach to thermodynamics, which is concerned directly with the structure of the matter, is known as statistical thermodynamics.
2. The analysis of macroscopic system requires simple mathematical formulae.	2. The behaviour of the system is found by using statistical methods, as the number of molecules is very large. so advanced statistical and mathematical methods are needed to explain the changes in the system.
3. The values of the properties of the system are their average values. For example, consider a sample of a gas in a closed container. The pressure of the **gas is the average value** of the pressure exerted by millions of individual molecules. Similarly the temperature of this gas is the average value of transnational kinetic energies of millions of individual molecules. these properties like pressure and temperature can be measured very easily. The changes in properties can be felt by our senses.	3. The properties like velocity, momentum, impulse, kinetic energy, and instruments cannot easily measure force of impact etc. that describe the molecule. Our senses cannot feel them.
4. In order to describe a system only a few properties are needed.	4. Large numbers of variables are needed to describe a system. So the approach is complicated.

Q. 8: Explain the concept of continuum and its relevance in thermodynamics. Define density and pressure using this concept. *(June, 01, March– 02, Jan–03)*

Or

Discuss the concept of continuum and its relevance. *(Dec–01)*

Or

Discuss the concept of continuum and its relevance in engineering thermodynamics. *(May–02)*

Or

What is the importance of the concept of continuum in engineering thermodynamics. *(May–03)*

Sol: Even the simplification of matter into molecules, atoms, electrons, and so on, is too complex a picture for many problems of thermodynamics. Thermodynamics makes no hypotheses about the structure of the matter of the system. The volumes of the system considered are very large compared to molecular dimensions. The system is regarded as a continuum. The system is assumed to contain continuous distribution of matter. There are no voids and cavities. The pressure, temperature, density and other properties are the average values of action of many molecules and atoms. Such idealization is a must for solving most problems. The laws and concepts of thermodynamics are independent of structure of matter.

According to this concept there is minimum limit of volume upto which the property remain continuum. Below this volume, there is sudden change in the value of the property. Such a region is called region of discrete particles and the region for which the property are maintain is called region of continuum. The limiting volume up to which continuum properties are maintained is called continuum limit.

For Example: If we measure the density of a substance for a large volume (υ_1), the value of density is (ρ_1). If we go on reducing the volume by $\delta v'$, below which the ratio $äm/äv$ deviates from its actual value and the value of $äm/äv$ is either large or small.

Thus according to this concept the design could be defined as

$$\rho = \lim \delta v{-} \delta v' \, [\delta m / \delta v]$$

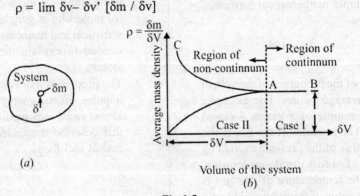

Fig 1.5

Q. 9: Define different types of properties?

Sol: For defining any system certain parameters are needed. Properties are those observable characteristics of the system, which can be used for defining it. For example pressure, temp, volume.

Properties further divided into three parts;

Intensive Properties
Intensive properties are those, which have same value for any part of the system or the properties that are independent of the mass of the system. EX; pressure, temp.

Extensive Properties
EXtensive properties are those, which dependent Upon the mass of the system and do not maintain the same value for any part of the system. EX; mass, volume, energy, entropy.

Specific Properties
The extensive properties when estimated on the unit mass basis result in intensive property, which is also known as specific property. EX; sp. Heat, sp. Volume, sp. Enthalpy.

Q. 10: Define density and specific volume.
Sol: DENSITY (ρ)
Density is defined as mass per unit volume;
Density = mass/ volume; ρ = m/v, kg/m^3
P for Hg = 13.6×10^3 kg/m^3
ρ for water = 1000 kg/m^3

Specific Volume (v)
It is defined as volume occupied by the unit mass of the system. Its unit is m^3/kg. Specific volume is reciprocal of density.
v = v/m; m^3/kg

Q. 11: Differentiate amongst gauge pressure, atmospheric pressure and absolute pressure. Also give the value of atmospheric pressure in bar and mm of Hg. *(Dec–02)*

Sol: While working in a system, the thermodynamic medium exerts a force on boundaries of the vessel in which it is contained. The vessel may be a container, or an engine cylinder with a piston etc. The exerted force F per unit area A on a surface, which is normal to the force, is called intensity of pressure or simply pressure p. Thus
$$P = F/A = \rho.g.h$$
It is expressed in Pascal (1 Pa = 1 Nm^2),
bar (1 bar = 10^5 Pa),
standard atmosphere (1 atm = 1.0132 bar),
or technical atmosphere (1 kg/cm^2 or 1 atm).
1 atm means 1 atmospheric absolute.
The pressure is generally represented in following terms.
1. Atmospheric pressure
2. Gauge pressure
3. Vacuum (or vacuum pressure)
4. Absolute pressure

Atmospheric Pressure (P_{atm})
It is the pressure exerted by atmospheric air on any surface. It is measured by a barometer. Its standard values are;
$$1\ P_{atm} = 760 \text{ mm of Hg i.e. column or height of mercury}$$
$$= \rho.g.h. = 13.6 \times 10^3 \times 9.81 \times 760/1000$$

= 101.325 kN/m² = 101.325 kPa
= 1.01325 bar

when the density of mercury is taken as 13.595 kg/m³ and acceleration due to gravity as 9.8066 m/s²

Gauge Pressure (P_{gauge})

It is the pressure of a fluid contained in a closed vessel. It is always more than atmospheric pressure. It is measured by an instrument called pressure gauge (such as Bourden's pressure gauge). The gauge measures pressure of the fluid (liquid and gas) flowing through a pipe or duct, boiler etc. irrespective of prevailing atmospheric pressure.

Vacuum (Or Vacuum pressure) (P_{vacc})

It is the pressure of a fluid, which is always less than atmospheric pressure. Pressure (i.e. vacuum) in a steam condenser is one such example. It is also measured by a pressure gauge but the gauge reads on negative side of atmospheric pressure on dial. The vacuum represents a difference between absolute and atmospheric pressures.

Absolute Pressure (P_{abs})

It is that pressure of a fluid, which is measured with respect to absolute zero pressure as the reference. Absolute zero pressure can occur only if the molecular momentum is zero, and this condition arises when there is a perfect vacuum. Absolute pressure of a fluid may be more or less than atmospheric depending upon, whether the gauge pressure is expressed as absolute pressure or the vacuum pressure.

Inter–relation between different types of pressure representations. It is depicted in Fig. 1.6, which can be expressed as follows.

$$P_{abs} = P_{atm} + P_{gauge}$$
$$P_{abs} = P_{atm} - P_{vace}$$

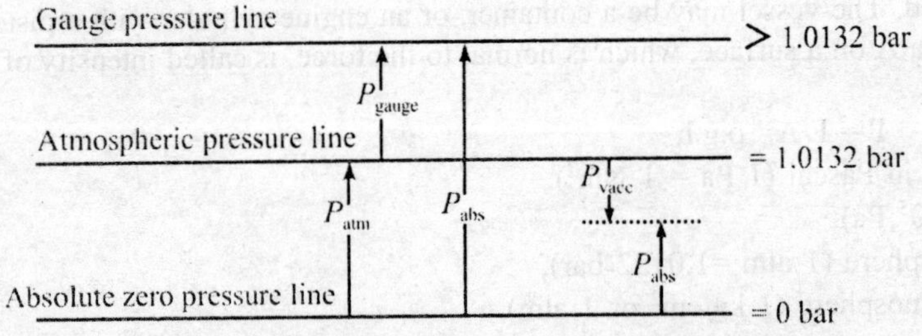

Fig 1.6 Depiction of atmospheric, gauge, vacuum, and absolute pressures and their interrelationship.

Hydrostatic Pressure

Also called Pressure due to Depth of a Fluid. It is required to determine the pressure exerted by a static fluid column on a surface, which is drowned under it. Such situations arise in water filled boilers, petrol or diesel filled tank in IC engines, aviation fuel stored in containers of gas turbines etc.

This pressure is also called 'hydrostatic pressure' as it is caused due to static fluid. The hydrostatic pressure acts equally in all directions on lateral surface of the tank. Above formula holds good for gases also. But due to a very small value of p (and w), its effect is rarely felt. Hence, it is generally neglected in thermodynamic calculations. One such tank is shown in Fig. 1.7. It contains a homogeneous liquid of weight density w. The pressure p exerted by it at a depth h will be given by

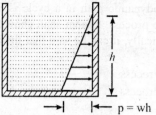

Fig 1.7 *Pressure under depth of a fluid increases with increase in depth.*

Q. 12: Write short notes on State, point function and path function.

STATE

The State of a system is its condition or configuration described in sufficient detail.

State is the condition of the system identified by thermodynamic properties such as pressure, volume, temperature, etc. The number of properties required to describe a system depends upon the nature of the system. However each property has a single value at each state. Each state can be represented by a point on a graph with any two properties as coordinates.

Any operation in which one or more of properties of a system change is called a change of state.

Point Function

A point function is a single valued function that always possesses a single – value is all states. For example each of the thermodynamics properties has a single – value in equilibrium and other states. These properties are called point function or state function.

Or

when two properties locate a point on the graph (coordinates axes) then those properties are called as point function.

For example pressure, volume, temperature, entropy, enthalpy, internal energy.

Path Function

Those properties, which cannot be located on a graph by a point but are given by the area or show on the graph.

A path function is different from a point function. It depends on the nature of the process that can follow different paths between the same states. For example work, heat, heat transfer.

Q. 13: Define thermodynamic process, path, cycle.

Sol: Thermodynamic system undergoes changes due to the energy and mass interactions. Thermody-namic state of the system changes due to these interactions.

The mode in which the change of state of a system takes place is termed as the **PROCESS** such as constant pressure, constant volume process etc. In fig 1.8, process 1–2 & 3–4 is constant pressure process while 2–3 & 4–1 is constant volume process.

Let us take gas contained in a cylinder and being heated up. The heating of gas in the cylinder shall result in change in state of gas as it's pressure, temperature etc. shall increase. However, the mode in which this change of state in gas takes place during heating shall he constant volume mode and hence the process shall be called constant volume heating process.

The **PATH** refers to the series of state changes through which the system passes during a process. Thus, path refers to the locii of various intermediate states passed through by a system during a process.

CYCLE refers to a typical sequence of processes in such a fashion that the initial and final states are identical. Thus, a cycle is the one in which the processes occur one after the other so as to finally, land the system at the same state. Thermodynamic path in a cycle is in closed loop form. After the occurrence of a cyclic process, system shall show no sign of the processes having occurred. Mathematically, it can be said that the cyclic integral of any property in a cycle is zero.

1–2 & 3–4 = Constant volume Process
2–3 & 4–1 = Constant pressure Process
1–2, 2–3, 3–4 & 4–1 = Path
1–2–3–4–1 = Cycle

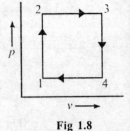

Fig 1.8

Q. 14: Define thermodynamic equilibrium of a system and state its importance. What are the conditions required for a system to be in thermodynamic equilibrium? Describe in brief.

(March–02, Dec–03)

Or

What do you known by thermodynamic equilibrium. *(Dec–02, Dec–04, may–05, Dec–05)*

Sol: Equilibrium is that state of a system in which the state does not undergo any change in itself with passage of time without the aid of any external agent. Equilibrium state of a system can be examined by observing whether the change in state of the system occurs or not. If no change in state of system occurs then the system can be said in equilibrium.

Let us consider a steel glass full of hot milk kept in open atmosphere. It is quite obvious that the heat from the milk shall be continuously transferred to atmosphere till the temperature of milk, glass and atmosphere are not alike. During the transfer of heat from milk the temperature of milk could be seen to decrease continually. Temperature attains some final value and does not change any more. This is the equilibrium state at which the properties stop showing any change in themselves.

Generally, ensuring the mechanical, thermal, chemical and electrical equilibriums of the system may ensure thermodynamic equilibrium of a system.

1. Mechanical Equilibrium: When there is no unbalanced force within the system and nor at its boundaries then the system is said to be in mechanical equilibrium.

For a system to be in mechanical equilibrium there should be no pressure gradient within the system i.e., equality of pressure for the entire system.

2. Chemical Equilibrium: When there is no chemical reaction taking place in the system it is said to be in chemical equilibrium.

3. Thermal equilibrium: When there is no temperature gradient within the system, the system is said to be in thermal equilibrium.

4. Electrical Equilibrium: When there is no electrical potential gradient within a system, the system is said to be in electrical equilibrium.

When all the conditions of mechanical, chemical thermal, electrical equilibrium are satisfied, the system is said to be in thermodynamic equilibrium.

Q. 15: What do you mean by reversible and irreversible processes? Give some causes of irreversibility.

(Feb–02, July–02)

Or

Distinguish between reversible and irreversible process *(Dec–01, May–02)*

Or

Briefly state the important features of reversible and irreversible processes. *(Dec–03)*

Sol: Thermodynamic system that is capable of restoring its original state by reversing the factors responsible for occurrence of the process is called reversible system and the thermodynamic process involved is called reversible process.

Thus upon reversal of a process there shall be no trace of the process being occurred, i.e., state changes during the forward direction of occurrence of a process are exactly similar to the states passed through by the system during the reversed direction of the process.

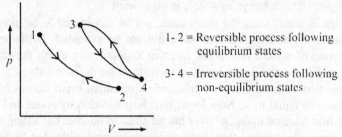

Fig. 1.9. Reversible and irreversible processes

It is quite obvious that such reversibility can be realised only if the system maintains its thermodynamic equilibrium throughout the occurrence of process.

Irreversible systems are those, which do not maintain equilibrium during the occurrence of a process. Various factors responsible for the non–attainment of equilibrium are generally the reasons responsible for irreversibility Presence of friction, dissipative effects etc.

Q. 16: What do you mean by cyclic and quasi – static process. *(March–02, Jan–03, Dec–01, 02, 05)*

Or

Define quasi static process. What is its importance in study of thermodynamics. *(May–03)*

Sol: Thermodynamic equilibrium of a system is very difficult to be realised during the occurrence of a thermodynamic process. 'Quasi–static' consideration is one of the ways to consider the real system as if it is behaving in thermodynamic equilibrium and thus permitting the thermodynamic study. Actually system does not attain thermodynamic equilibrium, only certain assumptions make it akin to a system in equilibrium for the sake of study and analysis.

Quasi–static literally refers to "almost static" and the infinite slowness of the occurrence of a process is considered as the basic premise for attaining near equilibrium in the system. Here it is considered that the change in state of a system occurs at infinitely slow pace, thus consuming very large time for completion of the process. During the dead slow rate of state change the magnitude *of* change in a state shall also be infinitely small. This infinitely small change in state when repeatedly undertaken one after the other results in overall state change but the number *of* processes required for completion of this state change are infinitely large. Quasi–static process is presumed to remain in thermodynamic equilibrium just because of infinitesimal state change taking place during the occurrence of the process. Quasi–static process can be understood from the following example.

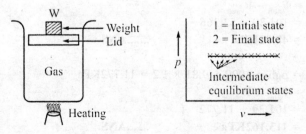

Fig 1.9 *Quasi static process*

Let us consider the locating of gas in a container with certain mass 'W' kept on the top lid (lid is such that it does not permit leakage across its interface with vessel wall) of the vessel as shown in Fig. 1.9. After certain amount of heat being added to the gas it is found that the lid gets raised up. Thermodynamic state change is shown in figure. The "change in state", is significant.

During the "change of state" since the states could not be considered to be in equilibrium, hence for unsteady state of system, thermodynamic analysis could not be extended. Difficulty in thermody-namic analysis of unsteady state of system lies in the fact that it is not sure about the state *of* system as it is continually changing and for analysis one has to start from some definite values.

Let us now assume that the total mass comprises of infinitesimal small masses of 'w' such that all 'w' masses put together become equal to w. Now let us start heat addition to vessel and as soon as the lifting of lid is observed put first fraction mass 'w' over the lid so as to counter the lifting and estimate the state change. During this process it is found that the state change is negligible. Let us further add heat to the vessel and again put the second fraction mass 'w' as soon as the lift is felt so as to counter it. Again the state change is seen to be negligible. Continue with the above process and at the end it shall be seen that all fraction masses 'w' have been put over the lid, thus amounting to mass 'w' kept over the lid of vessel and the state change occurred is exactly similar to the one which occurred when the mass kept over the lid was 'W'. In this way the equilibrium nature of system can be maintained and the thermodynamic analysis can be carried out. P–V representation for the series of infinitesimal state changes occurring between states 1 & 2 is also shown in figure 1.9.

Note:

In $PV = R_0 T$, $R_0 = 8314$ KJ/Kgk

And in $PV = mRT$; $R = R_0/M$; Where M = Molecular Weight

Q. 17: Convert the following reading of pressure to kPa, assuming that the barometer reads 760mm Hg.

(1) 90cm Hg gauge (2) 40cm Hg vacuum (3) 1.2m H$_2$O gauge

Sol: Given that h = 760mm of Hg for P_{atm}

$$P_{atm} = \rho g h = 13.6 \times 10^3 \times 9.81 \times 760/1000 = 101396.16$$

N/m² = 101396.16Pa = 101.39KPa ...(i)

(a) 90cm Hg gauge

$$P_{gauge} = \rho g h = 13.6 \times 10^3 \times 9.81 \times 90/100 = 120.07 KPa \quad \text{...(ii)}$$

$$P_{abs} = P_{atm} + P_{gauge} = 101.39 + 120.07$$

$$P_{abs} = \mathbf{221.46 KPa} \qquad \text{.......ANS}$$

(b) 40cm Hg vacuum

$$P_{vacc} = \rho g h = 13.6 \times 10^3 \times 9.81 \times 40/100 = 53.366 KPa \quad \text{...(iii)}$$

$$P_{abs} = P_{atm} - P_{vacc}$$
$$= 101.39 - 53.366$$

$$P_{abs} = \mathbf{48.02 KPa} \qquad \text{.......ANS}$$

(c) 1.2m Water gauge

$$P_{gauge} = \rho g h = 1000 \times 9.81 \times 1.2 = 11.772 KPa \quad \text{...(iv)}$$

$$P_{abs} = P_{atm} + P_{gauge}$$
$$= 101.39 + 11.772$$

$$P_{abs} = \mathbf{113.162 KPa} \qquad \text{.......ANS}$$

Fundamental Concepts, Definitions and Zeroth Law / 13

Q. 18: The gas used in a gas engine trial was tested. The pressure of gas supply is 10cm of water column. Find absolute pressure of the gas if the barometric pressure is 760mm of Hg.

Sol: Given that h = 760mm of Hg for P_{atm}

$P_{atm} = \rho g h = 13.6 \times 10^3 \times 9.81 \times 760/1000 = 101396.16$
$N/m^2 = 101.39 \times 10^3 \, N/m^2$ ...(i)
$P_{gauge} = \rho g h = 1000 \times 9.81 \times 10/100 = 981 \, N/m^2$ ...(ii)
$P_{abs} = P_{atm} + P_{gauge}$
$= 101.39 \times 10^3 + 981$
$P_{abs} = 102.37 \times 10^3 \, N/m^2$ ....ANS

Q. 19: A manometer shows a vacuum of 260 mm Hg. What will be the value of this pressure in N/m² in the form of absolute pressure and what will be absolute pressure (N/m²), if the gauge pressure is 260 mm of Hg. Explain the difference between these two pressures.

Sol: Given that $P_{Vacc} = 260$mm of Hg

$P_{Vacc} = \rho g h = 13.6 \times 10^3 \times 9.81 \times 260/1000$
$= 34.688 \times 10^3 \, N/m^2$ ....ANS ...(i)
$P_{atm} = \rho g h = 13.6 \times 10^3 \times 9.81 \times 760/1000 = 101396.16$
$N/m^2 = 101.39 \times 10^3 \, N/m^2$ ...(ii)
$P_{abs} = P_{atm} - P_{Vacc}$
$= 101.39 \times 10^3 - 34.688 \times 10^3$
$= 66.61 \times 10^3 \, N/m^2$ ....ANS

Now if $P_{gauge} = 260$mm of Hg =
$P_{gauge} = 260$mm of Hg $= 13.6 \times 10^3 \times 9.81 \times 260/1000 = 34.688 \times 10^3 \, N/m^2$
$P_{abs} = P_{atm} + P_{gauge}$
$= 101.39 \times 10^3 + 34.688 \times 10^3$
$= 136.07 \times 10^3 \, N/m^2$ ....ANS

ANS: $P_{vacc} = 34.7 \times 10^3 \, N/m^2$(vacuum), $P_{abs} = 66.6$kPa, 136kpa

Difference is because vacuum pressure is always Negative gauge pressure. Or vacuum in a gauge pressure below atmospheric pressure and gauge pressure is above atmospheric pressure.

Q. 20: Calculate the height of a column of water equivalent to atmospheric pressure of 1bar if the water is at 15⁰C. What is the height if the water is replaced by Mercury?

Sol: Given that P = 1bar = $10^5 N/m^2$

$P_{atm} = \rho g h$, for water equivalent
$10^5 = 1000 \times 9.81 \times h$
h = 10.19m ....ANS
$P_{atm} = \rho g h$, for Hg
$10^5 = 13.6 \times 10^3 \times 9.81 \times h$
h = 0.749m ....ANS

ANS: 10.19m, 0.75m

Q. 21: The pressure of a gas in a pipeline is measured with a mercury manometer having one limb open. The difference in the level of the two limbs is 562mm. Calculate the gas pressure in terms of bar.

Sol: The difference in the level of the two limbs = P_{gauge}

$P_{gauge} = P_{abs} - P_{atm}$

$$P_{abs} - P_{atm} = 562 \text{mm of Hg}$$
$$P_{abs} - 101.39 = \rho g h = 13.6 \times 10^3 \times 9.81 \times 562/1000 = 75.2 \times 10^3 \text{ N/m}^2 = 75.2 \text{ KPa}$$
$$P_{abs} = 101.39 + 75.2 = 176.5 \text{kPa}$$
ANS: P = 176.5kPa

Q. 22: Steam at gauge pressure of 1.5Mpa is supplied to a steam turbine, which rejects it to a condenser at a vacuum of 710mm Hg after expansion. Find the inlet and exhaust steam pressure in pascal, assuming barometer pressure as 76cm of Hg and density of Hg as 13.6×10^3 kg/m^3.

Sol: $P_{gauge} = 1.5 \times 10^6$ N/m^2

$P_{vacc} = 710$ mm of Hg

$P_{atm} = 76$ cm of Hg $= 101.3 \times 10^3$ N/m^2

$P_{inlet} = ?$

$P_{inlet} = P_{abs} = P_{gauge(inlet)} + P_{atm}$
$= 1.5 \times 10^6 + 101.3 \times 10^3$

$P_{inlet} = 1.601 \times 10^6$ Pa ....ANS

Since discharge is at vacuum i.e.;

$P_{exhaust} = P_{abs} = P_{atm} - P_{vacc}$
$= 101.3 \times 10^3 - 13.6 \times 10^3 \times 9.81 \times 710/1000$

$P_{exhaust} = 6.66 \times 10^6$ Pa ....ANS

ANS: $P_{inlet} = 1.6 \times 10^6$Pa, $P_{exhaust} = 6.66 \times 10^3$Pa

Q. 23: A U–tube manometer using mercury shows that the gas pressure inside a tank is 30cm. Calculate the gauge pressure of the gas inside the vessel. Take g = 9.78m/s^2, density of mercury =13,550kg/m^3. *(C.O.–Dec–03)*

Sol: Given that $P_{abs} = 30$mm of Hg

$P_{abs} = \rho g h = 13550 \times 9.78 \times 30/1000 = 39.755 \times 10^3$ N/m^2 ...(i)

$P_{atm} = \rho g h = 13550 \times 9.78 \times 760/1000 = 100714.44$ N/m^2 ...(ii)

$P_{gauge} = P_{abs} - P_{atm}$
$= 39.755 \times 10^3 - 100714.44$
$= -60958.74$ N/m^2 ....ANS

Q. 24: 12 kg mole of a gas occupies a volume of 603.1 m^3 at temperature of 140°C while its density is 0.464 kg/m^3. Find its molecular weight and gas constant and its pressure. *(Dec–03–04)*

Sol: Given data;

$V = 603.1$ m^3

$T = 140^0$C

$\rho = 0.464$ kg/m^3

Since $PV = nmR_0T$

$= 12$ Kg – mol

$= 12M$ Kg, M = molecular weight

Since $\rho = m/V$

$0.464 = 12M/603.1$

$M = 23.32$ ...(i)

Now Gas constant $R = R_0/M$, Where $R_0 = 8314$ KJ/kg–mol–k = Universal gas constant

$R = 8314/23.32 = 356.52$ J/kgk

$$PV = mR_0T$$
$$P = mR_0T/V, \text{ where m in kg, R} = 8314 \text{ KJ/kg–mol–k}$$
$$= [(12 \times 23.32) \times (8314/23.32)(273 + 140)]/ 603.1$$
$$P = 68321.04 \text{N/m}^2 \quad \text{........ANS}$$

Q. 25: An aerostat balloon is filled with hydrogen. It has a volume of 1000m³ at constant air temperature of 27⁰C and pressure of 0.98bar. Determine the load that can be lifted with the air of aerostat.

Sol: Given that:
$$V = 100 \text{m}^3$$
$$T = 300 \text{K}$$
$$P = 0.98 \text{bar} = 0.98 \times 10^5 \text{ N/m}^2$$
$$W = mg$$
$$PV = mR_0T$$

Where m = mass in Kg
R_0 = 8314 KJ/kg/mole K
But in Hydrogen; M = 2
  i.e.; $R = R_0/2 = 8314/2 = 4157$ KJ/kg.k
$$0.98 \times 10^5 \times 1000 = m \times 4157 \times 300$$
$$m = 78.58 \text{ kg}$$
$$W = 78.58 \times 9.81 = 770.11 \text{ N}$$
$$\text{....... ANS: 770.11N}$$

Q. 26: What is energy? What are its different forms? *(Dec— 02, 03)*

Sol: The energy is defined as the capacity of doing work. The energy possessed by a system may be of two kinds.

1. Stored energy: such as potential energy, internal energy, kinetic energy etc.

2. Transit energy: such as heat, work, flow energy etc.

The stored energy is that which is contained within the system boundaries, but the transit energy crosses the system boundary. The store energy is a thermodynamic property whereas the transit energy is not a thermodynamic property as it depends upon the path.

For example, the kinetic energy of steam issuing out from a steam nozzle and impinging upon the steam turbine blade is an example of stored energy. Similarly, the heat energy produced in combustion chamber of a gas turbine is transferred beyond the chamber by conduction/ convection and/or radiation, is an example of transit energy.

Form of Energy

1. Potential energy (PE)
The energy possessed by a body or system by virtue of its position above the datum (ground) level. The work done is due to its falling on earth's surface.
Potential energy, PE = Wh = mgh N.m
Where, W = weight of body, N ; m = mass of body, kg
h = distance of fall of body, m
g = acceleration due to gravity, = 9.81 m/s²

2. Kinetic Energy (KE)

The energy possessed by a system by virtue of its motion is called kinetic energy. It means that a system of mass m kg while moving with a velocity V_1 m/s, does $1/2mV1^2$ joules of work before coming to rest. So in this state of motion, the system is said to have a kinetic energy given as;

$$K.E. = 1/2mv_1^2 \text{ N.m}$$

However, when the mass undergoes a change in its velocity from velocity V_1 to V_2, the change in kinetic energy of the system is expressed as;

$$K.E. = 1/2mv_2^2 - 1/2mv_1^2$$

3. Internal Energy (U)

It is the energy possessed by a system on account of its configurations, and motion of atoms and molecules. Unlike the potential energy and kinetic energy of a system, which are visible and can be felt, the internal energy is invisible form of energy and can only be sensed. In thermodynamics, main interest of study lies in knowing the change in internal energy than to know its absolute value.

The internal energy of a system is the sum of energies contributed by various configurations and inherent molecular motions. These contributing energies are

(1) Spin energy: due to clockwise or anticlockwise spin of electrons about their own axes.

(2) Potential energy: due to intermolecular forces (Coulomb and gravitational forces), which keep the molecules together.

(3) Transitional energy: due to movement of molecules in all directions with all probable velocities within the system, resulting in kinetic energy acquired by the translatory motion.

(4) Rotational energy: due to rotation of molecules about the centre of mass of the system, resulting in kinetic energy acquired by rotational motion. Such form of energy invariably exists in diatomic and polyatomic gases.

(5) Vibrational energy: due to vibration of molecules at high temperatures.

(6) Binding energy: due to force of attraction between various sub–atomic particles and nucleus.

(7) Other forms of energies such as

Electric dipole energy and magnetic dipole energy when the system is subjected to electric and/or magnetic fields.

High velocity energy when rest mass of the system m_o changes to variable mass m in accordance with Eisenstein's theory of relativity).

The internal energy of a system can increase or decrease during thermodynamic operations.

The internal energy will increase if energy is absorbed and will decrease when energy is evolved.

4. Total Energy

Total energy possessed by a system is the sum of all types of stored energy. Hence it will be given by

$$E_{total} = PE + KE + U = mgh + 1/2mv^2 + U$$

It is expressed in the unit of joule (1 J = 1 N m)

Q. 27: State thermodynamic definition of work. Also differentiate between heat and work.

(May-02)

HEAT

Sol: Heat is energy transferred across the boundary of a system due to temperature difference between the system and the surrounding. The heat can be transferred by conduction, convection and radiation. The main characteristics of heat are:

1. Heat flows from a system at a higher temperature to a system at a lower temperature.
2. The heat exists only during transfer into or out of a system.
3. Heat is positive when it flows into the system and negative when it flows out of the system.
4. Heat is a path function.
5. It is not the property of the system because it does not represent an exact differential dQ. It is therefore represented as δQ.

Heat required to raise the temperature of a body or system, $Q = mc(T_2 - T_1)$
Where,   m = mass, kg
   T_1, T_2 = Temperatures in °C or K.
   c = specific heat, kJ/kg-K.

Specific heat for gases can be specific heat at constant pressure (C_p) and constant volume (c_v)

Also; mc = thermal or heat capacity, kJ.

mc = water equivalent, kg.

WORK

The work may be defined as follows:

"Work is defined as the energy transferred (without transfer of mass) across the boundary of a system because of an intensive property difference other than temperature that exists between the system and surrounding."

Pressure difference results in mechanical work and electrical potential difference results in electrical work.

Or

"Work is said to be done by a system during a given operation if the sole effect of the system on things external to the system (surroundings) can be reduced to the raising of a weight".

The work is positive when done by the system and negative if work is done on the system.

Q. 28: Compare between work and heat ?   *(May–01)*

Sol: There are many similarities between heat and work.

1. The heat and work are both transient phenomena. The systems do not possess heat or work. When a system undergoes a change, heat transfer or work done may occur.
2. The heat and work are boundary phenomena. They are observed at the boundary of the system.
3. The heat and work represent the energy crossing the boundary of the system.
4. The heat and work are path functions and hence they are inexact differentials.
5. Heat and work are not the properties of the system.
6. Heat transfer is the energy interaction due to temperature difference only. All other energy interactions may be called work transfer.
7. The magnitude of heat transfer or work transfer depends upon the path followed by the system during change of state.

Q. 29: What do you understand by flow work? It is different from displacement work? How.

(May–05)

FLOW WORK

Sol: Flow work is the energy possessed by a fluid by virtue of its pressure.

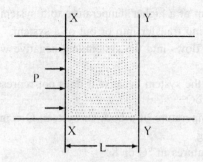

Fig 1.10

Let us consider any two normal sec-tions XX and YY of a pipe line through which a fluid is flowing in the direction as shown in Fig. 1.10.

Let
L = distance between sections XX and YY
A = cross–sectional area of the pipe line
p = intensity of pressure at section l.
Then, force acting on the volume of fluid of length 'L' and
cross–sectional area 'A' = p x A.
Work done by this force = p x A x L = p x V,
Where;
V = A x L = volume of the cylinder of fluid between sections XX and YY

Now, energy is the capacity for doing work. It is due to pressure that p x V amount of work has been done in order to cause flow of fluid through a length 'L',

So flow work = p x V mechanical unit

Displacement Work

When a piston moves in a cylinder from position 1 to position 2 with volume changing from V_1 to V_2, the amount of work W done by the system is given by $W_{1-2} = \int_{V_1}^{V_2} p\, dV$.

The value of work done is given by the area under the process 1 – 2 on diagram (Fig. 1.11)

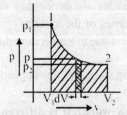

Fig 1.11 *Displacement work*

Q. 30: Find the work done in different processes?

(1) ISOBARIC PROCESS (PRESSURE CONSTANT)

$$W_{1-2} = \int_{V_1}^{V_2} p\, dV = p\,(V_2 - V_2)$$

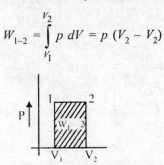

Fig. 1.12: *Constant pressure process*

Fig. 1.13: *Constant volume process*

(2) ISOCHORIC PROCESS (VOLUME CONSTANT)

$$W_{1-2} = \int_{V_1}^{V_2} p\, dV = 0 \quad (\because V_1 = V_2)$$

(3) ISOTHERMAL PROCESS (T or, PV = const)

$$W_{1-2} = \int_{V_1}^{V_2} p\, dV$$

$$pV = p_1 V_1 = p_2 V_2 = C. \qquad \frac{V_1}{V_2} = \frac{p_2}{p_1}$$

$$p = \frac{p_1 V_1}{V}.$$

$$W_{1-2} = p_1 V_1 \int_{V_1}^{V_2} \frac{dV}{V} = p_1 V_1 \ln \frac{V_2}{V_1}$$

$$= p_1 V_1 \ln \frac{P_1}{P_2}$$

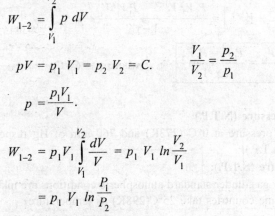

Fig. 1.14 : *Isothermal process*

(4) POLYTROPIC PROCESS (PV^n= C)

$$pV^n = p_1 V_1^n = P_2 V_2^n = C$$

$$p = \frac{(p_1 V_1^n)}{V^n}$$

$$W_{1-2} = \int_{V_1}^{V_2} p\, dV$$

$$= \int_{V_1}^{V_2} \frac{p_1 V_1^n}{V^n} dV = (p_1 V_1^n) \left| \frac{V^{-n+1}}{-n+1} \right|_{V_1}^{V_2}$$

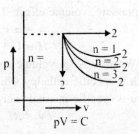

Fig. 1.15 *Polytropic process*

$$= \frac{p_1 V_1^n}{1-n}\left(V_2^{1-2} - V_1^{1-n}\right)$$

$$= \frac{p_2 V_2^n \times V_2^{1-n} - p_1 V_1^n \times V_1^{1-n}}{1-n}$$

$$= \frac{p_1 V_1 - p_2 V_2}{n-1} = \frac{p_1 V_1}{n-1}\left[1 - \left(\frac{p_2}{p_1}\right)^{\frac{n-1}{n}}\right]$$

(5) ADIABATIC PROCESS

$(PV^\gamma = C)$

Here δQ or $dQ = 0$

$\delta Q = dU + dW$

$0 = dU + dW$

$dW = dU = -c_v dT$

$dW = pdV$ $\quad [P_1 V_1^\gamma = P_2 V_2^\gamma = C]$

$$= \int_{v_1}^{v_2} \frac{C}{V^\gamma} dV = C \int_{v_1}^{v_2} V^{-\gamma} dV = C \frac{V^{-\gamma+1}}{-\gamma+1}$$

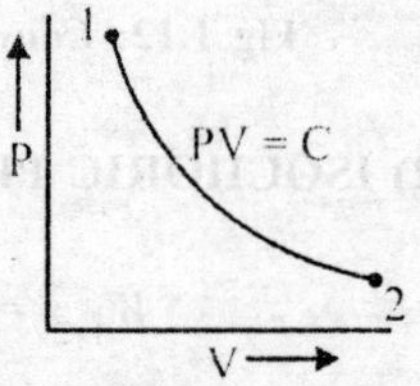

Fig. 1.16 *Adiabatic Process*

$$= \frac{C}{1-\gamma}\left[V_2^{1-\gamma} - V_1^{1-\gamma}\right] = \frac{P_2 V_2^\gamma V_2^{1-\gamma} - P_1 V_1^\gamma V_1^{1-\gamma}}{1-\gamma}$$

$$\boxed{W_{1-2} = \frac{P_2 V_2 - P_1 V_1}{1-\gamma} = \frac{P_1 V_1 - P_2 V_2}{\gamma-1}} \quad \text{where } \gamma = C_p/C_v$$

Q. 31: Define N.T.P. AND S.T.P.

Sol: Normal Temperature and Pressure (N.T.P.):

The conditions of temperature and pressure at 0°C (273K) and 760 mm of Hg respectively are called normal temperature and pressure (N.T.P.).

Standard Temperature and Pressure (S.T.P.):

The temperature and pressure of any gas, under standard atmospheric conditions are taken as $15^0C(288K)$ and 760 mm of Hg respectively. Some countries take $25^0C(298K)$ as temperature.

Q. 32: Define Enthalpy.

Sol: The enthalpy is the total energy of a gaseous system. It takes into consideration, the internal energy and pressure, volume effect. Thus, it is defined as:

$h = u + Pv$

$H = U + PV$

Where v is sp. volume and V is total volume of m Kg gas.

h is specific enthalpy while H is total enthalpy of m kg gas

u is specific internal energy while U is total internal energy of m kg gas. From ideal gas equation,

$Pv = RT$

$h = u + RT$

$h = f(T) + RT$

Therefore, h is also a function of temperature for perfect gas.
$$h = f(T)$$
$$\Rightarrow \quad dh \propto dt$$
$$\Rightarrow \quad dh = C_p dT$$
$$\Rightarrow \quad \int_1^2 dH = \int_1^2 mC_p dT$$
$$\Rightarrow \quad H_2 - H_1 = mC_p(T_2 - T_1)$$

Q. 33: Gas from a bottle of compressed helium is used to inflate an inelastic flexible balloon, originally folded completely flat to a volume of 0.5m³. If the barometer reads 760mm of Hg, What is the amount of work done upon the atmosphere by the balloon? Sketch the systems before and after the process.

Sol: The displacement work $W = \int_{bottle} Pdv + \int_{balloon} Pdv$

Since the wall of the bottle is rigid i.e.; $\int_{bottle} Pdv = 0$

$$W = \int Pdv; \text{ Here } P = 760mm \, Hg = 101.39 \, KN/m^2$$
$$dV = 0.5m^3$$
$$W = 101.39 \times 0.5 \, KN\text{–}m$$
$$W = 50.66 \, KJ \quad \ldots\ldots\text{ANS}$$

Q. 34: A piston and cylinder machine containing a fluid system has a stirring device in the cylinder the piston is frictionless, and is held down against the fluid due to the atmospheric pressure of 101.325kPa the stirring device is turned 10,000 revolutions with an average torque against the fluid of 1.275MN. Mean while the piston of 0.6m diameter moves out 0.8m. Find the net work transfer for the systems.

Sol: Given that
$$P_{atm} = 101.325 \times 10^3 \, N/m^2$$
Revolution = 10000
Torque = 1.275×10^6 N
Dia = 0.6m
Distance moved = 0.8m
Work transfer = ?

W.D by stirring device $W_1 = 2\Pi \times 10000 \times 1.275 \, J = 80.11 \, KJ$ ...(i)
This work is done on the system hence it is –ive.
Work done by the system upon surrounding
$$W_2 = F.dx = P.A.dx$$
$$= 101.32 \times \Pi/4 \times (0.6)^2 \times 0.8$$
$$= 22.92 \, KJ \quad \ldots(ii)$$
Net work done = $W_1 + W_2$
$$= -80.11 + 22.92 = -57.21 \, KJ \, (\text{–ive sign indicates that work is done on the system})$$
ANS: W_{net} = **57.21KJ**

Q. 35: A mass of 1.5kg of air is compressed in a quasi static process from 0.1Mpa to 0.7Mpa for which PV = constant. The initial density of air is $1.16 kg/m^3$. Find the work done by the piston to compress the air.

Sol: Given data:
$$m = 1.5 kg$$
$$P_1 = 0.1 MPa = 0.1 \times 10^6 Pa = 10^5 N/m^2$$
$$P_2 = 0.7 MPa = 0.7 \times 10^6 Pa = 7 \times 10^5 N/m^2$$
$$PV = c \text{ or Temp is constant}$$
$$\rho = 1.16 \, Kg/m^3$$
W.D. by the piston = ?
For PV = C;
$$WD = P_1 V_1 \log V_2/V_1 \text{ or } P_1 V_1 \log P_1/P_2$$
$$\rho = m/V; \text{ i.e.; } V_1 = m/\rho = 1.5/1.16 = 1.293 m^3 \quad ...(i)$$
$$W_{1-2} = P_1 V_1 \log P_1/P_2 = 10^5 \times 1.293 \log e \, (10^5/7 \times 10^5)$$
$$= 10^5 \times 1.293 \times (-1.9459)$$
$$= -251606.18 J = -251.6 \, KJ \, (\text{–ive means WD on the system})$$

ANS: – 251.6KJ

Q. 36: At a speed of 50km/h, the resistance to motion of a car is 900N. Neglecting losses, calculate the power of the engine of the car at this speed. Also determine the heat equivalent of work done per minute by the engine.

Sol: Given data:
$$V = 50 Km/h = 50 \times 5/18 = 13.88 \, m/sec$$
$$F = 900 N$$
Power = ?
Q = ?
$$P = F.V = 900 \times 13.88 = 12500 \, W = \textbf{12.5KW ANS}$$
Heat equivalent of W.D. per minute by the engine = power × 1 minute
$$= 12.5 \, KJ/sec \times 60 \, sec = 750 KJ$$

ANS: Q =750KJ

Q. 37: An Engine cylinder has a piston of area $0.12 m^2$ and contains gas at a pressure of 1.5Mpa the gas expands according to a process, which is represented by a straight line on a pressure volume diagram. The final pressure is 0.15Mpa. Calculate the work done by the gas on the piston if the stroke is 0.30m. *(Dec–05)*

Sol: Work done will be the area under the straight line which is made up of a triangle and a rectangle.
i.e.; WD = Area of Triangle + Area of rectangle
Area of Triangle = ½ × base × height = ½ × AC × AB
AC = base = volume = Area × stroke = 0.12 × 0.30
Height = difference in pressure = $P_2 - P_1 = 1.5 - 0.15$
$$= 1.35 MPa \text{ Area of Triangle}$$
$$= ½ \times (0.12 \times 0.30) \times 1.35 \times 10^6$$
$$= 24.3 \times 10^3 \, J = 24.3 \, KJ \quad ...(i)$$
Area of rectangle = AC × AD
$$= (0.12 \times 0.30) \times 0.15 \times 10^6$$
$$= 5400 \, J = 5.4 \, KJ$$

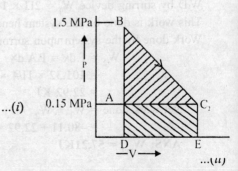

...(ii)

W.D. = (1) + (2) = 24.3 + 5.4 = 29.7KJ
ANS 29.7KJ

Q. 38: The variation of pressure with respect to the volume is given by the following equation $p = (3V^2 + V + 25)$ NM². Find the work done in the process if initial volume of gas is 3 m³ and final volume is 6 m³.

Sol: $P = 3V^2 + V + 25$
Where $V_1 = 3m^3$; $V_2 = 6m^3$

$$WD = \int PdV = \int_{V_1}^{V_2} PdV = \int_3^6 (3V^2 + V + 25) dV$$

$$= (3V^3/3 + V^2/2 + 25V)_3^6 = 277.5 J$$

ANS: 277.5×10⁵ N–m

Q. 39: One mole of an ideal gas at 1.0 Mpa and 300K is heated at constant pressure till the volume is doubled and then it is allowed to expand at constant temperature till the volume is doubled again. Calculate the work done by the gas. *(Dec–01–02)*

Sol: Amount of Gas = 1 mole
$P_1 = 1.0$ MPa
$T_1 = 300^0K$ Process 1–2: Constant pressure
$P_1V_1/T_1 = P_2V_2/T_2$ i.e.; $V_1/T_1 = V_2/T_2$
$V_2 = 2 V_1$; i.e.; $V_1/300 = 2V_1/T_2$
$T_2 = 600K$ ...(i)
For 1 mole, R = Universal gas constant
= 8.3143 KJ/kg mole K
= 8314.3 Kg–k

$$WD = \int_1^2 Pdv \text{ ; Since } PV = RT$$

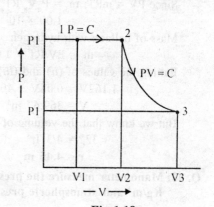

Fig 1.18

$= PV_2 - PV_1$
$= R(T_2 - T_1) = 8314.3 (600 - 300) = 2494.29$KJ ...(i)

Process 2 – 3: Isothermal process

$$W_{2-3} = \int_1^2 PdV = P_2V_2 \ln V_3/V_2 = RT_2 \ln 2V_2/V_2 = RT\ln 2 = 8314.3 \times 600 \ln 2 = 3457.82 \text{KJ}$$

Total WD = $WD_{1-2} + WD_{2-3}$
= 2494.29 + 3457.82 = 5952.11 KJ

ANS: 5952.11J

Q. 40: A diesel engine piston which has an area of 45 cm² moves 5 cm during part of suction stroke of 300 cm³ of fresh air is drawn from the atmosphere. The pressure in the cylinder during suction stroke is 0.9×10^5 N/m² and the atmospheric pressure is 1.01325 bar. The difference between suction pressure and atmospheric pressure is accounted for flow resistance in the suction pipe and inlet valve. Find the network done during the process. *(Dec–01)*

Sol: Net work done = work done by free air boundary + work done on the piston

The work done by free air is negative as boundary contracts and work done in the cylinder on the piston is positive as the boundary expands

Net work done = The displacement work W

$$= \int_{bottle} (PdV)\, Piston + \int_{balloon} (PdV)\, Freeboundary$$

$$= [0.9 \times 10^5 \times 45/(100)^2 \times 5/100] + [-1.01325 \times 10^5 \times 300/10^6]$$

$$= -10.14\ Nm \qquad \text{......ANS}$$

Q. 41: Determine the size of a spherical balloon filled with hydrogen at 30^0C and atmospheric pressure for lifting 400Kg payload. Atmospheric air is at temperature of 27^0C and barometer reading is 75cm of mercury. *(May–02)*

Sol: Given that:

Hydrogen temperature = 30^0C = 303K

Load lifting = 400Kg

Atmospheric pressure = $13.6 \times 10^3 \times 0.75 \times 9.81 = 1.00 \times 10^5\ N/m^2$ = 1.00 bar

Atmospheric Temperature = 27^0C = 300K

The mass that can be lifted due to buoyancy force,

So the mass of air displaced by balloon(m_a) = Mass of balloon hydrogen gas (m_b) + load lifted ...(i)

Since PV = mRT; $m_a = P_a V_a / RT_a$; R = 8314/29 = 287 KJ/Kgk For Air; 29 = Mol. wt of air

$$= 1.00 \times 10^5 \times V / 287 \times 300 = 1.162V\ Kg \qquad ...(ii)$$

Mass of balloon with hydrogen

$$m_b = PV/RT = 1.00 \times 10^5 \times V/(8314/2 \times 300) = 0.08V\ Kg \qquad ...(iii)$$

Putting the values of (ii) and (iii) in equation (i)

1.162V = 0.08V + 400

V = 369.67 m^3

But we know that the volume of a balloon (sphere) = $4/3 \Pi r^3$

$322 = 4/3 \Pi r^3$

$r = 4.45\ m$ ......ANS

Q. 42: Manometer measure the pressure of a tank as 250cm of Hg. For the density of Hg 13.6×10^3 Kg/m^3 and atmospheric pressure 101KPa, calculate the tank pressure in MPa. *(May–01)*

Sol: $P_{abs} = P_{atm} + P_{gauge}$

$$P_{abs} = P_{atm} + \tilde{n}.g.h$$
$$= 101 \times 10^3 + 13.6 \times 10^3 \times 9.81 \times 250 \times 10^{-2}$$
$$= 434.2 \times 10^3\ N/m^2$$
$$= 0.4342\ MPa \qquad \text{......ANS}$$

Q. 43: In a cylinder–piston arrangement, 2kg of an ideal gas are expanded adiabatically from a temperature of 125^0C to 30^0C and it is found to perform 152KJ of work during the process while its enthalpy change is 212.8KJ. Find its specific heats at constant volume and constant pressure and characteristic gas constant. *(May–03)*

Sol: Given data:

m = 2Kg

T_1 = 125^0C

T_2 = 30^0C

W = 152KJ

H = 212.8KJ

$C_P = ?,\ C_V = ?,\ R = ?$

We know that during adiabatic process is:
$$W.D. = P_1V_1 - P_2V_2/\gamma-1 = mR(T_1 - T_2)/\gamma-1$$
$$152 \times 10^3 = 2 \times R(125 - 30)/(1.4 - 1)$$
$$R = 320 J/Kg^0K = 0.32 \, KJ/Kg^0K \quad \ldots\ldots\text{ANS}$$
$$H = mc_p \, dT$$
$$212.8 = 2.C_p.(125 - 30)$$
$$C_p = 1.12 \, KJ/Kg^oK \quad \ldots\ldots\text{ANS}$$
$$C_p - C_v = R$$
$$C_v = 0.8 \, KJ/Kg^oK \quad \ldots\ldots\text{ANS}$$

Q. 44: Calculate the work done in a piston cylinder arrangement during the expansion process, where the process is given by the equation:

$P = (V^2 + 6V)$ bar, The volume changes from $1m^3$ to $4m^3$ during expansion. *(Dec–04)*

Sol: $P = (V^2 + 6V)$ bar

$V_1 = 1m^3$; $V_2 = 4m^3$

$$WD = \int P\,dV = \int_{V_1}^{V_2} P\,dV = \int_1^4 (V^2 + 6V)\,dV$$
$$= (V^3/3 + 6V^2/2)\Big|_1^4$$
$$= 66J \quad \ldots\ldots\text{ANS}$$

Q. 45: Define and explain Zeroth law of thermodynamics *(Dec–01,04)*

Or

State the zeroth law of thermodynamics and its applications. Also explain how it is used for temperature measurement using thermometers. *(Dec–00)*

Or

State the zeroth law of thermodynamics and its importance as the basis of all temperature measurement. *(Dec–02,05, May–03,04)*

Or

Explain with the help of a neat diagram, the zeroth law of thermodynamics. Dec–03

Concept of Temperature

The temperature is a thermal state of a body that describes the degree of hotness or coldness of the body.

If two bodies are brought in contact, heat will flow from hot body at a higher temperature to cold body at a lower temperature.

Temperature is the thermal potential causing the flow of heat energy.

It is an intensive thermodynamic property independent of size and mass of the system.

The temperature of a body is proportional to the stored molecular energy i.e. the average molecular kinetic energy of the molecules in a system. (A particular molecule does not have a temperature, it has energy. The gas as a system has temperature).

Instruments for measuring ordinary temperatures are known as thermometers and those for measuring high temperatures are known as pyrometers.

Equality of Temperature

Two systems have equal temperature if there are no changes in their properties when they are brought in thermal contact with each other.

Zeroth Law: Statement

When a body A is in thermal equilibrium with a body B, and also separately with a body C, then B and C will be in thermal equilibrium with each other. This is known as the zeroth law of thermodynamics.

This law forms the basis for all temperature measurement. The thermometer functions as body 'C' and compares the unknown temperature of body 'A' with a known temperature of body 'B' (reference temperature).

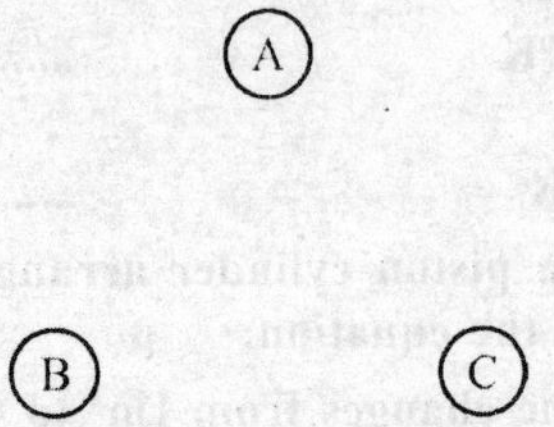

Fig. 1.21 *Zeroth Law*

This law was enunciated by R.H. Fowler in the year 1931. However, since the first and second laws already existed at that time, it was designated as Zeroth law so that it precedes the first and second laws to form a logical sequence.

Temperature Measurement Using Thermometers

In order to measure temperature at temperature scale should be devised assigning some arbitrary numbers to a known definite level of hotness. A thermometer is a measuring device which is used to yield a number at each of these level. Some material property which varies linearly with hotness is used for the measurement of temperature. The thermometer will be ideal if it can measure the temperature at all level.

There are different types of thermometer in use, which have their own thermometric property.

1. Constant volume gas thermometer   (Pressure P)
2. Constant pressure gas thermometer   (Volume V)
3. Electrical Resistance thermometer   (Resistance R)
4. Mercury thermometer   (Length L)
5. Thermocouple   (Electromotive force E)
6. Pyrometer   (Intensity of radiation J)

Q. 46: Express the requirement of temperature scale. And how it help to introduce the concept of temperature and provides a method for its measurement. *(Dec–01,04)*

Temperature Scales

The temperature of a system is determined by bringing a second body, a thermometer, into contact with the system and allowing the thermal equilibrium to be reached. The value of the temperature is found by measuring some temperature dependent property of the thermometer. Any such property is called thermometric property.

To assign numerical values to the thermal state of the system, it is necessary to establish a temperature scale on which the temperature of system can be read. This requires the selection of basic unit and reference state. Therefore, the temperature scale is established by assigning numerical values to certain easily reproducible states. For this purpose it is customary to use the following two fixed points:

(1) **Ice Point:** It is the equilibrium temperature of ice with air–saturated water at standard Atmospheric pressure.

(2) **Steam Point:** The equilibrium temperature of pure water with its own vapour of standard atmospheric pressure.

SCALE	ICE POINT	STEAM POINT	TRIPLE POINT
KELVIN	273.15K	373.15K	273.15K
RANKINE	491.67R	671.67R	491.69R
FAHRENHEIT	32°F	212°F	32.02°F
CENTIGRADE	0°C	100°C	0.01°C

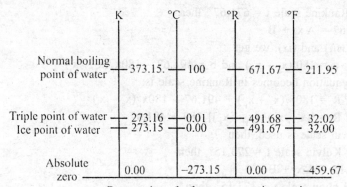

Compansion of references on various scales

Fig 1.22

Requirement of Temperature Scale

The temperature scale on which the temperature of the system can be read is required to assign the numerical values to the thermal state of the system. This requires the selection of basic unit & reference state.

Q. 47: Establish a correlation between Centigrade and Fahrenheit temperature scales. *(May–01)*

Sol: Let the temperature 't' be linear function of property x. (x may be length, resistance volume, pressure etc.) Then using equation of Line ;

$$t = A.x + B \qquad \ldots(i)$$

At Ice Point for Centigrade scale t = 0°, then

$$0 = A.x_i + B \qquad \ldots(ii)$$

At steam point for centigrade scale t = 100°, then

$$100 = A.x_s + B \qquad \ldots(iii)$$

From equation (iii) and (ii), we get

$$a = 100/(x_s - x_i) \text{ and } b = -100x_i/(x_s - x_i)$$

Finally general equation becomes in centigrade scale is;

$$t^0 C = 100x/(x_s - x_i) - 100x_i/(x_s - x_i)$$

$$t^0 C = [(x - x_i)/(x_s - x_i)]100 \qquad \ldots(iv)$$

Similarly if Fahrenheit scale is used, then

At Ice Point for Fahrenheit scale t = 32°, then

$$32 = A.x_i + B \qquad \ldots(v)$$

At steam point for Fahrenheit scale t = 212°, then

$$212 = A.x_s + B \qquad \ldots(vi)$$

From equation (v) and (vi), we get

$$a = 180/(x_s - x_i) \text{ and } b = 32 - 180x_i/(x_s - x_i)$$

Finally general equation becomes in Fahrenheit scale is;
$$t^0 F = 180x/(x_s - x_i) + 32 - 180x_i/(x_s - x_i)$$
$$t^0 F = [(x - x_i)/ (x_s - x_i)]180 + 32 \qquad \ldots(vii)$$

Similarly if Rankine scale is used, then
At Ice Point for Rankine scale t = 491.67°, then
$$491.67 = A.x_i + B \qquad \ldots(viii)$$
At steam point Rankine scale t = 671.67°, then
$$671.67 = A.x_s + B \qquad \ldots(ix)$$
From equation (viii) and (ix), we get
$$a = 180/(x_s - x_i) \text{ and } b = 491.67 - 180x_i/(x_s - x_i)$$
Finally general equation becomes in Rankine scale is;
$$t^0 R = 180\times/(x_s - x_i) + 491.67 - 180x_i/(x_s - x_i)$$
$$\mathbf{t^0 R = [(x - x_i)/ (x_s - x_i)] \ 180 + 491.67} \qquad \ldots(x)$$

Similarly if Kelvin scale is used, then
At Ice Point for Kelvin scale t = 273.15°, then
$$273.15 = A.x_i + B \qquad \ldots(xi)$$
At steam point Kelvin scale t = 373.15°, then
$$373.15 = A.x_s + B \qquad \ldots(xii)$$
From equation (xi) and (xii), we get
$$a = 100/(x_s - x_i) \text{ and } b = 273.15 - 100x_i/(x_s - x_i)$$
Finally general equation becomes in Kelvin scale is;
$$t^0 K = 100x/(x_s - x_i) + 273.15 - 100x_i/(x_s - x_i)$$
$$\mathbf{t^0 K = [(x - x_i)/ (x_s - x_i)] \ 100 + 273.15} \qquad \ldots(xiii)$$

Now compare between above four scales:
$$(x - x_i)/ (x_s - x_i) = C/100 \qquad \ldots(A)$$
$$= (F-32)/180 \qquad \ldots(B)$$
$$= (R-491.67)/180 \qquad \ldots(C)$$
$$= (K - 273.15)/100 \qquad \ldots(D)$$

Now joining all four values we get the following relation
$$K = C + 273.15$$
$$C = 5/9[F - 32]$$
$$= 5/9[R - 491.67]$$
$$F = R - 459.67$$
$$= 1.8C + 32$$

Q. 48: Estimate triple point of water in Fahrenheit, Rankine and Kelvin scales. *(May–02)*

Sol: The point where all three phases are shown of water is known as triple point of water.
Triple point of water T = 273.16°K
Let t represent the Celsius temperature then
$$t = T - 273.15 \ ^0C$$
Where t is Celsius temperature ^0C and Kelvin temperature $T(^0K)$
$$T^0{}_F = 9/5 T^0{}_C + 32 = 9/5 \times 0.01 + 32 = 32.018 \ ^0F$$

$$T^0_R = 9/5 T^0_K = 9/5 \times 273.16 = 491.7 \text{ R}$$
$$T^0_C = 9/5(T^0_K - 32)$$
$$T_K = t^0_C + 273.16$$
$$T_R = t^0_F + 459.67$$
$$\mathbf{T_R/T_K = 9/5}$$

Q. 49: During temperature measurement, it is found that a thermometer gives the same temperature reading in ^{0}C and in ^{0}F. Express this temperature value in ^{0}K. *(Dec–02)*

Sol: The relation between a particular value C on Celsius scale and F on Fahrenheit sacale is found to be as mentioned below.

$$C/100 = (F - 32)/180$$

As given, since the thermometer gives the same temperature reading say '×' in ^{0}C and in ^{0}F, we have from equation (i)

$$x/100 = (x - 32)/180$$
$$180x = 100(x - 32) = 100x - 3200$$
$$x = -40^0$$

Value of this temperature in ^{0}K = 273 + (– 40^0)
$$= 233^0 K \qquad \text{.......ANS}$$

CHAPTER 2

FIRST LAW OF THERMODYNAMICS

Q. 1: Define first law of thermodynamics?
Sol: The First Law of Thermodynamics states that work and heat are mutually convertible. The present tendency is to include all forms of energy. The First Law can be stated in many ways:
1. Energy can neither be created nor destroyed; it is always conserved. However, it can change from one form to another.
2. All energy that goes into a system comes out in some form or the other. Energy does not vanish and has the ability to be converted into any other form of energy.
3. If the system is carried through a cycle, the summation of work delivered to the surroundings is equal to summation of heat taken from the surroundings.
4. No machine can produce energy without corresponding expenditure of energy.
5. Total energy of an isolated system in all its form, remain constant

The first law of thermodynamics cannot be proved mathematically. Its validity stems from the fact that neither it nor any of its corollaries have been violated.

Q. 2: What is the first law for:
   (1) A closed system undergoing a cycle
   (2) A closed system undergoing a change of state

(1) First Law For a Closed system Undergoing a Change of State
According to first law, when a closed system undergoes a thermodynamic cycle, the net heat transfer is equal to the network transfer. The cyclic integral of heat transfer is equal to cyclic integral of work transfer.

$$\oint dQ = \oint dW.$$

where $\oint$ stands for cyclic integral (integral around complete cycle), dQ and dW are small elements of heat and work transfer and have same units.

(2) First Law for a Closed System Undergoing a Change of State
According to first law, when a system undergoes a thermodynamic process (change of state) both heat and work transfer take place. The net energy transfer is stored within the system and is called stored energy or total energy of the system.

When a process is executed by a system the change in stored energy of the system is numerically equal to the net heat interaction minus the net work interaction during the process.

$$dE = dQ - dW \quad \ldots(i)$$
$$E_2 - E_1 = Q_{1-2} - W_{1-2}$$

Where E is an extensive property and represents the total energy of the system at a given state, i.e., E = Total energy

$$dE = dPE + dKE + dU$$

If there is no change in PE and KE then, PE = KE = 0

$$dE = dU, \text{ putting in equation (1), we get}$$
$$dU = dQ - dW$$
or $$dQ = dU + dW$$

This is the first law of thermodynamics for closed system.
Where,

$$dU = \text{Change in Internal Energy}$$
$$dW = \text{Work Transfer} = PdV$$
$$dQ = \text{Heat Transfer} = mcdT$$

{Heat added to the system taken as positive and heat rejected/removal by the system taken as -ive}
For a cycle $\quad dU = 0; \; dQ = dW$

Q. 3: Define isolated system?

Sol: Total energy of an isolated system, in all its forms, remains constant. i.e., In isolated system there is no interaction of the system with the surrounding. i.e., for an isolated system, $dQ = dW = 0$; or, $dE = 0$, or E = constant i.e., Energy is constant.

Q. 4: What are the corollaries of first law of thermodynamics?

Sol: The first law of thermodynamics has important corollaries.

Corollary 1 : *(First Law for a process).*

There exists a property of a closed system, the change in the value of this property during a process is given by the difference between heat supplied and work done.

$$dE = dQ - dW$$

where E is the property of the system and is called total energy which includes internal energy (U), kinetic energy (KE), potential energy (PE), electrical energy, chemical energy, magnetic energy, etc.

Corollary 2: *(Isolated System).*

For an isolated system, both heat and work interactions are absent ($dQ = 0$, $dW = 0$) and E = constant. Energy can neither be created nor destroyed, however, it can be converted from one form to another.

Corollary 3 : *(PMM - 1).*

A perpetual motion machine of the first kind is impossible.

Q. 5: State limitations of first law of thermodynamics?

Sol: There are some important limitations of First Law of Thermodynamics.

1. When a closed system undergoes a thermodynamic cycle, the net heat transfer is equal to the net work transfer. The law does not specify the direction of flow of heat and work nor gives any condition under which energy transfers can take place.
2. The heat energy and mechanical work are mutually convertible. The mechanical energy can be fully converted into heat energy but only a part of heat energy can be converted into mechanical work. Therefore, there is a limitation on the amount of conversion of one form of energy into another form.

Q. 6: Define the following terms:
(1) Specific heat; (2) Joule's law; (3) Enthalpy

Specific Heat
The sp. Heat of a solid or liquid is usually defined as the heat required to raise unit mass through one degree temperature rise.

i.e., $dQ = mcdT$;

$dQ = mC_p dT$; For a reversible non flow process at constant pressure;

$dQ = mC_v dT$; For a reversible non flow process at constant volume;

C_p = Heat capacity at constant pressure

C_v = Heat capacity at constant volume

Joule's Law
Joules law experiment is based on constant volume process, and it state that the I.E. of a perfect gas is a function of the absolute temperature only.

i.e., $U = f(T)$

$dU = dQ - dW$; It define constant volume i.e $dw = 0$

$dU = dQ$; but $dQ = mC_v dT$, at constant volume

$dU = mC_v dT$; for a perfect gas

Enthalpy
It is the sum of I.E. (U) and pressure – volume product.

$H + pv$

For unit mass $pv = RT$

$$h = C_V T + RT = (C_V + R)T = C_p T = (dQ)_p$$
$$H = mC_p T$$
$$dH = mC_p dT$$

Q. 7: What is the relation between two specific heat?
Sol: $dQ = dU + dW$; for a perfect gas

dQ at constant pressure

dU at Constant volume; $= mC_v dT = mC_v(T_2 - T_1)$

dW at constant pressure $= PdV = P(V_2 - V_1) = mR(T_2 - T_1)$

Putting all the values we get

$$dQ = mC_v(T_2 - T_1) + mR(T_2 - T_1)$$
$$dQ = m(C_v + R)(T_2 - T_1)$$

but $\quad dQ = mC_p(T_2 - T_1)$

$mC_p(T_2 - T_1) = m(C_v + R)(T_2 - T_1)$

$C_p = C_v + R$; $C_p - C_v = R$ ...(i)

Now divided by C_v; we get

$C_p/C_v - 1 = R/C_v$; Since $C_p/C_v = y$ (gama = 1.41)

$y - 1 = R/C_v$;

or $\quad C_v = R/(y-1)$; $C_p = yR/(y-1)$; $C_p > C_v$; $y > 1$

Q. 8: Define the concept of process. How do you classify the process.

Sol: A process is defined as a change in the state or condition of a substance or working medium. For example, heating or cooling of thermodynamic medium, compression or expansion of a gas, flow of a fluid from one location to another. In thermodynamics there are two types of process; Flow process and Non-flow process.

Flow Process: The processes in open system permits the transfer of mass to and from the system. Such process are called flow process. The mass enters the system and leaves after exchanging energy. e.g. I.C. Engine, Boilers.

Non-Flow Process: The process occurring in a closed system where there is no transfer of mass across the boundary are called non flow process. In such process the energy in the form of heat and work cross the boundary of the system.

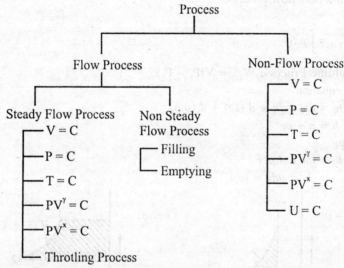

In steady flow fluid flow at a uniform rate and the flow parameter do not change with time. For example if the absorption of heat work output, gas flow etc. occur at a uniform rate (Not varying with time), the flow will be known as steady flow. But if these vary throughout the cycle with time, the flow will be known as non steady flow process e.g., flow of gas or flow of heat in an engine but if a long interval of time is chosen as criteria for these flows, the engine will be known to be operating under non – flow condition.

Q. 9: What is Work done, heat transfer and change in internal energy in free expansion or constant internal energy process.

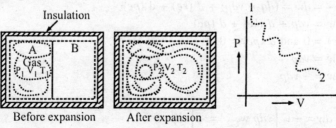

Fig 2.1

A free expansion process is such a process in which the system expands freely without experience any resistance. I.E. is constant during state change This process is highly irreversible due to eddy flow of fluid during the process and there is no heat transfer.

dU = 0; dQ = dW (For reversible process)

dQ = 0; dW = 0; $T_1 = T_2$; dU = 0

Q. 10: How do you evaluate mechanical work in different steady flow process?

work done by a steady flow process,

$$W_{1-2} = \int_1^2 vdp$$

and work done in a non–flow process,

$$W_{1-2} = \int_1^2 pdv$$

1. **Constant Volume Process;** $W_{1-2} = V(P_1 - P_2)$

   Steady flow equation

   $dq = du + dh + d(ke) + d(pv)$

   Now $\quad h = a + pr$

   Differentiating

   $dh = du + dtper$

   $= du = pdv = vdp.$

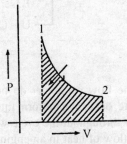

Non-flow process   Steady flow process

From First Law of Thermodynamics for a closed system.

$dq = du + pdv$

$db = dg + vdp$

∴ $\quad dq - = dw = (dq + vdp) + d(ke) + d(pe)$

∴ $\quad -dw = vdp + d(ke) + d(pe)$

if $\quad d(ke) = 0$ and $d(pe) = 0$

$\quad -dw = vdp$

or $\quad dw = -vdp$

Integrating, $\int_1^2 dw = -v \int_1^2 vdp \quad w_{1-2} = -\int_1^2 vdp$

2. **Constant Pressure process;** $W_{1-2} = V(P_1 - P_2) = 0$

$$w_{1-2} = -\int_1^2 v\,dp = -v\int_1^2 dp = v \qquad [\because p_1 = p_2]$$

3. **Constant temperature process;**

$$W_{1-2} = P_1V_1 \ln P_1/P_2 = P_1V_1 \ln V_2/V_1$$

$$w_{1-2} = \int_1^2 v\,dp = -\int_1^2 \frac{p_1 v_1}{p}\,dp \qquad \left[\begin{array}{c} \because pv " p_1 v_1 \\ v " \dfrac{p_1 v_1}{p} \end{array}\right]$$

$$= -p_1 v_1 \int_1^2 \frac{dp}{p} = -p_1 v_1 \ln \frac{p_2}{p_1} = p_2 v_2 \ln \frac{p_1}{p_2}$$

$$= p_1 v_1 \ln \frac{p_2}{v_1} \qquad \left(\because \frac{p_1}{p_2} = \frac{v_2}{v_1}\right)$$

4. **Adiabatic Process;** $W_{1-2} = \gamma(P_1V_1 - P_2V_2)/(\gamma - 1)$

$$pv^\gamma = p_1 v_1^\gamma = p_2 v_2^\gamma = \text{constant}$$

$$v = v_1 \left(\frac{p_1}{p}\right)^{\frac{1}{\gamma}}$$

$$w_{1-2} = -\int_1^2 v\,dp = -\int_1^2 v_1 \left(\frac{p_1}{p}\right)^{\frac{2}{\gamma}} dp$$

$$w_{1-2} = -v_1 p_1^{\frac{1}{\gamma}} \int_1^2 p^{-\frac{1}{\gamma}}\,dp = -v_1 p_2^{\frac{1}{\gamma}} \left|\frac{p^{-\frac{1}{\gamma}-1}}{-\frac{2}{\gamma}+1}\right|_1^2$$

$$= \frac{-v_1 p_1^{\frac{1}{\gamma}}}{\frac{\gamma-1}{\gamma}} \left[p_2^{\frac{\gamma-1}{\gamma}} - p_1^{\frac{\gamma-1}{\gamma}}\right]$$

$$w_{1-2} = \frac{\gamma}{\gamma - 1}(p_1 v_1 - p_2 v_2).$$

5. **Polytropic process;** $W_{1-2} = n(P_1V_1 - P_2V_2)/(n - 1)$

$$w_{1-2} = \frac{n}{n-1}(p_1 v_1 - p_2 v_2).$$

Non – Flow Process

S.No.	PROCESS	P-V-T RELATION	WORK DONE	dU	dQ	dH
1.	V = C	$P_1/T_1 = P_2/T_2$	0	$= mC_V(T_2 - T_1)$	$= mC_V(T_2 - T_1)$	$= mC_P(T_2 - T_1)$
		During Expansion and heating WD and Q is +ive while during Compression and cooling WD and Q is –ive				
2.	P = C	$V_1/T_1 = V_2/T_2$	$= P(V_2 - V_1)$ $= mR(T_2 - T_1)$	$= mC_V(T_2 - T_1)$	$= mC_P(T_2 - T_1)$	$= mC_P(T_2 - T_1)$
3.	T = C	$P_1V_1 = P_2V_2$	$= P_1V_1 \ln P_1/P_2$ $= P_1V_1 \ln V_2/V_1$ $= mRT_1 \ln V_2/V_1$	0	Q = W	0
4.	$Pv^\gamma = C$	$P_1V_1^\gamma = P_2V_2^\gamma = C$ $T_1/T_2 = (v_2/v_1)^{\gamma-1}$ $= (P_1/P_2)^{\gamma-1/\gamma}$ $V_1/V_2 = (P_2/P_1)^{1/\gamma}$	$= mR(T_2 - T_1)/\gamma-1$ $= (P_1V_1 - P_2V_2)/\gamma-1$	$= -dW$ $= mC_P(T_2 - T_1)$	0	$= mC_P(T_2 - T_1)$

6. Throttling Process

The expansion of a gas through an orifice or partly opened valve is called throttling.

$$q_{1-2} = 0 \text{ and } w_{1-2} = 0$$

Now $\quad h_1 + \dfrac{V_1^2}{2} + gz_1 + q_{1-2} = h_2 + \dfrac{V_2^2}{2} + gz_2 + w_{1-2}$

If $\qquad V_1 = V_2$ and $z_1 = z_2$, $h_1 = h_2$

The throttling process is a constant enthalpy process.
If the readings of pressure and temperature of Joule Thompson porous plug experiment are plotted,

$$h_1 = h_2 = h_3 = h_4 = h_5$$

The slope of this constant enthalpy curve is called Joule Thompson coefficient.

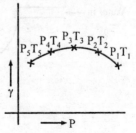

Constant Enthalpy Process

$$p = \left[\dfrac{dT}{dp}\right]_h$$

For a perfect gas, $p = 0$.

Q. 11: Define the following terms:
(1) Control surface
(2) Steam generator
(3) Flow work
(4) Flow Energy
(5) Mass flow rate

Control Surface : A control system has control volume which is separated from its surrounding by a real or imaginary control surface which is fixed in shape, position and orientation. Matter can continually flow in and out of control Volume and heat and work can cross the control surface. This is also an open system.

Steam Generator: The volume of generator is fixed. Water is Supplied. Heat is supplied. Steam comes out. It is a control system as well as open system.

The flow process can be analysed as a closed system by applying the concept of control volume. The control surface can be carefully selected and all energies of the system including flow energies can be considered inside the system. The changes of state of the working substance (mass) need not be considered during its passage through the system.

PE = force × Distance = $(p_1 A_1).x$.
$\qquad = p_1 V_1$ (J)

Now specific volume of working substance is p_1
$\qquad$ Fe = $p_1 v_1$ (J/kg).

Flow Work: The flow work is the energy required to move the working substance against its pressure It is also called flow or displacement energy

It a working substance with pressure p, flow through area A, (m²) and moves through a distance x. (m) work required to move the working substance.

Flow work = force X distance = (P.A).x = PV Joule

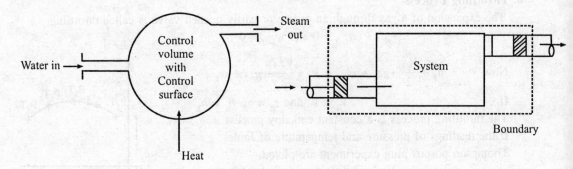

Fig. 2.2 *Control volume.*

Flow Energy: Flow work analysis is based on the consideration that there is no change in KE, PE, U. But if these energies are also considered in a flow process, The flow energy per unit mass will be expressed as

$$E = F.W + KE + PE + I.E.$$
$$E_{flow} = PV + V^2/2 + gZ + U$$
$$= (PV + U) + V^2/2 + gZ$$
$$E = h + V^2/2 + gZ$$

Mass Flow Rate (m_f)

In the absence of any mass getting stored the system we can write;

Mass flow rate at inlet = Mass flow rate at outlet

i.e., $m_{f1} = m_{f2}$

since m_f = density X volume flow rate = density X Area X velocity = $\rho.A.V$

$$\rho_1.A_1.V_1 = \rho_2.A_2.V_2$$

or, $m_f = A_1.V_1/v_1 = A_2.V_2/v_2$; Where: v_1, v_2 = specific volume

Q. 13: Derive steady flow energy equation *(May-05)*

Sol: Since the steady flow process is that in which the condition of fluid flow within a control volume do not vary with time, i.e. the mass flow rate, pressure, volume, work and rate of heat transfer are not the function of time.

i.e., for steady flow

$(dm/dt)_{entrance} = (dm/dt)_{exit}$; i.e, dm/dt = constant

$dP/dt = dV/dt = d\rho/dt = dE_{chemical} = 0$

Assumptions

The following conditions must hold good in a steady flow process.
   (a) The mass flow rate through the system remains constant.
   (b) The rate of heat transfer is constant.
   (c) The rate of work transfer is constant.
   (d) The state of working:; substance at any point within the system is same at all times.
   (e) There is no change in the chemical composition of the system.
   If any one condition is not satisfied, the process is called unsteady process.

Let;
A_1, A_2 = Cross sectional Area at inlet and outlet
ρ_1, ρ_2 = Density of fluid at inlet and outlet
m_1, m_2 = Mass flow rate at inlet and outlet
u_1, u_2 = I.E. of fluid at inlet and outlet
P_1, P_2 = Pressure of mass at inlet and outlet
v_1, v_2 = Specific volume of fluid at inlet and outlet
V_1, V_2 = Velocity of fluid at inlet and outlet
Z_1, Z_2 = Height at which the mass enter and leave
Q = Heat transfer rate
W = Work transfer rate

Consider open system; we have to consider mass balanced as well as energy balance.

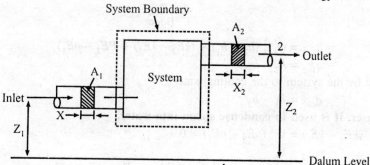

Fig 2.3

In the absence of any mass getting stored the system we can write;
Mass flow rate at inlet = Mass flow rate at outlet
i.e., $\quad m_{f1} = m_{f2}$
since m_f = density X volume flow rate = density X Area X velocity = $\rho.A.V$

$\quad \rho_1.A_1.V_1 = \rho_2.A_2.V_2$
or, $\quad A_1.V_1/v_1 = A_2.V_2/v_2$; v_1, v_2 = specific volume

Now total energy of a flow system consist of P.E, K.E., I.E., and flow work
Hence, $\quad E = PE + KE + IE + FW$
$\quad\quad\quad = h + V^2/2 + gz$

Now; Total Energy rate cross boundary as heat and work
= Total energy rate leaving at (2) - Total energy rate leaving at (1)

$\quad Q - W = m_{f2}[h_2 + V_2^2/2 + gZ_2] - m_{f1}[h_1 + V_1^2/2 + gZ_1]$

For steady flow process $m_f = m_{f1} = m_{f2}$

$\quad Q - W = m_f[(h_2 - h_1) + \tfrac{1}{2}(V_2^2 - V_1^2) + g(Z_2 - Z_1)]$

For unit mass basis

$\quad Q - W_s = [(h_2 - h_1) + \tfrac{1}{2}(V_2^2 - V_1^2) + g(Z_2 - Z_1)]$ J/Kg-sec

W_s = Specific heat work
May also written as

$\quad dq - dw = dh + dKE + dPE$

Or;

$h_1 + V_1^2/2 + gZ_1 + q_{1-2} = h_2 + V_2^2/2 + gZ_2 + W_{1-2}$

Q. 14: Write down different cases of steady flow energy equation?

1. Bolter

$(kE_2 - kE_1) = 0$, $(pE_2 - pE_1) = 0$, $w_{1-2} = 0$

Now,
$$q_{1-2} = w_{1-2}(h_2 - h_1) + (kE_2 - kE_1) + (pE_2 - pE_1)$$
$$q_{1-2} = h_2 - h_1$$

Heat supplied in a boiler increases the enthalpy of the system.

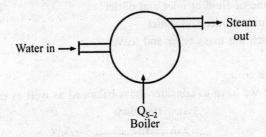
Boiler

$$q_{1-2} = w_{1-2}(h_2 - h_1) + (kE_2 - kE_1) + PE_2 - pE_1)$$
$$-q_{1-2} = h_2 - h_1$$

Heat is lost by the system to the cooling water
$$q_{1-2} = h_1 - h_2$$

2. Condenser. It is used to condense steam into water.

$(kE_2 - kE_1) = 0$, $(pE_2 - pE_1) = 0$
$$w_{1-2} = 0.$$
$$q_{1-2} - w_{1-2} = (h_2 - h_1) + (kE_2 - kE_1) + (pE_2 - pE_1)$$
$$-q_{1-2} = h_2 - h_1$$

Heat is lost by the system to the cooling water
$$q_{1-2} = h_1 - h_2$$

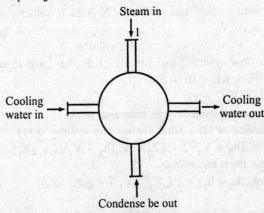
Condenser

3. Refrigeration Evaporator. It is used to evaporate refrigerant into vapour.

$(kE_2 - kE_1) = 0$, $(pE_2 - pE_1) = 0$
$$w_{1-2} = 0$$
$$q_{1-2} = h_2 - h_1$$

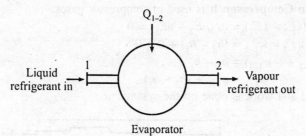

Evaporator

The process is reverse of that of condenser. Heat is supplied by the surrounding to increas3e the enthalpy of refrigerant.

4. Nozzle. Pressure energy is converted in to kinetic energy

$$q_{1-2} = V, w_{1-2} = 0$$
$$(pE_2 - pE_1) = 0$$

Now, $\quad q_{1-2} - w_{1-2} = (h_2 - h_1) + (kE_2 - kE_1) + 0$

$$\frac{V_2^2}{2} - \frac{V_1^2}{2} = (h_1 - h_2)$$

$$V_2^2 = V_1^2 + 2(h_1 - h_2)$$

$$V_2 = \sqrt{V_1^2 + 2(h_1 - h_2)}$$

If $V_1 << V_2$

$$V_2 = \sqrt{2(h_1 - h_2)}$$

Mass flow rate,

$$m = \frac{A_1 V_1}{v_1} = \frac{A_2 V_2}{v_2}$$

Nozzle

5. Turbine. It is used to produce work.

$$q_{1-2} = 0. (kE_2 - kE_1) = 0$$
$$(pE_2 - pE_1) = 0$$
$$-w_{1-2} = (h_2 - h_1)$$
$$w_{1-2} = (h_1 - h_2)$$

The work is done by the system due to decrease in enthalpy.

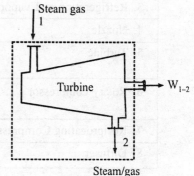

6. Rotary Compressor

$$q_{1-2} = 0, (kE_2 - kE_1) = 0$$
$$(pE_2 - pE_1) = 0$$
$$w_{1-2} = h_2 - h_1$$

Work is done to increase enthalpy.

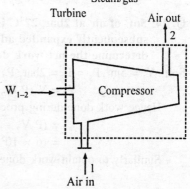

7. Reciprocatin Compressor. It is used to compressor gases.

$(kE_2 - kE_1) = 0, (pE_2 - pE_1) = 0$

$q_{1-2} - w_{1-2} = (h_2 - h_1) + 0 + 0$

$- q_{1-2} - (-w_{1-2}) = h_2 - h_1$

$w_{1-2} = q_{1-2} + (h_2 - h_1)$

Heat is rejected and work is done on the system.

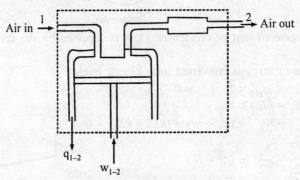

Cooling of cylinder (by air or water)
Reciprocating compressor

Dirrerent Cases of Sfee	Sfee
1. Boiler	$q = h_2 - h_1$
2. Condenser	$q = h_1 - h_2$
3. Refrigeration or Evaporator	$q = h_1 - h_2$
4. Nozzle	$V_2^2/2 - V_1^2/2 = h_1 - h_2$
5. Turbine	$W_{1-2} = h_1 - h_2$; WD by the system due to decrease in enthalpy
6. Rotary compressor	$W_{1-2} = h_2 - h_1$; WD by the system due to increae in enthalpy
7. Reciprocating Compressor	$W_{1-2} = q_{1-2} + (h_2 - h_1)$
8. Diffuser	$q - w = (h_2 - h_1) + \frac{1}{2}(V_2^2 - V_1^2)$

Q. 14: 5m³ of air at 2bar, 27⁰C is compressed up to 6bar pressure following $PV^{1.3}$ = constant. It is subsequently expanded adiabatically to 2 bar. Considering the two processes to be reversible, determine the net work done, also plot the processes on T – S diagrams. *(May – 02)*

Sol: $V_1 = 5m^3, P_1 = P_3 = 2bar, P_2 = 6bar,$ and $n = 1.3$

$V_2 = V_1(P_1/P_2)^{1/1.3} = 5(2/6)^{1/1.3} = 2.147 m^3$

Hence work done during process 1 – 2 is W_{1-2}

$= (P_2V_2 - P_1V_1)/(1- n)$

$= (6 \times 10^5 \times 2.47 - 2 \times 10^5 \times 5)/(1-1.3) = - 9.618 \times 10^5$ J

Similarly to obtain work done during processes 2 – 3, we apply

$W_{2-3} = (P_3V_3 - P_2V_2)/(1 - \gamma)$; where $\gamma = 1.4$

And $V_3 = V_2(P_2/P_3)^{1/\gamma} = 2.147(6/2)^{1/1.4} = 4.705 m^3$

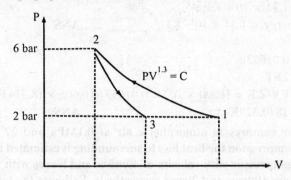

Fig 2.4

Thus $W_{2-3} = (2 \times 10^5 \times 4.705 - 2 \times 10^5 \times 2.147)/(1 - 1.4) = 8.677 \times 10^5$ J

Net work done

$W_{net} = W_{1-2} + W_{2-3} = -9.618 \times 10^5 + 8.677 \times 10^5 = -0.9405 \times 10^5$ J

$W_{net} = -94.05$ KJ   ...ANS

Q. 15: The specific heat at constant pressure of a gas is given by the following relation: $C_p = 0.85 + 0.00004T + 5 \times 10T^2$ where T is in Kelvin. Calculate the changes in enthalpy and internal energy of 10 kg of gas when its temperature is raised from 300 K to 2300 K. Take that the ratio of specific heats to be 1.5. A steel cylinder having a volume of 0.01653 m³ contains 5.6 kg of ethylene gas C_2H_4 molecular weight 28. Calculate the temperature to which the cylinder may be heated without the pressure exceeding 200 bar; given that compressibility factor Z = 0.605. *(Dec-03-04)*

Sol: $C_p = 0.85 + 0.00004T + 5 \times 10T^2$

$dh = m.C_p.dT$

$dh = m. \int_{T_2 = 300}^{T_2 = 2300} (0.85 + 0.00004T + 5 \times 10T^2)\, dT$

$= 10 \times \left[0.85(T_2 - T_1) + (4 \times 10^5/2)(T_2^2 - T_1^2) + (5 \times 10/3)(T_2^3 - T_1^3) \right]_{T_1 = 300}^{T_2 = 2300}$

$= 10 \times [0.85(2300 - 300) + 4 \times 10^{-5}/2(2300^2 - 300^2) + 5 \times 10/3(2300^3 - 300^3)]$

$= 2.023 \times 10^{12}$ KJ

Change in Enthalpy = 2.023×10^{12} KJ   ...ANS

$C_V = C_P/\gamma$

$du = mC_V dT$

$= m.C_P/\gamma \cdot dT$

$= m/\gamma. \int_{T_2 = 300}^{T_2 = 2300} (0.85 + 0.00004T + 5 \times 10T^2)\, dT$

$= (10/1.5)$
$= (10/1.5) \times [0.85(2300 - 300) + (4 \times 10^{-5}/2)(2300^2 - 300^2) + (5 \times 10/3)(2300^3 - 300^3)]$
$= 1.34 \times 10^{12}$ KJ

Change in Internal Energy $= 1.34 \times 10^{12}$ KJ ....ANS

Now; $v = 0.01653 m^3$
$Pv = ZRT$
$T = P.V/Z.R = [\{200 \times 10^5 \times 0.01653\}/\{0.605 \times (8.3143 \times 10^3/28)\}]$
$T = 1840.329 K$ ....ANS

Q. 16: An air compressor compresses atmospheric air at 0.1MPa and 27°C by 10 times of inlet pressure. During compression the heat loss to surrounding is estimated to be 5% of compression work. Air enters compressor with velocity of 40m/sec and leaves with 100m/sec. Inlet and exit cross section area are 100cm² and 20cm² respectively. Estimate the temperature of air at exit from compressor and power input to compressor. *(May–02)*

Sol: Given that;
At inlet: $P_1 = 0.1$MPa; $T_1 = 27 + 273 = 300$K; $V_1 = 40$m/sec;
$A_1 = 100$cm²
At exit: $P_2 = 10P_1 = 1.0$MPa; $V_2 = 100$m/sec; $A_2 = 20$cm²
Heat lost to surrounding = 5% of compressor work
Since Mass flow rate $m_f = A_1 \cdot V_1/v_1 = A_2 \cdot V_2/v_2$;
Where: v_1, v_2 = specific volume ...(i)
$(100 \times 10^{-4} \times 40)/v_1 = (20 \times 10^{-4} \times 100)/v_2$ ...(ii)
or; $v_2/v_1 = 0.5$
Also $P_1 v_1 = RT_1$ & $P_2 V_2 = RT_2$
Or; $P_1 v_1/T_1 = P_2 v_2/T_2 = R$ ...(iii)
$T_2/T_1 = (P_2 v_2/P_1 v_1)$
$T_2 = T_1(P_2 v_2/P_1 v_1) = (10 P_1 \times 0.5/P_1) \times 300 = 1500$K
Also $v_1 = RT_1/P_1 = \{(8.3143 \times 10^3/29) \times 300\}/(0.1 \times 10^6) = 0.8601$ m³/kg
From equation (2) $m_f = (100 \times 10^{-4} \times 40)/0.8601 = 0.465$kg/sec

$m_f = 0.465$kg/sec ....ANS

Applying SFEE to control volume:
$Q - W_S = m_f[(h_2 - h_1) + \frac{1}{2}(V_2^2 - V_1^2) + g(Z_2 - Z_1)]$
$Q = 5\%$ of $W_S = 0.05(-W_S)$
– ve sign is inserted because the work is done on the system
$-0.05(-W_S) - W_S = 0.465[1.005(1500 - 300) + \frac{1}{2}(100^2 - 40^2)/1000]$
(Neglecting the change in potential energy)
$W_S = -592.44$ KJ/sec ....ANS
–ive sign shows work done on the system
–Power input required to run the compressor is 592.44KW

Q. 17: A steam turbine operating under steady state flow conditions, receives 3600Kg of steam per hour. The steam enters the turbine at a velocity of 80m/sec, an elevation of 10m and specific enthalpy of 3276KJ/kg. It leaves the turbine at a velocity of 150m/sec. An elevation of 3m and a specific enthalpy of 2465 KJ/kg. Heat losses from the turbine to the surroundings amount to 36MJ/hr. Estimate the power output of the turbine. *(May – 01(C.O.))*

Sol: Steam flow rate = 3600Kg/hr = 3600/3600 = 1 Kg/sec
Steam velocity at inlet V_1 = 80m/sec
Steam velocity at exit V_2 = 150m/sec
Elevation at inlet Z_1 = 10m
Elevation at exit Z_2 = 3m
Sp. Enthalpy at inlet h_1 = 3276KJ/kg
Sp. Enthalpy at exit h_2 = 2465KJ/kg
Heat losses from the turbine to surrounding Q = 36MJ/hr = 36 × 10⁶/3600 = 10KJ/sec
Turbine operates under steady flow condition, so apply SFEE
For unit mass basis:
$$Q - W_s = (h_2 - h_1) + \tfrac{1}{2}(V_2^2 - V_1^2) + g(Z_2 - Z_1) \text{ J/Kg-sec}$$
$$-10 - W_s = [(2465 - 3276) + (150^2 - 80^2)/2 \times 1000 + 9.81(3-10)/1000]$$
$$W_s = 793 \text{ KJ/Kg-sec} = 793 \text{ KW} \quad \text{.......ANS}$$

Q. 18: In an isentropic flow through nozzle, air flows at the rate of 600Kg/hr. At inlet to the nozzle, pressure is 2Mpa and temperature is 127°C. The exit pressure is 0.5Mpa. If initial air velocity is 300m/sec. Determine
(i) Exit velocity of air, and
(ii) Inlet and exit area of the nozzle. *(Dec – 01)*

Sol:

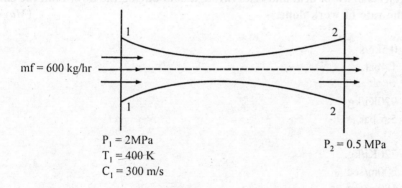

Fig. 2.5

Rate of flow of air m_f = 600Kg/hr
Pressure at inlet P_1 = 2MPa
Temperature at inlet T_1 = 127 + 273 = 400K
Pressure at exit P_2 = 0.5MPa
The velocity at inlet V_1 = 300m/sec
Let the velocity at exit = V_2
And the inlet and exit areas be A_1 and A_2
Applying SFEE between section 1 – 1 & section 2 – 2
$$Q - W_S = m_f[(h_2 - h_1) + \tfrac{1}{2}(V_2^2 - V_1^2) + g(Z_2 - Z_1)]$$
$$Q = W_S = 0 \text{ and } Z_1 = Z_2$$

For air $h_2 - h_1 = C_p(T_2 - T_1)$
$$0 = C_p(T_2 - T_1) + \tfrac{1}{2}(V_2^2 - V_1^2)$$
$$V_2^2 = 2C_p(T_2 - T_1) + V_1^2 \qquad \ldots(i)$$
Now $\quad T_2/T_1 = (P_2/P_1)^{\gamma-1/\gamma}$
For air $\quad \gamma = 1.4$
$$T_2 = 400(0.5/2.0)^{1.4-1/1.4} = 269.18K$$
from equation 1
$$V_2 = [2 \times 1.005 \times 10^3 (400 - 269.18) + (300)^2 \,]^{1/2}$$
$$V_2 = 594 \text{ m/sec} \qquad \ldots\text{ANS}$$
Since $\quad P_1v_1 = RT_1$
$\quad v_1 = 8.314 \times 400/\,29 \times 2000 = 0.05733$ m³/kg
Also $\quad m_f.v_1 = A_1v_1$
$\quad A_1 = 600 \times 0.05733/3600 \times 300 = 31.85$ mm² $\qquad \ldots\text{ANS}$
$\quad P_2v_2 = RT_2$
$\quad v_2 = 8.314 \times 269.18/\,29 \times 500 = 0.1543$ m³/kg
Now $\quad m_f.v_2 = A_2v_2$
$\quad A_2 = 600 \times 0.1543/3600 \times 594 = 43.29$ mm² $\qquad \ldots\text{ANS}$

Q. 19: 0.5kg/s of a fluid flows in a steady state process. The properties of fluid at entrance are measured as p_1 = 1.4bar, density = 2.5kg/m³, u_1 = 920Kj/kg while at exit the properties are p_2 = 5.6 bar, density = 5 kg/m³, u_2 = 720Kj/kg. The velocity at entrance is 200m/sec, while at exit it is 180m/sec. It rejects 60kw of heat and rises through 60m during the flow. Find the change of enthalpy and the rate of work done. *(May-03)*

Sol: Given that:
$\quad m_f = 0.5$kg/s
$\quad P_1 = 1.4$bar,
density = 2.5kg/m³,
$\quad u_1 = 920$Kj/kg
$\quad P_2 = 5.6$ bar,
density = 5 kg/m³,
$\quad u_2 = 720$Kj/kg.
$\quad V_1 = 200$m/sec
$\quad V_2 = 180$m/sec
$\quad Q = -60$kw
$\quad Z_2 - Z_1 = 60$m
$\quad \Delta h = ?$
$\quad W_S = ?$
Since $\quad h_2 - h_1 = \Delta U + \Delta Pv$
$\quad h_2 - h_1 = [U_2 - U_1 + (P_2/\rho_2 - P_1/\rho_1)]$
$\quad\quad\quad = [(720 - 920) \times 10^3 + (5.6/5 - 1.4/2.5) \times 10^5]$
$\quad\quad\quad = [-200 \times 10^3 + 0.56 \times 10^5] = -144$KJ/kg
$\quad \Delta H = m_f \times (h_2 - h_1) = 0.5 \times (-144)$ Kj/kg = -72KJ/sec $\qquad \ldots\text{ANS}$

Now Applying SFEE

$$-Q - W_S = m_f[(h_2 - h_1) + \tfrac{1}{2}(V_2^2 - V_1^2) + g(Z_2 - Z_1)]$$
$$60 \times 10^3 - W_S = 0.5[-144 \times 10^3 + (180^2 - 100^2)/2 + 9.81 \times 60]$$
$$W_S = 13605.7 \text{ W} = 136.1 \text{KW} \quad \text{.......ANS}$$

Q. 20: Carbon dioxide passing through a heat exchanger at a rate of 100kg/hr is cooled down from 800^0C to 50^0C. Write the steady flow energy equation. Assuming that the change in pressure, kinetic and potential energies and flow work interaction are negligible, determine the rate of heat removal. (Take Cp = 1.08Kj/kg-K) *(Dec-03)*

Sol: Given data:
$$m_f = 100 \text{Kg/hr} = 100/3600 \text{ Kg/sec} = 1/36 \text{ Kg/sec}$$
$$T_1 = 800^0\text{C}$$
$$T_2 = 50^0\text{C}$$
$$C_p = 1.08 \text{Kj/kg-K}$$
Rate of heat removal = Q = ?
Now Applying SFEE
$$Q - W_S = m_f[(h_2 - h_1) + \tfrac{1}{2}(V_2^2 - V_1^2) + g(Z_2 - Z_1)]$$
Since change in pressure, kinetic and potential energies and flow work interaction are negligible, i.e.;
$$W_S = \tfrac{1}{2}(V_2^2 - V_1^2) = g(Z_2 - Z_1) = 0$$
Now
$$Q = m_f[(h_2 - h_1)] = m_f[C_p.dT] = (1/36) \times 1.08(800 - 50)$$
$$Q = 22.5 \text{ KJ/sec} \quad \text{.......ANS}$$

Q. 22: A reciprocating air compressor takes in 2m³/min of air at 0.11MPa and 20^0C which it delivers at 1.5MPa and 111^0C to an after cooler where the air is cooled at constant pressure to 25^0C. The power absorbed by the compressor is 4.15KW. Determine the heat transfer in (a) Compressor and (b) cooler. C_P for air is 1.005KJ/Kg-K.

Sol: $v_1 = 2\text{m}^3/\text{min} = 1/30 \text{ m}^3/\text{sec}$
$$P_1 = 0.11 \text{MPa} = 0.11 \times 10^6 \text{ N/m}^2$$
$$T_1 = 20^0\text{C}$$
$$P_2 = 1.5 \times 10^6 \text{ N/m}^2$$
$$T_2 = 111^0\text{C}$$
$$T_3 = -25^0\text{C}$$
$$W = 4.15 \text{KW}$$
$$C_P = 1.005 \text{KJ/kgk}$$
$$Q_{1-2} = ? \text{ and } Q_{2-3} = ?$$
From SFEE
$$Q - W_S = m_f[(h_2 - h_1) + \tfrac{1}{2}(V_2^2 - V_1^2) + g(Z_2 - Z_1)]$$
There is no data about velocity and elevation so ignoring KE and PE
$$Q_{1-2} - W_{1-2} = m[cp(T_2 - T_1)] \quad \text{...(i)}$$
Now $P_1v_1 = mRT_1$
$$m = (0.11 \times 10^6 \times 1/30)/(287 \times 293) = 0.0436 \text{Kg/sec}; R = 8314/29 = 287 \text{ For Air}$$
From equation *(i)*
$$Q_{1-2} - 4.15 \times 10^3 = 0.0436[1.005 \times 10^3 (111 - 20)]$$
$$Q_{1-2} = 8.137 \text{KJ/sec} \quad \text{.......ANS}$$

For process 2 – 3; $W_{2-3} = 0$

$Q_{2-3} - W_{2-3} = m[cp(T_2 - T_1)]$

$Q_{2-3} - 0 = 0.0436[1.005 \times 10^3 (-111 + 25)]$

$Q_{2-3} = -3.768 KJ/sec$ ....ANS

Q. 22: A centrifugal air compressor delivers 15Kg of air per minute. The inlet and outlet conditions are

At inlet: Velocity = 5m/sec, enthalpy = 5KJ/kg

At out let: Velocity = 7.5m/sec, enthalpy = 173KJ/kg

Heat loss to cooling water is 756KJ/min find:
(1) The power of motor required to drive the compressor.
(2) Ratio of inlet pipe diameter to outlet pipe diameter when specific volumes of air at inlet and outlet are $0.5 m^3/kg$ and $0.15 m^3/kg$ respectively. Inlet and outlet lines are at the same level.

Sol: Device: Centrifugal compressor

Mass flow rate m_f = 15Kg/min

Condition at inlet:

V_1 = 5m/sec; h_1 = 5KJ/kg

Condition at exit:

V_2 = 7.5m/sec; h_3 = 173KJ/kg

Heat loss to cooling water Q = –756 KJ/min

From SFEE

$Q - W_S = m_f[(h_2 - h_1) + \frac{1}{2}(V_2^2 - V_1^2) + g(Z_2 - Z_1)]$

$-756 - W_S = 15[(173 - 5) + \frac{1}{2}(7.5^2 - 5^2)/1000 + 0]$

W_S = –3276.23KJ/min = - 54.60KJ/sec ....ANS   Fig 2.6

(-ive sign indicate that work done on the system) Thus the power of motor required to drive the compressor is 54.60KW.

Mass flow rate at inlet = Mass flow rate at outlet = 15kg/min = 15/60 kg/sec

Mass flow rate at inlet = $m_{f1} = A_1.V_1/v_1$

$15/60 = A_1 \times 5/0.5$

$A_1 = 0.025 m^2$

Now; Mass flow rate at outlet = $m_{f2} = A_2.V_2/v_2$

$15/60 = A_2 \times 7.5/0.15$

$A_2 = 0.005 m^2$

$A_1/A_2 = 5$

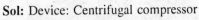

$\Pi d_1^2/\Pi d_1^2 = 5$

$d_1/d_2 = 2.236$ ....ANS

Thus the ratio of inlet pipe diameter to outlet pipe diameter is 2.236

Q. 23: 0.8kg/s of air flows through a compressor under steady state condition. The properties of air at entrance are measured as p_1 = 1bar, velocity 10m/sec, specific volume $0.95 m^3/kg$ and internal energy u_1 = 30KJ/kg while at exit the properties are p_2 = 8 bar, velocity 6m/sec, specific volume $0.2 m3/kg$ and internal energy u_2 = 124KJ/kg. Neglecting the change in potential energy. Determine the power input and pipe diameter at entry and exit.   *(May-05(C.O.))*

Sol: Device: Centrifugal compressor
Mass flow rate $m_f = 0.8$ Kg/sec
Condition at inlet:
$P_1 = 1$ bar $= 1 \times 10^5$ N/m² $V_1 = 10$ m/sec;
$u_1 = 30$ KJ/kg $v_1 = 0.95$ m³/kg
Condition at exit:
$P_2 = 8$ bar $= 8 \times 10^5$ N/m²
$V_2 = 6$ m/sec;
$u_2 = 124$ KJ/kg
$v_2 = 0.2$ m³/kg

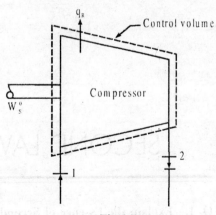

Fig 2.7

The change in enthalpy is given by
$$h_2 - h_1 = (u_2 + P_2 U_2) - (u_1 + P_1 U_1)$$
$$= (124 \times 10^3 + 8 \times 10^5 \times 0.2)$$
$$- (30 \times 10^3 + 1 \times 10^5 \times 0.95)$$
$$= 159000 \text{ J/Kg} = 159 \text{ KJ/kg} \qquad \ldots(i)$$

Heat loss to cooling water
$$Q = -(dU + dW) = -(U_2 - U_1) - W_s \text{ KJ/sec}$$
$$Q = -(30 - 124) - W_s = -96 - W_s \qquad \ldots(ii)$$

From Sfee
$$Q - W_s = m_f [(h_2 - h_1) + 1/2 (v_2^2 - V_1^2) + g(Z_2 - Z_1)]$$
$$-96 - W_s - W_s = 0.8 [159 + 1/2 (6^2 - 10^2)]$$
$$-96 - 2W_s = 0.8 [159 + 1/2 (6^2 - 10^2)]$$
$$-96 - 2W_s = 101.6$$
$$W_s = -98.8 \text{ KJ/sec} \qquad \text{.......ANS}$$

(–ive sign indicate that work done on the system)
Thus the power of motor required to drive the compressor is 54.60KW
Mass flow rate at inlet = Mass flow rate at outlet
$$= 0.8 \, A_1 \times 10/0.95$$
$$A_1 = 0.076 \text{ m}^2$$
$$\Pi/4 . d_{intel}^2 = 0.076$$
$$d_{inlet} = 0.096 \text{ m} = 96.77 \text{ mm} \qquad \text{.......ANS}$$

Now; Mass flow rate of outlet = $m_{f2} = A_2 . V_2/u_2$
$$0.8 = A_2 \times 6/0.2$$
$$A_2 = 0.0266 \text{ m}^2$$
$$\Pi/4 . d_{outlet}^2 = 0.0266$$
$$d_{outlet} = 0.03395 \text{ m} = 33.95 \text{ mm} \qquad \text{.......ANS}$$

CHAPTER 3

SECOND LAW OF THERMODYNAMICS

Q. 1: Explain the Essence of Second Law?
Sol: First law deals with conservation and conversion of energy. But fails to state the conditions under which energy conversion are possible. The second law is directional law which would tell if a particular process occurs or not and how much heat energy can be converted into work.

Q. 2: Define the following terms:
1. Thermal reservoir,
2. Heat engine,
3. Heat pump

(Dec-05)

Or

Write down the expression for thermal efficiency of heat engine and coefficient of performance (COP) of the heat pump and refrigerator. *(Dec-02,04)*

Sol: Thermal Reservoir. A thermal reservoir is the part of environment which can exchange heat energy with the system. It has sufficiently large capacity and its temperature is not affected by the quantity of heat transferred to or from it. The temperature of a heat reservoir remain constant. The changes that do take place in the thermal reservoir as heat enters or leaves are so slow and so small that processes within it are quasistatic. The reservoir at high temperature which supplies heat to the system is called HEAT SOURCE. For example: Boiler Furnace, Combustion chamber, Nuclear Reactor. The reservoir at low temperature which receives heat from the system is called HEAT SINK. For example: Atmospheric Air, Ocean, river.

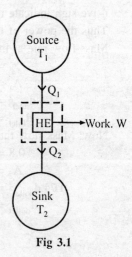

Fig 3.1

HEAT ENGINE. A heat engine is such a thermodynamics system that operates in a cycle in which heat is transferred from heat source to heat sink. For continuous production of work. Both heat and work interaction take place across the boundary of the engine. It receive heat Q_1 from a higher temperature reservoir at T_1. It converts part of heat Q_1 into mechanical work W_1. It reject remaining heat Q_2 into sink at T_2. There is a working substance which continuously flow through the engine to ensure continuous/cyclic operation.

Performance of HP: Measured by thermal efficiency which is the degree of useful conversion of heat received into work.

η_{th} = Net work output/ Total Heat supplied = $W/Q_1 = (Q_1 - Q_2) / Q_1$
$\eta_{th} = 1 - Q_2/Q_1 = 1 - T_2/T_1$; Since $Q_1/Q_2 = T_1/T_2$

Or, Thermal efficiency is defined as the ratio of net work gained (output) from the system to the heat supplied (input) to the system.

Heat Pump: Heat pump is the reversed heat engine which removes heat from a body at low temperature and transfer heat to a body at higher temperature. It receive heat Q_2 from atmosphere at temperature T_2 equal to atmospheric temperature.

It receive power in the form of work 'W' to transfer heat from low temperature to higher temperature. It supplies heat Q_1 to the space to be heated at temperature T_1.

Performance of HP: is measured by coefficient of performance (COP). Which is the ratio of amount of heat rejected by the system to the mechanical work received by the system.

$(COP)_{HP} = Q_1/W = Q1/(Q_1 - Q_2) = T_1/(T_1 - T_2)$

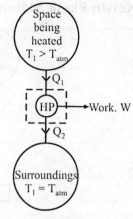

Fig 3.2

Refrigerator

The primary function of a heat pump is to transfer heat from a low temperature system to a high temperature system, this transfer of heat can be utilized for two different purpose, either heating a high temperature system or cooling a low temperature system. Depending upon the nature of use. The heat pump is said to be acting either as a heat pump or as a refrigerator. If its purpose is to cause heating effect it is called operating as a H.P. And if it is used to create cold effect, the HP is known to be operating as a refrigerator.

$(COP)_{ref}$ = Heat received/ Work Input
$= Q_2/W = Q_2/(Q_1 - Q_2)$
$(COP)_{ref} = Q_2/(Q_1 - Q_2) = T_2/(T_1 - T_2)$
$(COP)_{HP} = (COP)_{ref} + 1$

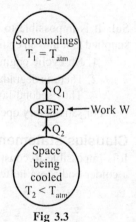

Fig 3.3

COP is greater when heating a room than when cooling it.

Q. 3: State and explain the second law of thermodynamics? (Dec-02)

Sol: There are many different way to explain second law; such as
1. Kelvin Planck Statement
2. Clausius statement
3. Concept of perpetual motion m/c of second kind
4. Principle of degradation of energy
5. Principle of increase of entropy

Among these the first and second are the basic statements while other concept/principle are derived from them.

Kelvin Plank is applicable to HE while the clausius statement is applicable to HP.

Kelvin Plank Statement

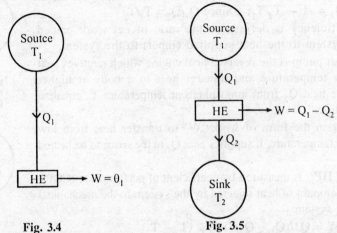

Fig. 3.4 Fig. 3.5

Sol: It is impossible to construct such a H.E. that operates on cyclic process and converts all the heat supplied to it into an equivalent amount of work. The following conclusions can be made from the statement
1. No cyclic engine can converts whole of heat into equivalent work.
2. There is degradation of energy in a cyclic heat engine as some heat has to be degraded or rejected. Thus second law of thermodynamics is called the law of degradation of energy.

For satisfactory operation of a heat engine there should be a least two heat reservoirs source and sink.

Clausius statement

It is impossible to construct such a H.P. that operates on cyclic process and allows transfer of heat from a colder body to a hotter body without the aid of an external agency.

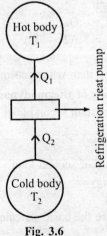

Fig. 3.6

Equivalent of Kalvin Plank and Clausius statement

The Kalvin plank and clausius statements of the second law and are equivalent in all respect. The equivalence of the statement will be proved by the logic and violation of one statement leads to violation of second statement and vice versa.

Violation of Clausius statement

A cyclic HP transfer heat from cold reservoir(T_2) to a hot reservoir(T_1) with no work input. This violates clausius statement.

A Cyclic HE operates between the same reservoirs drawing a heat Q_1 and producing W as work. As HP is supplying Q_1 heat to hot reservoir, the hot reservoir can be eliminated. The HP and HE constitute a HE operating in cycle and producing work W while exchanging heat with one reservoir(Cold) only. This violates the K-P statement

Violation of K-P Statement

A HE produce work 'W' by exchanging heat with one reservoir at temperature T_1 only. The K-P statement is violated.

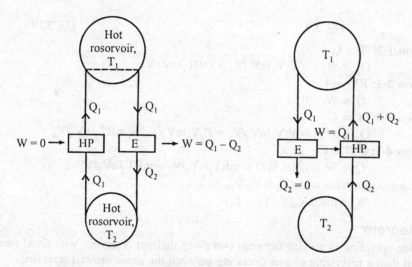

Fig. 3.7 Violation of Clascius Statement   Fig. 3.8 Violation of K-P Statement

H.P. is extracting heat Q_2 from low temperature (T_2) reservoir and discharging heat to high temperature (T_1) reservoir and getting work 'W'. The HE and HP together constitute a m/c working in a cycle and producing the sole effect of transmitting heat from a lower temperature to a higher temperature. The clausius statement is violated.

Q.No-4: State and prove the Carnot theorem   *(May – 02, Dec-02)*

Sol: Carnot Cycle: Sadi carnot; based on second law of thermodynamics introduced the concepts of reversibility and cycle in 1824. He show that the temperature of heat source and heat sink are the basis for determining the thermodynamics efficiency of a reversible cycle. He showed that all such cycles must reject heat to the sink and efficiency is never 100%. To show a non existing reversible cycle, Carnot invented his famous but a hypothetical cycle known as Carnot cycle.Carnot cycle consist of two isothermal and two reversible adiabatic or isentropic operation. The cycle is shown in P-V and T-S diagrams

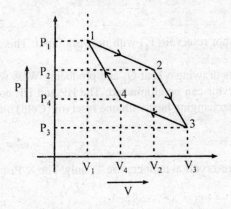

Fig. 3.9 Fig. 3.10

Operation 1-2: T = C
$$Q_1 = W_{1-2} = P_1 V_1 \ln V_2/V_1 = mRT_1 \ln V_2/V_1$$
Operation 2-3: $PV^\gamma = C$
$$Q = W = 0$$
Operation 3-4: T = C
$$Q_2 = W_{3-4} = P_3 V_3 \ln V_4/V_3 = P_3 V_3 \ln V_4/V_3 = mRT_2 \ln V_3/V_4$$
Operation 4-1: $PV^\gamma = C$
$$Q = W = 0 \text{Net WD} = mRT_1 \ln V_2/V_1 - mRT_2 \ln V_3/V_4 \ ;$$
Since compression ratio = $V_3/V_4 = V_2/V_1$, $T_2 = T_3$
$$W = mR \ln V_3/V_4 (T_1 - T_3)$$

Carnot Theorem

No heat engine operating in a cycle between two given thermal reservoir, with fixed temperature can be more efficient than a reversible engine operating between the same thermal reservoir.

- Thermal efficiency η_{th} = Work out/Heat supplied
- Thermal efficiency of a reversible engine (η_{rev})
$$\eta_{rev} = (T_1 - T_2)/T_1 \ ;$$
- No engine can be more efficient than a reversible carnot engine i.e $\eta_{rev} > \eta_{th}$

Carnot Efficiency

$$\eta = (\text{Heat added} - \text{Heat rejected}) / \text{Heat added} = [mRT_2 \ln V_2/V_1 - mRT_4 \ln V_3/V_4]/ mRT_2 \ln V_2/V_1$$
$$\eta = 1 - T_1/T_2$$

Condition:
1. If $T_1 = T_2$; No work, $\eta = 0$
2. Higher the temperature diff, higher the efficiency
3. For same degree increase of source temperature or decrease in sink temperature carnot efficiency is more sensitive to change in sink temperature.

Q.No-5. Explain Clausius inequality *(Dec-02, 05)*

Sol: When ever a closed system undergoes a cyclic process, the cyclic integral $\oint dQ/T$ is less than zero (i.e., negative) for an irreversible cyclic process and equal to zero for a reversible cyclic process.

The efficiency of a reversible H.E. operating within the temperature T_1 & T_2 is given by:

$$\eta = (Q_1 - Q_2)/Q_1 = (T_1 - T_2)/T_1 = 1 - (T_2/T_1)$$
i.e., $\quad 1 - Q_2/Q_1 = 1 - T_2/T_1$; or
$$Q_2/Q_1 = T_2/T_1 ; \text{ or; } Q_1/T_1 = Q_2/T_2$$
Or; $\quad Q_1/T_1 - (-Q_2/T_2) = 0$; Since Q_2 is heat rejected so –ive
$$Q_1/T_1 + Q_2/T_2 = 0;$$

or $\oint dQ/T = 0$ for a reversible engine. ....(i)

Now the efficiency of an irreversible H.E. operating within the same temperature limit T_1 & T_2 is given by
$$\eta = (Q_1 - Q_2)/Q_1 < (T_1 - T_2)/T_1$$
i.e., $\quad 1 - Q_2/Q_1 < 1 - T_2/T_1$;
or; $\quad -Q_2/Q_1 < T_2/T_1$;
or; $\quad Q_1/T_1 < Q_2/T_2$
Or; $\quad Q_1/T_1 - (-Q_2/T_2) < 0$;
Since Q_2 is heat rejected so –ive
$$Q_1/T_1 + Q_2/T_2 < 0;$$

or $\oint dQ/T < 0$ for an irreversible engine. ...(ii)

Combine equation (i) and (ii); we get
$$\oint dQ/T \, d \leq 0$$

The equation for irreversible cyclic process may be written as:
$$\oint dQ/T + I = 0$$

I = Amount of irreversibility of a cyclic process.

Q. 6: Heat pump is used for heating the premises in winter and cooling the same during summer such that temperature inside remains 25°C. Heat transfer across the walls and roof is found 2MJ per hour per degree temperature difference between interior and exterior. Determine the minimum power required for operating the pump in winter when outside temperature is 1°C and also give the maximum temperature in summer for which the device shall be capable of maintaining the premises at desired temperature for same power input. *(May-02)*

Sol: Given that:
Temperature inside the room $T_1 = 25^0C$
Heat transferred across the wall = 2MJ/hr°C
Outside temperature $T_2 = 1^0C$
To maintain the room temperature 25°C the heat transferred to the room = Heat transferred across the walls and roof.
$Q_1 = 2 \times 10^6 \times (25 - 1)/3600 = 1.33 \times 10^4$ J/sec = 13.33KW

For heat pump
COP= $T_1/(T_1 - T_2) = 298/(298 - 274) = 12.4167$
Also COP = Heat delivered / Net work done = Q_1/W_{net}
$\quad 12.4167 = 1.333 \times 10^4/W_{net}$
$\quad W_{net} = 1073.83$ J/sec = 1.074KW

Thus the minimum power required by heat pump = 1.074KW

Again, if the device works as refrigerator (in summer)

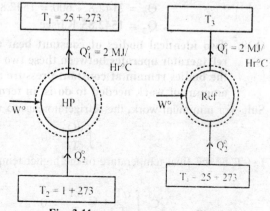

Fig. 3.11   Fig. 3.12

Heat transfer $Q_1 = \{2 \times 10^6/(60 \times 60)\} \times (T_3 - 298)$ Watt
Now $\quad$ COP $= Q_1/W_{net} = T_4/(T_3 - T_4)$
$$[2 \times 10^6 \times (T_3 - 298)]/[60 \times 60 \times 1073.83] = 298/(T_3 - 298)$$
On solving
$$T_3 = 322K = 49^0C \qquad \text{.......ANS}$$

Q. 7: A reversible heat engine operates between temperature 800^0C and 500^0C of thermal reservoir. Engine drives a generator and a reversed carnot engine using the work output from the heat engine for each unit equality. Reversed Carnot engine abstracts heat from 500^0C reservoir and rejected that to a thermal reservoir at 715^0C. Determine the heat rejected to the reservoir by the reversed engine as a fraction of heat supplied from 800^0C reservoir to the heat engine. Also determine the heat rejected per hour for the generator output of 300KW. *(May-01)*

Sol: Given that
$$T_1 = 800^0C = 1073K$$
$$T_2 = 500^0C = 773K$$
$$T_3 = 800^0C = 988K$$
$$\eta_{rev} = (Q_1 - Q_2)/Q_1$$
$$= (T_1 - T_2)/T_1 = W/Q_1$$
$$= (1073 - 773)/1073 = W/Q_1$$
$$W = 0.28Q_1 \qquad ...(i)$$

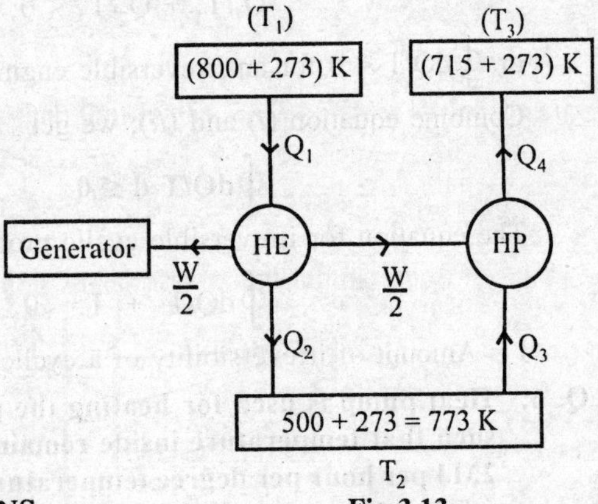

Now for H.P.
$$Q_4/(Q_3 - Q_4) = T_3/(T_3 - T_2)$$
$$Q_4/(W/2) = T_3/(T_3 - T_2)$$
$$Q_4 = (W/2)[T_3/(T_3 - T_2)]$$
$$= (0.28Q_1/2)\,[T_3/(T_3 - T_2)]$$
$$Q_4 = (0.28Q_1/2)\,[988/(988 - 773)]$$
$$= 0.643Q_1$$
$$\mathbf{Q_4 = 0.643Q_1} \qquad \text{.......ANS}$$

Now if W/2 = 300; W = 600KW
$$0.28Q_1 = 600$$
$$\mathbf{Q_1 = 2142.8\,KJ/sec} \qquad \text{.......ANS}$$
Since $\quad Q_1 = W + Q_2$
$$Q_2 = 2142.8 - 600 = 1542.85\text{ KJ/sec}$$
$$\mathbf{Q_2 = 1542.85\text{ KJ/sec}} \qquad \text{.......ANS}$$

Fig 3.13

Q. 8: Two identical bodies of constant heat capacity are at the same initial temperature T_1. A refrigerator operates between these two bodies until one body is cooled to temperature T_2. If the bodies remain at constant pressure and undergo no change of phase, find the minimum amount of work needed to do is, in terms of T_1, T_2 and heat capacity. *(Dec – 02, May - 05)*

Sol: For minimum work, the refrigerator has to work on reverse Carnot cycle.
$$\oint \frac{dQ}{T} = 0$$
Let T_f be the final temperature of the higher temperature body and let 'C' be the heat capacity.
$$C \int_{T_i}^{T_f} \frac{\alpha T}{T} + C\,\frac{\alpha T}{T} = 0$$

$$C \log_e \frac{T_f}{T_i} + C\log_e \frac{T_2}{T_i} = 0$$

$$\log_e \frac{T_f T_2}{T_i^2} = \log_e 1 \Rightarrow \frac{T_f T_2}{T_i^2} = 1$$

$$T_f = \frac{T_i^2}{T_2}$$

work required (minimum)

$$= C \int_{T_i}^{T_f} dT - C \int_{T_2}^{T_i} dT$$
$$= C (T_f - T_i) - C (T_i - T_2)$$
$$= C [T_f + T_2 - 2T_i]$$

$$w = C \frac{T_i^2}{T_2} \left[\frac{T_i^2}{T_2} + T_2 - 2T_i \right]$$

Q. 9: A reversible heat engine operates between two reservoirs at temperature of 600^0C and 40^0C. The Engine drives a reversible refrigerator which operates between reservoirs at temperature of 40^0C and -20^0C. The heat transfer to the heat engine is 2000KJ and net work output of combined engine refrigerator plant is 360KJ. Evaluate the heat transfer to the refrigerator and the net heat transfer to the reservoir at 40^0C. *(Dec – 05)*

Sol : $T_1 = 600 + 273 = 873K$

$T_2 = 40 + 273 = 313K$

$T_3 = -20 + 273 = 253K$

Heat transfer to engine = 200KJ
Net work output of the plant = 360KJ
Efficiency of heat engine cycle,

$\eta = 1 - T_2/T_1 = 1 - 313/873 = 0.642$
$W_1/Q_1 = 0.642 W_1 = 0.642 \times 2000 = 1284KJ$ ...(i)
C.O.P. $= T_3/(T_2 - T_3) = 253/(313 - 253) = 4.216$
$Q_4/W_2 = 4.216$ ...(ii)
$W_1 - W_2 = 360; W_2 = W_1 - 360$
$W_2 = 1284 - 360 = 924KJ$

From equation (ii)

$Q_4 = 4.216 \times 924 = 3895.6KJ$ ....ANS
$Q_3 = Q_4 + W_2 = 3895.6 + 924$
$Q_3 = 4819.6KJ$ ....ANS
$Q_2 = Q_1 - W_1 = 2000 - 1284$
$Q_2 = 716KJ$ ....ANS

Heat rejected to reservoir at $40^0C = Q_2 + Q_3 = 716 + 4819.6$
Heat rejected to reservoir at $40^0C = 5535.6KJ$ ....ANS
Heat transfer to refrigerator, $Q_4 = 3895.6KJ$ ....ANS

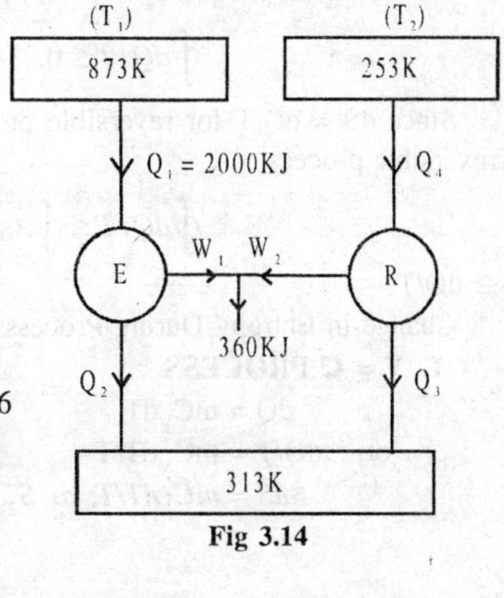

Fig 3.14

Q. 10: A cold storage of 100Tonnes of refrigeration capacity runs at 1/4th of its carnot COP. Inside temperature is – 15⁰C and atmospheric temperature is 35⁰C. Determine the power required to run the plant. Take one tonnes of refrigeration as 3.52KW. *(Dec – 03(C.O.))*

Sol: Given that T_{atm} = 35 + 273 = 308K

T_{inside} = -15 + 273 = 258K

COP = T_{inside} /(T_{atm} - T_{inside}) = 258 /(308 - 258) = 5.16 ...(i)

Again COP = Q/W

5.16 x ¼ = 100 x 3.52/ W

W = 272.87 KW ....ANS

Power required to run the plant is 272.87KW

Q. 11: Define entropy and show that it is a property of system. *(Dec-05)*

Sol: Entropy is a thermodynamics property of a system which can be defined as the amount of heat contained in a substance and its interaction between two state in a process. Entropy increase with addition of heat and decrease when heat is removed.

dQ = T.dS; T = Absolute Temperature and dS = Change in entropy.

dS = dQ/ T

T-S Diagrams

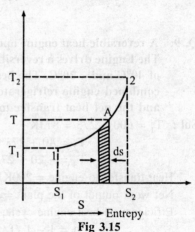

Fig 3.15

$$\int_1^2 dS = \int_1^2 dQ/T$$

The area under T-S diagram represent the heat added or rejected. Entropy is a point function From first las

dQ = dU + dW

T.dS = dU + P.dV

Carnot efficiency η = (T_1– T_2)/T_1 = dW/dQ

dW = η.dQ; If T_1 – T_2 = 1; η =1/T

dW = dQ/T = dS; if Temperature difference is one.

dS represents maximum amount of work obtainable per degree in temperature. Unit of Entropy = KJ/K

Principle of Entropy

From claucius inequality

$$\oint dQ/T \le 0$$

Since dS = dQ/T for reversible process and dS > dQ/T for irreversible process

$$\oint dQ/T \le \oint dS;\ or\quad dQ/T\ d \le dS\ or\ dS$$

e ≥ dQ/T

Change in Entropy During Process

1. V = C PROCESS

dQ = $mC_V dT$

or, dQ/T = $mC_V dT/T$

dS = $mC_V dT/T$; or $S_2 - S_1 = mC_V ln P_2/P_1$

2. P = C; PROCESS
$$dQ = mC_p dT$$
or, $dQ/T = mC_p dT/T$
$$dS = mC_p dT/T; \text{ or } S_2 - S_1 = mC_p \ln T_2/T_1 = mC_p \ln V_2/V_1$$

3. T = C; PROCESS
$$dQ = mRT \ln V_2/V_1$$
or, $dQ/T = (mRT/T) \ln V_2/V_1$
or $S_2 - S_1 = mR \ln V_2/V_1 = m(C_P - C_V) \ln V_2/V_1 = mR \ln P_1/P_2$

4. $PV^\gamma = C$; PROCESS
$$dQ = 0; \; dS = 0$$

5. $PV^n = C$; PROCESS
$$dQ = [(\gamma - n)/(\gamma - 1)] \, dW$$
$$= [(\gamma - n)/(\gamma - 1)] \, PdV$$
$$dQ/T = [(\gamma - n)/(\gamma - 1)] \, PdV/T$$
$$dS = [(\gamma - n)/(\gamma - 1)] \, mRdV/V$$
or $S_2 - S_1 = [(\gamma - n)/(\gamma - 1)] \, mR \ln V_2/V_1$

Q. 12: Show that the entropy change in a process when a perfect gas changes from state 1 to state 2 is given by $S_2 - S_1 = C_p \ln T_2/T_1 + R \ln P_1/P_2$. *(May–02, 03)*

Using clausius equality for reversible cycle,, we have

$$\oint \left(\frac{\delta q}{T} \right)_{rev.} = 0 \quad \ldots(i)$$

Let a control mass system undergoes a reversible process from state 1 to state 2 along path A and let the cycle be completed by returning back through path C, which is also reversible, then

$$\int_1^2 \left(\frac{\delta q}{T} \right)_A + \int_1^2 \left(\frac{\delta q}{T} \right)_C = 0 \quad \ldots(ii)$$

Also we can move through path B and C then

$$\int_1^2 \left(\frac{\delta q}{T} \right)_B + \int_1^2 \left(\frac{\delta q}{T} \right)_C = 0$$

From (*ii*) and (*iii*)

$$\int_1^2 \left(\frac{\delta q}{T} \right)_A - \int_1^2 \left(\frac{\delta q}{T} \right)_B = 0$$

$$\int_1^2 \left(\frac{\delta q}{T} \right)_A = \int_1^2 \left(\frac{\delta q}{T} \right)_B$$

The quantity $\int \left(\frac{\delta q}{T} \right)$ is independent of path A and B but depends on end states 1 and 2. Therefore this is point function and not a path function, and hence a property of the system.

$$\int_1^2 \left(\frac{\delta q}{T} \right)_{rev} = \int_1^2 ds$$

where; s is specific entropy.

or
$$S_2 - S_1 = \int_1^2 \left(\frac{\delta q}{T} \right)_{rev}$$

Also; $\delta q = T \cdot ds$ (for reversible process)
From first law
$$\delta q = du + P\,dv$$
$$h = u + Pv$$
$$dh = du + Pdv + vdP$$
$$dh - v\,dP = du + P\,dv$$
Using equations
$$T\,ds = du + P\,dv$$
$$\int_1^2 ds = \int_1^2 \frac{du}{T} + \int_1^2 \frac{P}{T}\,dv$$
$$s_2 - s_1 = C_v \int_1^2 \frac{dT}{T} + \int_1^2 \frac{R}{v}\,dv$$
$$s_2 - s_1 = C_v \ln(T_2/T_1) + R \ln(v_2/v_1)$$
$$T\,ds = dh - v\,dP$$
$$\int_1^2 ds = \int_1^2 \frac{dh}{T} + \int_1^2 \frac{v}{T}\,dP$$
$$s_2 - s_1 = \int_1^2 C_p \frac{dT}{T} - \int_1^2 \frac{R}{P}\,dP$$
$$s_2 - s_1 = C_p \ln(T_2/T_1, - R \ln P_2\, P_1)$$

Q. 13: 5Kg of ice at -10^0C is kept in atmosphere which is at 30^0C. Calculate the change of entropy of universe when if melts and comes into thermal equilibrium with the atmosphere. Take latent heat of fusion as 335KJ/kg and sp. Heat of ice is half of that of water. *(Dec- 05)*

Sol: Mass of ice, m = 5Kg
Temperature of ice = -10^0C = 263K
Temperature of atmosphere = 30^0C = 303K
Heat absorbed by ice from atmosphere = Heat in solid phase + latent heat + heat in liquid phase
$$= m_i C_i dT + M_i L_i + m_w C_w dT$$
$$= 5 \times 4.187/2\,(0 + 10) + 5 \times 335 + 5 \times 4.187 \times (30 - 0)$$
$$= 104.675 + 1675 + 628.05$$
$$Q = 2407.725 KJ$$
Entropy change of atmosphere $(\Delta s)_{atm} = -Q/T = -2407.725/303$
$$(\Delta s)_{atm} = -7.946 KJ/k$$
Entropy change of ice $(\Delta s)_{ice}$
= Entropy change as ice gets heated from -10^0C to 0^0C + Entropy change as ice melts at 0^0C to water at 0^0C + Entropy change of water as it gets heated from 0^0C to 30^0C
$$= \int dQ/T + \int dQ/T$$
$$= m\left[\int_{263}^{273} C_{ice}\cdot dt/T + L/273 + \int_{273}^{303} C_W \cdot dT/T\right]$$
$$= 5[(4.18/2)\ln 273/263 + 335/273 + 4.18\ln 303/273]$$

$$= 5 \times 1.7409 = 8.705 \text{KJ}$$

Entropy of universe = Entropy change of atmosphere $(\Delta s)_{atm}$ + Entropy change of ice $(\Delta s)_{ice}$

$$= -7.946 \text{KJ/k} + 8.705 \text{KJ}$$
$$= 0.7605329 \text{KJ/kg} \quad \text{.......ANS}$$

Q. 14: $0.05 m^3$ of air at a pressure of 8bar and $280^0 C$ expands to eight times its original volume and the final temperature after expansion is $25^0 C$. Calculate change of entropy of air during the process. Assume $C_P = 1.005 \text{KJ/kg} - k$; $C_V = 0.712 \text{KJ/kg} - k$. *(Dec–01)*

Sol: $V_1 = 0.05 m^3$

$P_1 = 8 \text{bar} = 800 \text{KN/m}^2$
$T_1 = 280^0 C = 553 K$
$V_2 = 8V_1 = 0.4 m^3$
$T_2 = 298 K$
$dS = ?$
$C_P = 1.005 \text{KJ/kg} - k$;
$C_V = 0.712 \text{KJ/kg} - k$.
$R = C_P - C_V = 0.293 \text{KJ/kg}$
$P_1 V_1 = mRT_1$
$m = P_1 V_1 / RT_1 = (800 \times 0.05)/(0.293 \times 553) = 0.247 \text{Kg}$ ...(i)
$S_2 - S_1 = mC_V \ln T_2/T_1 + mR \ln V_2/V_1$
$= 0.247 \times 0.712 \ln(298/553) + 0.247 \times 0.293 \ln 8$
$= -0.108 + 0.15049$
$S_2 - S_1 = 0.04174 \text{KJ}$ ....ANS

Q. 15: Calculate the change in entropy and heat transfer through cylinder walls, if $0.4 m^3$ of a gas at a pressure of 10bar and $200^0 C$ expands by the law $PV^{1.35}$ = Constant. During the process there is loss of 380KJ of internal energy. (Take $C_P = 1.05 \text{KJ/kg k}$ and $C_V = 0.75 \text{KJ/kgK}$)

(May – 01)

Sol: $ds = ?$

$dQ = ?$
$V_1 = 0.4 m^3$
$P_1 = 10 \text{bar} = 1000 \text{KN/m}^2$
$T_1 = 200^0 C = 473 K$
$PV^{1.35} = C$
$dU = 380 \text{KJ}$
$C_P = 1.05, C_V = 0.75$

Since $P_1 V_1 = mRT$
$m = 1000 \times 0.4/[(1.005 - 0.75) \times 473]$
$= 2.82 \text{kg}$ ...(i)
$dU = mC_V (T_2 - T_1)$
$-380 = 2.82 \times 0.75 (T_2 - 473)$
$T_2 = 292 K$ ...(ii)
$W_{1-2} = mR(T_1 - T_2)/(n-1)$
$= [2.82 \times 0.3 (473 - 292)]/(1.35 - 1)$
$= 437.5 \text{KJ}$ ...(iii)

$y = C_P/C_V = 1.05/0.75 = 1.4$
$Q_{1-2} = [(\gamma - n)W_{1-2}]/(\gamma - 1)$
$= [(1.4 - 1.35) \times 437.5]/(1.4 - 1)$
$= \mathbf{54.69 KJ}$ ....ANS

$S_2 - S_1 = [(\gamma - n)mR\ln V_2/V_1]/(\gamma - 1)$ ...(iv)

Since in isentropic process $T_1/T_2 = (V_2/V_1)^{n-1}$
$473/292 = (V_2/V_1)^{1.35-1}$
$V_2/V_1 = 3.96$; Putting in equation 4
$S_2 - S_1 = [(1.4 - 1.35) \times 2.82 \times 0.3 \times \ln 3.96]/(1.4-1)$
$S_2 - S_1 = \mathbf{0.145\ KJ/K}$ ....ANS

Q. 16: 5 m³ of air at 2 bar, 27°C is compressed up to 6 bar pressure following $PV^{1.3} = C$. It is subsequently expanded adiabatically to 2 bar. Considering the two processes to be reversible, determine the net work. Also plot the processes on T - S diagram. *(Dec–01)*

Sol: Given that :
Initial volume of air $V_1 = 5$ m³
Initial pressure of air $P_1 = 2$ bar
Final pressure = 6 bar
Compression : Rev. Polytropic process ($PV^n = C$)
Expansion :Rev. adiabatic process ($PV^{1.4} = C$)
Now, work done during process (1-2)

$${}_1W_2 = \frac{P_1V_1 - P_2V_2}{n-1}$$

also $\dfrac{V_2}{V_1} = \left(\dfrac{P_1}{P_2}\right)^{\frac{1}{n}}$ $V_2 = 5\left(\dfrac{2}{6}\right)^{\frac{1}{1.3}} = 2.148$ m³

$${}_1W_2 = \frac{2 \times 100 \times 5 - 6 \times 100 \times 2.148}{1.3 - 1} = -962.67\ kJ$$

Now, work done during expansion process (2-3)

$${}_2W_3 = \frac{P_2V_2 - P_3V_3}{\gamma - 1}$$

$\dfrac{V_3}{V_2} = \left(\dfrac{P_2}{P_3}\right)^{\frac{1}{\gamma}}$

$\Rightarrow \quad V_3 = 2.148\left(\dfrac{6}{2}\right)^{\frac{1}{1.4}} = 4.708$ m³

$${}_2W_3 = \frac{6 \times 100 \times 2.148 - 2 \times 100 \times 4.708}{1.4 - 1}$$
$= 868\ kJ$

Net work output = $W_{1-2} + W_{2-3} = -962.67 + 868 = \mathbf{-94.67}$ ....ANS

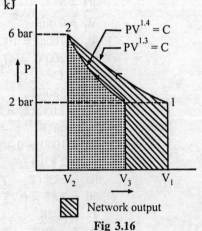
Fig 3.16

–ve sign shows that work input required for compression is more that work output obtained during expansion.

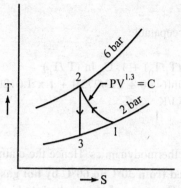

Fig. 3.17

Q. 17: One inventor claims that 2 kg of air supplied to a magic tube at 4 bar and 20°C and two equal mass streams at 1 bar are produced, one at -20°C and other at 80°C. Another inventor claims that it is also possible to produce equal mass streams, one at -40°C and other at 40°C. Whose claim is correct and why? Consider that it is an adiabatic system. (Take C_p air 1.012 kJ/kg K) *(Dec–02)*

Sol: Given that :

Air supplied to magic tube = 2 kg

Inlet condition at magic tube = 4 bar, 20°C

Exit condition : Two equal mass streams, one at – 20°C and other at 80°C for inventor 1. One at -40°C and other at 40°C for inventor-II.

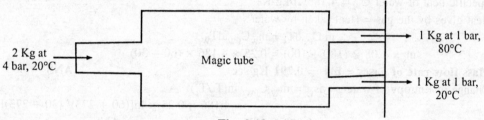

Fig. 3.18

Assume ambient condition 0°C i.e. To = 0°C

This is an irreversible process, the claim will be correct if net entropy of the universe (system surroundings) increases after the process.

Inventor I :

Total entropy at inlet condition is

$$S_1 = {}_mC_p \ln(T_1/T_0) = 2 \times 1.012 \ln\left(\frac{20+273}{273}\right) = 0.143 \text{ kJ/K}$$

Total entropy at exit condition is :

$$S_2 = 1.C_p \ln(T_1/T_0) + 1.C_p \ln(T_2/T_0)$$

$$S_2 = 1 \times 1.012 \ln\left(\frac{-20+273}{273}\right) + 1 \times 1.012 \ln\left(\frac{80+273}{273}\right)$$

$S_2 = 0.183$
$S_2 = S_1$
$\Rightarrow S_2 - S_1 > 0$

Thus the claim of inventor is acceptable

For Inventor 2

$S_2 = 1 \cdot C_P \cdot \ln(T_1/T_0) + 1 \cdot C_p \ln(T_2/T_0)$
$= 1 \times 1.012 \ln[(-40 + 273)/273)] + 1 \times 1.012 \ln[(40 + 273)/273]$
$= -0.0219 \, KJ/K$

Since $S_2 < S_1$
$S_2 - S_1 < 0$

This violates the second law of thermodynamics. Hence the claim of inventor is false. – ANS

Q. 18: **0.25Kg/sec of water is heated from 30°C to 60°C by hot gases that enter at 180°C and leaves at 80°C. Calculate the mass flow rate of gases when its $C_P = 1.08 KJ/kg\text{-}K$. Find the entropy change of water and of hot gases. Take the specific heat of water as 4.186KJ/kg-k.**

(May – 03)

Sol: Given that

Mass of water $m_w = 0.25 \, Kg/sec$
Initial temperature of water $T_{W1} = 30°C$
Final temperature of water $T_{W2} = 60°C$
Entry Temperature of hot gas $= T_{g1} = 180°C$
Exit Temperature of hot gas $= T_{g2} = 80°C$
Mass flow rate $m_f = ?$
Specific heat of gas $C_{Pg} = 1.08 \, KJ/kg\text{-}K$
Specific heat of water $C_W = 4.186 \, KJ/kg\text{-}K$
Heat gives by the gas = Heat taken by water

$$m_s C_{Ps} \cdot dT_s = m_W C_{PW} \cdot dT_W$$

$m_s \times 1.08 \times (180 - 100) = 0.25 \times 4.186 \times (60 - 30)$

Mass flow rate of gases = ms = 0.291 Kg/sec ....ANS

Change of Entropy of water = $ds_W = m_W \cdot C_{PW} \cdot \ln(T_2/T_1)$
$= 4.186 \times 0.25 \times \ln[(60 + 273)/(30 + 273)]$

Change of Entropy of water = 0.099 KJ/°K ....ANS

Change of Entropy of Hot gases = $ds_g = m_g \cdot C_{Pg} \cdot \ln(T_2/T_1)$
$= 0.291 \times 1.08 \times \ln[(80 + 273)/(180 + 273)]$

Change of Entropy of Hot gas = -0.0783 KJ/°K ....ANS

CHAPTER 4

INTRODUCTION TO I.C. ENGINE

Q. 1: What do you mean by I.C. Engine? how are they classified?
Sol.: Internal combustion engine more popularly known as I.C. engine, is a heat engine which converts the heat energy released by the combustion of the fuel inside the engine cylinder, into mechanical work. Its versatile advantages such as high efficiency light weight, compactness, easy starting, adaptability, comparatively lower cost has made its use as a prime mover universal

Classification of I.C. Engines
IC engines are classified according to:
1. **Nature of thermodynamic cycles as:**
    1. Otto cycle engine;
    2. Diesel cycle engine
    3. Dual combustion cycle engine
2. **Type of the fuel used:**
    1. Petrol engine
    2. Diesel engine.
    3. Gas engine
    4. Bi-fuel engine
3. **Number of strokes as**
    1. Four stroke engine
    2. Two stroke engine
4. **Method of ignition as:**
    1. Spark ignition engine, known as SI engine
    2. Compression ignition engine, known as C.I. engine
5. **Number of cylinder as:**
    1. Single cylinder engine
    2. Multi cylinder engine
6. **Position of the cylinder as:**
    1. Horizontal engine
    2. Vertical engine.
    3. Vee engine

4. In-line engine.
5. Opposed cylinder engine
 7. **Method of cooling as:**
 1. Air cooled engine
 2. Water cooled engine

Q. 2: Differentiate between SI and CI engines. *(May–02)*

Or

What is C.I. Engine, Why it has more compression ratio compared to S.I. Engine. *(May–05)*

Spark Ignition Engines (S.I. Engine)

It works on otto cycle. In Otto cycle, the energy supply and rejection occur at constant volume process and the compression and expansion occur isentropically. The engines working on Otto cycle use petrol as the fuel and incorporate a carburetor for the preparation of mixture of air fuel vapor in correct proportions for rapid combustion and a spark plug for the ignition of the mixture at the end of compression. The compression ratio is kept 5 to 10.5. Engine has generally high speed as compared to C.I. engine. Low maintenance cost but high running cost. These engines are also called spark ignition engines or simply S.I. Engine.

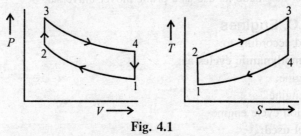

Fig. 4.1

Compression Ignition Engines (C.I. Engine)

It works on dieses cycle. In diesel engines, the energy addition occurs at constant pressure but energy rejection at constant volume. Here spark plug is replaced by fuel injector. The compression ratio is from 12 to 25. Engine has generally low speed as compared to S.I. engine. High maintenance cost but low running cost. These are known as compression ignition engines, (C.I) as the ignition is accomplished by heat of compression.

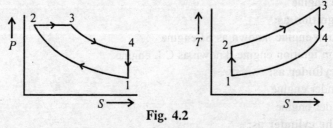

Fig. 4.2

The upper limit of compression ratio in S.I. Engine is fixed by anti knock quality of fuel. While in C.I. Engine upper limit of compression ratio is limited by thermal and mechanical stresses of cylinder material. That's way the compression ratio of S.I. engine has more compression ratio as compared to S.I. Engine.

Dual cycle is a combination of the above two cycles, where part or the energy is given a constant volume and rest at constant pressure.

Q. 3: Define Bore, stroke, compression Ratio, clearance ratio and mean effective pressure.

(Dec–01)

Or

Define clearance volume, mean effective pressure, Air standard cycle, compression Ratio.

(May–02)

Or

Air standard cycle, Cycle efficiency, mean effective pressure. *(May–03)*

Bore
The inner diameter of the engine cylinder is known as bore. It can be measured precisely by a vernier calliper or bore gauge. As the engine cylinder wears out with the passage of time, so the bore diameter changes to a larger value, hence the piston becomes lose in the cylinder, and power loss occurs. To correct this problem reboring to the next standard size is done and a new piston is placed. Bore is denoted by the letter 'D'. It is usually measured in mm (S.I. units) or inches (metric units). It is used to calculated the engine capacity (cylinder volume).

Stroke
The distance traveled by the piston from its topmost positions (also called as Top dead centre TDC), to its bottom most position (or bottom dead centre BDC) is called stroke it will be two times the crank radius. It is denoted by letter h. Units mm or inches (S.L, Metric). Now we can calculate the swept volume as follows: ($L = 2r$)

$$V_S = \left[\frac{\pi D^2}{4}\right] L$$

If D is in cm and L is also in cm than the units of V will be cm^3 which is usually written as cubic centimeter or c.c.

Clearance Volume
The volume above the T.D.C is called as clearance volume, this is provided so as to accommodate engine valves etc. this is referred as (V_C). Then total volume of the engine cylinder

$$V = V_S + V_C$$

Compression Ratio
It is calculated as follows

$$r_k = \frac{\text{Total volume}}{\text{Clearance volume}}$$

$$r_k = \frac{V_S + V_C}{V_C}$$

Mean Effective Pressure (P_m or P_{mef})
Mean effective pressure is that hypothetical constant pressure which is assumed to be acting on the piston during its expansion stroke producing the same work output as that from the actual cycle.

Mathematically,

$$P_m = \frac{\text{Work Output}}{\text{Swept volume}} = \frac{W_{net}}{(V_1 - V_2)}$$

It can also be shown as

$$P_m = \frac{\text{Area of Indicator diagram}}{\text{Length of diagram}} \times \text{constant}$$

The constant depends on the mechanism used to get the indicator diagram and has the units bar/m.

Indicated Mean Effective Pressure (P_{im})

Indicated power of an engine is given by

$$ip = \frac{P_{im} L A N K}{60,000} \quad \Rightarrow \quad P_{im} = \frac{60,000 \times i\,p}{L A N K}$$

Break Mean Effective Pressure (P_{bm})

Similarly, the brake mean effective pressure is given by

$$P_{bm} = \frac{60,000 \times b\,p}{L A N K}$$

where;
- ip = indicated power (kW)
- bp = Break Powder (kW)
- P_{im} = indicated mean effective pressure (N/m^2)
- Pbm = Break mean effective Pressure (N/m^2)
- L = length of the stroke
- A = area of the piston (m^2)
- N = number of power strokes
  = rpm for 2-stroke engines = rpm/2 for 4-stroke
- K = no. of cylinder.

Q. 4: Write short notes on Indicator diagram and indicated power. *(Dec–03)*

Sol.: An indicated diagram is a graph between pressure and volume. The former being taken on vertical axis and the latter on the horizontal axis. This is obtained by an instrument known as indicator. The indicator diagram are of two types;

(a) Theoretical or hypothetical

(b) Actual.

The theoretical or hypothetical indicator diagram is always longer in size as compared to the actual one. Since in the former losses are neglected. The ratio of the area of the actual indicator diagram to the theoretical one is called diagram factor.

Q. 5: Explain the working of any air standard cycle (by drawing it on *P-V* diagram) known to you. Why is it known as 'Air standard cycle.'? *(Dec–01)*

Or

Draw the Diesel cycle on *P-V* coordinates and explain its functioning. *(Dec–02)*

Or

Show Otto and diesel cycle on *P-V* and *T-S* diagram. *(May–03)*

Or

Stating the assumptions made, describe air standard otto cycle. *(Dec–04)*

Or

Derive a relation for the air standard efficiency of diesel cycle. Also show the cycle on P-V and T-S diagram. *(Dec–04)*

AIR STANDARD CYCLES

Most of the power plant operates in a thermodynamic cycle i.e. the working fluid undergoes a series of processes and finally returns to its original state. Hence, in order to compare the efficiencies of various cycles, a hypothetical efficiency called air standard efficiency is calculated.

If air is used as the working fluid in a thermodynamic cycle, then the cycle is known as "Air Standard Cycle".

To simplify the analysis of I.C. engines, air standard cycles are conceived.

Assumptions
1. The working medium is assumed to be a perfect gas and follows the relation
$$pV = mRT \quad \text{or} \quad P = \rho RT$$
2. There is no change in the mass of the working medium.
3. All the processes that constitute the cycle are reversible.
4. Heat is added and rejected with external heat reservoirs.
5. The working medium has constant specific heats.

Otto Cycle (1876) (Used S. I. Engines)

This cycle consists of two reversible adiabatic processes and two constant volume processes as shown in figure on P-V and T-S diagrams.

The process 1-2 is reversible adiabatic compression, the process 2-3 is heat addition at constant volume, the process 3-4 is reversible adiabatic expansion and the process 4-1 is heat rejection at constant volume.

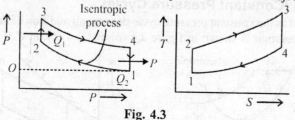

Fig. 4.3

The cylinder is assumed to contain air as the working substance and heat is supplied at the end of compression, and heat is rejected at the end of expansion to the sink and the cycle is repeated.

Process

0-1 = suction
1-2 = isentropic compression
2-3 = heat addition at constant volume
3-4 = isentropic expansion
4–1 = constant volume heat rejection 1-0 = exhaust

Heat supplied : $\quad Q_1 = mc_v (T_3 - T_2)$

Heat rejected : $\quad Q_2 = mc_v (T_4 - T_1)$

Efficiency : $\quad \eta = 1 - \dfrac{Q_2}{Q_1} = 1 - \dfrac{mc_v (T_4 - T_1)}{mc_v (T_3 - T_2)}$

$$\eta = 1 - \frac{T_4 - T_1}{T_3 - T_2}$$

Process 1 – 2 : $T_1 V_1^{\gamma-1} = T_2 V_2^{\gamma-1}$

$$\frac{T_1}{T_2} = \left(\frac{V_2}{V_1}\right)^{\gamma-1} \quad \text{or} \quad \left(\frac{V_1}{V_2}\right)^{\gamma-1} = \frac{T_2}{T_1}$$

Process 3 – 4 : $T_3 V_3^{\gamma-1} = T_4 V_4^{\gamma-1}$

$$\left(\frac{V_4}{V_3}\right)^{\gamma-1} = \frac{T_2}{T_1} \quad \text{also} \quad \frac{V_4}{V_3} = \frac{V_1}{V_2}$$

$$\Rightarrow \quad \frac{T_3}{T_4} = \frac{T_2}{T_1} \Rightarrow \frac{T_3}{T_2} = \frac{T_4}{T_1}$$

$$\Rightarrow \quad \frac{T_3}{T_2} - 1 = \frac{T_4}{T_1} - 1 \quad \text{(subtracting 1 from both sides)}$$

$$\Rightarrow \quad \frac{T_3 - T_2}{T_2} = \frac{T_4 - T_1}{T_1}$$

$$\Rightarrow \quad \frac{T_1}{T_2} = \frac{T_4 - T_1}{T_3 - T_2} = \left(\frac{V_2}{V_1}\right)^{\gamma-1} = \left(\frac{1}{r_k}\right)^{\gamma-1}$$

Substituting in eq. (i) $\text{hotto} = 1 - \dfrac{1}{r_k^{\gamma-1}}$

where r_k = compression ratio.

Diesel Cycle (1892) (Constant Pressure Cycle)

Diesel cycle is also known as the constant pressure cycle because all addition of heat takes place at constant pressure. The cycle of operation is shown in figure 2.4 (a) and 2.4 (b) on P-V and T-S diagrams.

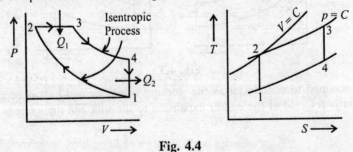

Fig. 4.4

The sequence of operations is as follows :
1. The air is compressed isentropically from condition '1' to condition '2'.
2. Heat is supplied to the compressed air from external source at constant pressure which is represented by the process 2-3.
3. The air expands isentropically until it reaches condition '4'.
4. The heat is rejected by the air to the external sink at constant volume until it reaches condition T and the cycle is repeated.

The air standard efficiency of the cycle can be calculated as follows:

Heat supplied: $Q_1 = Q_{2-3} = mc_p(T_3 - T_2)$

Heat rejected: $Q2 = Q_{4-1} = mc_v(T_4 - T_1)$

$$\eta = 1 - \frac{Q_2}{Q_1} = 1 - \frac{mc_v(T_4 - T_1)}{mc_p(T_3 - T_2)}$$

$$\eta = 1 - \frac{(T_4 - T_1)}{(T_3 - T_2)}$$

Compression ratio : $r_k = V_1/V_2$ ...(i)

Expansion ratio : $r_e = V_4/V_3$ ...(ii)

Cut of ratio : $r_c = V_3/V_2$ ...(iii)

It is seen that $r_k = r_e r_c$

Process 1-2 : $T_1 V_1^{\gamma-1} = T_2 V_2^{\gamma-1}$

$$\left(\frac{V_1}{V_2}\right)^{\gamma-1} = \frac{T_2}{T_1} \Rightarrow T_2 = T_1 (r_k)^{\gamma-1}$$

Process 2-3 : $\dfrac{P_2 V_2}{T_2} = \dfrac{P_3 V_3}{T_3} \Rightarrow \dfrac{V_3}{V_2} = \dfrac{T_3}{T_2} = r_c$ (As $P_2 = P_3$)

Process 3-4 : $T_3 V_3^{\gamma-1} = T_4 V_4^{\gamma-1} \Rightarrow \left(\dfrac{V_4}{V_3}\right)^{\gamma-1} = \dfrac{T_3}{T_4} \Rightarrow T_3 = T_4 (r_c)^{\gamma-1}$

Substituting

$$\eta = 1 - \frac{(T_4 - T_1)}{\gamma T_4 (r_c)^{\gamma-1} - T_1(r_k)^{\gamma-1}}$$

$$\frac{T_3}{T_1} = r_c (\eta)^{\gamma-1} \quad \left[\because \frac{T_3}{T_2} = r_c \text{ and } \frac{T_2}{T_1} = (r_k)^{\gamma-1}\right]$$

$$\frac{T_3}{T_4} = r_c^{\gamma-1}$$

$$\frac{T_4}{T_1} = \frac{r_c \cdot r_k^{\gamma-1}}{\left(\dfrac{r_k}{r_c}\right)^{\gamma-1}} = r_c^{\gamma}$$

$$\eta = 1 - \frac{T_1 \left[r_c^{\gamma} - 1\right]}{\gamma T_2 \left[r_c - 1\right]}$$

$$\gamma = 1 - \frac{1}{\gamma (r_k)^{\gamma-1}} \frac{\left[r_c^{\gamma} - 1\right]}{\left[r_c - 1\right]}$$

As $r_c > 1$, so $\dfrac{1}{\gamma}\left[\dfrac{r_c^{\gamma}-1}{r_c - 1}\right]$

is also > 1, therefore for the same compression ratio the efficiency of the diesel cycle is less than that of the otto cycle.

Q. 6: Compare otto cycle with Diesel cycle?

Sol.: These two cycles can be compared on the basis of either the same compression ratio or the same maximum pressure and temperature.

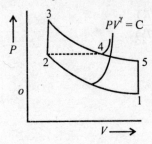 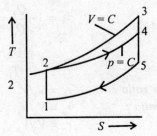

Fig. 4.5

1 - 2 - 3 - 5 = Otto Cycle,
for the same heat rejection Q_2 the higher the
1 - 2 - 4 - 5 = Diesel Cycles, heat given Q_1,
the higher is the cycle efficiency.

So from T-S diagram for cycle 1 - 2 - 3 - 5, Q_1 is more than that for 1 - 2 - 7 - 5 (area under the curve represents Q_1).

Hence $\eta_{Otto} > \eta_{Diesel}$
For the same heat rejection by both otto and diesel cycles.
Again both can be compared on the basis of same maximum pressure and temperature.

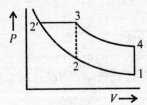 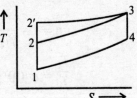

Fig. 4.6

1 - 2 - 3 - 4 = Otto Cycle; Here area under the curve
1 - 2' - 3 - 4 = Diesel Cycle
1 - 2' - 3 - 4 is more than 1 - 2 - 3 - 4
So $\eta diesel > \eta otto$; for the same T_{max} and P_{max}

Q. 7: Describe the working of four stroke SI engine. Illustrate using line diagrams.

(May–02, May–03, Dec–03)

Or

Explain the working of a 4 stroke petrol engine. *(Dec–02)*

Four Stroke Engine

Figure. shows the working of a 4 stroke engine. During the suction stroke only air (in case of diesel engine) or air with petrol (in case of petrol engine) is drawn into the cylinder by the moving piston.

Introductioon To I.C. Engine / 73

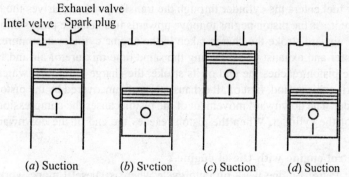

(a) Suction (b) Svction (c) Suction (d) Suction

Fig. 4.7. *Cycle of events in a four stroke petrol engine*

The charge enters the engine cylinder through the inlet valve which is open. During this stroke, the exhaust valve is closed. During the compression stroke, the charge is compressed in the clearance space. On completion of compression, if only air is taken in during the suction stroke, the fuel is injected into the engine cylinders at the end of compression. The mixture is ignited and the heat generated, while the piston is nearly stationary, sets up a high pressure. During the power stroke, the piston is forced downward by the high pressure. This is the important stroke of the cycle. During the exhaust stroke the products of combustion are swept out through the open exhaust valve while the piston returns. This is the scavenging stroke. All the burnt gases are completely removed from the engine cylinder and the cylinder is ready to receive the fresh charge for the new cycle.

Thus, in a 4-stroke engine there is one power stroke and three idle strokes. The power stroke supplies the necessary momentum to keep the engine running.

Q. 8: Describe the working of two stroke *SI* engine. Illustrate using line diagrams.

(May–03, 04, Dec–05)

Two Stroke Engine

In two stroke engine, instead of valves ports are provided, these are opened and closed by the moving piston. Through the inlet port, the mixture of air and fuel is taken into the crank case of the engine cylinder and through the transfer port the mixture enters the engine cylinder from the crank case. The exhaust ports serve the purpose of exhausting the gases from the engine cylinder. These ports are more than one in number and are arranged circumferentially.

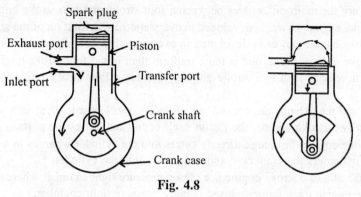

Fig. 4.8

A mixture of air fuel enters the cylinder through the transfer ports and drives the burnt gases from the previous stroke before it. As the piston begins to move upwards fresh charge passes into the engine cylinder. For the remainder of upward stroke the charge taken in the engine cylinder is compressed after the piston has covered the transfer and exhaust ports. During the same time mixture of air and fuel is taken into the crank case. When the piston reaches the end of its stroke, the charge is ignited, which exerts pressure on top of the piston. During this period, first of all exhaust ports are uncovered by the piston and so the exhaust gases leave the cylinder. The downward movement of the piston causes the compression of the charge taken into the crank case of the cylinder. When the piston reaches the end of the downward stroke. The cycle repeats.

Q. 9: Compare Petrol engine with Diesel engine.?
Sol.: (*i*) **Basic cycle:** Petrol Engine work on Otto cycle whereas Diesel Engine work on diesel cycles.

(*ii*) **Induction of fuel:** During the suction stroke in petrol engine, the air fuel mixture is sucked in the cylinder while in diesel engine only air is sucked into the cylinder during its suction stroke.

(*iii*) **Compression Ratio:** In petrol engine the compression ratio in the range of 5:1 to 8:1 while in diesel engine it is in the range of 15:1 to 20:1.

(*iv*) **Thermal efficiency:** For same compression ratio, the thermal efficiency of diesel engine is lower than that of petrol engine.

(*v*) **Ignition:** In petrol engine the charge (A/F mixture) is ignited by the spark plug after the compression of mixture while in diesel engine combustion of fuel due to its high temperature of compressed air.

Two Stroke Engine
In two stroke engine all the four operation i.e. suction, compression, ignition and exhaust are completed in one revolution of the crank shaft.

Four Stroke Engine
In four stroke engine all the four operation are completed in two revolutions of crank shaft.

Application of 2-stroke Engines
2 stroke engine are generally used where low cost, compactness and light weight are the major considerations

Q. 10: compare the working of 4 stroke and 2 – stroke cycles of internal combustion engines.
(Dec–01, 04)

Sol.: The following are the main differences between a four stroke and two stroke engines.
1. In a four stroke engine, power is developed in every alternate revolution of the crankshaft whereas; in a two stroke engine power is developed in every revolution of the crankshaft.
2. In a two stroke engine, the torque is more uniform than in the four stroke engine hence a lighter flywheel is necessary in a two stroke engine, whereas a four stroke engine requires a heavier flywheel.
3. The suction and the exhaust are opened and closed by mechanical valves in a four stroke engine, whereas in a two stroke engine, the piston itself opens and closes the ports.
4. In a four stroke engine the charge directly enters into the cylinder whereas in a two stroke engine the charge first enters the crankcase and then flows into the cylinder.
5. The crankcase of a two stroke engine is a closed pressure tight chamber whereas the crankcase of a four stroke engine even though closed is not a pressure tight chamber.

6. In a four stroke engine the piston drives out the burnt gases during the exhaust stroke, whereas, in a two stroke engine the high pressure fresh charge scavenges out the burnt gases.
7. The lubricating oil consumption in a two stroke engine is more than in four stroke engine.
8. A two stroke engine produces more noise than a four stroke engine.
9. Since the fuel burns in every revolution of the crankshaft in a two stroke engine the rate of cooling is more than in a four stroke engine.
10. A valve less two stroke engines runs in either direction, whereas a four stroke engine cannot run in either direction.

Q. 11: What are the advantage of a two stroke engine over a four stroke engine.?
Sol.: The following are the advantages of a two stroke engine over a four stroke engine:
1. A two stroke engine has twice the number of power stroke than a four strokes engine at the same speed. Hence theoretically a two stroke engine develops double the power per cubic meter of the swept volume than the four stroke engine.
2. The weight of the two stroke engine is less than four stroke engine because of the lighter flywheel due to more uniform torque on the crankshaft.
3. The scavenging is more complete in low-speed two stroke engines, since exhaust gases are not left in the clearance volume as in the four stroke engine.
4. Since there are only two strokes in a cycle, the work required to overcome the friction and the exhaust strokes is saved.
5. Since there are no mechanical valves and the valve gears, the construction of two stroke engine is simple which reduces its initial cost.
6. A two stroke engine can be easily reversed by a simple reversing gear mechanism.
7. A two stroke engine can be easily started than a four stroke engine:
8. A two stroke engine occupies less space.
9. A lighter foundation will be sufficient for two stroke engine.
10. A two stroke engine has less maintenance cost since it requires only few parts.

Q. 12: What are the disadvantages of two stroke engine?
Sol.: The following are some of the disadvantages of two stroke engine when compared with four stroke engine:
1. Since the firing takes place in every revolution, the time available for cooling will be less than in a four stroke engine.
2. Incomplete scavenging results in mixing of exhaust gases with the fresh charge which will dilute it, hence lesser power output.
3. Since the transfer port is kept open only during a short period, less quantity of the charge will be admitted into the cylinder which will reduce the power output.
4. Since both the exhaust and the transfer ports are kept open during the same period, there is a possibility of escaping of the fresh charge through the exhaust port which will reduce the thermal efficiency.
5. For a given stroke and clearance volume, the effective compression stroke is less in a two stroke engine than in a four stroke engine.
6. In a crankcase compressed type of two stroke engine, the volume of charge down into the crankcase is less due to the reduction in the crankcase volume because of rotating parts.
7. A fan scavenged two stroke engine has less mechanical efficiency since some power is required to run the scavenged fan.

8. A two stroke engine needs better cooling arrangement because of high operating temperature.
9. A two stroke engine consumes more lubricating oil.
10. The exhaust in a two stroke engine is noisy due to sudden release of the burnt gases.

Q. 13: Calculate the thermal efficiency and compression ratio for an automobile working on otto cycle. If the energy generated per cycle is thrice that of rejected during the exhaust. Consider working fluid as an ideal gas with $\gamma = 1.4$ *(May–01)*

Sol.: Since we have
$$\eta_{otto} = (Q_1 - Q_2)/Q_1$$
Where
Q_1 = Heat supplied
Q_2 = Heat rejected
Given that $Q_1 = 3Q_2$
$$\eta_{otto} = (3Q_2 - Q_2)/3Q_2 = 2/3 = 66.6\% \qquad ...(i)$$
We also have;
$$\eta_{otto} = 1 - 1/(r)^{\gamma - 1}$$
$$0.667 = 1/(r)^{1.4 - 1}$$
$$r = (3)^{1/0.4} = 15.59 \qquad \text{.......ANS}$$

Q. 14: A 4 stroke diesel engine has length of 20 cm and diameter of 16 cm. The engine is producing indicated power of 25 KW when it is running at 2500 RPM. Find the mean effective pressure of the engine. *(May–03)*

Sol.: Length or stroke = 20 cm = 0.2 m
Diameter or Bore = 16 cm = 0.16 m
Indicating power = 25 KW
Speed = 2500 RPM
Mean effective pressure = ?
$K = 1$
Indicated power = $P_{ip} = (P_{mef}.L.A.N.K)/60$
Where $N = N/2$ = 1250 RPM (for four stroke engine)
$$25 \times 10^3 = \{P_{mef} \times 0.2 \times (\pi/4)(0.16)^2 \times 1250 \times 1\}/60$$
$$P_{mef} = 298.415 \text{KN/m}^2 \qquad \text{.......ANS}$$

Q. 15: A 4 stroke diesel engine has L/D ratio of 1.25. The mean effective pressure is found with the help of an indicator equal to 0.85MPa. The engine produces indicated power of 35 HP. While it is running at 2500 RPM. Find the dimension of the engine. *(Dec–03)*

Sol.:
$L/D = 1.25$
$P_{mef} = 0.85$ MPa = 0.85×10^6 N/m^2
$P_{IP} = 35$ HP = 35/1.36 KW (Since 1KW = 1.36HP or 1HP = 1/1.36 KW)
$N = 2500$ RPM = 1250 RPM for four stroke engine ($N = N/2$ for four stroke)
Indicated power = $P_{ip} = (P_{mef}.L.A.N.K)/60$
$$(35/1.36) \times 10^3 = \{0.85 \times 10^6 \times 1.25D \times (\pi/4)(D)^2 \times 1250 \times 1\}/60$$
$$D = 0.11397 \ m = 113.97 \text{ mm}$$
$$L = 1.25 \ D = 142.46 \text{ mm}$$
$$\mathbf{D = 113.97 \text{ mm}, L = 142.46 \text{ mm}} \qquad \text{.......ANS}$$

Q. 16: An engine of 250 mm bore and 375 mm stroke works on otto cycle. The clearance volume is 0.00263 m³. The initial pressure and temperature are 1 bar and 50°C. If the maximum pressure is limited to 25 bar. Find
  (1) The air standard efficiency of the cycle.
  (2) The mean effective pressure for the cycle. *(Dec–00)*

Sol.: Given that:
  Bore diameter d = 250mm
  Stroke length L = 375mm
  Clearance volume V_C = 0.00263m³
  Initial pressure P_1 = 1bar
  Initial temperature P_3 = 25 bar
  We know that, swept volume

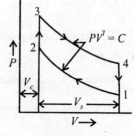

Fig. 4.9

$$V_s = \frac{\pi}{4} d^2 \cdot L = \frac{\pi}{4} \times (0.25)^2 \times 0.375 = 0.0184077 \text{ m}^3$$

Compression ratio 'r' $= \dfrac{V_c + V_s}{V_c} = \dfrac{0.0184077 + 0.0263}{0.00263} = 8$

∴ The air standard efficiency for Otto cycle is given by

$$\eta_{otto} = 1 - \frac{1}{(r)^{\gamma-1}} = 1 - \frac{1}{(8)^{1.4-1}} = 0.5647 \text{ or } 56.57\%$$

$$\frac{T_2}{T_1} = (r)^{\gamma-1} = (8)^{1.4-1} = 2.297; \ T_2 = (50 + 273) \times 2.297 = 742.06 \text{ K}$$

$$\frac{P_2}{P_1} = \left(\frac{V_1}{V_2}\right)^{\gamma} = (8)^{1.4} = 18.38; \ P_2 = 1 \times 18.38 = 18.38 \text{ bar}$$

Process (2 – 3)

$$V_2 = V_3; \ \frac{P_2}{T_2} = \frac{P_3}{T_3}$$

$$T_3 = \frac{25}{18.38} \times 742.06 = 1009.38$$

$$q_s = C_p (T_3 - T_2) = 1.005 (1009.38 - 742.06) = 268.65 \text{ kJ/kg}$$

$$\eta_{otto} = \frac{w}{q_s}; \ w = q_s \times \eta_{otto} = 268.65 \times 0.5647 = 151.70 \text{ kJ/kg}$$

Mean effective pressure $P_m = \dfrac{W}{V_2 - V_2}; \ P_m = \dfrac{151.70 \times m}{0.021 - 0.00263}$

$$m = \frac{P_1 V_1}{RT_1} = \frac{1 \times 10^5 \times 0.021}{0.287 \times 10^3 \times (50 + 273)} = 0.02265$$

$$P_m = \frac{151.70 \times 0.02265}{0.021 - 0.00263} = 187 \text{ kPa} = 1.87 \text{ bar}$$

Q. 17: An Air standard otto cycle has a compression ratio of 8. At the start of compression process the temperature is 26°C and the pressure is 1 bar. If the maximum temperature of the cycle is 1080K. Calculate

(1) Net out put

(2) Thermal efficiency. Take $C_V = 0.718$ (Dec–04)

Sol.: Compression Ratio $(R_c) = 8$

$$T_1 = 26°C = 26 + 273 = 299K = 1 \text{ bar}$$
$$T_3 = 1080 \text{ k}$$

(i) Net output = work done per kg of air = $\oint \delta w = \oint \delta q$

Process (1 – 2) Isentropic compression process $\dfrac{T_2}{T_1} = \left(\dfrac{V_1}{V_2}\right)^{\gamma-1}$

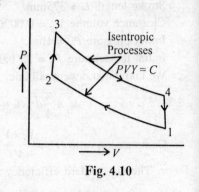

Fig. 4.10

$$T_2 = T_1 \left(\dfrac{P_2}{P_1}\right)^{\gamma-1}$$

$$T_2 = T_1 (R_c)^{\gamma-1} \quad \left(\because Rc = \dfrac{P_2}{P_1}\right)$$

$$T_2 = 299\,(8)^{1.4-1} = 299\,(8)^{0.4} = 299 \times 2.29 = 686.29\ K$$

$$\dfrac{T_3}{T_4} = \left(\dfrac{V_4}{V_3}\right)^{\gamma-1} ; \ T_4 = \dfrac{T_3}{R_C^{\gamma-1}} = \dfrac{10.30}{8^{1.4-1}} = \dfrac{1080}{(8)^{0.4}} ; \ T_4 = \dfrac{1080}{2.29} = 471.62\ k$$

Net output = work done per kg of air = $\oint \delta w$

$$\oint \delta w = C_v (T_3 - T_2) - Cv (T_4 - T_1)$$
$$= 0.718\,(1080 - 686.92) - 0.718\,(471.62 - 299)$$
$$= 0.718 \times 393.08 - 0.718 \times 172.62 = 282.23 - 123.94$$

Net Output = 158.28 KJ/Kg   ....ANS

(ii) $\eta_{thermal} = \dfrac{\oint \delta w}{qs} \times 100 = \dfrac{\text{work done per kg of air}}{\text{heat suplied per kg of air}}$

$$q_s = Cv\,(T_3 - T_2) = 0.718\,(1080 - 686.29) = 282.23 \text{ KJ/kg}$$

$$\eta \text{ thermal} = \dfrac{\oint \delta w}{q_s} \times 100 = \dfrac{158.28}{282.23} \times 100$$

η thermal = 56.08%   ....Ans

Q. 18: A diesel engine operating on Air Standard Diesel cycle operates on 1 kg of air with an initial pressure of 98kPa and a temperature of 36°C. The pressure at the end of compression is 35 bar and cut off is 6% of the stroke. Determine (i) Thermal efficiency (ii) Mean effective pressure.   (May–05)

Sol.: Given that :

$$\dot{m} = 1 \text{ kg},$$
$$P_1 = 98 \text{ kPa} = 98 \times 10^3 \text{ Pa};$$
$$T_1 = 36°C = 36 + 273 = 309 \text{ K},$$
$$P_2 = 35 \text{ bar} = 35 \times 10^5 \text{ Pa}$$
$$V_3 - V_2 = 0.06 V_s$$

For air standard cycle $P_1 V_1 = mRT_1$
$$98 \times 10^3 \times V_1 = 1 \times 287 \times 309$$
$$V_1 = 1.10 \text{ m3}; \ V_1 = V_2 + V_3 = 1.10$$

As process 1-2 is adiabtic compression process,

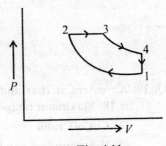

Fig. 4.11

$$\frac{T_2}{T_1} = \left(\frac{P_2}{P_1}\right)^{\frac{\gamma-1}{\gamma}}$$

$$\frac{T_2}{309} = \left(\frac{35 \times 10^5}{98 \times 10^3}\right)^{\frac{1.4-1}{1.4}} \Rightarrow T_2 = 858.28 \text{ K}$$

$$P_2 V_2 = mRT_2$$
$$35 \times 10^5 \times V_2 = 1 \times 287 \times 858.28;$$
$$V_C = V_2 = 0.07 \text{m}^3$$
$$V_s = V_1 = 1.10 \text{m}^3$$

However, $V_3 - V_2 = 0.06 V_s$
$$V_3 - 0.07 = 0.06 \times 1.10; \ V_3 = 0.136 \text{m}^3$$

Compression ratio $\quad R_c = \dfrac{V_1}{V_2} = \dfrac{1.10}{0.07} = 15.71$

$$\rho = \frac{V_1}{V_2} = \frac{0.136}{0.07} = 1.94$$

$$\gamma_{thermal} = 1 - \frac{1}{(R_c)^{\gamma-1}} \left[\frac{(\rho^\gamma - 1)}{\gamma(\rho - 1)}\right]$$

$$= 1 - \frac{1}{(15.71)^{1.4-1}} \left[\frac{(1.94)^{1.4} - 1}{1.4(1.94 - 1)}\right] = 1 - \frac{1}{(15.71)^{0.4}} \left[\frac{253 - 1}{1.4 \times 0.94}\right]$$

$$= 1 - \frac{1}{3.01} \left[\frac{153}{1.32}\right] = 1 - \frac{1}{3.01}(1.16) = 1 - 0.39 = 0.61$$

P_{mef} is given by $= P_1 \cdot R_c \left[\dfrac{\gamma(R_c)^\gamma (\rho - 1) - (\rho^\gamma - 1)}{(R_c - 1)(\gamma - 1)}\right]$

$$= 98 \times 10^3 \times 15.71 \left[\frac{1.4(15.71)^{1.4-1}(1.94-1) - (1.94^{1.4}-1)}{(15.71-1)(1.4-1)}\right]$$

$$= 98 \times 103 \times 15.71 \left[\frac{1.4(15.71)^{0.4}(0.94) - (1.94^{1.4}-1)}{14.71 \times 0.4}\right]$$

$$= 1539580 \left[\frac{1.32 \times 3.01 - (253 - 1)}{5.88} \right]$$

$$= \left[\frac{3.97 - 1.53}{5.88} \right] = 1539580 \left[\frac{2.44}{5.88} \right]$$

$$= 1539580 \times 0.415 \text{ Pa} = 6389257 \text{ Pa} = 6389.3 \text{ KPa} \quad \text{.......ANS}$$

Q. 19: Air enters at 1bar and 230°C in an engine running on diesel cycle whose compression ratio is 18. Maximum temperature of cycle is limited to 1500°C. Compute

(1) Cut off ratio
(2) Heat supplied per kg of air
(3) Cycle efficiency. *(Dec–05)*

Sol.: Given that:

$$P_1 = 1\text{bar}$$
$$T_1 = 230 + 273 = 503\text{K}$$
$$T_3 = 1500 + 273 = 1773\text{K}$$

Compression ratio $r = 18$
Since $T_2/T_1 = (r)^{\gamma-1}$

$$T_2 = T_1 \times (r)^{\gamma-1}$$
$$= 503(18)^{1.4-1} = 1598.37\text{K}$$

(1) Cut off ratio $(\rho) = V_3/V_2 = T_3/T_2$
$$T_3/T_2 = \rho$$
$$\rho = 1773/1598.37$$
$$\boldsymbol{\rho = 1.109} \quad \text{.......ANS}$$

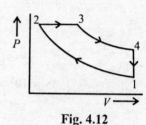
Fig. 4.12

(2) Heat supplied per kg of air
$$Q = C_P (T_3 - T_2) = 1.005 (1773 - 1598.37)$$
$$\boldsymbol{Q = 175.50 \text{ KJ/kg}} \quad \text{.......ANS}$$

(3) Cycle efficiency

$$\eta_{\text{diesel}} = \{1 - 1/[\gamma(r)^{\gamma-1}]\}\{(\rho^\gamma - 1)/(\rho - 1)\}$$
$$\eta_{\text{diesel}} = \{1 - 1/[1.4(18)^{1.4-1}]\}\{(1.109^{1.4} - 1)/(1.109 - 1)\}$$
$$\eta_{\text{diesel}} = \{1 - 0.225\}\{(0.156)/(0.109)\}$$
$$\eta_{\text{diesel}} = 0.678$$
or $\quad \boldsymbol{\eta_{\text{diesel}} = 67.8\%} \quad \text{.......ANS}$

CHAPTER 5

PROPERTIES OF STEAM AND THERMODYNAMICS CYCLE

Q. 1: Discuss the generation of steam at constant pressure. Show various process on temperature volume diagram. *(Dec–04)*

Sol.: Steam is a pure substance. Like any other pure substance it can be converted into any of the three states, i.e., solid, liquid and gas. A system composed of liquid and vapour phases of water is also a pure substance. Even if some liquid is vaporised or some vapour get condensed during a process, the system will be chemically homogeneous and unchanged in chemical composition.

Assume that a unit mass of steam is generated starting from solid ice at -10°C and 1atm pressure in a cylinder and piston machine. The distinct regimes of heating are as follows :

Regime (A-B) : The heat given to ice increases its temperature from -10°C to 0°C. The volume of ice also increases with the increase in temperature. Point B shows the saturated solid condition. At B the ice starts to melt (Fig. 5.1, Fig. 5.3).

Regime (B-C): The ice melts into water at constant pressure and temperature. At C the melting - process ends. There is a sudden decrease in volume at 0°C as the ice starts to melt. It is a peculiar property of water due to hydrogen bonding (Fig. 5.3).

Regime (C-D): The temperature of water increases an heating from 0°C to 100°C (Fig. 5.1). The volume of water first decreases with the increase in temperature, reaches to its minimum at 4°C (Fig. 5.3) and again starts to increase because of thermal expansion.

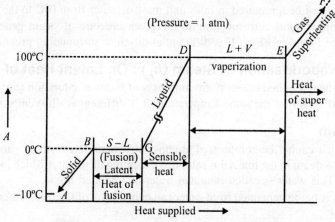

Fig 5.1 *Generation of steam at 1 atm pressure.*

Point D shows the saturated liquid condition.

Regime (D-E): The water starts boiling at D. The liquid starts to get converted into vapour. The boiling ends at point E. Point E shows the saturated vapour condition at 100°C and 1 bar.

Regime (E-F): It shows the superheating of steam above saturated steam point. The volume of vapour increases rapidly and it behaves as perfect gas. The difference between the superheated temperature and the saturation temperature at a given pressure is called degree of superheat.

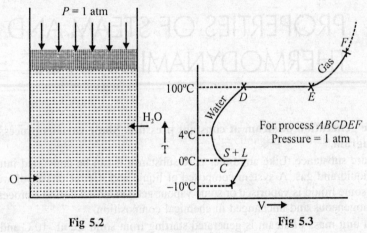

Fig 5.2 Fig 5.3

Point B, C, D, E are known as saturation states. State B : Saturated solid state.

State C & D : Both saturated liquid states.

State C is for hoar frost and state D is for vaporization. State E : Saturated vapour state.

At saturated state the phase may get changed without change in pressure or temperature.

Q. 2: Write some important term in connection with properties of steam.

Or

Short notes on Dryness fraction measurement. *(May-03)*

sensible Heat of Water or Heat of the Liquid or Enthalpy of Liquid (h_f)

Sol.: It is the quantity of heat required to raise unit mass of water from 0°C to the saturation temperature (or boiling point temperature) corresponding to the given pressure of steam generation. In Fig 5.5, 'h_f' indicates enthalpy of liquid in kJ/kg. It is different at different surrounding pressures.

Laten Heat of Vapourisation of Steam (h_{fg}) : Or, Latent Heat of Evaporation

It is the quantity of heat required to transform unit mass of water at saturation temperature to unit mass of steam (dry saturated steam) at the same temperature. It is different at different surrounding pressures.

Saturated Steam

It is that steam which cannot be compressed at constant temperature without partially condensing it. In Fig. 5.5, condition of steam in the line AB is saturated excepting the point A which indicates water at boiling point temperature. This water is called saturated water or saturated liquid.

The steam as it is being generated from water can exist in any of the three different states given below.
  (1) Wet steam
  (2) Dry (or dry saturated) steam
  (3) Superheated steam.

Amongst these, the superheated state of steam is most useful as it contains maximum enthalpy (heat) for doing useful work. Dry steam is also widely utilized, but the wet steam is of least utility. Different states of steam and sequential stages of their evolution are shown in Fig. 5.4 a-e. Their corresponding volumes are also shown therein.

WET SATURATED STEAM Wet steam is a two-phase mixture comprising of boiling water particles and dry steam in equilibrium state. Its formation starts when water is heated beyond its boiling point, thereby causing start of evaporation. A wet steam may exist in different proportions of water particles and dry steam. Accordingly, its qualities are also different. Quality of wet steam is expressed in terms of dryness fraction which is explained below.

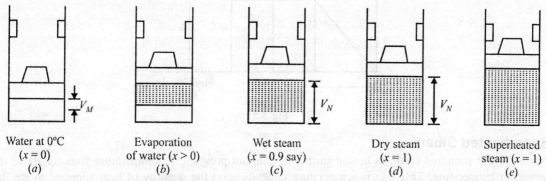

Fig 5.4: *Different states of steam and the stages of their evolution.*

Dryness Fraction of Steam

Dryness fraction of steam is a factor used to specify the quality of steam. It is defined as the ratio of weight of dry steam W_{ds} present in a known quantity of wet steam to the total weight of Wet steam W_{ws}. It is a unit less quantity and is generally denoted by x. Thus

$$x = \frac{W_{ds}}{W_{ds} + W_{ws}}$$

it is evident from the above equation that $x = 0$ in pure water state because $W_{ds} = 0$. It can also be seen, in Fig. 5.4a that $W_{ds} = 0$ in water state. But for presence of even a very small amount of dry steam i.e. $W_{ds} \ne 0$, x will be greater than zero as shown in Fig. 5.4b. On the other hand for no water particles at all in a sample of steam, $W_{ws} = 0$. Therefore x can acquire a maximum value of 1. It cannot be more than 1. The values of dryness fraction for different states of steam are shown in Fig. 5.4, and are as follows.

(i) Wet steam $\quad\quad\quad\quad 1 > x > 0$
(ii) Dry saturated steam $\quad x = 1$
(iii) Superheated steam $\quad x = 1$

The dryness fraction of a sample of steam can be found experimentally by means of calorimeters.

Dry (Or Dry Saturated) Steam

A dry saturated steam is a single-phase medium. It does not contain any water particle. It is obtained on complete evaporation of water at a certain saturation temperature. The saturation temperature differs for different pressures. It means that if water to be evaporated is at higher pressure, it will evaporate at higher temperature. As an illustration, the saturation temperatures at different pressures are given below for a ready reference.

p (bar)	0.025	0.30	2.0	9.0	25.0	80.0	150.0	200.0
t_{sat} (°C)	21.094	69.12	120.23	175.35	223.93	294.98	342.11	365.71

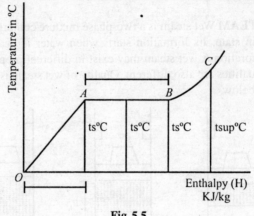

Fig 5.5

Superheated Steam

When the dry saturated steam is heated further at constant pressure, its temperature rises-up above the saturation temperature. This rise in temperature depends upon the quantity of heat supplied to the dry steam. The steam so formed is called superheated steam and its temperature is known as superheated temperature t_{sup} °C or T_{sup} K. A superheated steam behaves more and more like a perfect gas as its temperature is raised. Its use has several advantages. These are

(i) It can be expanded considerably (to obtain work) before getting cooled to a lower temperature.

(ii) It offers a higher thermal efficiency for prime movers since its initial temperature is higher.

(iii) Due to high heat content, it has an increased capacity to do work. Therefore, it results in economy of steam consumption.

In actual practice, the process of superheating is accomplished in a super heater, which is installed near boiler in a steam (thermal) power plant.

Degree of Super Heat

It is the difference between the temperature of superheated steam and saturation temperature corresponding to the given pressure.

So, degree of superheated = $t_{sup} - t_s$

Where;

t_{sup} = Temperature of superheated steam

t_s = Saturation temperature corresponding to the given pressure of steam generation.

Super Heat

It is the quantity of heat required to transform unit mass of dry saturated steam to unit mass of superheated steam at constant pressure so,

Super heat = $1 \times C_p \times (t_{sup} - t_s)$ KJ/Kg

Saturated Water

It is that water whose temperature is equal to the saturation temperature corresponding to the given pressure.

Q. 3: How you evaluate the enthalpy of steam, Heat required, specific volume of steam, Internal energy of steam?

(1) Evaluation the Enthalpy of Steam

Let
h_f = Heat of the liquid or sensible heat of water in KJ/kg
h_{fg} = Latent heat of vapourisation of steam in KJ/kg
t_s = Saturation temperature in 0°C corresponding to the given pressure.
t_{sup} = Temperature of superheated steam in °C
x = dryness fraction of wet saturated steam
C_p = Sp. Heat of superheated steam at constant pressure in KJ/kg.k.

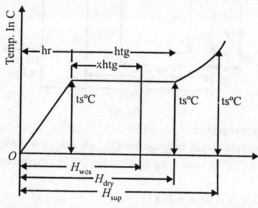

Fig 5.6

(a) Enthalpy of dry saturated steam

1 kg of water will be first raised to saturation temperature (t_s) for which h_f (sensible heat of water) quantity of heat will be required. Then 1 kg of water at saturation temperature will be transformed into 1 kg of dry saturated steam for which h_{fg} (latent heat of steam) will be required. Hence enthalpy of dry saturated steam is given by

$$H_{dry} \text{ (or } h_g) = h_f + h_{fg} \text{ kJ/kg}$$

(b) Enthalpy of wet saturated steam

1 kg of water will be first raised to saturation temperature (t_s) for which h_f (sensible heat of water) will be required. Then 'x' kg of water at saturation temperature will be transformed into 'x' kg of dry saturated steam at the same temperature for which $x.h_{fg}$ amount of heat will be required. Hence enthalpy of wet saturated steam is given by

$$H_{wet} = h_f + x.h_f g \text{ kJ/kg}$$

(c) Enthalpy of superheated steam

1 kg of water will be first raised to saturation temperature (t_s) for which h_f (sensible heat of water) will be required. Then, 1kg of water at saturation temperature will be transformed into 1kg dry saturated steam at the same temperature for which h_{fg} (latent heat of steam) will be required. Finally, 1kg dry saturated steam will be transformed into 1kg superheated steam at the same pressure for which heat required is

$$1 \times C_p(t_{sup} - t_s) = Cp\ (t_{sup} - t_s) \text{ kJ}$$

Hence enthalpy of superheated steam is given by

$$H_{sup} = h_f + h_{fg} + C_p\ (t_{sup} - t_s) \text{ kJ/kg}$$

(2) Evaluation of heat Required

Heat required to generate steam is different from 'total heat' or enthalpy of steam. Heat required to generate steam means heat required to produce steam from water whose initial temperature is $t°C$ (say) which is always greater than 0°C. Total heat or enthalpy of steam means heat required to generate steam from water whose initial temperature is 0°C. If, however, initial temperature of water is actually 0°C, then of course heat required to generate steam becomes equal to total heat or enthalpy of steam.

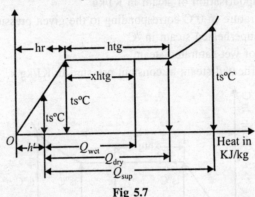

Fig 5.7

(a) **When steam is dry saturated**

heat required to generate steam is given by $Q_{Dry} = h_f + h_{fg} - h'$ kJ/kg,

where h' = heat required to raise 1 kg water from 0°C to the given initial temperature (say $t°C$) of water

$= mst = 1 \times 4.2 \times (t - 0) = 4.2t$ kJ

[sp. heat of water = 4.2 kJ/kg K].

(b) **When steam is wet saturated, heat required to generate steam is given by**

$Q_{wet} = h_f + x.h_{fg} - h'$ kJ/kg.

(c) **When steam is superheated, heat required is given by**

$Q_{sup} = h_f + h_{fg} + C_p(t_{sup} - t_s) - h'$ kJ/kg.

(3) Evaluation of Specific Volume of Steam

Specific volume of steam means volume occupied by unit mass of steam. It is expressed in m^3/kg. Specific volume of steam is different at different pressure. Again, corresponding to a given pressure specific volume of dry saturated steam, wet saturated steam and superheated steam will be different from one another.

(a) Sp. volume of dry saturated steam (V_g or V_{Dry})

It is the volume occupied by unit mass of dry saturated steam corresponding to the given pressure of steam generation.

Sp. volume of dry saturated steam corresponding to a given pressure can be found out by experiment. However, sp. volume of dry saturated steam corresponding to any pressure of steam generation can be found out directly from steam table. In the steam table, 'v_g' denotes the sp. volume of dry saturated steam in "m^3/kg",

(b) Specific volume of wet saturated steam (V_{wet})

It is the volume occupied by unit mass of wet saturated steam corresponding to the given pressure of steam generation.

Sp. volume of wet saturated steam is given by

V_{wet} = volume occupied by 'x' kg dry saturated steam + volume occupied by $(1 - x)$ kg. water,
where,

x = dryness fraction of wet saturated steam.

Let v_g = sp. volume of dry saturated steam in m³/kg corresponding to given pressure of wet saturated steam.

v_f = sp. volume of water in m³/kg corresponding to the given pressure of wet saturated steam.
Then,

$V_{wet} = x.vg + (1 - x) v_f$ m³/kg.

Since $(1 - x) v_f$ is very small compared to $x.v_g$, it is neglected.
[Average value of $v_f = 0.001$ m³/kg upto atmospheric pressure].
So,

$V_{wet} = xv_g$ m³/kg.

(c) Specific volume of superheated steam

It is the volume occupied by unit mass of superheated steam corresponding to the given pressure of superheated steam generation. Superheated steam behaves like a perfect gas. Hence the law

$$\frac{P_1 V_1}{T_1} = \frac{P_2 V_2}{T_2}$$

is applicable to superheated steam. Let

v_g = sp. volume of dry saturated steam corresponding to given pressure of steam generation is m³/kg.

T_S = absolute saturation temperature corresponding to the given pressure of steam generation.

T_{sup} = absolute temperature of superheated steam

P = pressure of steam generation

V_{sup} = required specific volume of superheated steam.

Then, in the above formula,

$P_1 = P_2$
$V_1 = v_g$, $V_2 = V_{sup}$
$T_1 = T_s$, $T_2 = T_{sup}$

$$\frac{v_g}{T_s} = \frac{V_{sup}}{V_{sup}}$$

$V_{sup} = v_g \times T_{sup}/T_s$ m³/kg

(4) Evaluation of Internal Energy of Steam

It is the actual heat energy stored in steam above the freezing point of water.

We know that enthalpy = internal energy + pressure energy
$$= U + PV,$$

where

U = internal energy of the fluid
PV = pressure energy of the fluid
P = pressure of the fluid
V = volume of the fluid.

If 'U' is in kJ/kg, 'P' is in kN/m² and 'V' is in m³/kg,

then $U + PV$ denotes specific enthalpy is kJ/kg.   [sp. enthalpy means enthalpy per unit mass]
From the above equation, we get
u = enthalpy – PV = H – PV kJ/kg,
where H = enthalpy per unit steam in kJ/kg.

(a) Internal energy of dry saturated steam is given by

$U_{Dry} = H_{Dry} - P.v_g$ kJ/kg,
where H_{Dry} (or h_g) = enthalpy of dry saturated steam in kJ/kg.
v_g = sp. volume of dry saturated steam in m³/kg, and
P = pressure of steam generation in kN/m²

(b) Internal energy of wet saturated steam is given by

$u_{wet} = H_{wet} - P.V_{wet}$ kJ/kg,
where
H_{wet} = enthalpy of wet saturated steam in kJ/kg
V_{wet} = sp. volume of wet saturated steam in m³/kg

(c) Internal energy of superheated steam is given by

$u_{sup} = H_{sup} - P.v_{sup}$ kJ/kg,
where
H_{sup} = enthalpy of superheated steam in kJ/kg.
v_{sup} = sp. volume of superheated steam in m³/kg.

Q. 4: Write short notes on Steam table.

Sol.: Steam table provides various physical data regarding properties of saturated water and steam. This table is very much helpful in solving the problem on properties of steam. It should be noted that the pressure in this table is absolute pressure.

In this table, various symbols used to indicate various data are as stated below:

(1) 'P' indicates absolute pressure in bar
(2) 't' indicates saturation temperature corresponding to any given pressure. This has been often denoted by 't_s'.
(3) 'v_f' indicates specific volume of water in m³/kg corresponding to any given pressure.
(4) 'v_g' indicates specific volume of dry saturated steam corresponding to any given pressure.
(5) 'h_f' indicates heat of the liquid in kJ/kg corresponding to any given pressure.
(6) 'h_{fg}' indicates latent heat of evaporation in kJ/kg corresponding to any given pressure.
(7) 'h_g' indicates enthalpy of dry saturated steam in kJ/kg corresponding to any given pressure.
(8) 'S_f' indicates entropy of water in kJ/kg.K corresponding to any given pressure.
(9) 'S_g' indicates entropy of dry saturated steam in kJ/kg.K corresponding to any given pressure.
(10) "S_{fg}" indicates entropy of evaporation corresponding to any pressure. There are two types of steam tables :

One steam table is on the basis of absolute pressure of steam and another steam table is on the basis of saturation temperature. Extracts of two types of steam tables are given below.

Table 5.1. On the Basis of Pressure

Absolute pressure (P) bar	Saturation temperature (t) °C	Sp. volume in m3/kg		Specific enthalpy in kJ/kg			Specific entropy in kJ/kg K	
		Water (vf)	Steam (vg)	Water (hf)	Latent heat (hfg)	Steam (hg)	Water (Sf)	Steam (Sg)
1.00	99.63	0.001	1.69	417.5	2258	2675.5	1.303	6.056
1.10	102.3	0.00104	1.59	428.8	2251	2679.8	1.333	5.994
1.20	104.8	0.00104	1.428	439.4	2244	2683.4	1.361	5.937
1.50	111.4	0.00105	1.159	467.1	2226	2693.1	1.434	5.790

Table 5.2. On the Basis of Saturation Temperature

Saturation temperature (t) in °C	Absolute pressure (P) in bar	Sp. volume in m3/kg		Specific enthalpy in kJ/kg			Specific entropy in kJ/kg K	
		Water (vf)	Steam (vg)	Water (hf)	Latent heat (hfg)	Steam (hg)	Water (Sf)	Steam (Sg)
10	0.0123	0.001	106.4	42.0	2477	2519	0.151	8.749
20	0.0234	0.001	57.8	83.9	2454	2537.9	0.296	8.370
40	0.0738	0.001	19.6	167.5	2407	2574.5	0.572	7.684

Ques No-5: Explain Mollier diagram and Show different processes on mollier diagram.?

Sol.: A Mollier diagram is a chart drawn between enthalpy H (on ordinate) and entropy Φ or S (on abscissa). it is also called H-Φ diagram. It depicts properties of water and steam for pressures up to 1000 bar and temperatures up to 800°C. In it the specific volume, specific enthalpy, specific entropy, and dryness fraction are given in incremental steps for different pressures and temperatures. A Mollier diagram is very convenient in predicting the states of steam during compression and expansion, during heating and cooling, and during throttling and isentropic processes directly. It does not involve any detailed calculations as is required while using the steam tables. Sample of a Mollier chart is shown in Fig. 5.8 for a better understanding.

There is a thick saturation line that indicates 'dry and saturated state' of steam. The region below the saturation line represents steam 'in wet conditions' and above the saturation line, the steam is in 'superheated state'. The lines of constant dryness fraction and of constant temperature are drawn in wet and superheated regions respectively. It should be noted that the lines of constant pressure are straight in wet region but curved in superheated region.

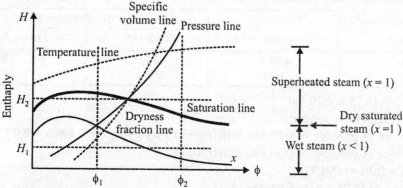

Fig 5.8: *A sample Mollier diagram (H-Φ chart) showing its details.*

Q. 6: 10 kg of wet saturated steam at 15 bar pressure is superheated to the temperature of 290°C at constant pressure. Find the heat required and the total heat of steam. Dryness fraction of steam is 0.85.

Sol.: From steam table, we obtain the following data:

Absolute pressure (P) bar	Saturation temperature (t) °C	Specific enthalpy kJ/kg	
		Water (hf)	Latent heat (hfg)
15	198.3	844.6	1947

Total heat of 10 kg wet saturated steam
$$= 10 \times 2499.55 = 24995.5 \text{ kJ} \quad \text{.......ANS}$$

Total heat of 1 kg superheated steam is given by $H_{sup} = h_f + h_{fg} + Cp(t_{sup} - t_s)$ kJ
$$= 844.6 + 1947 + 2.1 \times (290 - 198.3) \text{ kJ}$$
$$= 2984.17 \text{ kJ} \quad \text{.......ANS}$$

Total heat of 10 kg superheated steam $= 10 \times 2984.17 = \mathbf{29841.7 \text{ kJ}}$ ....ANS

$$= h_f + h_{fg} + C_p(t_{sup} - t_s) - (h_f + xh_{fg})$$
$$= h_{fg} + C_p(t_{sup} - t_s) - xh_{fg}$$
$$= 1947 + 2.1 \times (290 - 198.3) - 0.85 \times 1947 \text{ kJ} = 484.62 \text{ kJ}$$

Heat required to convert 10 kg wet saturated steam into 10 kg superheated steam
$$= 10 \times 484.62 = \mathbf{4846.2 \text{ kJ}} \quad \text{.......ANS}$$

Total heat of 1 kg wet saturated steam is given by $H_{wet} = h_f + xh_{fg}$ kJ
$$= 844.6 + 0.85 \times 1947 \text{ kJ} = 2499.55 \text{ kJ}$$

Heat required to convert 1 kg wet saturated steam into 1 kg superheated steam
$$= H_{sup} - H_{wet},$$

where H_{sup} = enthalpy of 1 kg superheated steam = $h_f + h_{fg} + C_p(t_{sup} - t_s)$ kJ

H_{wet} = enthalpy of 1 kg wet saturated steam = $h_f + xh_{fg}$ kJ

Heat required to convert 1 kg wet saturated steam into 1 kg superheated steam

Q. 7: Steam is being generated in a boiler at a pressure of 15.25 bar. Determine the specific enthalpy when
   (i) Steam is dry saturated
   (ii) Steam is wet saturated having 0.92 as dryness fraction, and
   (iii) Steam is superheated, the temperature of steam being 270°C.

Sol.: Note. Sp. enthalpy means enthalpy per unit mass. From steam table, we get the following data:

Absolute pressure (P) bar	Saturation temperature (t) °C	Specific enthalpy kJ/kg	
		Water (hf)	Latent heat (hfg)
15	198.3	844.6	1947
15.55	200.0	852.4	1941

Now 15.55 − 15.25 = 0.30 bar
      15.55 − 15 = 0.55 bar

For a difference of pressure of 0.55 bar, difference of t (or t_s) = 200 − 198.0 = 2.0°C
For a difference of pressure of 0.30 bar, difference of
$t = (2/0.55) \times 0.30 = 1.091°C.$
Corresponding to 15.25 bar, exact value of

$t = 200 - 1.091 = 198.909°C$

For a difference of pressure of 0.55 bar, difference of h_f (heat of the liquid) = 852.4 – 844.6 = 7.8 kJ/kg
For a difference of pressure of 0.30 bar, difference of h_f = (7.8/0.55) × 0.30 = 4.255 kJ/kg.
Corresponding to 15.25 bar, exact value of
$$h_f = 852.4 - 4.255 = 848.145 \text{ kJ/kg.}$$
Again, for a difference of pressure of 0.55 bar, difference of h_{fg} (latent heat of evaporation)
$$= 1947 - 1941 = 6 \text{ kJ/kg.}$$
For a difference of 0.30 bar, difference of
$$h_{fg} = (6/0.55) \times 0.30 = 3.273 \text{ kJ/kg.}$$
Corresponding to 15.25 bar, exact value of
$$h_{fg} = 1941 + 3.273 = 1944.273 \text{ kJ/kg}$$
[Greater the pressure of steam generation, less is the latent heat of evaporation.]
The data calculated above are written in a tabular form as below :

Absolute pressure (P) bar	Saturation temperature (t) °C	Specific enthalpy kJ/kg	
		Water (hf)	Latent heat (hfg)
15.25	198.909	848.145	1944.273

(i) When steam is dry saturated, its enthalpy is given by
$$H_{Dry} = h_f + h_{fg} \text{ kJ/kg}$$
$$= 848.145 + 1944.273 = \mathbf{2792.418 \text{ kJ/kg}} \quad \text{.......ANS}$$

(ii) When steam is wet saturated, its enthalpy is given by
$$H_{wet} = h_f + x h_{fg} \text{ kJ/kg}$$
$$= 848.145 + 0.92 \times 1944.273 = \mathbf{2636.876 \text{ kJ/kg}} \quad \text{.......ANS}$$

(iii) When steam is superheated, its enthalpy is given by
$$H_{sup} = h_f + h_{fg} + C_p(t_{sup} - t_s) \text{ kJ/kg}$$
$$= 848.145 + 1944.273 + 2.1 \times (270 - 198.909) = \mathbf{2941.71 \text{ kJ/kg}} \quad \text{.......ANS}$$

Q. 8: **200 litres of water is required to be heated from 30°C to 100°C by dry saturated steam at 10 bar pressure. Find the mass of steam required to be injected into water. Sp. heat of water is 4.2 kj/kg.K.**

Sol.: From steam table, we obtain the following data:

Absolute pressure (P) bar	Saturation temperature (t) °C	Specific enthalpy kJ/kg	
		Water (hf)	Latent heat (hfg)
10	1799	702.6	2015

Heat lost by 1 kg dry steam = $H_{Dry} - h'$ kJ,
where
H_{Dry} = enthalpy (or total heat) of 1 kg dry saturated steam
h' = heat required to raise 1 kg water from 0°C to 100°C
(i.e. h'= total heat of 1 kg water at 100°C)
Now,
$$H_{Dry} = h_f + h_{fg} = 702.6 + 2015 = 2717.6 \text{ kJ/kg}$$
$$h' = 1 \times 4.2 \times (100 - 0) = 420 \text{kJ/kg}$$

Heat lost by 1 kg dry saturated steam = 2717.6 − 420 = 2297.6 kJ
Let m = required mass of steam in kg.
Then, heat lost by m kg dry saturated steam = m × 2297.6 kJ
Now, 200 litres of water has a mass of 200 kg.
Heat gained by 200 kg water
$$= 200 \times 4.2 \times (100 - 30) \text{ kJ} = 58800 \text{ kJ}$$
Heat lost by m kg steam = heat gained by 200 kg water
m × 2297.6 = 58800
or,   m = **25.592 kg**   ....ANS

Q. 9: One Kg of steam at 1.5MPa and 400°C in a piston – cylinder device is cooled at constant pressure. Determine the final temperature and change in volume. If the cooling continues till the condensation of two – third of the mass.   *(May – 01)*

Sol.: Given that
Mass of steam m = 1kg
Pressure of steam P_1 = 1.5MPa = 15bar
Temperature of steam T_1 = 400°C
From superheated steam table
At P_1 = 15bar, T_1 = 400°C
$Å_1$ = 0.1324 m³/kg
$Å_2$ = (2 × 0.1324)/3 = 0.0882 m³/kg
Change in volume "Å = $Å_1$ - $Å_2$ = 0.1324 – 0.0882 = 0.0441 m³/kg
The steam is wet at 15bar, therefore, the temperature will be 198.32°C.

Q. 10: A closed metallic boiler drum of capacity 0.24m³ contain steam at a pressure of 11bar and a temperature of 200°C. Calculate the quantity of steam in the vessel. At what pressure in the vessel will the steam be dry and saturated if the vessel is cooled?   *(May–01)(C.O.)*

Sol.: Given that:
Capacity of drum V_1 = 0.24m³
Pressure of steam P_1 = 11bar
Temperature of steam T_1 = 200°C
At pressure 11bar from super heated steam table
At 10 bar and T = 200°C; Å = 0.2060 m³/kg
At 12 bar and T = 200°C; Å = 0.1693 m³/kg
Using linear interpolation:
(Å − 0.2060)/(0.1693 − 0.206) = (11 − 10)/(12 − 10)
Å = 0.18765 m³/kg
Quantity of steam = V/Å = 0.24/0.18765 = 1.2789 kg
From Saturated steam table
At 11bar; T_{sat} = 184.09°C
200°C > 184.09°C
i.e. steam is superheated
If the vessel is cooled until the steam becomes dry saturated, its volume will remain the same but its pressure will change.
From Saturated steam table; corresponding to $Å_g$ = 0.18765, the pressure is 1122.7KPa   ....ANS

Q. 11: (*a*) Steam at 10 bar absolute pressure and 0.95 dry enters a super heater and leaves at the same pressure at 250°C. Determine the change in entropy per kg of steam. Take C_{ps} = 2.25 kJ/kg K

(*b*) Find the internal energy of 1 kg of superheated steam at a pressure of 10 bar and 280°C. If this steam is expanded to a pressure of 1.6 bar and 0.8 dry, determine the change in internal energy. Assume specific heat of superheated steam as 2.1 kJ/kg-K. *(Dec–01)*

Sol.: (*a*) Given that :

$$P = 10 \text{ bar}$$
$$x = 0.95$$
$$t_{sup} = 250°C$$

From Saturated steam table

$$t_{sat} = 179.9$$

Now, entropy of steam at the entry of the superheater

$$s_1 = s_{f1} + x_1 s_{fg1}$$
$$= 2.1386 + 0.95 \times 4.4478 = 6.3640 \text{ kJ/kg K}$$

entropy of the steam at exit of superheater

$$s_2 = sgf + C_{ps} \ln\left(\frac{T_{sup}}{T_{sat}}\right)$$

$$= 6.5864 + 2.25 \ln\left(\frac{250 + 273}{179.9 + 273}\right)$$

$$= 6.9102 \text{ kJ/kg K}$$

Change in entropy = $s_2 - s_1$ = 6.9102 – 6.3640
= 0.5462 kJ/kg K

(*b*) Given that
State 1 : 10 bar 280°C
State 2 : 1.6 bar, 0.8 dry
Specific heat of superheated steam = 2.1 kJ/kg K
Internal energy at state 1 is :

$$u_1 = u_g + m.c.(T_1 - T_{sat}) = (h_g - pv_g) + m.c. (T_1 - T_{sat})$$
$$= (2776.2 - 1000 \times 0.19429) + 2.1(280 - 179.88) = 2792.16 \text{ kJ/kg}$$

Internal energy at state 2;

$$u_2 = uf_2 + xu_{g2}$$
$$= (h_f - Pv_f)_2 + x\,[h_{fg} - P\,(v_{fg})]_2$$
$$= (h_f - Pv_f) + x\,(h_g - hf) - P\,(v_g - v_f)]$$
$$= (h_f - Pv_f) + x\,((h_g - Pvg) - (h_f - P_{vf}))$$
$$= [475.38 + 160\,(0.0010547)] + [(2696.2 - 160 * (1.0911))$$
$$\quad - (475.38 - 160\,(0.0010547))]$$
$$= 475.21 + 0.8\,(2521.62 - 475.21)$$
$$= 2112.34 \text{ kJ/kg}$$

Change in internal energy = 211234 - 2792.16 = - 679.82 kJ/kg   ....**ANS**
-ve sign shows the reduction in internal energy.

Q. 12: A cylindrical vessel of 5m³ capacity contains wet steam at 100KPa. The volumes of vapour and liquid in the vessel are 4.95m³ and 0.05m³ respectively. Heat is transferred to the vessel until the vessel is filled with saturated vapour. Determine the heat transfer during the process. *(Dec–00)*

Sol.: Given that:
Volume of vessel $V = 5m^3$
Pressure of steam $P = 100KPa$
Volume of vapour $Vg = 4.95m^3$
Volume of liquid $V_f = 0.05m^3$
Since, the vessel is a closed container, so applying first law analysis, we have:

$$_1Q_2 - _1W_2 = U_2 - U_1$$
$$_1W_2 = \int PdV = 0$$
$$_1Q_2 = U_2 - U_1$$
$$U_1 = m_{f_1} \cdot u_{f_1} + m_{g_1} \cdot u_{g_1}$$

$$m_f = \frac{V_f}{v_{f_1}} = \frac{0.05}{0.001043} = 47.94 \quad \text{(using table B – 2)}$$

$$m_g = \frac{V_g}{v_{g_1}} = \frac{4.95}{1.694} = 2.922 \text{ kg}$$

The final condition of the steam is dry and saturated but its mass remains the same.
The specific volume at the end of heat transfer = v_{g2}

But $\quad v_2 = \frac{V}{m} = \frac{5.0}{(47.94 + 2.922)} = 0.0983$

Now $\quad v_2 = v_{g2} = 0.0983$

The pressure corresponding to $v_g = 0.0983$ from saturated steam table is 2030kPa or 2.03 bar.
At 2.03 bar $U_2 = u_{g2} \cdot m = (47.94 + 2922) \times 2600.5 = 132.26$ MJ
$$_1Q_2 = U_2 - U_1 = 132.26 - 27.33 = 104.93 \text{ MJ} \quad \text{........ANS}$$

Q. 13: Water vapour at 90kPa and 150°C enters a subsonic diffuser with a velocity of 150m/s and leaves the diffuser at 190kPa with a velocity of 55m/s and during the process 1.5 kJ/kg of heat is lost to surroundings. Determine
  (i) The final temperature
  (ii) The mass flow rate.
  (iii) The exit diameter, assuming the inlet diameter as 10cm and steady flow. *(May-01)*

Sol.: Given that :
Pressure at inlet = 90kPa = P_1
Temperature at inlet = 150°C = t_1
Velocity at inlet = 150 m/s = V_1
Pressure at exit = 190kPa = P_2
Velocity at exit = 55 m/s = V_2
Working substance-steam
Type of Process: Flow type
Governing Equation : S.F.E.E.

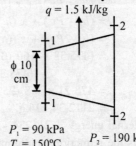

$q = 1.5$ kJ/kg
ϕ 10 cm
$P_1 = 90$ kPa
$T_1 = 150°C$
$C_1 = 150$ m/s
$P_2 = 190$ kPa
$C_2 = 55$ k/s

Fig 5.9

(i) Calculation for Final Temperature

From steady flow energy equation:
$$Q - W_s = m_f[(h_2 - h_1) + \tfrac{1}{2}(V_2^2 - V_1^2) + g(z_2 - z_1)]$$
$$W_S = g(z_2 - z_1) = 0$$
(since, there is no shaft and no change in datum level takes place)
$$Q = m_f[(h_2 - h_1) + \tfrac{1}{2}(V_2^2 - V_1^2)] \qquad \ldots(i)$$
since, the working substance is steam the properties of working substance at inlet and exit should be obtained from steam table.

At stage (1) for $P_1 = 90$ kPa and $t_1 = 150°C$
$t_1 > t_{sat}$ i.e.; superheated vapour
The steam thus behaves as perfect gas.
since $y = 1.3$ for superheated vapour and $R = 8.314/18$ kJ/kg K $= 0.4619$ kJ/kg K

$$C_p = \left(\frac{\gamma}{\gamma-1}\right) \cdot R = \left(\frac{1.3}{1.3-1}\right) \times 0.4619 = 2.00 \text{ kJ/kg K}$$

From equation (1)
$$-1.5 = 2(T_2 - T_1) + \left(\frac{(55)^2 - (155)^2}{2}\right) \times 10^{-3}$$
$$4.118 = T_2 - T_1$$
$$T_2 = 4.118 + 150 = 154.12°C \qquad \ldots\text{ANS}$$

(ii) Calculation for Mass Flow Rate

Now using ideal gas equation, assuming that superheated vapour behaves as ideal gas:
$$v_1 = \frac{RT_1}{P_1} = \frac{0.4619 \times (150+273) \times 10^3}{90 \times 10^3} \text{ m}^3/\text{kg}$$
$$v_1 = 2.170 \text{ m3/kg}$$
$$v_2 = \frac{RT_2}{P_2} = \frac{0.4619 \times (154.12+273) \times 10^3}{90 \times 10^3} \text{ m}^3/\text{kg}$$
$$v_2 = 1.038 \text{ m}^3/\text{kg}$$

Mass flow rate can be obtained by using continuity equation
$$mf \cdot v = A_1 V_1 = A_2 V_2$$
$$m_f = \frac{A_1 V_1}{v_1} = \frac{\frac{\pi}{4} \times (0.10)^2 \times 150}{2.170} = 0.543 \text{ kg/sec} \qquad \ldots\text{ANS}$$

(iii) Calculation for Exit Diameter

$$A_2 = \frac{A_1 V_1}{v_1} \times \frac{v_2}{V_2} = \frac{\frac{\pi}{4} \times (0.10)^2 \times 150 \times 1.038}{2.170 \times 55}$$
$$A_2 = 0.010246 \text{ m}^2$$
$$d_2 = \sqrt{\frac{A_2 \times 4}{\pi}} = 0.1142 = 11.42 \text{ cm} \qquad \ldots\text{ANS}$$

Q. 14: A turbine in a steam power plant operating under steady state conditions receives superheated steam at 3MPa and 350°C at the rate of 1kg/s and with a velocity of 50m/s at an elevation of 2m above the ground level. The steam leaves the turbine at 10kPa with a quality of 0.95 at an elevation of 5m above the ground level. The exit velocity of the steam is 120m./s. The energy losses as heat from the turbine are estimated at 5kJ/s. Estimate the power output of the turbine. How much error will be introduced, if the kinetic energy and the potential energy terms are ignored? *(Dec–01)*

Sol.: Given that; the turbine is running under steady state condition. At inlet: $P_1 = 3$MPa; $T_1 = 350°C$; $m_f = 1$ kg/sec; $V_1 = 50$ m/s; $Z_1 = 2$m At exit: $P_2 = 10$kPa; $x = 0.95$; $Z_2 = 5$ m; $V_2 = 120$ m/s Heat exchanged during expansion = $Q = -5$ kJ/sec From superheat steam table, at 3MPa and 350°C $h_1 = 3115.25$ kJ/kg $h_2 = h_{f2} + xh_{fg2}$ From saturated steam table; at 10kPa $h_2 = 191.81 + 0.95(2392.82) = 2464.99$ kJ/kg

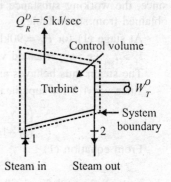

Fig. 5.10

From steady flow energy equation:

$Q - W_s = m_f [(h_2 - h_1) + \frac{1}{2}(V_2^2 - V_1^2) + g(z_2 - z_1)]$

$-5 - W_S = 1 \times [(2464.99 - 3115.25) + \{(120)^2 - (50)^2\}/2 \times 1000 + 9.8(5 - 2)/1000] - W_S = -639.28$ kJ/sec

$W_S = 639.28$ kJ/sec ....**ANS**

If the changes in potential and kinetic energies are neglected; then SFEE as; $Q - W_s = m_f(h_2 - h_1) - 5 - {}_1W_2 = 1 \times (2464.99 - 3115.25) {}_1W_2 = 645.26$ KJ/sec

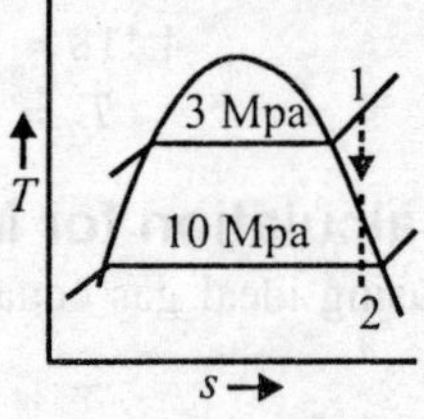

% Error introduced if the kinetic energy and potential energy terms are ignored:

% Error = $[(W_s - {}_1W_2)/W_s] \times 100$
= $[(639.28 - 645.26)/639.28] \times 100$
= -0.935%

So **Error is = 0.935%** ....**ANS**

Fig. 5.11

Q. 15: 5kg of steam is condensed in a condenser following reversible constant pressure process from 0.75 bar and 150°C state. At the end of process steam gets completely condensed. Determine the heat to be removed from steam and change in entropy. Also sketch the process on T-s diagram and shade the area representing heat removed. *(May–02)*

Sol.: Given that :

At state 1: $P_1 = 0.75$ bar and $T_1 = 150°C$ Applying SF'EE to the control volume $Q - W_s = m_f[(h_2 - h_1) + \frac{1}{2}(V_2^2 - V_1^2) + g(z_2 - z_1)]$ neglecting the changes in kinetic and potential energies.i.e.; $Ws = \frac{1}{2}(V_2^2 - V_1^2) = g(z_2 - z_1) = 0$ i.e.; $Q = m_f [(h_2 - h_1)]$ From super heat steam table at $P_1 = 0.75$ bar and $T_1 = 150°C$ We have

$(75 - 50)/(100 - 50) = (h_1 - 2780.08)/(2776.38 - 2780.08)$

$h_1 = 2778.23$ KJ/kg

Also, entropy at state (1)

$(75 - 50)/(100 - 50) = (s_1 - 7.94)/(7.6133 - 7.94)$

$s_1 = 7.77665$ KJ/kgK

at state (2), the condition is saturated liquid.

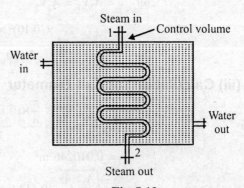

Fig 5.12

From saturated steam table $h_2 = h_f = 384.36$ kJ/kg $s_2 = s_f = 1.2129$ kJ/kgK

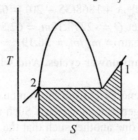

Fig. 5.13

$Q = 384.36 - 2778.23 = -2393.87$ kJ/kg -ve sign shows that heat is rejected by system Total heat rejected = $5 \times 2393.87 = 11.9693$ MJ similarly, total change in entropy
$$= m(s_2 - s_1) = 5(1.2129 - 7.7767) = -32.819 \text{ kJ/K} \qquad \text{......ANS}$$

Q. 16: In a steam power plant, the steam 0.1 bar and 0.95 dry enters the condenser and leaves as saturated liquid at 0.1 bar and 45°C. Cooling water enters the condenser in separate steam at 20°C and leaves at 35°C without any loss of its pressure and no phase change. Neglecting the heat interaction between the condenser and surroundings and changes in kinetic energy and potential energy, determine the ratio of mass-flow rate of cooling water to condensing steam. *(Dec–02)*

Sol.: Given that:

Inlet condition of steam : Pressure 'P_1' = 0.1 bar dryness fraction x = 0.95 Exit condition of steam :saturated water at 0.1 bar and 45°C Inlet temperature of cooling water = 20°C Exit temperature of cooling water = 35°C Applying SFEE to control volume $Q - W_s = m_f[(h_2 - h_1) + \frac{1}{2}(V_2^2 - V_1^2) + g(z_2 - z_1)]$

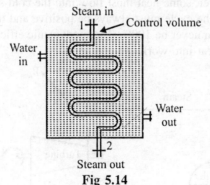

Fig 5.14

neglecting the changes in kinetic and potential energies. And there is no shaft work i.e.; $W_s = 0$

$Q = m_f(h_2 - h_1)$
$h_1 = h_{f1} + xhf_{g1}$
$\quad = 191.81 + 0.95(2392.82) = 2464.99$ kJ/kg
$h_2 = h_{f2}$ at 0.1 bar (since the water is saturated liquid)
$h_2 = 191.81$ kJ/kg
$Q = m_f(191.81 - 2464.99)$
$\quad = -2273.18 m_f$ kJ/kg ...(i)

-ve sign shows the heat rejection.

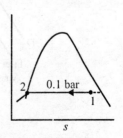

Fig. 5.15

By energy balance; Heat lost by steam = Heat gained by cooling water

$$Q = m_{fw} \cdot C_w (T_4 - T_3) = m_{fw} \times 4.1868(35 - 20) = 62.802 m_{fw} \quad ...(ii)$$

Equate equation (i) and (ii); We get $Q = -2273.18 m_f = 62.802 m_{fw}$, $m_{fw}/m_f = 36.19$ Ratio of mass flow rate of cooling water to condensing steam = $m_{fw}/m_f = \mathbf{36.19}$ ....ANS

Q. 17: What do you know about Steam power cycles. And what are the main component of a steam power plant.?

Sol.: Steam power plant converts heat energy Q from the combustion of a fuel into mechanical work W of shaft rotation which in turn is used to generate electricity. Such a plant operates on thermodynamic cycle in a closed loop of processes following one another such that the working fluid of steam and water repeats cycles continuously. If the first law of thermodynamics is applied to a thermodynamic cycle in which the working fluid returns to its initial condition, the energy flowing into the fluid during the cycle must be equal to that flowing out of the cycle.

$$Q_{in} + W_{in} = Q_{out} + W_{out}$$

Or, $\quad Q_{in} - Q_{out} = W_{out} - W_{in}$

Where;

Q = Rate of Heat transfer

W = Rate of work transfer, i.e. power

Heat and work are mutually convertible. However, although all of a quantity of work energy can be converted into heat energy (by a friction process), the converse is not true. A quantity of heat cannot all be converted into work.

Heat flows by virtue of a temperature difference, and which means that in order to flow, two heat reservoirs must be present; a hot source and a cold sink. During a heat flow from the hot source to the cold sink, a fraction of the flow may be converted into work energy, and the function of a power plant is to produce this conversion. However, some heat must flow into the cold sink because of its presence. Thus the rate of heat transfer Q_{out} of the cycle must always be positive and the efficiency of the conversion of heat energy into work energy can never be 100%. Thermodynamic efficiency for a cycle, nt_h is a measure of how Well a cycle converts heat into work.

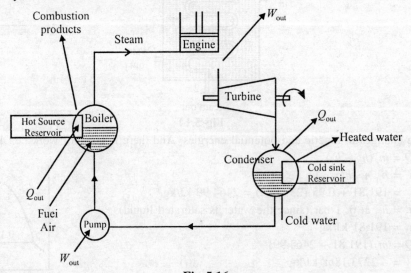

Fig 5.16

Components of a Steam Power Plant

There are four components of a steam power plant:
1. The boiler: Hot-source reservoir in which combustion gases raise steam.
2. Engine/Turbine: The steam reciprocating engine or turbine to convert a portion of the heat energy into mechanical work.
3. Condenser: Cold sink into which heat is rejected.
4. Pump: Condensate extraction pump or boiler feed pump to return the condensate back into the boiler.

Q. 18: Define Carnot vapour Cycle. Draw the carnot vapour cycle on *T-S* diagram and make the different thermodynamics processes. *(Dec–01, Dec–03)*

Sol.: It is more convenient to analyze the performance of steam power plants by means of idealized cycles which are theoretical approximations of the real cycles. The Carnot cycle is an ideal, but non-practising cycle giving the maximum possible thermal efficiency for a cycle operating on selected maximum and minimum temperature ranges. It is made up of four ideal processes: 1 - 2 : Evaporation of water into saturated steam within the boiler at the constant maximum cycle temperature $T_1 (= T_2)$

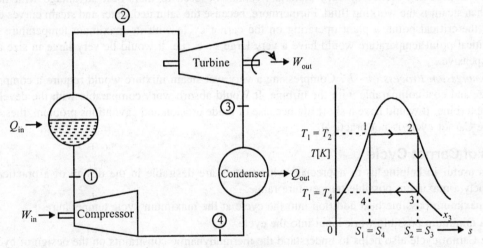

Fig. 5.17 *Carnot Cycle*

2 - 3 : Ideal (i.e., constant-entropy) expansion within the steam engine or turbine i.e., $S_2 = S_3$.

3 - 4 : Partial condensation within the condenser at the constant minimum cycle temperature $T_3 (= T_4)$.

4 - 1 : Ideal (i.e., constant-entropy) compression of very wet steam within the compressor to complete the cycle, i.e., $S_4 = S_1$.

$$S_2 - S_1 = \frac{Q_{in}}{mT_t}$$
$$Q_{in} = (m\,T_1)(S_2 - S_1)$$
$$Q_{out} = (m\,T_3)(S_3 - S_4)$$
$$(S_2 - S_1) = (S_3 - S_4)$$
$$Q_{out} = (v_1\,T_3)(S_2 - S_1)$$
$$\eta_s = 1 - \frac{Q_{out}}{Q_{in}} = 1 - \frac{(mT_3)(S_2 - S_1)}{(mT_1)(S_2 - S_1)} = 1 - \frac{T_3}{T_1}$$

Q. 19: What are the limitations and uses of Carnot vapour cycle.?

Limitations
This equation shows that the wider the temperature range, the more efficient is the cycle.
(a) T_3: In practice T_3 cannot be reduced below about 300 K (27°C), corresponding to a condenser pressure of 0.035 bar. This is due to two tractors:
   (i) Condensation of steam requires a bulk supply of cooling water and such a continuous natural supply below atmospheric temperature of about 15°C is unavailable.
   (ii) If condenser is to be of a reasonable size and cost, the temperature difference between the condensing steam and the cooling water must be at least 10°C.
(b) T_1: The maximum cycle temperature T_1 is also limited to about 900 K (627°C) by the strength of the materials available for the highly stressed parts of the plant, such as boiler tubes and turbine blades. This upper limit is called the metallurgical limit.
(c) *Critical Point* : In fact the steam Carnot cycle has a maximum cycle temperature of well below this metallurgical limit owing to the properties of steam; it is limited to the critical-point temperature of 374°C (647 K). Hence modern materials cannot be used to their best advantage with this cycle when steam is the working fluid. Furthermore, because the saturated water and steam curves converge to the critical point, a plant operating on the carnot cycle with its maximum temperature near the critical-point temperature would have a very large s.s.c., i.e. it would be very large in size and very expensive.
(d) *Compression Process (4 - 1* : Compressing a very wet steam mixture would require a compressor of size and cost comparable with the turbine. It Would absorb work comparable with the developed by the turbine. It would have a short life because of blade erosion and cavitations problem. these reasons the Carnot cycle is not practical.

Uses of Carnot Cycle
1. It is usefui in helping us to appreciate what factors are desirable in the design of a practical cycle; namely a maximum possible temperature range.
   - maximum possible heat addition into the cycle at the maximum cycle temperature
   - a minimum possible work input into the cycle.
2. The Carnot cycle also helps to understand the thermodynamic constraints on the design of cycles. For example, even if such a plant were practicable and even if the maximum cycle temperature could be 900K the cycle thermal efficiency would be well below 100%. This is called Cartrot lintitation.

$$\eta_{th} = 1 - \frac{T_3}{T_1} = 1 - \frac{300}{900} = 66.7\%$$

A hypothetical plant operating on such a cycle would have a plant efficiency lower than this owing to the inefficiencies of the individual plant items.

$$\eta_{plant} = \eta_{th} \times \eta_{item\ 1} \times \eta_{item\ 2} \times \eta_{item\ 3} \times \ldots$$

Q. 20: What is the performance criterion of a steam power plant.?
Sol.: The design of a power plant is determined largely by the consideration of capital cost and operating cost; the former depends mainly on the plant size and latter is primarily a function of the overall efficiency of the plant. In general the efficiency can usually be improved, but only by increasing the capital cost of the plant, hence a suitable compromise must be reached between capital costs and operating costs.

1. Specific steam consumption (S.S.C.). The plant capital cost is mainly dependent upon the size of the plant components. These sizes will themselves depend on the flow rate of the steam which is passed through them.

Hence, an indication of the relative capital cost of different steam plant is provided by the mass flow rate m of the steam required per unit power output, i.e., by the specific steam consumption (s.s.c.) or steam rate

$$s.s.c. = \frac{\dot{m}}{\dot{W}}\frac{kg/s}{kW} = \frac{\dot{m}}{\dot{W}}\frac{kg}{kWs} = \frac{3600\dot{m}}{\dot{W}}\frac{kg}{kWh} = \frac{3600}{\dot{W}/\dot{m}}\frac{kg}{kWh}$$

In M.K.S. system.

1 horsepower hour ≈ 632 k cal

1 kilowatt hour ≈ 860 k cal.

$$\therefore \quad s.s.c. = \frac{632}{\dot{W}} \text{ kg/HP-hr} = \frac{860}{\dot{W}} \text{ kg/k Wh}$$

3. Work ratio is defined as the ratio of net plant output to the gross (turbine) output.

$$\text{Work ratio} = \frac{\dot{W}_{out} \dot{W}_{in}}{\dot{W}_{out}} = \frac{\dot{W}_t - \dot{W}_c}{\dot{W}_t}$$

Q. 21: Explain Rankine cycle with the help of P-V, T-s and H-s diagrams. *(May–05)*

Or

Write a note on Rankine cycle. *(Dec–01, Dec–05)*

Sol.: One of the major problems of Carnot cycle is compressing a very wet steam mixture from the condenser pressure upto the boiler pressure. The problem can be avoided by condensing the steam completely in the condenser and then compressing the water in a comparatively small feed pump. The effect of this modification is to make the cycle practical one. Furthermore, far less work is required to pump a liquid than to compress a vapour and therefore this modification also has the result that the feed pump's work is only one or two per cent of the work developed by the turbine. We can therefore neglect this term in our cycle analysis.

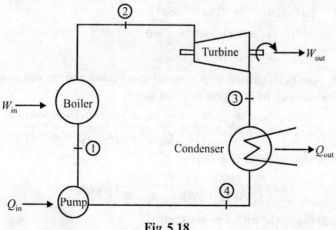

Fig. 5.18

The idealized cycle for a simple steam power plant taking into account the above modification is called the Rankine cycle shown in the figure, Fig. 5.18. It is made up of four practical processes:

(a) 1 - 2 : Heat is added to increase the temperature of the high-pressure water up to its saturation value (process 1 to A). The water is then evaporated at constant temperature and pressure (process A to 2). Both processes occur within the boiler, but not all of the heat supplied is at the maximum cycle temperature. Thus, the mean temperature at which heat is supplied is lower than that in the equivalent Carnot cycle. Hence, the basic steam cycle thermal efficiency is inherently lower. Applying the first law of thermodynamics to this process:

$$(\dot{Q}_{in} - \cancel{\dot{Q}_{out}}^0) + (\cancel{\dot{W}_{in}}^0 - \cancel{\dot{W}_{out}}^0) = \dot{m}_{fluid}(h_{final} - h_{initial})$$

$$\dot{Q}_{out} = 0; \dot{W}_{in} = 0, \dot{W}_{out} = 0$$

$$\therefore \quad \dot{Q}_{in} = \dot{m}_f(h_2 - h_1)$$

(b) 2 - 3: The high pressure saturated steam is expanded to a low pressure within a reciprocating engine or a turbine.

If the expansion is ideal (i.e., one of constant entropy), the cycle is called the Rankine cycle. However, in actual plant friction takes place in the flow of steam through the engine or turbine which results in the expansion with increasing entropy. Applying first law to this process:

$$(\cancel{\dot{Q}_{in}}^0 - \dot{Q}_{out}) + (\cancel{\dot{W}_{in}}^0 - \cancel{\dot{W}_{out}}^0) = \dot{m}_f(h_{final} - h_{initial})$$

$$\dot{Q}_{in} = 0, \dot{W}_{in} = 0, \dot{W}_{out} = 0.$$

$$\dot{Q}_{out} = \dot{m}_f(h_3 - h_4)$$

(c) 3 - 4: The low-pressure wet steam is completely condensed at constant condenser pressure back into saturated water. The latent heat of condensation is thereby rejected to the condenser cooling water which, in turn, rejects this heat to the atmosphere. Applying first law of the thermodynamics,

$$(\cancel{\dot{Q}_{in}}^0 - \dot{Q}_{out}) + (\cancel{\dot{W}_{in}}^0 - \cancel{\dot{W}_{out}}^0) = \dot{m}_f(h_{final} - h_{initial})$$

(d) 4 - 1: The low pressure saturated water is pumped back up to the boiler pressure and, in doing so, it becomes sub-saturated. The water then reenters the boiler and begins a new cycle. Applying the first law:

$$(\dot{Q}_{in} - \dot{Q}_{out}) + (\dot{W}_{in} - \dot{W}_{out}) = \dot{m}_f(h_{final} - h_{initial})$$

$$\dot{Q}_{in} = 0, \dot{Q}_{out} = 0, \dot{W}_{out} = 0$$

$$\dot{W}_{in} = \dot{m}_f(h_1 - h_4).$$

However, W_{in} can be neglected with reasonable accuracy and we can assume $h_1 = h_4$. The thermal efficiency of the cycle is given by:

$$\eta_{th} = \frac{\dot{W}_{out} - \dot{W}_{in}}{\dot{Q}_{in}} = \frac{\dot{W}_{out}}{\dot{Q}_{in}} = \frac{m_f(h_2 - h_3)}{m_f(h_2 - h_1)} = \frac{h_2 - h_3}{h_2 - h_3}$$

Specific steam consumption is given by:

$$s.s.c = \frac{3600}{\dot{W}/m}\left[\frac{kg}{kWh}\right] = \frac{3600}{h_2 - h_3} \text{ kg/kWh}$$

Q. 22: Compare Rankine cycle with Carnot cycle

Sol.: Rankine cycle without superheat : 1 – A – 2 – 3 – 4 – 1.
Rankine cycle with superheat : 1 – A – 2 – 2' –3' – 4 – 1.
Carnor cyle without superheat : A – 2 – 3 –4" – A.

Carnot cycle with superheat : $A - 2'' - 3' - 4^2 - A$.

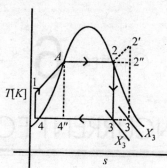

Fig. 5.20

(1) The thermal efficiency of a Rankine cycle is lower than the equivalent Carnot cycle. Temperature of heat supply to Carnot cycle = T_A; Mean temperature of heat supply to Rankine

$$\text{cycle} = \frac{T_1 + T_2}{2}, \quad T_A > \frac{T_1 + T_2}{2}$$

(2) Carnot cycle needs a compressor to handle wet steam mixture whereas in Rankine cycle, a small pump is used.

(3) The steam can be easily superheated at constant pressure along 2 - 2' in a Rankine cycle. Superheating of steam in a Carnot cycle at constant temperature along A - ?" is accompanied by a fall of pressure which is difficult to achieve in practice because heat transfer and expansion process should go side by side. Therfore Rankine cycle is used as ideal cycle for steam power plants.

Chapter 6

FORCE: CONCURRENT FORCE SYSTEM

Q. 1: Define Engineering Mechanics
Sol.: Engineering mechanics is that branch of science, which deals the action of the forces on the rigid bodies. Everywhere we feel the application of Mechanics, such as in railway station, where we seen the railway bridge, A car moving on the road, or simply we are running on the road. Everywhere we saw the application of mechanics.

Q. 2: Define matter, particle and body. How does a rigid body differ from an elastic body?
Sol.: Matter is any thing that occupies space, possesses mass offers resistance to any stress, example Iron, stone, air, Water.

A body of negligible dimension is called a particle. But a particle has mass.

A body consists of a No. of particle, It has definite shape.

A rigid body may be defined as the combination of a large no. of particles, Which occupy fixed position with respect to another, both before and after applying a load.

Or, A rigid body may be defined as a body, which can retain its shape and size even if subjected to some external forces. In actual practice, no body is perfectly rigid. But for the shake of simplicity, we take the bodies as rigid bodies.

An elastic body is that which regain its original shape after removal of the external loads.

The basic difference between a rigid body and an elastic body is that the rigid body don't change its shape and size before and after application of a force, while an elastic body may change its shape and size after application of a load, and again regain its shape after removal of the external loads.

Q. 3: Define space, motion.
Sol.: The geometric region occupied by bodies called space.

When a body changes its position with respect to other bodies, then body is called as to be in motion.

Q. 4: Define mass and weight.
Sol.: The properties of matter by which the action of one body can be compared with that of another is defined as mass.

$$m = \rho \cdot v$$

Where,
ρ = Density of body and v = Volume of the body

Weight of a body is the force with which the body is attracted towards the center of the earth.

Q. 5: Define Basic S.I. Units and its derived unit.
Sol.: S.I. stands for "System International Units". There are three basic quantities in S.I. Systems as concerned to engineering Mechanics as given below:

Sl.No.	Quantity	Basic Unit	Notation
1	Length	Meter	m
2	Mass	Kilogram	kg
3	Time	Second	s

Meter: It is the distance between two given parallel lines engraved upon the polished surface of a platinum-Iridium bar, kept at 00C at the "International Bureau of Weights and Measures" at Serves, near Paris.

Kilogram: It is the mass of a particular cylinder made of Platinum Iridium kept at "International Bureau of Weights and Measures" at Serves, near Paris.

Second: It is $1/(24 \times 60 \times 60)$th of the mean solar day. A solar day is defined as the time interval between the instants at which the sun crosses the meridian on two consecutive days.

With the help of these three basic units there are several units are derived as given below.

Sl.No.	Derived Unit	Notation
1	Area	m^2
2	Volume	m^3
3	Moment of Inertia	m^4
4	Force	N
5	Angular Acceleration	Rad/sec^2
6	Density	kg/m^3
7	Moment of Force	N.m
8	Linear moment	kg.m/sec
9	Power	Watt
10	Pressure/stress	$Pa(N/m^2)$
11	Mass moment of Inertia	$kg.m^2$
12	Linear Acceleration	m/s^2
13	Velocity	m/sec
14	Momentum	kg-m/sec
15	Work	N-m or Jule
16	Energy	Jule

Q. 6: What do you mean by 1 Newton's? State Newton's law of motion

Sol.: 1-Newton: It is magnitude of force, which develops an acceleration of 1 m/s^2 in 1 kg mass of the body.

The entire subject of rigid body mechanics is based on three fundamental law of motion given by an American scientist Newton.

Newton's first law of motion: A particle remains at rest (if originally at rest) or continues to move in a straight line (If originally in motion) with a constant speed. If the resultant force acting on it is Zero.

Newton's second law of motion: If the resultant force acting on a particle is not zero, then acceleration of the particle will be proportional to the resultant force and will be in the direction of this force.

$$F = m.a$$

Newton's s third law of motion: The force of action and reaction between interacting bodies are equal in magnitude, opposite in direction and have the same line of action.

Q. 7: Differentiate between scalar and Vector quantities. How a vector quantity is represented?

Sol.: A quantity is said to be scalar if it is completely defined by its magnitude alone. Ex: Length, area, and time. While a quantity is said to be vector if it is completely defined only when its magnitude and direction are specified. For Ex: Force, velocity, and acceleration.

Vector quantity is represented by its magnitude, direction, point of application. Length of line is its magnitude, inclination of line is its direction, and in the fig 6.1 point C is called point of application.

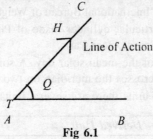

Fig 6.1

Here AC represent the vector acting from A to C
T = Tail of the vector
H = Head of the vector
Q = Direction of the vector
Arrow represents the Sense.

Q. 8: What are the branches of mechanics, differentiate between static's, kinetics and kinematics.

Sol.: Mechanics is mainly divided in to two parts Static's and Dynamics, Dynamics further divided in kinematics and kinetics

Statics: It deals with the study of behavior of a body at rest under the action of various forces, which are in equilibrium.

Dynamics: Dynamics is concerned with the study of object in motion

Kinematics: It deals with the motion of the body with out considering the forces acting on it.

Kinetics: It deals with the motion of the body considering the forces acting on it.

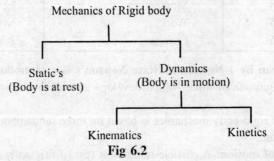

Fig 6.2

Q. 9: Define force and its type?

Sol.: Sometime we push the wall, then there are no changes in the position of the wall, but no doubt we apply a force, since the applied force is not sufficient to move the wall, i.e no motion is produced. So this is clear that a force may not necessarily produce a motion in a body. But it may simply tend to do, So we can say

The force is the agency, which change or tends to change the state of rest or motion of a body. It is a vector quantity.

A force is completely defined only when the following four characteristics are specified- Magnitude, Point of application, Line of action and Direction.

OR:

The action of one body on another body is defined as force.

In engineering mechanics, applied forces are broadly divided in to two types. Tensile and compressive force.

Tensile Forces

A force, which pulls the body, is called as tensile force. Here member AB is a tension member carrying tensile force P. (see fig 6.3)

Compressive Force

A force, which pushes the body, is called as compressive force. (see fig 6.4)

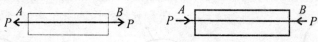

Fig 6.3: *Tensile Force*   **Fig 6.4:** *Compressive force*

Q. 10: Define line of action of a force?

Sol.: The direction of a force along a straight line through its point of application, in which the force tends to move a body to which it is applied. This line is called the line of action of the force.

Q. 11: How do you classify the force system?

Sol.: Single force is of two types i.e.; Tensile and compressive. Generally in a body several forces are acting. When several forces of different magnitude and direction act upon a rigid body, then they are form a System of Forces, These are

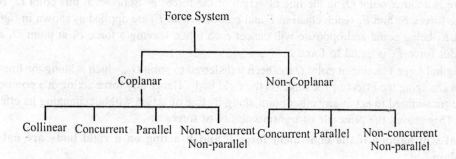

Fig 6.5

Coplanar Force System: The forces, whose lines of action lie on the same plane, are known as coplanar forces.

Non-Coplanar Force System: The forces, whose lines of action not lie on the same plane, are known as non-coplanar force system.

Concurrent Forces: All such forces, which act at one point, are known as concurrent forces.

Coplanar-Concurrent System: All such forces whose line of action lies in one plane and they meet at one point are known as coplanar-concurrent force system.

Coplanar-Parallel Force System: If lines of action of all the forces are parallel to each other and they lie in the same plane then the system is called as coplanar-parallel forces system.

Coplanar-Collinear Force System: All such forces whose line of action lies in one plane also lie along a single line then it is called as coplanar-collinear force system.

Non-concurrent Coplanar Forces System: All such forces whose line of action lies in one plane but they do not meet at one point, are known as non-concurrent coplanar force system.

CONCURRENT FORCE SYSTEM

Q. 12: State and explain the principle of transmissibility of forces?

(Dec-00, May-01, May(B.P.)-01, Dec-03)

Sol.: It state that if a force acting at a point on a rigid body, it may be considered to act at any other point on its line of action, provided this point is rigidly connected with the body. The external effect of the force on the body remains unchanged. The problems based on concurrent force system (you study in next article) are solved by application of this principle.

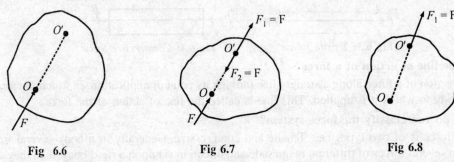

Fig 6.6　　　　　Fig 6.7　　　　　Fig 6.8

For example, consider a force 'F' acting at point 'O' on a rigid body as shown in fig(6.6). On this rigid body," There is another point O_1 in the line of action of the force 'F' Suppose at this point O_1 two equal and opposite forces F_1 and F_2 (each equal to F and collinear with F) are applied as shown in fig(6.7). The force F and F_2 being equal and opposite will cancel each other, leaving a force F_1 at point O_1 as shown in fig(6.7). But force F_1 is equal to force F.

The original force F acting at point O has been transferred to point O_1, which is along the line of action of F without changing the effect of the force on the rigid body. Hence any force acting at a point on a rigid body can be transmitted to act at any other point along its line of action without changing its effect on the rigid body. This proves the principle of transmissibility of forces.

Q. 13: What will happen if the equivalent force F and F acting on a rigid body are not in line? Explain.

Sol.: If the equivalent force of same magnitude 'F' acting on a rigid body are not in line, then no change of the position of the body, Because the resultant of both two forces is the algebraic sum of the two forces which is $F - F = 0$ or $F + F = 2F$.

Q. 14: What will happen if force is applied to (*i*) Rigid body (*ii*) Non- Rigid body?

Sol.: (*i*) Since Rigid body cannot change its shape on application of any force, so on application of force "It will start moving in the direction of applied force without any deformation."

(*ii*) Non-Rigid body change its shape on application of any force, So on application of force on Non-Rigid body " It will start moving in the direction of applied force with deformation."

Q. 15: Define the term resultant of a force system? How you find the resultant of coplanar concurrent force system?

Sol.: Resultant is a single force which produces the same effect as produced by number of forces jointly in a system. In equilibrium the magnitude of resultant is always zero.

There are many ways to find out the resultant of the force system. But the first thing to see that how many forces is acting on the body,

1. If only one force act on the body then that force is the resultant.
2. If two forces are acting on the rigid body then there are two methods for finding out the resultant, i.e. 'Parallelogram law' (Analytical method) and 'triangle law' (Graphical method).
3. If more than two forces are acting on the body then the resultant is finding out by 'method of resolution' (Analytical method) and 'Polygon law' (Graphical method).

So we can say that there are mainly two type of method for finding the resultant.

1. Analytical Method.
2. Graphical Method

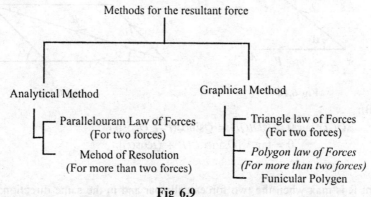

Fig 6.9

Finally; The resultant force, of a given system of forces, may be found out analytically by the following methods:

(a) Parallelogram law of forces

(b) Method of Resolution.

Q. 16: State and prove parallelogram law of forces

Sol.: This law is used to determine the resultant of two forces acting at a point of a rigid body in a plane and is inclined to each other at an angle of a.

It state that "If two forces acting simultaneously on a particle, be represented in magnitude and direction by two adjacent sides of a parallelogram then their resultant may be represented in magnitude and direction by the diagonal of the parallelogram, which passes through their point of intersection."

Let two forces P and Q act at a point 'O' as shown in fig (6.10).The force P is represented in magnitude and direction by vector OA, Where as the force Q is represented in magnitude and direction by vector OB, Angle between two force is 'a'. The resultant is denoted by vector OC in fig. 6.11. Drop perpendicular from C on OA.

Let,

P,Q = Forces whose resultant is required to be found out.

θ = Angle which the resultant forces makes with one of the forces

α = Angle between the forces P and Q

110 / *Problems and Solutions in Mechanical Engineering with Concept*

Now $\angle CAD = \alpha$ :: because $OB // CA$ and OA is common base.

In $\triangle ACD$ :: $\cos\alpha = AD/AC \Rightarrow AD = AC\cos\alpha$

:: But $AC = Q$; i.e., $AD = Q\cos\alpha$ ...(i)

And $\sin\alpha = CD/AC \Rightarrow CD = AC\sin\alpha$

$\Rightarrow CD = Q\sin\alpha$ ...(ii)

Now in $\triangle OCD \Rightarrow OC^2 = OD^2 + CD^2$

$\Rightarrow R^2 = (OA + AD)^2 + CD^2$

$= (P + Q\cos\alpha)^2 + (Q\sin\alpha)^2$

$\Rightarrow = P^2 + Q^2\cos^2\alpha + 2PQ\cos\alpha + Q^2\sin\alpha$

$$R = \sqrt{(P^2 + Q^2 + 2PQ\cos\alpha)}$$

It is the magnitude of resultant 'R'

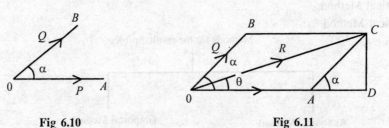

Fig 6.10 Fig 6.11

Direction (θ):

in $\triangle OCD$ $\tan\theta = CD/OD = Q\sin\alpha/(P + Q\cos\alpha)$

i.e., $\theta = \tan^{-1}[Q\sin\alpha / (P + Q\cos\alpha)]$

Conditions

(i) Resultant R is max when the two forces collinear and in the same direction.

i.e., $\alpha = 0°$ $\Rightarrow$ $R_{max} = P + Q$

(ii) Resultant R is min when the two forces collinear but acting in opposite direction.

i.e., $\alpha = 180°$ $\Rightarrow$ $R_{min} = P - Q$

(iii) If $\alpha = 90°$, i.e when the forces act at right angle, then

$R = \sqrt{P^2 + Q^2}$

(iv) If the two forces are equal i.e., when $P = Q \Rightarrow R = 2P.\cos(\theta/2)$

Q. 17: A 100N force which makes an angle of 45° with the horizontal x-axis is to be replaced by two forces, a horizontal force F and a second force of 75N magnitude. Find F.

Sol.: Here 100N force is resultant of 75N and F Newton forces, Draw a Parallelogram with Q = 75N and $P = F$ Newton

$\theta = 45°$ and α is not given.

We know that

$\tan\theta = Q\sin\alpha/(P + Q\cos\alpha)$

$\tan 45° = 75\sin\alpha/(F + 75\cos\alpha)$

since $\tan 45° = 1 \rightarrow 75\sin\alpha = F + 75\cos\alpha$

or, $F = 75(\sin\alpha - \cos\alpha)$ ...(i)

Now, $R = (P^2 + Q^2 + 2PQ\cos\alpha)^{1/2}$

$(100)^2 = F^2 + 75^2 + 2.F.75.\cos\alpha$

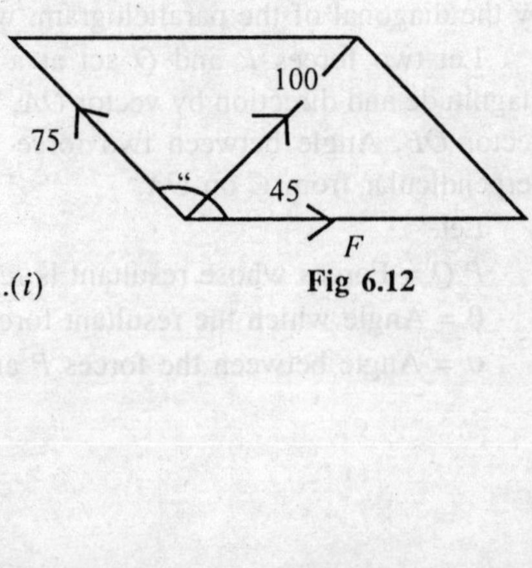

Fig 6.12

$$F^2 + 150 \cdot F \cdot \cos\alpha = 4375$$
$$F(F + 75\cos\alpha + 75\cos\alpha) = 4375$$
$$F(75\sin\alpha + 75\cos\alpha) = 4375$$
$$F(\sin\alpha + \cos\alpha) = 58.33 \qquad \qquad ...(ii)$$

Value of 'F' from eq(i) put in equation(ii), we get
$$75(\sin\alpha - \cos\alpha)(\sin\alpha + \cos\alpha) = 58.33$$
$$\sin^2\alpha - \cos^2\alpha = 0.77$$
$$-\cos 2\alpha = 0.77 \rightarrow \alpha = 70.530 \qquad \qquad ...(iii)$$

Putting the value of a in equation (i), we get
$$F = 45.71N \qquad \qquad ....ANS$$

Q. 18: Find the magnitude of two forces such that if they act at right angle their resultant is $\sqrt{10}$ KN, While they act at an angle of 60°, their resultant is $\sqrt{13}$ KN.

Sol.: Let the two forces be P and Q, and their resultant be 'R'

Since $\qquad R = \sqrt{(P^2 + Q^2 + 2PQ\cos\alpha)}$

Case–1: If $\qquad \alpha = 90°$, than $R = (10)^{1/2}$ KN
$$10 = P^2 + Q^2 + 2PQ\cos 90°$$
$$10 = P^2 + Q^2, \quad \cos 90° = 0 \qquad \qquad ...(i)$$

Case–2: If $\qquad \alpha = 600$, than $R = (13)^{1/2}$ KN
$$13 = P^2 + Q^2 + 2PQ\cos 60°$$
$$13 = P^2 + Q^2 + PQ, \quad \cos 60° = 0.5 \qquad \qquad ...(ii)$$

From equation (i) and (ii)
$$PQ = 3 \qquad \qquad ...(iii)$$

Now $\qquad (P + Q)^2 = P^2 + Q^2 + 2PQ = 10 + 2.3 = 16$
$$P + Q = 4 \qquad \qquad ...(iv)$$
$$(P - Q)^2 = P^2 + Q^2 - 2PQ = 10 - 2 \times 3$$
$$P - Q = 2 \qquad \qquad ...(v)$$

From equation (v) and (iv)
$$P = 3KN \text{ and } Q = 1KN \qquad \qquad ....ANS$$

Q. 19: Two forces equal to $2P$ and P act on a particle. If the first force be doubled and the second force is increased by 12KN, the direction of their resultant remain unaltered. Find the value of P.

Sol.: In both cases direction of resultant remain unchanged, so we used the formula,
$$\tan\theta = Q\sin\alpha/(P + Q\cos\alpha)$$

Case-1: $\qquad P = 2P, Q = P$
$$\tan\theta = P\sin\alpha/(2P + P\cos\alpha) \qquad \qquad ...(i)$$

Case-2: $\qquad P = 4P, Q = P + 12$
$$\tan\theta = (P + 12)\sin\alpha/(4P + (P + 12)\cos\alpha) \qquad \qquad ...(ii)$$

Equate both equations:
$$P\sin\alpha/(2P + P\cos\alpha) = (P + 12)\sin\alpha/(4P + (P + 12)\cos\alpha)$$
$$4P^2\sin\alpha + P^2\sin\alpha\cos\alpha + 12P\sin\alpha\cos\alpha$$
$$= 2P^2\sin\alpha + 24P\sin\alpha + P^2\sin\alpha\cos\alpha + 12P\sin\alpha\cos\alpha$$
$$2P^2\sin\alpha = 24P\sin\alpha$$
$$P = 12KN \qquad \qquad ....ANS$$

Q. 20: The angle between the two forces of magnitude 20KN and 15KN is 60°, the 20KN force being horizontal. Determine the resultant in magnitude and direction if

(*i*) the forces are pulls

(*ii*) the 15KN force is push and 20KN force is a pull.

Sol.: Since there are two forces acting on the body, So we use Law of Parallelogram of forces.

Case-1: $P = 20$ KN, $Q = 15$ KN, $\alpha = 60°$

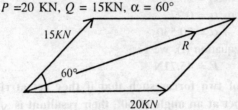

Fig 6.13

$R^2 = P^2 + Q^2 + 2PQ\cos\alpha = 20^2 + 15^2 + 2 \times 20 \times 15\cos 60°$

R = 30.41KN ....ANS

$\tan\theta = Q\sin\alpha/(P + Q\cos\alpha) = 15\sin 60°/(20 + 15\cos 60°)$

θ = 25.28° ....ANS

Case-2: Now angle between two forces is 120°, $P = 20$KN, $Q = 15$KN, $\alpha = 120°$

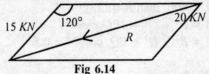

Fig 6.14

$R^2 = P^2 + Q^2 + 2PQ\cos\alpha = 20^2 + 15^2 + 2 \times 20 \times 15\cos 120°$

R = 18.027KN ....ANS

$\tan\theta = Q\sin\alpha/(P + Q\cos\alpha) = 15\sin 120°/(20 + 15\cos 120°)$

θ = –46.1° ....ANS

Q. 21: Explain composition of a force. How you make component of a single force?

Sol.: When a force is split into two parts along two directions not at right angles to each other, those parts are called component of a force. And process is called composition of a force.

In *BOAC*, angle *BOC* = angle *OCA* = β

(Because // lines *OB* and *AC*)

Angle *CAO* = 180 - (α + β)

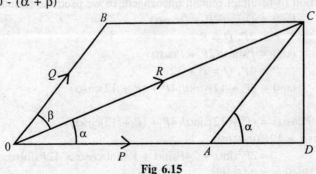

Fig 6.15

Using sine rule in Triangle OCA

$$OA/\sin\beta = OC/\sin(\alpha + \beta) = AC/\sin\alpha \rightarrow P/\sin\beta = R/\sin(\alpha + \beta) = Q/\sin\alpha$$

Or we can say that; $\quad P = R.\sin\beta/\sin(\alpha + \beta)$

$$Q = R.\sin\alpha/\sin(\alpha + \beta)$$

Here P and Q are component of the force 'R' in any direction.

Q. 22: A 100N force which makes as angle of 45° with the horizontal x-axis is to be replaced by two forces, a horizontal force F and a second force of 75N magnitude. Find F.

Sol.: given $Q = 75N$ and $P = F\ N$

$\theta = 45°$ and α is not given.

We know that

$Q = R.\sin\alpha/\sin(\alpha + \beta)$

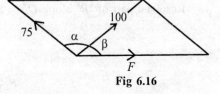

Fig 6.16

$75 = 100\sin 45°/\sin(45 + \beta)$

on solving, $\quad \beta = 25.530$

$P = R.\sin\beta/\sin(\alpha + \beta)$

$F = 100\sin 25.53°/\sin(45° + 25.53°)$

$F = \mathbf{45.71N}$ ....ANS

Q. 23: What is resolution of a force? Explain principle of resolution.

Sol.: When a force is resolved into two parts along two mutually perpendicular directions, without changing its effect on the body, the parts along those directions are called resolved parts. And process is called resolution of a force.

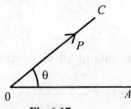 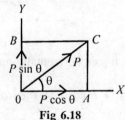

Fig 6.17 Fig 6.18

Horizontal Component $(\Sigma H) = P\cos\theta$

Vertical Component $(\Sigma V) = P\sin\theta$

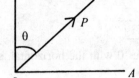 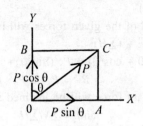

Fig 6.19 Fig 6.20

Horizontal Component $(\Sigma H) = P\sin\theta$

Vertical Component $(\Sigma V) = P\cos\theta$

Principle of Resolution: It states, "The algebraic sum of the resolved parts of a number of forces in a given direction is equal to the resolved part of their resultant in the same direction."

Q.No-24: What is the method of resolution for finding out the resultant force.

Or

How do you find the resultant of coplanar concurrent force system?

Sol.: The resultant force, of a given system of forces may be found out by the method of resolution as discussed below:

Let the forces be P_1, P_2, P_3, P_4, and P_5 acting at 'o'. Let OX and OY be the two perpendicular directions. Let the forces make angle a_1, a_2, a_3, a_4, and a_5 with Ox respectively. Let R be their resultant and inclined at angle θ. with OX.

Resolved part of 'R' along OX = Sum of the resolved parts of P_1, P_2, P_3, P_4, P_5 along OX.

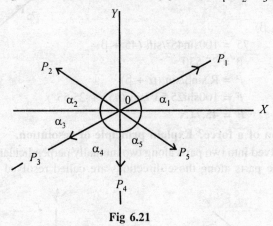

Fig 6.21

i.e.,

Resolve all the forces horizontally and find the algebraic sum of all the horizontally components (i.e., ΣH)

$R\cos\theta = P_1\cos\alpha_1 + P_2\cos\alpha_2 + P_3\cos\alpha_3 + P_4\cos\alpha_4 + P_5\cos\alpha_5$
$= X$ (Let)

Resolve all the forces vertically and find the algebraic sum of all the vertical components (i.e., ΣV)

$R\sin? = P_1\sin\alpha_1 + P_2\sin\alpha_2 + P_3\sin\alpha_3 + P_4\sin\alpha_4 + P_5\sin\alpha_5$
$= Y$ (Let)

The resultant R of the given forces will be given by the equation:

$R = \sqrt{(\Sigma V)^2 + (\Sigma H)^2}$

We get $R^2(\sin^2\theta + \cos^2\theta) = P_1^2(\sin^2\alpha_1 + \cos^2\alpha_1) + \text{------}$

i.e., $\qquad R^2 = P_1^2 + P_2^2 + P_3^2 + \text{------}$

And The resultant force will be inclined at an angle 'θ' with the horizontal, such that

$\tan\theta = \Sigma V / \Sigma H$

NOTE:

1. Some time there is confusion for finding the angle of resultant (θ), The value of the angle θ will be very depending upon the value of ΣV and ΣH, for this see the sign chart given below, first for ΣH and second for ΣV.

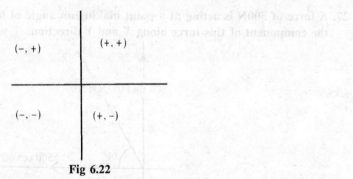

Fig 6.22

a. When ΣV is +ive, the resultant makes an angle between 0° and 180°. But when ΣV is –ive, the resultant makes an angle between 180° and 360°.

b. When ΣH is +ive, the resultant makes an angle between 0° and 90° and 270° to 360°. But when ΣH is -ive, the resultant makes an angle between 90° and 270°.

2. Sum of interior angle of a regular Polygon
$$= (2.n - 4).90°$$
Where, n = Number of side of the polygon
For Hexagon, n = 6; angle = (6 X 2–4) X 90 = 720°
And each angle = total angle/n = 720/6 = 120°

3. It resultant is horizontal, then θ = 0°
i.e. $\Sigma H = R$, $\Sigma V = 0$

4. It Resultant is vertical, then θ = 90°; i.e., $\Sigma H = 0$, V = R

Q. 25: What are the basic difference between components and resolved parts?

Sol.: 1. When a force is resolved into two parts along two mutually perpendicular directions, the parts along those directions are called resolved parts. When a force is split into two parts along two directions not at right angles to each other, those parts are called component of a force. And process is called composition of a force.

2. All resolved parts are components, but all components are not resolved parts.

3. The resolved parts of a force in a given direction do not represent the whole effect of the force in that direction.

Q. 26: What are the steps for solving the problems when more than two coplanar forces are acting on a rigid body.

Sol.: The steps are as;

1. Check the Problem for concurrent or Non concurrent
2. Count Total No. of forces acting on the body.
3. First resolved all the forces in horizontal and vertical direction.
4. Make the direction of force away from the body.
5. Take upward forces as positive, down force as negative, Left hand force as negative, and Right hand force as positive
6. Take sum of all horizontal parts i.e., ΣH
7. Take sum of all vertical parts i.e., ΣV
8. Find the resultant of the force system using,
$$R = \sqrt{(\Sigma V)^2 + (\Sigma H)^2}$$
9. Find angle of resultant by using $\tan\theta = \Sigma V/\Sigma H$
10. Take care about sign of ΣV and ΣH.

Q. 27: A force of 500N is acting at a point making an angle of 60° with the horizontal. Determine the component of this force along X and Y direction.

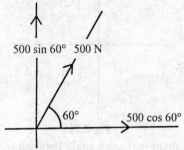

Fig 6.23

Sol.: The component of 500N force in the X and Y direction is

ΣH = Horizontal Component = $500\cos60°$

ΣV = Vertical Component = $500\sin60°$

$\Sigma H = 500\cos60°$, $\Sigma V = 500\sin60°$ ....ANS

Q. 28: A small block of weight 300N is placed on an inclined plane, which makes an angle 600 with the horizontal. What is the component of this weight?

(*i*) Parallel to the inclined plane

(*ii*) Perpendicular to the inclined plane. As shown in fig(6.24)

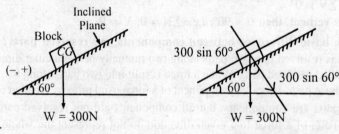

Fig 6.24   Fig 6.25

Sol.: First draw a line perpendicular to inclined plane, and parallel to inclined plane

ΣH = Sum of Horizontal Component

= Perpendicular to plane

= $300\cos60° = 150N$ ....ANS

ΣV = Sum of Vertical Component

= Parallel to plane

= $300\sin600 = 259.81N$ ....ANS

NOTE: There is no confusion about $\cos\theta$ and $\sin\theta$, the angle 'θ' made by which plane, the component of force on that plane contain $\cos\theta$, and other component contain $\sin\theta$.

Q. 29: The 100N force is applied to the bracket as shown in fig(6.26). Determine the component of F in,

(*i*) the x and y directions

(*ii*) the x' and y' directions

(*iii*) the x and y' directions

Force: Concurrent Force System / 117

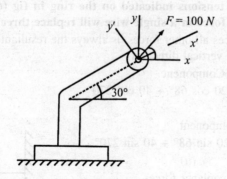

Fig 6.26

Sol.:

(1) Components in x and y directions
$\Sigma H = 100\cos 500 = \mathbf{64.2N}$ ....ANS
$\Sigma V = 100\sin 500 = \mathbf{76.6N}$ ....ANS

(2) Components in x' and y' directions
$\Sigma H' = 100\cos 200 = \mathbf{93.9N}$ ....ANS
$\Sigma V' = 100\sin 200 = \mathbf{34.2N}$ ....ANS

(3) Components in x and y' directions
$\Sigma H = 100\cos 500 = \mathbf{64.2N}$ ....ANS
$\Sigma V' = 100\sin 200 = \mathbf{34.2N}$ ....ANS

Q. 30: Determine the x and y components of the force exerted on the pin at A as shown in fig (6.27).

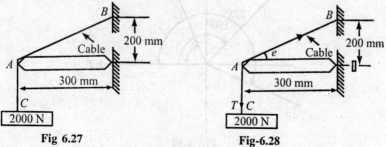

Fig 6.27   Fig-6.28

Sol.: Since there is a single string, so the tension in the string throughout same, Let 'T' is the tension in the string.

At point C, there will be an equal and opposite reaction, so
$$T = 2000N \qquad \ldots(i)$$
Now $\tan\theta = 200/300 \Rightarrow \theta = 33.69°$

Horizontal component of T is;
$\Sigma H = T\cos\theta = 2000\cos 33.69°$
$= \mathbf{1664.3N}$ ....ANS

Vertical component of T is;
$\Sigma V = T\sin\theta = 2000\sin 33.69°$
$= \mathbf{1109.5N}$ ....ANS

Q. 31: Three wires exert the tensions indicated on the ring in fig (6.29). Assuming a concurrent system, determine the force in a single wire will replace three wires.

Sol.: Single force, which replaces all other forces, is always the resultant of the system, so first resolved all the forces in horizontal and vertical direction

ΣH = Sum of Horizontal Component
= 60 cos 0° + 20 cos 68° + 40 cos 270°
= 67.49N ...(i)

ΣV = Sum of Vertical Component
= 60 sin 0° + 20 sin 68° + 40 sin 270°
= –21.46N ...(ii)

Let R be the resultant of coplanar forces
$R = (\Sigma H^2 + \Sigma V^2)^{1/2}$
= $(67.49^2 + (-21.46)^2)^{1/2}$
R = 70.81N ....ANS
$\theta = \tan^{-1}(R_V/R_H)$
= $\tan^{-1}(-21.45/67.49)$
θ = **–17.63°** ....ANS

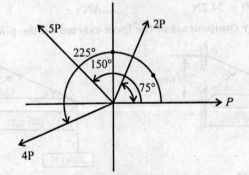

Fig 6.29

Angle made by resultant (70.81),–17.63° and lies in forth coordinate.

Q. 32: Four forces of magnitude P, $2P$, $5P$ and $4P$ are acting at a point. Angles made by these forces with x-axis are 0°, 75°, 150° and 225° respectively. Find the magnitude and direction of resultant force.

Fig. 6.30

Sol.: first resolved all the forces in horizontal and vertical direction

ΣH = Sum of Horizontal Component
= P cos 0° + $2P$cos 75° + $5P$cos 150° + $4P$cos 225°
= –5.628P ...(i)

ΣV = Sum of Vertical Component
= Psin 0° + $2P$sin 75° + $5P$sin 150° + $4P$sin 225°
= 1.603P ...(ii)

$R = ((-5.628P)^2 + (1.603P)^2)^{1/2}$
R = 5.85P ....ANS
$\theta = \tan^{-1}(R_V/R_H)$

$= \tan^{-1}(1.603P/-5.628P)$
$\theta = -15.89°$ ....ANS

Angle made by resultant (5.85P),−15.890 and lies in forth coordinate.

Q. 33: Four coplanar forces are acting at a point. Three forces have magnitude of 20, 50 and 20N at angles of 45°, 200° and 270° respectively. Fourth force is unknown. Resultant force has magnitude of 50N and acts along x-axis. Determine the unknown force and its direction from x-axis.

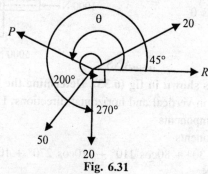

Fig. 6.31

Sol.: Let unknown force be 'P' which makes an angle of 'θ' with the x-axis, If R_H and R_V be the sum of horizontal and vertical components of the resultant, and resultant makes an angle of θ' with the horizontal. Then;

$\Sigma H = R\cos\theta$ = Horizontal component of resultant
$\Sigma V = R\sin\theta$ = Vertical component of resultant

Since Resultant make an angle of 00 (Since acts along x-axis) with the X-axis so
$\Sigma H = R\cos 0° = R$
$\Sigma V = R\sin 0° = 0$

i.e., $\Sigma H = R$ and $\Sigma V = 0$
i.e., $R = \Sigma H = 50$

$\Sigma H = 20\cos 45° + 50\cos 200° + P\cos\theta + 20\cos 270° = 50$

On solving $P\cos\theta = 82.84$ ...(i)

As the same,
$\Sigma V = 20\sin 45° + 50\sin 200° + P\sin\theta + 20\sin 270° = 0$

On solving $P\sin\theta = 22.95$ ...(ii)

Now, square both the equation and add
$P^2\cos^2\theta + P^2\sin^2\theta = 22.95^2 + 82.84^2$

P = 85.96N ....ANS

Let angle made by the unknown force be ?
$\tan\theta = P\sin\theta/P\cos\theta$
$= 22.95/82.84$
$\theta = 15.48°$ ....ANS

Angle made by unknown force is 15.48° and lies in first coordinate.

Q. 34: Determine the resultant 'R' of the four forces transmitted to the gusset plane if θ = 45° as shown in fig(6.32).

Sol.: First resolved all the forces in horizontal and vertical direction, Clearly note that the angle measured by x-axis,

$\Sigma H = 4000\cos 45° + 3000\cos 90° + 1000\cos 0° + 5000\cos 225°$
= 292.8N   ...(i)
$\Sigma V = 4000\sin 45° + 3000\sin 90° + 1000\sin 0° + 5000\sin 225°$
= 2292.8N   ...(ii)
$R^2 = R_H^2 + R_V^2$
$R^2 = (292.8)^2 + (22923.8)^2$
R = 2311.5N   ....ANS

Let angle made by resultant is θ
$\tan\theta = \Sigma V / \Sigma H$
= 2292.8/292.8
θ = 82.72°   ....ANS

Q. 35: Four forces act on bolt as shown in fig (6.33). Determine the resultant of forces on the bolt.
Sol.: First resolved all the forces in vertical and horizontal directions; Let
ΣH = Sum of Horizontal components
ΣV = Sum of Vertical components
$\Sigma H = 150\cos 30° + 80\cos 110° + 110\cos 270° + 100\cos 345°$
= 199.13N   ...(i)
$\Sigma V = 150\sin 30° + 80\sin 110° + 110\sin 270° + 100\sin 345°$
= 14.29N   ...(ii)

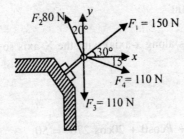

Fig. 6.33   Fig. 6.34

$R = (\Sigma H^2 + \Sigma V^2)^{1/2}$
= $\{(199.13)^2 + (14.29)^2\}^{1/2}$
R = 199.6N   ....ANS

Let angle made by resultant is θ
$\tan\theta = \Sigma V/\Sigma H$
= 14.29/199.13
θ = 4.11°   ....ANS

Q. 36: Determine the resultant of the force acting on a hook as shown in fig (6.35).
Sol.: First resolved all the forces in vertical and horizontal directions Let
ΣH = Sum of Horizontal components
$\Sigma H = 80\cos 25° + 70\cos 50° + 50\cos 315°$
= 152.86N   ...(i)
ΣV = Sum of Vertical components

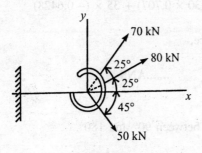

Fig. 6.35   Fig. 6.36

$$\Sigma V = 80\sin 25° + 70\sin 50° + 50\sin 315°$$
$$= 52.07 N \qquad \qquad \ldots(ii)$$
$$R = (R_H^2 + R_V^2)^{1/2} = \{(152.86)^2 + (52.07)^2\}^{1/2}$$
$$R = 161.48 N \qquad \qquad \ldots\ldots\text{ANS}$$

Let angle made by resultant is θ
$$\tan\theta = \Sigma V/\Sigma H \Rightarrow = 52.07/152.86$$
$$\theta = 18.81° \qquad \qquad \ldots\ldots\text{ANS}$$

Q. 37: The following forces act at a point:

(i) 20N inclined at 300 towards North of east

(ii) 25N towards North

(iii) 30N towards North west,

(iv) 35N inclined at 400 towards south of west.

Find the magnitude and direction of the resultant force.

Sol.: Resolving all the forces horizontally i.e. along East-West, line,
$$\Sigma H = 20\cos 30° + 25\cos 90° + 30\cos 135° + 35\cos 220°$$
$$= (20 \times 0.886) + (25 \times 0) + \{-30(-0.707) + 35(-0.766)\} N$$
$$= -30.7 N \qquad \qquad \ldots(i)$$

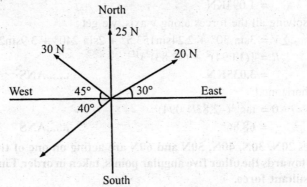

Fig. 6.37

And now resolving all the forces vertically i.e., along North-South line,
$$\Sigma V = 20\sin 30° + 25\sin 90° + 30\sin 135° + 35\sin 220°$$

= (20 × 0.5) + (25 × 1.00) + (30 × 0.707) + 35 × (– 0.6428)
= 33.7N ...(ii)

We know that the magnitude of the resultant force,
$R = \sqrt{\Sigma H^2 + \Sigma V^2}$
On solving, $R = 45.6$ N ....ANS

Direction of the resultant force;
$\tan\theta = \Sigma V / \Sigma H$

Since ΣH is –ve and ΣV is +ve, therefore θ lies between 90° and 180°.

Actual $\theta = 180° - 47° 42'$
= 132.18° ....ANS

Q. 38: Determine the resultant of four forces acting on a body shown in fig (6.38).

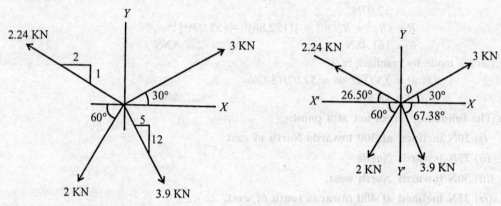

Fig. 6.38   Fig. 6.39

Sol.: Here 2.24KN makes an angle $\tan^{-1}$ (1/2) with horizontal. Also 3.9KN makes an angle of $\tan^{-1}$ (12/5) with horizontal.

Let the resultant R makes an angle θ with x-axis. Resolving all the forces along x-axis, we get,
$\Sigma H = 3\cos 30° + 2.24\cos 153.5° + 2\cos 240° + 3.9\cos 292.62°$
= 1.094KN ...(i)

Similarly resolving all the forces along y-axis, we get
$\Sigma V = 3\sin 30° + 2.24\sin 153.5° + 2\sin 240° + 3.9\sin2 92.62° = -2.83KN =$ ...(ii)

Resultant $R = \{(1.094)^2 + (-2.83)^2\}^{1/2}$
= 3.035KN ....ANS

Angle with horizontal
$\theta = \tan^{-1}(-2.83/1.094)$
= 68.86° ....ANS

Q. 39: The forces 20N, 30N, 40N, 50N and 60N are acting on one of the angular points of a regular hexagon, towards the other five angular points, taken in order. Find the magnitude and direction of the resultant force.

Sol.: In regular hexagon each angle is equal to 120°, and if each angular point is joint together, then each section makes an angle of 30°.

First resolved all the forces in vertical and horizontal directions Let

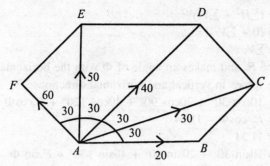

Fig. 6.40

ΣH = Sum of Horizontal components
ΣH = 20cos 0° + 30cos 30° + 40cos 60° + 50cos 90° + 60cos 120°
= 35.98N   ...(i)

ΣV = Sum of Vertical components
ΣV = 20sin 0° + 30sin 30° + 40sin 60° + 50sin 90° + 60sin 120°
= 151.6N   ...(ii)

$R = (\Sigma H^2 + \Sigma V^2)^{1/2} = \{(35.98)^2 + (151.6)^2\}^{1/2}$
R = 155.81N   ....ANS

Let angle made by resultant is θ
Tan θ = $\Sigma V / \Sigma H$ = 151.6/35.98
θ = 76.64°   ....ANS

Q. 40: The resultant of four forces, which are acting at a point, is along Y-axis. The magnitudes of forces F_1, F_3, F_4 are 10KN, 20KN and 40KN respectively. The angle made by 10KN, 20KN and 40KN with X-axis are 300, 900 and 1200 respectively. Find the magnitude and direction of force F_2, if resultant is 72KN.

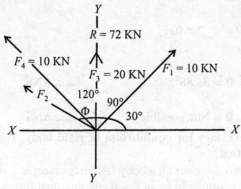

Fig. 6.41

Sol.: Given that resultant is along Y-axis that means resultant(R) makes an angle of 90° with the X-axis, *i.e.,* horizontal component of R is zero, and Magnitude of resultant is equal to vertical component, Let
ΣH = Sum of Horizontal components = 0
ΣV = Sum of Vertical components

$$R = (\Sigma H^2 + \Sigma V^2)^{1/2}$$
$$= (0 + \Sigma V^2)^{1/2}$$
$$R = \Sigma V;$$

Let unknown force be F_2 and makes an angle of Φ with the horizontal X-axis;
Now resolved all the forces in vertical and horizontal directions;

$$\Sigma H = 10\cos 30° + 20\cos 90° + 40\cos 120° + F_2\cos\Phi$$
$$0 = F_2 \cos\Phi - 11.34$$
$$F_2\cos\Phi = 11.34 \qquad \qquad \qquad ...(i)$$
$$72 = 10\sin 30° + 20\sin 90° + 40\sin 120° + F_2\sin\Phi$$
$$72 = F_2\sin\Phi + 59.64$$
$$F_2\sin\Phi = 12.36 \qquad \qquad \qquad ...(ii)$$

Divide equation (ii) by (i), we get
$$\tan\Phi = 12.36/11.34$$
$$\Phi = 47.460 \qquad \qquad ....ANS$$

Putting the value of Φ in equation (i) we get
$$F_2\cos 47.46 = 11.34 \Rightarrow F_2 = 16.77 KN \qquad ....ANS$$

Q. 41: A body is subjected to the three forces as shown in fig 6.42. If possible, determine the direction θ of the force F so that the resultant is in X-direction when:

(1) $F = 5000N$;

(2) $F = 3000N$. *(Dec(C.O)-03)*

Sol.: Since Resultant is in X direction, i.e., Vertical component of resultant is zero.
$$\Sigma V = 0$$
$$R = \Sigma H$$

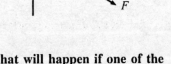

Resolve the forces in X and Y direction
$$\Sigma V = 2000\cos 60° + 3000 - F\cos\theta = 0$$
$$4000 - F\cos\theta = 0$$
or, $\qquad F\cos\theta = 4000 \qquad ...(i)$
Now
(i) If $F = 5000$
$$\cos\theta = 4/5, \quad \theta = 36.86° \qquad ....ANS$$
(ii) If $F = 3000$
$$\cos\theta = 4/3, \quad \theta = \text{Not possible} \qquad ....ANS$$

Q. 42: State the condition necessary for equilibrium of rigid body. What will happen if one of the conditions is not satisfied?

Sol.: When two or more than two force act on a body (all forces meet at a single point) in such a way that body remain in state of rest or continue to be in linear motion, than forces are said to be in equilibrium.

According to Newton's law of motion it means that the resultant of all the forces acting on a body in equilibrium is zero. i.e.,
$$R = 0,$$
$$\Sigma V = 0,$$
$$\Sigma H = 0$$

When body is in equilibrium, then there are two types of forces applied on the body
- Applied forces
- None applied forces

┌── Self weight (W = m.g. act vertically downwards)
│
└── Contact reaction (Action = reaction

NOTE
- If the resultant of a number of forces acting on a particle is zero, the particle will be in equilibrium.
- Such a set of forces, whose resultant is zero, are called equilibrium forces.
- The force, which brings the set of forces in equilibrium, is called an equilibrant. As a matter of fact, the equilibrant is equal to the resultant force in magnitude, but opposite in nature.

Q. 43: Explain 'action' and 'reaction' with the help of suitable examples.

Sol.: Two body A and B are in contact at point 'O'. Body A

Press against the body B. Hence action of body A on the body B is F. Reaction of Body B on body A is R. From Newton's third law of motion (*i.e.*, action = reaction), both these forces are equal there for $F = R$

i.e., Action = Reaction

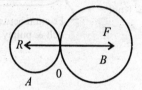

Fig 6.43

Or, "Any pressure on a support causes an equal and opposite pressure from the support so that action and reaction are two equal and opposite forces."

Q. 44: Describe the different uses of strings. Illustrate the tension in the strings.

Sol.: When a weight is attached to a string then it will be in tension. Various diagrams are shown below to describe this concept.

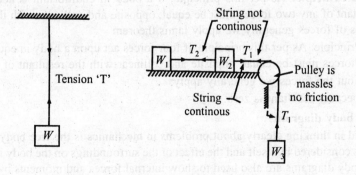

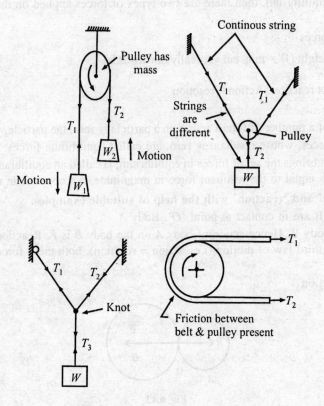

Fig 6.44 *Different uses of String*

Q. 45: What is the principle of equilibrium:
Sol.: Principle of equilibrium may be divided in to three parts;

(1) **Two Force Principle:** Since Resultant is zero when body is in equilibrium, so if two forces are acting on the body, then they must be equal, opposite and collinear.

(2) **Three Force Principle:** As per this principle, if a body in equilibrium is acted upon by three forces, then the resultant of any two forces must be equal, opposite and collinear with the third force. For finding out the values of forces generally we apply lamis theorem

(3) **Four Force Principle:** As per this principle, if four forces act upon a body in equilibrium, then the resultant of any two forces must be equal, opposite and collinear with the resultant of the other two.

And for finding out the forces we generally apply;

$\Sigma H = \Sigma V = 0$, because resultant is zero.

Q. 46: What is free body diagram?
Sol.: An important aid in thinking clearly about problems in mechanics is the free body diagram. In such a diagram, the body is considered by itself and the effect of the surroundings on the body is shown by forces and moments. Free body diagrams are also used to show internal forces and moments by cutting away the unwanted portion of a body.

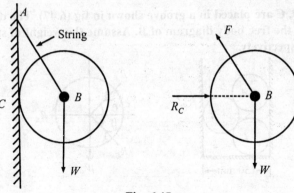

Fig. 6.45

Such a diagram of the body in which the body under consideration is freed from all the contact surface, and all the forces acting on it.(Reaction) are drawn is called a free body diagram.

Q. 47: Explain lami's theorem?
Sol.: It states that "If three coplanar forces acting at a point be in equilibrium, then each force is proportional to the sine of the angle between the other two." Mathematically,
$$P/\sin\beta = Q/\sin\gamma = R/\sin\alpha$$

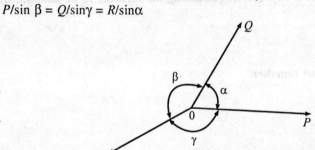

Fig 6.46

Q. 48: Explain law of superposition?
Sol.: When two forces are in equilibrium (equal, opposite and collinear), their resultant is zero and their combined action on a rigid body is equivalent to that of no force at all., Thus

"The action of a given system of forces on a rigid body will in no way be changed if we add to or subtract from them another system of forces in equilibrium.", this is called law of superposition.

Q. 49: What are the steps for solving the problems of equilibrium in concurrent force system.
Sol.: The steps are as following:
1. Draw free body diagram of the body.
2. Make the direction of the forces away from the body.
3. Count how many forces are acting on the body.
2. If there is three forces are acting then apply lamis theorem. And solved for unknown forces.
3. If there are more then three forces are acting then first resolved all the forces in horizontal and vertical direction, Make the direction of the forces away from the body.
4. And then apply equilibrium condition as $R_H = R_V = 0$.

Q. 50: Three sphere A, B, C are placed in a groove shown in fig (6.47). The diameter of each sphere is 100mm. Sketch the free body diagram of B. Assume the weight of spheres A, B, C as 1KN, 2KN and 1KN respectively.

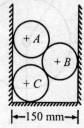

Fig 6.47   Fig 6.48

Sol.: For θ,
$$\cos\theta = 50/100, \cos\theta = .5, \theta = 60°$$
FBD of block B is given in fig 9.47

Q. 51: Two cylindrical identical rollers A and B, each of weight W are supported by an inclined plane and vertical wall as shown in fig 6.49. Assuming all surfaces to be smooth, draw free body diagrams of

(i) roller A,

(ii) roller B

(iii) Roller A and B taken together.

Sol.: Let us assumed

W = Weight of each roller

R = Radius of each roller

R_A = Reaction at point A

R_B = Reaction at point B

R_C = Reaction at point C

R_D = Reaction at point D

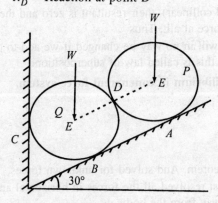

Fig 6.49

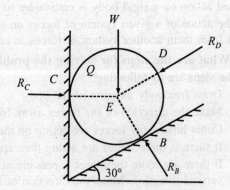

Fig 6.50 *FBD of Roller 'B'*

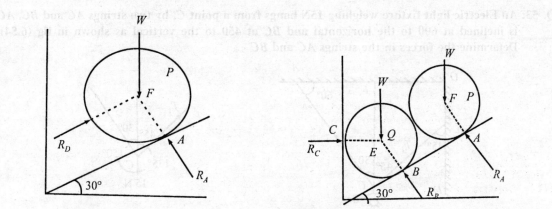

Fig 6.51 *FBD of Roller 'A'*   **Fig 6.52** *FBD of Roller 'B' & 'A' taken together*

Q. 52: Three forces act on a particle 'O' as shown in fig(6.53). Determine the value of 'P' such that the resultant of these three forces is horizontal. Find the magnitude and direction of the fourth force which when acting along with the given three forces, will keep 'O' in equilibrium.

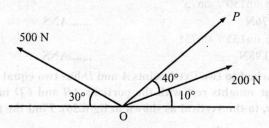

Fig 6.53

Sol.: Since resultant(R) is horizontal so the vertical component of resultant is zero, i.e.,

$\Sigma V = 0, \Sigma H = R$

$\Sigma V = 200\sin 10° + P\sin 50° + 500\sin 150° = 0$

On solving, $P = -371.68$N   ...(i)

$\Sigma H = 200\cos 10° + P\cos 50° + 500\cos 150° = 0$

Putting the value of 'P', we get

$\Sigma H = -474.96$N   ...(ii)

Let Unknown force be 'Q' and makes an angle of ? with the horizontal X-axis. Additional force makes the system in equilibrium Now,

$\Sigma H = Q\cos\theta - 474.96$N $= 0$

i.e., $Q\cos\theta = 474.96$N------(3)

Since ΣV already zero, Now on addition of force Q, the body be in equilibrium so again ΣV is zero.

$\Sigma V = 200\sin 10° - 371.68\sin 50° + 500\sin 150° + Q\sin\theta = 0$

But $200\sin 10° - 371.68\sin 500 + 500\sin 1500 = 0$ by equation (1)

So, $Q\sin\theta = 0$, that means $Q = 0$ or $\sin\theta = 0$,

Q is not zero so $\sin\theta = 0$, $\theta = 0$

Putting $\theta = 0$ in equation (*iii*),

$Q = 474.96$N, $\theta = 0°$ ....ANS

Q. 53: An Electric light fixture weighing 15N hangs from a point C, by two strings AC and BC. AC is inclined at 600 to the horizontal and BC at 450 to the vertical as shown in fig (6.54), Determine the forces in the strings AC and BC

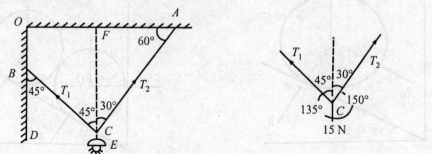

Fig 6.54   Fig 6.55

Sol.: First draw the F.B.D. of the electric light fixture,

Apply lami's theorem at point 'C'

$T_1/\sin 150° = T_2/\sin 135° = 15/\sin 75°$

$T_1 = 15.\sin 150°/ \sin 75°$

$T_1 = 7.76N$ ....ANS

$T_2 = 15.\sin 135°/ \sin 75°$

$T_2 = 10.98N$ ....ANS

Q. 54: A string ABCD, attached to two fixed points A and D has two equal weight of 1000N attached to it at B and C. The weights rest with the portions AB and CD inclined at an angle of 300 and 600 respectively, to the vertical as shown in fig(6.56). Find the tension in the portion AB, BC, CD

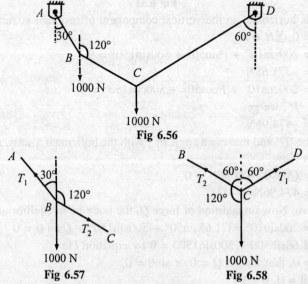

Fig 6.56

Fig 6.57   Fig 6.58

Sol.: First string ABCD is split in to two parts, and consider the joints B and C separately

Let,

T_1 = Tension in String AB

T_2 = Tension in String BC
T_3 = Tension in String CD
Since at joint B there are three forces are acting. SO Apply lamis theorem at joint B,
$T_1/\sin 60° = T_2/\sin 150° = 1000/\sin 150°$
$T_1 = \{\sin 60° \times 1000\}/\sin 150°$
= 1732N ....ANS
$T_2 = \{\sin 150° \times 1000\}/\sin 150°$
= 1000N ....ANS
Again Apply lamis theorem at joint C,
$T_2/\sin 120° = T_3/\sin 120° = 1000/\sin 120°$
$T_3 = \{\sin 120° \times 1000\}/\sin 120°$
= 1000N ....ANS

Q. 55: A fine light string $ABCDE$ whose extremity A is fixed, has weights W_1 and W_2 attached to it at B and C. It passes round a small smooth peg at D carrying a weight of 40N at the free end E as shown in fig(6.59). If in the position of equilibrium, BC is horizontal and AB and CD makes 150° and 120° with BC, find (*i*) Tension in the portion AB,BC and CD of the string and (*ii*) Magnitude of W_1 and W_2.

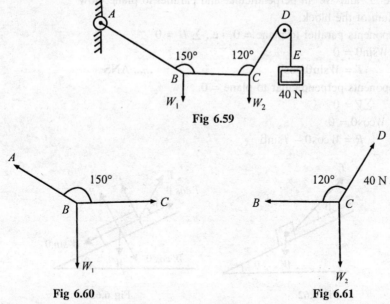

Fig 6.59

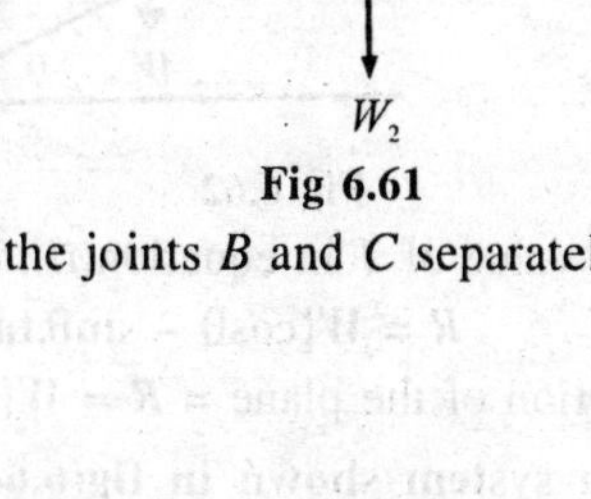

Fig 6.60   Fig 6.61

Sol.: First string $ABCD$ is split in to two parts, and consider the joints B and C separately
Let,
T_1 = Tension in String AB
T_2 = Tension in String BC
T_3 = Tension in String CD
T_4 = Tension in String DE
$T_4 = T_3 = 40N$

Since at joint B and C three forces are acting on both points. But at B all three forces are unknown and at point C only two forces are unknown SO Apply lamis theorem first at joint C,

$T_2/\sin 150° = W_2/\sin 120° = 40/\sin 90°$

$T_2 = \{\sin 150° \times 40\}/\sin 90°$
$= 20N$ ....ANS

$W_2 = \{\sin 120° \times 40\}/\sin 90°$
$= 34.64N$ ....ANS

Now for point B, We know the value of T_2 So, Again Apply lamis theorem at joint B,

$T_1/\sin 90° = W_1/\sin 150° = T_2/\sin 120°$

$T_1 = \{\sin 90° \times 20\}/\sin 120°$
$= 23.1N$ ....ANS

$W_1 = \{\sin 150° \times 20\}/\sin 120°$
$= 11.55N$ ....ANS

Q. 56: Express in terms of θ, β and W the force T necessary to hold the weight in equilibrium as shown in fig (6.62). Also derive an expression for the reaction of the plane on W. No friction is assumed between the weight and the plane.

Sol.: Since block is put on the inclined plane, so plane give a vertical reaction on the block say 'R'. Also resolved the force 'T' and 'W' in perpendicular and parallel to plane, now

For equilibrium of the block,

Sum of components parallel to plane = 0, i.e., $\Sigma H = 0$

$T\cos\beta - W\sin\theta = 0$ ...(i)

Or $\quad T = W\sin\theta/\cos\beta$ ....ANS

Sum of components perpendicular to plane = 0,

i.e., $\quad \Sigma V = 0$

$R + T\sin\beta - W\cos\theta = 0$

Or $\quad R = W\cos\theta - T\sin\beta$ ...(ii)

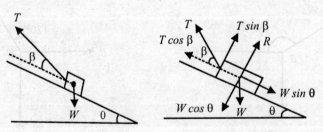

Fig 6.62   Fig 6.63

Putting the value of T in equation(ii), We get

$R = W\{\cos\theta - \sin\theta.\tan\beta\}$ ....ANS

Hence reaction of the plane = $R = W\{\cos\theta - \sin\theta.\tan\beta\}$ ....ANS

Q. 57: For the system shown in fig(6.64), find the additional single force required to maintain equilibrium.

Sol.: Let α and β be the angles as shown in fig. Resolved all the forces horizontal and in vertical direction. When we add a single force whose magnitude is equal to the resultant of the force system and direction is opposite the the direction of resultant. Let;

ΣH = Sum of horizontal component
ΣV = Sum of vertical component
First
$\Sigma H = 20\cos\alpha + 20\cos(360° - \beta)$
$\Sigma H = 20\cos\alpha - 20\cos\beta$ ...(i)
Now
$\Sigma V = 20\sin\alpha + 20\sin(3600 - \beta)$
$-50 = -50 + 20\sin\alpha + 20\sin\beta$ ...(ii)
Hence the resultant of the system = $R = (\Sigma H^2 + \Sigma V^2)^{1/2}$
Let additional single force be 'R' and its magnitude is equal to
$R' = R = [(20\cos\alpha - 20\cos\beta)^2 + (-50 + 20\sin\alpha + 20\sin\beta)^2]^{1/2}$ ....ANS
This force should act in direction opposite to the direction of force 'R'.

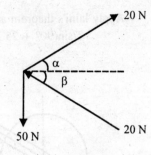

Fig 6.64

Q. 58: A lamp of mass 1Kg is hung from the ceiling by a chain and is pulled aside by a horizontal chord until the chain makes an angle of 600 with ceiling. Find the tensions in chain and chord.

Sol.: Let,
T_{chord} = Tension in chord
T_{chain} = Tension in chain

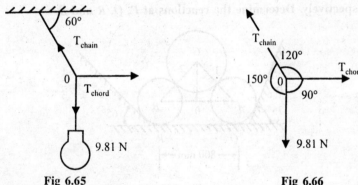

Fig 6.65   Fig 6.66

W = weight of lamp = $1 \times g$ = 9.81N
Consider point 'C', there are three force acting, so apply lamis
theorem at point 'C', as point C is in equilibrium
$T\text{chord}/\sin150° = T_{chain}/\sin90° = 9.81/\sin120°$
$T_{chord} = 9.81 \times \sin150° /\sin120°$
T_{chord} = **5.65N** ....ANS
$T_{chain} = 9.81 \times \sin90° /\sin120°$
T_{chain} = **11.33N** ....ANS

Q. 59: A roller shown in fig(6.67) is of mass 150Kg. What force T is necessary to start the roller over the block A?

Sol.: Let R be the reaction given by the block to the roller, and supposed to act at point A makes an angle of ? as shown in fig,
For finding the angle θ,
$\sin\theta = 75/175 = 0.428$
$\theta = 25.37°$

Apply lami's theorem at 'A', Since the body is in equilibrium

$T/\sin(90° + 25.37°) = 150 \times g / \sin(64.63° + 65°)$

$T = [150 \times g \times \sin(90° + 25.37°)] / \sin(64.63° + 65°)$

$T = 1726.33 N$

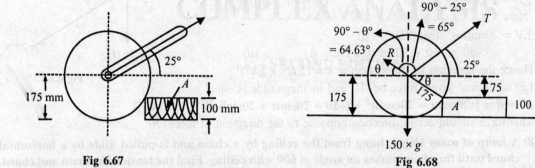

Fig 6.67   Fig 6.68

Q. 60: Three sphere A, B and C having their diameter 500mm, 500mm and 800mm respectively are placed in a trench with smooth side walls and floor as shown in fig(6.69). The center to center distance of spheres A and B is 600mm. The weights of the cylinders A, B and C are 4KN, 4KN and 8KN respectively. Determine the reactions at P, Q, R and S.

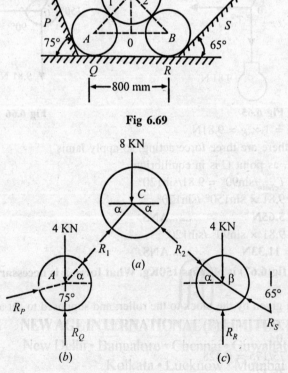

Fig 6.69

Fig 6.70

Sol.: From triangle ABC in fig 6.69

$$Cos\alpha = AD/AC = 300/(250 + 400)$$
$$Cos\alpha = 62.51°$$

Consider FBD of sphere C(Fig 6.70(a))
Consider equilibrium of block C

$$\Sigma H = R_1 Cos\alpha - R_2 Cos\alpha = 0$$

i.e.,
$$R_1 = R_2 \quad ...(i)$$
$$\Sigma H = R_1 sin\alpha - R_2 sin\alpha - 8 = 0 \Rightarrow \text{putting } R_1 = R_2$$
$$\Sigma V = R_1 sin\alpha - R_1 sin\alpha = 8 \Rightarrow 2R_1 = 8/sin\alpha$$
$$= R_1 = 8/2sin\alpha = 4.509$$

i.e., **$R_1 = R_2 = 4.509$ KN**

Consider equilibrium of block A

$$\Sigma H = R_p sin75° - R_1 cos\alpha = 0$$
$$=> R_p = R_1 cos\alpha/sin75° = 4.5cos62.51°/sin75°$$
$R_p = 2.15$ KN ....ANS
$$\Sigma V = R_p cos75° - R_1 sin62.50 + R_Q - W_A = 0$$
$R_Q = 7.44$ KN ....ANS

Consider equilibrium of block B

$$\Sigma H = R_S sin65° - R_2 cos\alpha = 0$$
$$=> R_S = R_2 cos\alpha/sin65° = 4.5cos62.51°/sin65°$$
$R_S = 2.29$ KN ....ANS
$$\Sigma V = R_S cos65° - R_2 sin\alpha + R_R - W_B = 0$$
$$\Sigma V = 2.29cos65° - 4.509sin62.5° + R_R - 4 = 0$$
$R_R = 7.02$ KN ....ANS

Q. 61: Determine the magnitude and direction of smallest force P required to start the wheel over the block. As shown in fig(6.71).

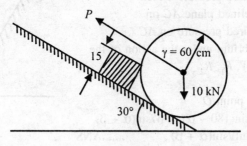

Fig. 6.71

Sol.: Let the reaction of the block be R. The least force P is always perpendicular in the reaction R. When the wheel is just on the point of movement up, then it loose contact with inclined plane and reaction at this point becomes zero.

Consider triangle OMP

$$OM = 60 cm$$
$$OP = 60 - 15 = 45 cm$$
$$MP = \{(OM)^2 - (OP)^2\}^{1/2}$$

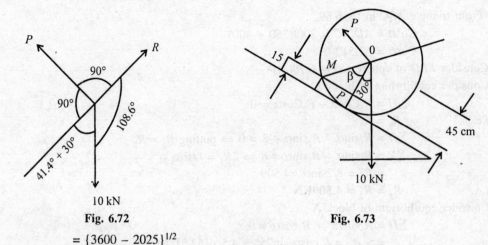

Fig. 6.72

Fig. 6.73

$= \{3600 - 2025\}^{1/2}$
$= 39.68 \text{cm}$
$\tan \beta = MP/OP = 39.68/45, \quad \beta = 41.400$

Using lamis theorem at point O
$P/\sin 108.6° = 10/\sin 90° = R/\sin 161.4°$
$P = (10 \times \sin 108.6°)/\sin 90° = 9.4 \text{KN}$

Hence smallest force $P = 9.4 \text{KN}$

Q. 62: A heavy spherical ball of weight W rests in a V shaped trough whose sides are inclined at angles α and β to the horizontal. Find the pressure on each side of the trough. If a second ball of equal weight be placed on the side of inclination α, so as to rest above the first, find the pressure of the lower ball on the side of inclination β.

Sol.: Let
R_1 = Reaction of the inclined plane AB on the sphere or required pressure on AB
R_2 = Reaction of the inclined plane AC on the sphere or required pressure on AC

The point O is in equilibrium under the action of the following three forces: W, R_1, R_2

Case – 1:

Apply lami's theorem at point O
$R_1/\sin\beta = R_2/\sin(180 - \alpha) = W/\sin(\alpha + \beta)$
or $R_1 = W\sin\beta/\sin(\alpha + \beta)$ ....ANS
and $R_2 = W\sin\alpha/\sin(\alpha + \beta)$ ....ANS

Fig 6.74

Case – 2: Let

R_3 = Reaction of the inclined plane AC on the bottom sphere or required pressure on AC

Since the two spheres are equal, the center line O_1O_2 is parallel to the plane AB.

When the two spheres are considered as a single unit, the action and reaction between them at the point of contact cancel each other. Considering equilibrium of two spheres taken together and resolving the forces along the Line O_1O_2, we get

Force: Concurrent Force System / 137

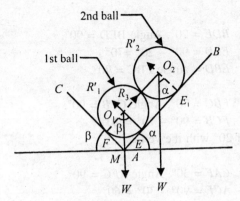

Fig. 6.75

$$R_3\cos\{90° - (\alpha + \beta)\} = W\sin\alpha + W\sin\alpha$$
$$R_3\sin(\alpha + \beta) = 2W\sin\alpha$$
Or, $\quad R_3 = 2W\sin\alpha/\sin(\alpha + \beta) \quad$ ....ANS

Q. 63: A right circular roller of weight 5000N rests on a smooth inclined plane and is held in position by a cord AC as shown in fig 6.76. Find the tension in the cord if there is a horizontal force of magnitude 1000N acting at C. *(May–02-03)*

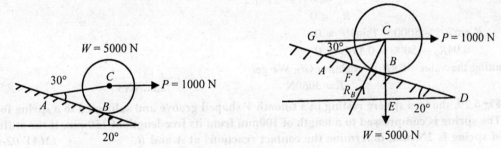

Fig 6.76   Fig 6.77

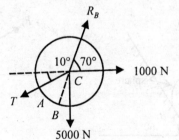

Fig 6.78

Sol.: Let R_B be the contact reaction at point B. This reaction makes an angle of 20° with the vertical Y-axis.

Let Tension in string AC is 'T', which makes an angle of 100 with the horizontal X-axis as shown in fig (6.78).

See fig(6.77)

In Triangle *EBD*
Angle $BDE = 20°$, Angle $BED = 90°$,
Angle $EBD = 90° - 20° = 70°$
Since Angle $EBD =$ Angle $FBC = 70°$,
Now In Triangle *FBC*
Angle $FBC = 70°$, Angle $CFB = 90°$,
Angle $FCB = 90° - 70° = 20°$
i.e., R_B makes an angle of 20° with the vertical
Now In Triangle *ACF*
Angle $CAF = 30°$, Angle $AFC = 90°$,
Angle $ACF = 90° - 30° = 60°$
Now Angle $GCB = 90°$,
Angle $GCA = 90° - 20° - 60° = 10°$
i.e., Tension T makes an angle of 10° with the Horizontal
Consider Fig(3), The body is in equilibrium, SO apply condition of equilibrium
$$R_H = 0$$
$$1000 + R_B\cos70° - T\cos10° = 0$$
$$1000 + 0.34 RB - 0.985T = 0$$
$$R_B = 2.89T - 2941.2 \quad \ldots(i)$$
$$R_V = 0$$
$$R_B\sin70° - 5000 - T\sin10° = 0$$
$$0.94 R_B - 5000 - 0.174T = 0 \quad \ldots(ii)$$
Putting the value of *RB* in equation (*ii*), We get
$$T = 3060 N \quad \ldots\ldots\text{.ANS}$$

Q. 64: Fig 6.79, shows a sphere resting in a smooth *V* shaped groove and subjected to a spring force. The spring is compressed to a length of 100mm from its free length of 150mm. If the stiffness of spring is 2N/mm, determine the contact reactions at *A* and *B*.   *(MAY 02-03)*

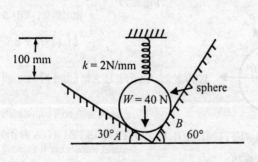

Fig 6.79   Fig 6.80

Sol.: The spring is compressed from 150mm to 100mm. So it is exiting a compressive force, which is acting vertically downward on the sphere.
Since,
Spring force$(F) = K.x$

Given that $K = 2N/mm$
$$x = 150 - 100 = 50mm$$
$$F = 2 \times 50 = 100N \qquad ...(i)$$

Let R_A and R_B be the contact reaction at Pont A and B.

Here wt of sphere and F are collinear force, both act down ward so the net force is = 100 + 40, acting down ward.

Apply lamis theorem at point 'O'
$$R_A/\sin(90° + 30°) = R_B/\sin(90° + 60°)$$
$$= 140/\sin(180° - 90°)$$

On solving
$$R_A = 121N \qquad ....ANS$$
$$R_B = 70N \qquad ....ANS$$

Q. 65: Three sphere A, B and C weighing 200N, 400N and 200N respectively and having radii 400mm, 600mm and 400mm respectively are placed in a trench as shown in fig 6.81. Treating all contact surfaces as smooth, determine the reactions developed.

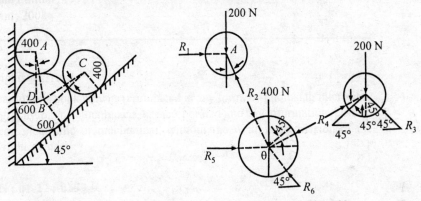

Fig 6.81　　　　　　　　　　　　Fig 6.82

Sol.: From the fig 6.81
$$\sin\alpha = BD/AB = (600 - 400)/(400 + 600) = 0.2$$
$$\alpha = 11.537°$$

Referring to FBD of sphere A (Fig a)
$$R_2\cos\alpha = 200$$
$$R_2 = 200/\cos 11.537° = \textbf{204.1 N} \qquad ....ANS$$
And $R_1 - R_2\sin\alpha = 0$
$$R_1 = \textbf{40.8N} \qquad ....ANS$$

Referring to the FBD of sphere C [Fig. 6.82(b)],
Sum of forces parallel to inclined plane = 0
$$R_4\cos\alpha - 200\cos 45° = 0$$
$$R_4 = \textbf{144.3 N} \qquad ....ANS$$
Sum of forces perpendicular to inclined plane = 0
$$R_4\cos(45 - \alpha) - R_3\cos 45° = 0$$
$$R_3 = \textbf{170.3N} \qquad ....ANS$$

Referring to *FBD* of cylinder B (Fig. 6.82(c)]

$$\Sigma V = 0$$
$$R_6 \sin 45° - 400 - R_2 \cos\alpha - R_4 \cos(45 + \alpha) = 0$$
$$R_6 \sin 45° = 400 + 204.1 \cos 11.537° + 144.3 \cos 56.537°$$
$$\mathbf{R_6 = 961.0 \text{ N}} \qquad \text{.......ANS}$$
$$\Sigma H = 0$$
$$R_5 - R_2 \sin\alpha - R_4 \sin(45 + \alpha) - R_6 \cos 45° = 0$$
$$R_5 = 204.1 \sin 11.537 + 144.3 \sin 56.537 + 961.0 \cos 45°$$
$$\mathbf{R_5 = 840.7 \text{ N}} \qquad \text{.......ANS}$$

CHAPTER 7

FORCE: NON - CONCURRENT FORCE SYSTEM

Q. 1: Define Non-concurrent force system. Why we find out the position of Resultant in Non-concurrent force system?

Sol.: In Equilibrium of concurrent force system, all forces are meet at a point of a body. But if the forces acting on the body are not meet at a point, then the force system is called as Non-concurrent force system.

In concurrent force system we find the resultant and its direction. Because all the forces are meet at one point so the resultant will also pass through that point, i.e. the position of resultant is already clear. But in non-concurrent force system we find the magnitude, direction and distance of the resultant from any point of the body because forces are not meet at single point they act on many point of the body, so we don't know the exact position of the resultant. For finding out the position of resultant we used the concept of moment.

Q. 2: Define Moment of a Force? What is moment center and moment arm? Also classify the moment.

Sol.: It is the turning effect produced by a force, on the body, on which it acts. The moment of a force is equal to the product of the force and the perpendicular distance of the point about which the moment is required, and the line of action of the force.

The force acting on a body causes linear displacement, while moment causes angular displacement.

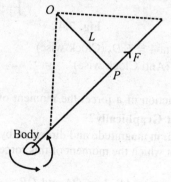

Fig. 7.1

If M = Moment
F = Force acting on the body, and
L = Perpendicular distance between the point about which the moment is required and the line of action of the force. Then $M = F.L$

The point about which the moment is considered is called **Moment Center**. And the Perpendicular distance of the point from the line of action of the force is called **moment Arm**.

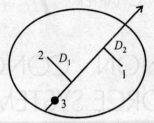

Fig. 7.2

The moment is of the two types:

Clockwise moment:

It is the moment of a force, whose effect is to turn or rotate the body, in the clockwise direction. It takes +ive.

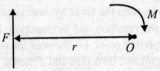

Fig. 7.3

Anticlock wise Moment:

It is the moment of a force, whose effect is to turn or rotate the body, in the anticlockwise direction. It take -ive.

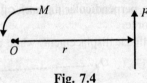

Fig. 7.4

In Fig. 7.2; Moment about Point 1 = $F.D_2$ (Clock wise)

Moment about Point 2 = $F.D_1$ (Anti Clock wise)

Moment about Point 3 = 0

i.e. if point lie on the line of action of a force, the moment of the force about that point is zero.

Q. 3: How you represent moment Graphically?

Sol.: Consider a force F represented, in magnitude and direction, by the line AB. Let 'O' be a point about which the moment of this force is required to be found out.

From 'O' draw OC perpendicular to AB. Join OA and OB.

Now moment of the force P about O = F X OC = AB.OC

But AB.OC is equal to twice the area of the triangle ABO.

Thus the moment of a force about any point is geometrically equal to twice the area of the triangle, whose base is the line representing the force and whose vertex is the point, About which the moment is taken.

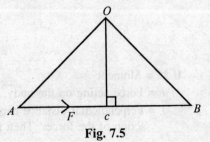

Fig. 7.5

Mo = 2.Area of Triangle OAB

Unit of moment = N-m

Q. 4: State Varignon's theorem. How it can help on determination of moments? In what condition is it used?

Sol.: Varignon's theorem also called Law of Moment.

The practical application of varignon's theorem is to find out the position of the resultant from any point of the body.

It states *"If a number of coplanar forces are acting simultaneously on a particle, the algebraic sum of the moments of all the forces about any point is equal to the moment of their resultant force about the same point."*

Proof: Let us consider, for the sake of simplicity, two concurrent forces P and Q represented in magnitude and direction by AB and AC as shown in fig. 7.6.

Let 'O' be the point, about which the moment are taken, through O draw a line OD parallel to the direction of force P, to meet the line of action of the force Q at C. Now with AB and AC as two adjacent sides, complete the Parallelogram $ABDC$ as shown in fig. 7.6. Joint the diagonal AD of the parallelogram and OA and OB. From the parallelogram law of forces, We know that the diagonal AD represents in magnitude and direction, the resultant of two forces P and Q. Now we see that the moment of the force P about O: = 2. Area of the triangle AOB ...(i)

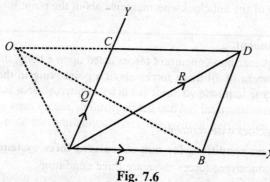

Fig. 7.6

Similarly, moment of the force Q about O: = 2. Area of the triangle AOC ...(ii)
And moment of the resultant force R about O: = 2.Area of the triangle AOD ...(iii)
But from the geometry of the fig.ure, we find that
Area of triangle AOD = Area of triangle AOC + Area of triangle ACD
But Area of triangle ACD = Area of triangle ABD = Area of triangle AOB
(Because two "AOB and ADB are on the same base AB and between the same // lines)
Now Area of triangle AOD = Area of triangle AOC + Area of triangle AOB
Multiply both side by 2 we get;
2. Area of triangle AOD = 2.Area of triangle AOC + 2. Area of triangle ACD, i.e.
Moment of force R about O = Moment of force P about O + Moment of force Q about O
or,
Where $R \cdot d = \Sigma M$
ΣM = Sum of the moment of all forces
d = Distance between the resultant force and the point where moment of all forces are taken.
This principle is extended for any number of forces.

Q. 5: How do you find the resultant of Non - coplanar concurrent force system?

Sol.: The resultant of non-concurrent force system is that force, which will have the same rotational and translation effect as the given system of forces, It may be a force, a pure moment or a force and a moment.

$$R = \{(\Sigma H)^2 + (\Sigma V)^2\}^{1/2}$$
$$\tan \theta = \Sigma V / \Sigma H$$
$$\Sigma M = R \cdot d$$

Where,

ΣH = Sum of all horizontal component
ΣV = Sum of all vertical component
ΣM = Sum of the moment of all forces
d = Distance between the resultant force and the point where moment of all forces are taken.

Q. 6: How you find the position of resultant force by moments?

Sol.: First of all, find the magnitude and direction of the resultant force by the method of resolution. Now equate the moment of the resultant force with the algebraic sum of moments of the given system of forces about any point or **simply using Varignon's theorem**. This may also be found out by equating the sum of clockwise moments and that of the anticlockwise moments about the point through which the resultant force will pass.

Q. 7: Explain principle of moment.

Sol.: If there are number of coplanar non-concurrent forces acted upon a body, then for equilibrium of the body, the algebraic sum of moment of all these forces about a point lying in the same plane is zero.

i.e. $\quad \Sigma M = 0$

Or we can say that,

clock wise moment = Anticlockwise moment

Q. 8: What are the equilibrium conditions for non-concurrent force system?

Sol.: For Equilibrium of non-concurrent forces there are three conditions:

1. Sum of all the horizontal forces is equal to zero, i.e
   $$\Sigma H = 0$$
2. Sum of all the horizontal forces is equal to zero, i.e
   $$\Sigma V = 0$$
3. Sum of the moment of all the forces about any point is equal to zero, i.e
   $$\Sigma M = 0$$

If any one of these conditions is not satisfied then the body will not be in equilibrium.

Q. 9: Define equilibrant.

Sol.: The force, which brings the set of forces in equilibrium, is called an equilibrant. As a matter of fact, the equilibrant is equal to the resultant force in magnitude, but opposite in nature.

Q. 10: What are the cases of equilibrium?

Sol.: As the result of the acting forces, the body may have one of the following states:

(1) The body may move in any one direction:

It means that there is resultant force acting on it. A little consideration will show, that if the body is to be at rest or in equilibrium, the resultant force causing movement must be zero. or ΣH and ΣV must be zero.

$$\Sigma H = 0 \quad \text{and} \quad \Sigma V = 0$$

(2) The body may rotate about itself without moving:

It means that there is single resultant couple acting on it with no resultant force. A little consideration will show, that if the body is to be at rest or in equilibrium, the moment of the couple causing rotation must be zero. or

$$\Sigma M = 0$$

(3) The body may move in any one direction, ant at the same time it may also rotate about itself:

It means that there is a resultant force and also resultant couple acting on it. A little consideration will show, that if the body is to be at rest or in equilibrium, the resultant force causing movement and the resultant moment of the couple causing rotation must be zero. i.e.

$$\Sigma H = 0, \quad \Sigma V = 0 \quad \text{and} \quad \Sigma M = 0$$

(4) The body may be completely at rest:

It means that there is neither a resultant force nor a couple acting on it. A little consideration will show, that in this case the following condition are already satisfied:

$$\Sigma H = 0, \quad \Sigma V = 0 \quad \text{and} \quad \Sigma M = 0$$

Q. 11: Determine the resultant of four forces tangent to the circle of radius 3m shown in fig. (7.7). What will be its location with respect to the center of the circle? *(Dec–03-04)*

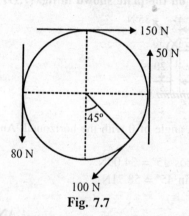

Fig. 7.7

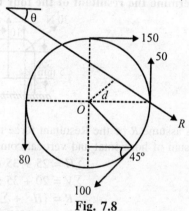

Fig. 7.8

Sol: Let resultant be 'R' which makes an angle of θ with the horizontal X axis. And at a distance of x from point 'O'. Let ΣH and ΣV be the horizontal and vertical component.

$$\Sigma H = 150 - 100 \cos 45° = 79.29 N \qquad ...(i)$$
$$\Sigma V = 50 - 100 \sin 45° - 80 = -100.7 N \qquad ...(ii)$$
$$R = \{\Sigma H^2 + \Sigma V^2\}^{1/2}$$
$$R = \{(79.29)^2 + (100.7)^2\}^{1/2}$$
$$R = 128.17 N \qquad \text{........ANS}$$

Calculation For angle θ

$$\tan \theta = \Sigma V / \Sigma H$$
$$= -100.71 / 79.28$$
$$\theta = -51.78° \qquad \text{........ANS}$$

Calculation For distance 'd'
According to Varignon's theorem, $R \cdot d = \Sigma M$
(Taking moment about point 'O')
i.e. $\quad 128.17 \times d = 150 \times 3 - 50 \times 3 + 100 \times 3 - 80 \times 3$
$$d = 2.808 \, m \qquad \text{........ANS}$$

Q. 12: Determine the moment of the 50N force about the point A, as shown in fig. (7.9).

Sol.: Taking moment about point A,

$$\Sigma M_A = 50 \cos 150° \times 150 - 50 \sin 150° \times 200$$

(Negative sign because, both moments are anticlockwise)

$$\Sigma M_A = -11475.19 \text{N} - \text{mm} \quad \quad \text{......ANS}$$

Hence moment about A = 11475.19N –mm (Anticlockwise)

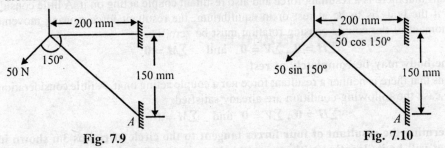

Fig. 7.9   Fig. 7.10

Q. 13: Determine the resultant of the four forces acting on the plate shown in fig. (7.11)

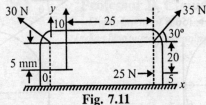

Fig. 7.11

Sol.: Let us assume R be the Resultant force is acting at an angle of θ with the horizontal. And ΣH and ΣV be the sum of horizontal and vertical components.

$$\Sigma H = 25 + 35 \cos 30° - 30 \cos 45° = 34.09 \text{N} \quad \quad ...(i)$$
$$\Sigma V = 20 + 35 \sin 30° + 30 \sin 45° = 58.71 \text{N} \quad \quad ...(ii)$$
$$R = \{H^2 + \Sigma V^2\}^{1/2}$$
$$R = 67.89 \text{N} \quad \quad \text{......ANS}$$

For direction of resultant

$$\tan \theta = \Sigma V / \Sigma H$$
$$= 58.71/34.09$$
$$\theta = 59.85° \quad \quad \text{......ANS}$$

Q. 14: A beam AB (fig. 7.12) is hinged at A and supported at B by a vertical cord, which passes over two frictionless pulleys C and D. If pulley D carries a vertical load Q, find the position x of the load P if the beam is to remain in equilibrium in the horizontal position.

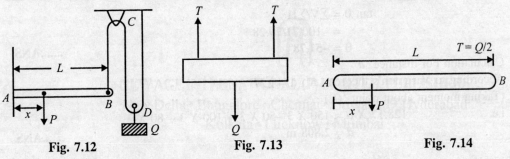

Fig. 7.12   Fig. 7.13   Fig. 7.14

Sol.: First consider the free body diagram of block Q,
From the fig. 7.13,
$$2T = Q, T = Q/2$$
i.e tension in the rope = $Q/2$

Now consider the F.B.D. of the beam as shown in fig. 7.14, Here two forces are acting force 'P' at a distance 'X' from point 'A' and $T = Q/2$ at a distance 'l' from point 'A'

Taking moment about point 'A', i.e. $\Sigma M_A = 0$
$$P \times x = Q/2 \times l$$
$$X = \frac{QL}{2P} \qquad \text{.......ANS}$$

Q. 15: A uniform wheel of 600 mm diameter, weighing 5KN rests against a rigid rectangular block of 150mm height as shown in fig. 7.15. Find the least pull, through the center of the wheel, required just to turn the wheel over the corner A of the block. Also find the reaction of the block. Take the entire surface to be smooth.

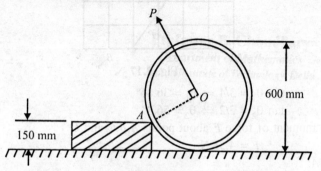

Fig. 7.15

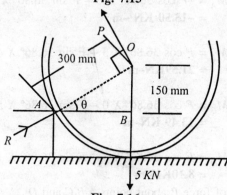

Fig. 7.16

Sol.: Let P = least pull required just to turn the wheel

Least pull must be applied normal to AO. F.B.D of wheel is shown in fig. 7.16, from the fig.,
$$\sin \theta = 150/300, \theta = 30°$$
$$AB = \{(300)^2 - (150)^2\}^{1/2} = 260 \text{ mm}$$
Now taking moment about point A, considering body is in equilibrium
$$P \times 300 - 5 \times 260 = 0$$
$$P = 4.33 \text{ KN} \qquad \text{.......ANS}$$

148 / *Problems and Solutions in Mechanical Engineering with Concept*

Calculation for reaction of the block
Let R = Reaction of the block
Since body is in equilibrium, resolving all the force in horizontal direction and equate to zero,
$$R \cos 30° - P \sin 30° = 0$$
$$R = 2.5 KN \qquad \text{.......ANS}$$

Q. 16: In the fig. (7.17) assuming clockwise moment as positive, compute the moment of force $F = 4.5$ KN and of force $P = 3.61$ KN about points A, B, C and D. Each block is of $1m^2$.

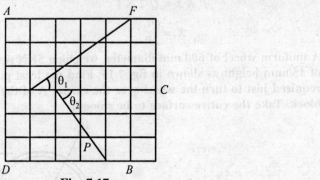

Fig. 7.17

Sol.: Here $\tan \theta_1 = 3/4 \Rightarrow \theta_1 = 36.86°$
$\tan \theta_2 = 3/2 \Rightarrow \theta_2 = 56.3°$

First we find the moment of force F about points A, B, C and D
$$F = 4.5KN$$

(1) About point A:
$$M_A = -F \cos 36.86° \times 3 - F \sin 36.86° \times 1$$
$$= -13.50 \text{ KN–m} \qquad \text{.......ANS}$$

(2) About point B:
$$M_B = F \cos 36.86° \times 3 + F \sin 36.86° \times 4$$
$$= 21.59 KN–m \qquad \text{.......ANS}$$

(3) About point C:
$$M_C = F \cos 36.86° \times 0 - F \sin 36.86° \times 5$$
$$= 13.49 \text{ KN–m} \qquad \text{.......ANS}$$

(4) About point D:
$$M_D = F \cos 36.86° \times 3 - F \sin 36.86° \times 1$$
$$= 8.10 KN–m \qquad \text{.......ANS}$$

Now we find the moment of force P about points A, B, C and D
$$P = 3.61KN$$

(1) About point A:
$$M_A = -P \cos 56.3° \times 3 + P \sin 56.3° \times 2$$
$$= 0.002 KN–m \qquad \text{.......ANS}$$

(2) About point B:
$$M_B = P \cos 56.3° \times 3 - P \sin 56.3° \times 3$$
$$= -7.007 KN–m \qquad \text{.......ANS}$$

(3) About point C:

$$M_C = -P\cos 56.3° \times 0 - P\sin 56.3° \times 4$$
$$= -12.0134 \text{ KN-m} \quad \text{.......ANS}$$

(4) About point D:

$$M_D = -P\cos 56.3° \times 3 - P\sin 56.3° \times 2$$
$$= -11.998 \text{ KN-m} \quad \text{.......ANS}$$

Q. 17: A uniform wheel of 60 cm diameter weighing 1000 N rests against rectangular obstacle 15 cm high. Find the least force required which when acting through center of the wheel will just turn the wheel over the corner of the block. Find the angle of force with horizontal.

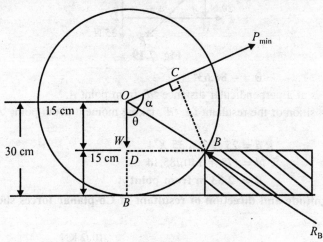

Fig. 7.18

Sol.: Let,

P_{min} = Least force applied as shown in fig. 7.18
α = Angle of the least force
From triangle OBC, $BC = BO\sin\alpha$
$$BC = 30\sin\alpha$$
In Triangle BOD, $BD = \{(BO)^2 - (OD)^2\}^{1/2}$
$$BD = (30^2 - 15^2)^{1/2} = 25.98$$
Taking moment of all forces about point B, We get
$$P_{min} \times BC - W \times BD = 0$$
$$P_{min} = W \times BD/BC$$
$$P_{min} = 1000 \times 25.98 / 30\sin\alpha$$
We get minimum value of P when α is maximum and maximum value of α is at 90° i.e. 1, putting $\sin\alpha = 1$
$$P_{min} = 866.02 \text{ N} \quad \text{.......ANS}$$

Q. 18: A system of forces is acting at the corner of a rectangular block as shown in fig. 7.19. Determine magnitude and direction of resultant.

Sol.: Let R be the resultant of the given system. And ΣH and ΣV be the horizontal and vertical component of the resultant.

$$\Sigma H = 25 - 20 = 5 \text{ KN} \quad \text{...(i)}$$
$$\Sigma V = -50 - 35 = -85 \text{ KN} \quad \text{...(ii)}$$

$$R^2 = \Sigma H^2 + \Sigma V^2$$
$$R^2 = (5)^2 + (-85)^2$$
$$R = 85.14 N \quad \text{........ANS}$$

Let Resultant makes an angle of , with the horizontal
$$\tan \theta = \Sigma V/\Sigma H = -85/5$$

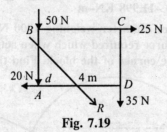

Fig. 7.19

$$\theta = -86.63° \quad \text{........ANS}$$

Let resultant 'R' is at a perpendicular distance 'd' from point A,

For finding the position of the resultant i.e. 'd', taking moment about point 'A'., or apply varignon's theorem

$$R.d = 25 \times 3 + 35 \times 4$$
$$d = (75 + 140)/85.14$$
$$d = 2.53 \text{ m from point } A \quad \text{........ANS}$$

Q. 19: Find the magnitude and direction of resultant of Co-planar forces shown in fig. 7.20.

(Dec–00–01)

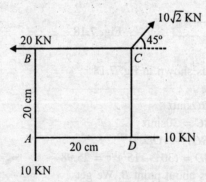

Fig. 7.20

Sol.: Using the equation of equilibrium,
$$\Sigma H = -20 + 10 + 10\sqrt{2} \cos 45°$$
$$\Sigma H = 0 \quad \text{...(i)}$$
$$\Sigma V = -10 + 10\sqrt{2} \sin 45°$$
$$\Sigma V = 0 \quad \text{... (ii)}$$

Since ΣH and ΣV both are zero, but in non concurrent forces system, the body is in equilibrium when
$$\Sigma H = \Sigma V = \Sigma M = 0$$

So first we check the value of ΣM, if it is zero then body is in equilibrium, and if not then that moment is the resultant.

Taking moment about point A,

$\Sigma M_C = 0$
$= -10 \times 20 - 10 \times 20 = -400$ KN–cm,

Since moment is not zero i.e. Body is not in equilibrium, Hence the answer is
$M = 400$ KN–cm (Anticlockwise)

Q. 20: Three similar uniform slabs each of length '$2a$' are resting on the edge of the table as shown in fig. 7.21. If each slab is overhung by maximum possible amount, find amount by which the bottom slab is overhanging. *(Dec–00-01)*

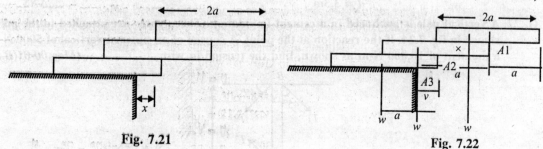

Fig. 7.21   Fig. 7.22

Sol.: The maximum overhang of top beam is 'a',

Now taking moment about point A_2, considering all load acting on middle beam.
$$-W(a - X) + W.X = 0$$
on solving $X = a/2$ ...(i)

Now taking moment about point A_3
$$-W(a - Y) + W[Y - (a - X)] + W(X + Y) = 0$$
$$-Wa + WY + WY - Wa + WX + WX + WY = 0$$
$$3Y - 2a + 2X = 0$$
$$Y = a/3 \quad\quad\quad\quad\quad ....\text{ANS}$$

Since bottom beam overhang by $a/3$ amount

Q. 21: Determine the resultant of force system acting tangential to the circle of radius 1m as shown in fig. 7.23. Also find its direction and line of action *(May–00-01)*

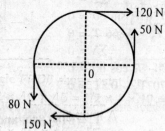

Fig. 7.23

Sol.: $\Sigma H = 120 - 150 = -30$N   ...(i)
$$\Sigma V = 50 - 80 = -30\text{N} \quad\quad\quad ...(ii)$$
$$R = (\Sigma H^2 + \Sigma V^2)^{1/2}$$
$$R = ((-30)^2 + (-30)^2)^{1/2}$$
$$R = 42.43 \text{ N} \quad\quad\quad\quad\quad ....\text{ANS}$$

$\tan \theta = -30/-30$

$\theta = 45°$ ....ANS

Now for finding the position of the resultant Let the perpendicular distance of the resultant from center 'O' be 'd'.

Apply varignon's theorem, taking moment about point O.

$R.d = -80 \times 1 + 150 \times 1 + 120 \times 1 - 50 \times 1$

$42.42 \times d = -80 \times 1 + 150 \times 1 + 120 \times 1 - 50 \times 1$

$d = 3.3m$ ....ANS

Q. 22: A vertical pole is anchored in a cement foundation. Three wires are attached to the pole as shown in fig. 7.24. If the reaction at the point. A consist of an upward vertical of 5000 N and a moment of 10,000 N-m as shown, find the tension in wire. *(May 00-01(B.P.))*

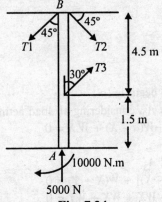

Fig. 7.24

Sol.: Resolve all the forces in horizontal and vertical direction. From the condition of equilibrium

Taking moment about point B, We get

$T_3 \sin 30° \times 4.5 - 10000 = 0$

$T_3 = 4444.44 N$ ....ANS

$\Sigma H = 0$

$T_3 \sin 30° + T_2 \cos 45° - T_1 \sin 60° = 0$

$2222.22 + 0.707 T_2 - 0.866 T_1 = 0$ ...(i)

$\Sigma V = 0$

$T_3 \cos 30° + 5000 - T_2 \sin 45° - T_1 \cos 60° = 0$

$8849 - 0.707 T_2 - 0.5 T_1 = 0$ ...(ii)

From equation (i) and (ii)

$T_1 = 8104.84 N$ and $T_2 = 6783.44 N$ ....ANS

Q. 23: A man raises a 10 Kg joist of length 4m by pulling on a rope, Find the tension T in the rope and reaction at A for the position shown in fig. 7.25. *(May 00-01(B.P.))*

Force: Non-Concurrent Force System / 153

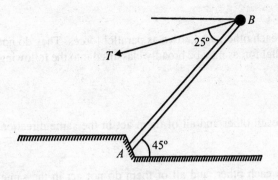

Fig. 7.25

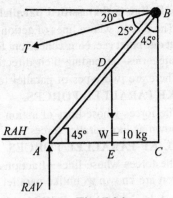

Fig. 7.26

Sol.: Apply condition of equilibrium

$\Sigma H = 0 \quad R_{AH} - T\cos 20° = 0 \quad R_{AH} = T\cos 20°$ ...(i)
$\Sigma V = 0 \quad R_{AV} - 10 - T\sin 20° = 0 \quad R_{AV} = 10 + T\sin 20°$ ...(ii)

Now taking moment about point A

$T\sin 20° \times AC + 10\sin 45° \times AE - T\cos 20 \times BC = 0$,

$AC = 4\cos 45° = 2.83$ m
$BC = 4\sin 45° = 2.83$ m
$AE = 2\cos 45° = 1.41$ m

$T \times 0.34 \times 2.83 + 10 \times 0.71 \times 1.41 - T \times 0.94 \times 2.83 = 0$,
$0.9622\ T + 10.011 - 2.66\ T = 0$
$T = 5.9$ Kg ....ANS

Putting the value of T in equation (i) and (ii)

$R_{AH} = 5.54$ Kg ....ANS
$R_{AV} = 15.89$ Kg ....ANS

Q. 24: The 12m boom AB weight 1 KN, the distance of the center of gravity G being 6 m from A. For the position shown, determine the tension T in the cable and the reaction at B.

(Dec–03-04)

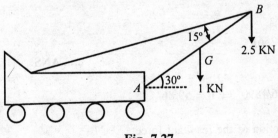

Fig. 7.27

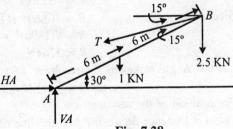

Fig. 7.28

Sol.: The free body diagram of the boom is shown in fig. 7.28

$\Sigma M_A = 0$

$T\sin 15° \times 12 - 2.5 \times 12\cos 30° - 1 \times 6\cos 30° = 0$

$T = 10.0382$ KN ....ANS

Reaction at $B = (2.5^2 + 10^2 + 10 \times 2.5 \times \cos 75°)^{1/2}$

$R_B = 10.61$ KN ....ANS

Q. 25: Define and classified parallel forces?

Sol.: The forces, whose lines of action are parallel to each other, are known as parallel forces. They do not meet at one point (i.e. Non-concurrent force). The parallel forces may be broadly classified into the following two categories, depending their direction.

There are two types of parallel force

1. LIKE PARALLEL FORCES

The forces whose lines of action are parallel to each other and all of them act in the same direction are known as like parallel forces.

2. UNLIKE PARALLEL FORCES

The forces whose lines of actions are parallel to each other, and all of them do not act in the same direction are known as unlike parallel forces.

Q. 26: A horizontal line PQRS is 12 m long, where PQ = QR = RS = 4m. Forces of 1000, 1500, 1000 and 500 N act at P, Q, R and S respectively with downward direction. The lines of action of these make angle of 90°, 60°, 45° and 30° respectively with PS. Find the magnitude, direction and position of the resultant force.

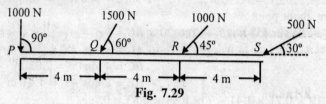

Fig. 7.29

The system of given forces is shown in fig. 7.29

Let R be the resultant of the given system. And R_H and R_V be the horizontal and vertical component of the resultant.

Resolving all the forces horizontally

$$\Sigma H = -1000 \cos 90° - 1500 \cos 60° - 1000 \cos 45° - 500 \cos 30°$$
$$\Sigma H = -1890 \text{ N} \qquad \ldots(i)$$

Resolving all the forces vertically

$$\Sigma V = -1000 \sin 90° - 1500 \sin 60° - 1000 \sin 45° - 500 \sin 30°$$
$$\Sigma V = -3256 \text{N} \qquad \ldots(ii)$$

Since,
$$R = \sqrt{(\Sigma H)^2 + (\Sigma V)^2}$$
$$R = \sqrt{(1890)^2 + (3256)^2}$$
$$R = 3764\text{N} \qquad \text{......ANS}$$

Let θ = Angle makes by the resultant

$$\tan \theta = \Sigma V/\Sigma H = 3256/1890 \Rightarrow \theta = 59.86°$$

For position of the resultant

Let, d = Distance between P and the line of action of the resultant force.

Apply varignon's theorem

$$R.d = 1000 \sin 90° \times 0 + 1500 \sin 60° \times 4 + 1000 \sin 45° \times 8 + 500 \sin 30° \times 12$$
$$3256.d = 13852$$
$$d = 3.67 \text{ m} \qquad \text{......ANS}$$

Q. 27: Replace the two parallel forces acting on the control lever by a single equivalent force R.
Sol.: Since single equivalent force is resultant.
Let ΣH and ΣV be the horizontal and vertical component of the resultant.
Resolving all the forces horizontally
$$\Sigma H = 50 - 80 = -30N \qquad ...(i)$$
Since there is no vertical force i.e. the resultant is horizontal. Now for finding out the point of application of resultant, Let resultant is at a distance of 'd' from point 'O'. Apply varignon's theorem, and taking moment about point 'O'
$$R.d = 50 \times 80 - 80 \times 50 = 0$$
But $\quad R = \Sigma H = -30N$, so $d = 0$
$d = 0$, means point of application of resultant is 'O'
Hence an equivalent force 30N acts in –ive x-axis at point 'O' which replace the given force system.

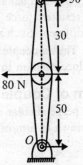

Fig. 7.30

Q. 28: A system of loads acting on a beam is shown in fig. 7.31. Determine the resultant of the loads.
Sol.: Let R be the resultant of the given system. And ΣH and ΣV be the horizontal and vertical component of the resultant. And resultant makes an angle of θ with the horizontal.
Resolving all the forces horizontally
$$\Sigma H = 20 \cos 60°$$
$$\Sigma H = 10 KN \qquad ...(i)$$
Resolving all the forces vertically

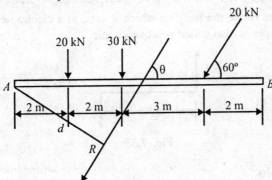

Fig. 7.31

$$\Sigma V = 20 + 30 + 20 \sin 60°$$
$$\Sigma V = 67.32 KN \qquad ...(ii)$$
Since, $\quad R = \sqrt{(\Sigma H)^2 + (\Sigma V)^2} \Rightarrow \sqrt{(10)^2 + (67.32)^2}$
$\quad \mathbf{R = 68.05 KN}$ ....ANS

Let θ = Angle makes by the resultant
$$\tan \theta = \Sigma V / \Sigma H = 67.32/10 \Rightarrow \theta = 81.55°$$
For position of the resultant
Let, d = Distance between Point A and the line of action of the resultant force.
Apply varignon's theorem
$$R.d = 20 \times 2 + 30 \times 4 + 20 \sin 30° \times 7$$
$$68.05.d = 281.2$$
$$\mathbf{d = 4.132 \ m} \qquad \text{........ANS}$$

Q. 29: Define couple and Arm of couple?

Sol.: If two equal and opposite parallel forces (i.e. equal and unlike) are acting on a body, they don't have any resultant force. That is no single force can replace two equal and opposite forces, whose line of action are different. Such a set of two equal and opposite forces, whose line of action are different, form a couple.

Thus a couple is unable to produce any translatory motion (motion in a straight line). But a couple produce rotation in the body on which it acts.

Arm of Couple

The perpendicular distance (d) between the lines of action of the two equal and opposite parallel forces, is known as arm of couple.

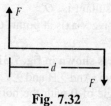

Fig. 7.32

Q. 30: Define different types of couple?

Sol.: There are two types of couples:

1. Clockwise Couple

A couple whose tendency is to rotate the body on which it acts, in a clockwise direction, is known as a clockwise couple. Such a couple is also called positive couple.

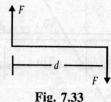

Fig. 7.33

2. Anticlockwise Couple

A couple whose tendency is to rotate the body on which it acts, in a anticlockwise direction, is known as a anticlockwise couple. Such a couple is also called Negative couple.

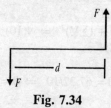

Fig. 7.34

Q 31: What is the moment of a couple?

Sol.: The moment of a couple is the product of the force (i.e. one of the forces of the two equal and opposite parallel forces) and the arm of the couple.

Mathematically:

Moment of a couple = $F.d$ N-m or N-mm
- The moment of couple may be clockwise or anticlockwise.
- The effect of the couple is unchanged if::
1. The couple is shifted to any other position.
2. The couple is rotated by an angle.
3. Any pair of force whose rotation effect is the same replaces the couple.
4. Sum of forces forming couple in any direction is zero.

Q. 32: What are the main characteristics of couple?
Sol.: The main characteristics of a couple
1. The algebraic sum of the forces, consisting the couple, is zero.
2. The algebraic sum of the moment of the forces, constituting the couple, about any point is the same, and equal to the moment of the couple itself.
3. A couple cannot be balanced by a single force, but can be balanced only by a couple, but of opposite sense.
4. Any number of coplanar couples can be reduced to a single couple, whose magnitude will be equal to the algebraic sum of the moments of all the couples.

Q. 33: Define magnitude of a couple.
Sol.: For a system, magnitude of a couple is equal to the algebraic sum of the moment about any point
If the system is reduces to a couple, the resultant force is zero, **(i.e. $\Sigma H = \Sigma V = 0$)** but $\Sigma M \neq 0$, i.e. the moment of the force system is the resultant.

Q. 34: A rectangle $ABCD$ has sides $AB = CD = 80$ mm and $BC = DA = 60$ mm. Forces of 150 N each act along AB and CD, and forces of 100 N each act along BC and DA. Make calculations for the resultant of the force system.

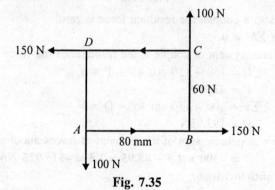

Fig. 7.35

Sol.: Let R be the resultant of the given system. And ΣH and ΣV be the horizontal and vertical component of the resultant. And resultant makes an angle of θ with the horizontal.

Resolving all the forces horizontally
$$\Sigma H = 150 - 150$$
$$\Sigma H = 0 \text{KN} \qquad \qquad ...(i)$$

Resolving all the forces vertically
$$\Sigma V = 100 - 100$$

$$\Sigma V = 0 KN \qquad ...(ii)$$

Since ΣH and ΣV both are 0, then resultant of the system is also zero.

But in Non-concurrent forces system, the resultant of the system may be a force, a couple or a force and a couple

i.e. in this case if couple is not zero then couple is the resultant of the force system.

For finding Couple, taking moment about any point say point 'A'.

$$M_A = -150 \times 60 - 100 \times 80, \text{ both are anticlockwise}$$

Then, Resultant moment = couple = –17000N–mm ....ANS

Q. 35: A square block of each side 1.5 m is acted upon by a system of forces along its sides as shown in the adjoining fig.ure. If the system reduces to a couple, determine the magnitude of the forces P and Q, and the couple.

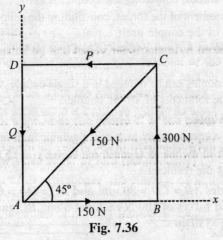

Fig. 7.36

Sol.: If the system is reduces to a couple, the resultant force is zero,

(i.e. $\Sigma H = \Sigma V = 0$) but $\Sigma M \neq 0$,

i.e. the moment of the force system or couple is the resultant of the force system.

$$\Sigma H = 150 - 150 \cos 45° - P = 0$$
$$P = 43.95 N \qquad ....ANS$$
$$\Sigma V = 300 - 150 \sin 45° - Q = 0$$
$$Q = 193.95 N \qquad ....ANS$$

Now moment of couple = Algebraic sum of the moment of forces about any corner, say A

$$= -300 \times 1.5 - 43.95 \times 1.5 = -515.925 \text{ Nm} \qquad ....ANS$$

-ive means moment is anticlockwise.

Q. 36: Resolve a force system in to a single force and a couple system. Also explain Equivalent force couple system.

Or

'Any system of co-planer forces can be reduced to a force – couple system at an arbitrary point'. Explain the above statement by assuming a suitable system

Sol.: A given force 'F' applied to a body at any point A can always is replaced by an equal force applied at another point B together with a couple which will be equivalent to the original force.

Let us given force F is acting at point 'A' as shown in fig. (7.37).

This force is to be replaced at the point 'B'. Introduce two equal and opposite forces at B, each of magnitude F and acting parallel to the force at A as shown in fig. (7.38). The force system of fig. (7.38) is equivalent to the single force acting at A of fig. (7.37). In fig. (7.38) three equal forces are acting. The two forces i.e. force F at A and the oppositely directed force F at B (i.e. vertically downwards force at B) from a couple. The moment of this couple is $F \times x$ clockwise where x is the perpendicular distance between the lines of action of forces at A and B. The third force is acting at B in the same direction in which the force at A is acting.

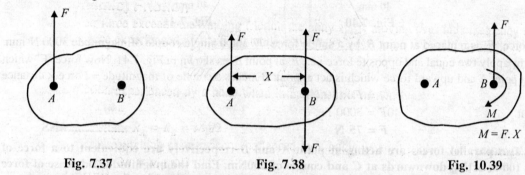

Fig. 7.37 **Fig. 7.38** **Fig. 10.39**

In fig. (7.39), the couple is shown by curved arrow with symbol M. The force system of fig. (7.39) is equivalent to fig. (7.38). Or in other words the Fig. (7.39) is equivalent to Fig. (7.37). Hence the given force F acting at A has been replaced by an equal and parallel force applied at point B in the same direction together with a couple of moment $F \times x$.

Thus force acting at a point in a rigid body can be replaced by an equal and parallel force at any other point in the body, and a couple.

Equivalent force System

An equivalent system for a given system of coplanar forces is a combination of a force passing through a given point and a moment about that point. The force is the resultant of all forces acting on the body. And the moment is the sum of all the moments about that point.

Hence equivalent system consists of:
1. A single force R passing through the given point, and
2. A single moment (ΣM)

Where,
R = the resultant of all force acting on the body
ΣM = Sum of all moments of all the forces about point P.

Q. 37: In designing the lifting hook, the forces acting on a horizontal section through B may be determined by replacing F by a equivalent force at B and a couple. If the couple is 3000 N-mm, determine F. Fig. (7.40).

160 / *Problems and Solutions in Mechanical Engineering with Concept*

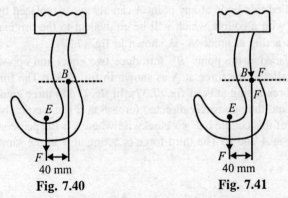

Fig. 7.40   Fig. 7.41

Sol.: Force 'F' is replaced at point B, by a single force 'F' and a single couple of magnitude 3000 N-mm.

Now apply two equal and opposite force i.e. 'F' at point B. as shown in Fig. 7.41. Now force 'F' which is act at point E and upward force which is act at point B makes a couple of magnitude = Force × distance

$$= F \times 40$$

But $40F = 3000$ i.e.

$$F = 75 \text{ N} \qquad \qquad \qquad \text{........ANS}$$

Q. 38: Two parallel forces are acting at point A and B respectively are equivalent to a force of 100 N acting downwards at C and couple of 200Nm. Find the magnitude and sense of force F_1 and F_2 shown in Fig. 7.42.

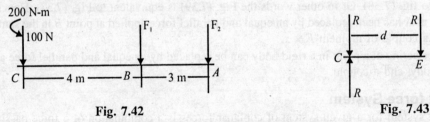

Fig. 7.42   Fig. 7.43

Sol.: The given system is converts to a single force and a single couple at C. Let R be the resultant of F_1 and F_2.

$$R_H = 0$$
$$R_V = -F_1 - F_2$$
$$R_V = -(F_1 + F_2) \qquad \qquad \qquad ...(i)$$

Since $R = (R^2_V)^{1/2} = (F_1 + F_2)$

Let resultant R act at a distance 'd' from the point C.

Now the single force i.e. R is converted in to a single force and a couple at C

Now apply two equal and opposite force i.e. 'R' at point C. as shown in Fig. 7.43. Now force 'R' which is act at point E and upward force which is act at point C makes a couple of magnitude = Force × distance

$$= R \times d$$

But $R.d = 200$ N-m

And single force R which is downward direction = 100 N   (-ive for downward)

i.e $(F_1 + F_2) = 100$ or $F_1 + F_2 = 100$ ...(ii)

Now taking moment about point C., or apply varignon's theorem.

$$R.d = 4F_1 + 7F_2, \text{ but } R.d = 200,$$

$4F_1 + 7F_2 = 200$ ...(iii)

solving equation (ii) and (iii)

$F_1 = 500/3$ N ....ANS

$F_2 = -200/3$ N ....ANS

Q. 39: A system of parallel forces is acting on a rigid bar as shown in Fig. 7.44. Reduce this system to
  (i) a single force
  (ii) A single force and a couple at A
  (iii) A single force and a couple at B.

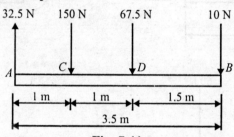

Fig. 7.44

Sol.: (i) A single force: A single force means just to find out the resultant of the system.

Since there are parallel force i.e resultant is sum of vertical forces,

$R = 32.5 - 150 + 67.5 - 10 = -60$

$R = (\Sigma V^2)^{1/2}$

R = 60N (downward) ....ANS

Let d = Distance of resultant from A towards right.

To find out location of resultant apply varignon's theorem :

$R.d = 150 \times 1 - 67.5 \times 2 + 10 \times 3.5$

$60.d = 150 \times 1 - 67.5 \times 2 + 10 \times 3.5$

$d = 0.833$ m

i.e resultant is at a distance of 0.83 m from A.

(ii) A single force and a couple at A: It means the whole system is to convert in to a single force and a single couple.

Since we convert all forces in to a single force i.e. resultant.

Now apply two equal and opposite force i.e. 'R' at point A. Now force 'R' which is act at point E and upward force which is act at point A makes a couple of magnitude,

Magnitude = Force × distance = 60 × 0.833

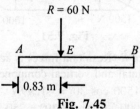

Fig. 7.45

= 49.98 Nm ....ANS and a single force of magnitude = 60N ....ANS

(*iii*) **A single force and a couple at *B*:** Since $AE = 0.833$ m, then $BE = 3.5 - 0.833 = 2.67$ m
Now, the force $R = -60N$ is moved to the point B, by a single force $R = -60N$ and a couple of magnitude $= R \times BE = -60 \times 2.67 = 160$ Nm

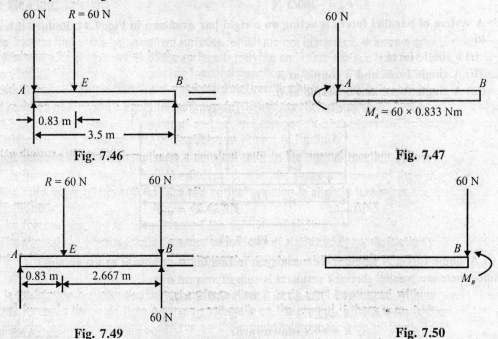

Fig. 7.46

Fig. 7.47

Fig. 7.49

Fig. 7.50

Hence single force is 60 N and couple is 160 Nm

Q. 40: The two forces shown in Fig. (7.51), are to be replaced by an equivalent force R applied at the point P. Locate P by finding its distance x from AB and specify the magnitude of R and the angle θ it makes with the horizontal.

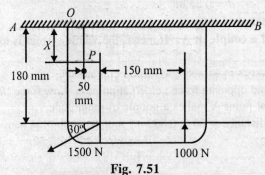

Fig. 7.51

Sol.: Let us assume the equivalent force R (Resultant force) is acting at an angle of θ with the horizontal. And ΣH and ΣV be the sum of horizontal and vertical components.

$$\Sigma H = -1500 \cos 30° = -1299N \qquad \ldots(i)$$
$$\Sigma V = 1000 - 1500 \sin 30° = 250 \text{ N} \qquad \ldots(ii)$$
$$R = \{\Sigma H^2 + \Sigma V^2\}^{1/2}$$

$$R = 1322.87 \text{ N} \quad \text{.......ANS}$$

For direction of resultant
$$\tan \theta = \Sigma V/\Sigma H$$
$$= 250/-1299$$
$$\theta = -10.89° \quad \text{.......ANS}$$

Now for finding the position of the resultant, we use Varignon's theorem,
i.e $R \times d = \Sigma M$, Take moment about point 'O'
$$1322.878 \times x = 1500 \cos 30° \times 180 + 1500 \sin 30° \times 50 - 1000 \times 200$$
on solving x = **53.92 mm** ....ANS

Q. 41: Fig. 7.52 shows two vertical forces and a couple of moment 2000 Nm acting on a horizontal rod, which is fixed at end A.
1. Determine the resultant of the system
2. Determine an equivalent system through A. *(May 00-01(B.P.))*

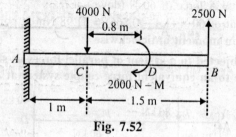

Fig. 7.52

Sol.: *(i)* Resultant of the system
$$\Sigma V = -4000 + 2500 = -1500 \text{N}$$
$$R = (\Sigma V^2)^{1/2}$$
$$= 1500 \text{N (acting downwards)}$$

for finding the position of the resultant, apply varignon's theorem
i.e Moment of resultant = sum of moment of all the forces about any point.
Let from point be A, and distance of resultant is 'd' m from A
$$R.d = 4000 \times 1 + 2000 - 2500 \times 2.5$$
$-1500 \times d = -250 \Rightarrow d = 0.166$ m from point A

(ii) **Equivalent system through A**
Equivalent system consist of:
1. A single force R passing through the given point, and
2. A single moment (ΣM)
Where,
R = the resultant of all force acting on the body
ΣM = Sum of all moments of all the forces about point A.
Hence single force is = 1500 N; And couple = 250 Nm

Q. 42: A rigid body is subjected to a system of parallel forces as shown in Fig. 7.53. Reduce this system to,
   (i) **A single force system**
   (ii) **A single force moment system at B** *(May–01-02)*

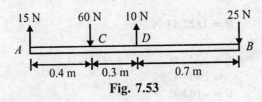

Fig. 7.53

Sol.: It is the equivalent force system

$$R = 15 - 60 + 10 - 25 = -60 \text{ N}$$

(Acting downward)

Now taking moment about point A, apply varignon's theorem

$$R.X = 60 \times 0.4 - 10 \times 0.7 + 25 \times 1.4$$
$$60.X = 52, \quad x = 0.867 \text{ m}$$

Where X is the distance of resultant from point A.

(1) A single force be 60 N acting downward

(2) Now a force of 60 N = A force of 60 N (down) at B

and anticlockwise moment of $60 \times (1.4 - 0.866) = 31.98$ Nm at point B.

60 N force and 31.98 Nm moment anticlockwise ....ANS

Q. 43: A rigid bar *CD* is subjected to a system of parallel forces as shown in Fig. 7.54. Reduce the given system of force to an equivalent force couple system at *F*. *(Dec–03-04)*

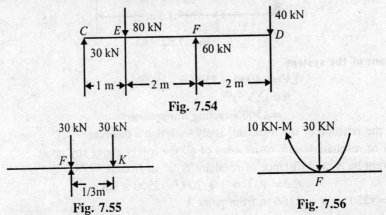

Fig. 7.54

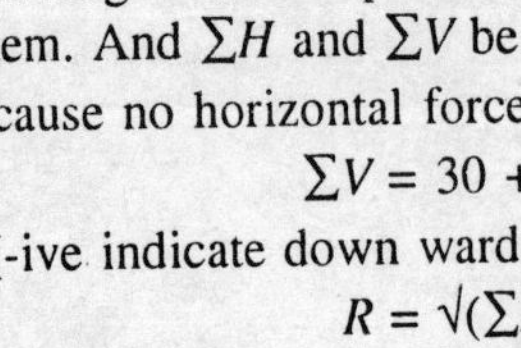

Fig. 7.55 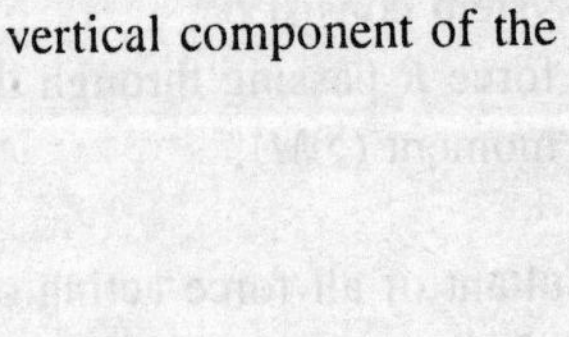   Fig. 7.56

Sol.: First find the magnitude and point of application of the resultant of the system, Let *R* be the resultant of the given system. And ΣH and ΣV be the horizontal and vertical component of the resultant.

$\Sigma H = 0$, because no horizontal force

$$\Sigma V = 30 + 60 - 80 - 40$$

⇒ –30KN (-ive indicate down ward force.)

Since, $\qquad R = \sqrt{(\Sigma H)^2 + (\Sigma V)^2}$

⇒ $\sqrt{(0)^2 + (-30)^2}$

R = 30KN (Downwards) ....ANS

For position of the resultant

Let, *d* = Distance between Point *F* and the line of action of the resultant force.

Apply varignon's theorem, take moment about point '*F*'

$R.d = 30 \times 3 - 80 \times 2 + 40 \times 2$

$30.d = 10$

$d = 1/3\,m$

Now it means resultant is acting at a distance of 1/3m from point F. Now the whole system is converted to a single force i.e. resultant, which is act at a point 'K'. Now apply two equal and opposite forces at point F. as shown in Fig. 7.55. Now resultant force which is act at point K and upward force which is act at point F makes a couple of magnitude = Force × distance

$$= 30 \times 1/3 = 10\,KN\text{--}m \text{ (clockwise)}$$

So two force replace by a couple at point F.

Now the system contains a single force of magnitude 30 KN and a couple of magnitude 10 KN–m.

Q. 44: **What force and moment is transmitted to the supporting wall at A? (Refer Fig. 7.57)**

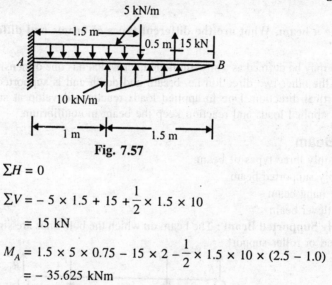

Fig. 7.57

$\Sigma H = 0$

$\Sigma V = -5 \times 1.5 + 15 + \dfrac{1}{2} \times 1.5 \times 10$

$\quad\quad = 15\,kN$

$M_A = 1.5 \times 5 \times 0.75 - 15 \times 2 - \dfrac{1}{2} \times 1.5 \times 10 \times (2.5 - 1.0)$

$\quad\quad = -35.625\,kNm$

A force of 15 KN (vertical) is transmitted to the wall along with an anticlockwise moment of 35.625 kNm.

CHAPTER 8

FORCE : SUPPORT REACTION

Q. 1: Define a beam. What are the different types of beams and different types of loading?

(Dec–05)

Sol.: A beam may be defined as a structural element which has one dimension (length) considerable larger compared to the other two direction i.e. breath and depth and is supported at a few points. It is usually loaded in vertical direction. Due to applied loads reactions develop at supports. The system of forces consisting of applied loads and reaction keep the beam in equilibrium.

Types of Beam

There are mainly three types of beam:
1. Simply supported beam
2. Over hang beam
3. Cantilever beam

1. Simply Supported Beam : The beam on which the both ends are simply supported, either by point load or hinged or roller support.

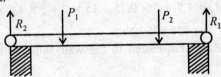

Fig 8.1

Fig 8.2

2. Over–Hanging Beam: The beam on which one end or both ends are overhang (or free to air.) are called overhanging beam.

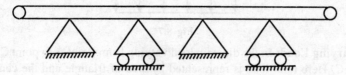

Fig 8.3

3. Cantilever Beam: If a beam is fixed at one end and is free at the other end, it is called cantilever beam, In cantilever beam at fixed end, there are three support reaction a horizontal reaction (R_H), a vertical reaction(R_V), and moment(M)

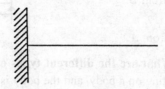

Fig 8.4

Types of Loading
Mainly three types of load acting on any beam;
   1. Concentrated load
   2. Uniformly distributed load
   3. Uniformly varying load

1. Concentrated load (or point load): If a load is acting on a beam over a very small length. It is called point load.

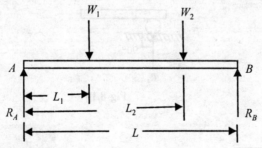

Fig 8.5

2. Uniformly Distributed Load: For finding reaction, this load may be assumed as total load acting at the center of gravity of the loading (Middle point).

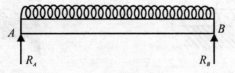

Fig 8.6

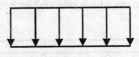

Fig 8.7

3. Uniformly Varying Load: In the diagram load varying from Point A to point C. Its intensity is zero at A and 900N/M at C. Here total load is represented by area of triangle and the centroid of the triangle represents the center of gravity.

Thus total load = $\frac{1}{2} \cdot AB \cdot BC$

And $\quad C.G. = \frac{1}{3} \cdot AB$ meter from B.

$\quad\quad\quad\quad = \frac{2}{3} \cdot AB$ meter from A.

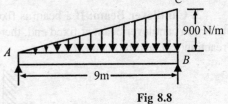

Fig 8.8

Q. 2: Explain support reaction? What are the different types of support and their reactions?

Sol.: When a number of forces are acting on a body, and the body is supported on another body, then the second body exerts a force known as reaction on the first body at the points of contact so that the first body is in equilibrium. The second body is known as support and the force exerted by the second body on the first body is known as support reaction.

There are three types of support;
1. Roller support
2. Hinged Support
3. Fixed Support

1. Roller Support: Beams end is supported on rollers. Reaction is at right angle. Roller can be treated as frictionless. At roller support only one vertical reaction.

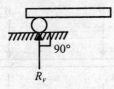

Fig 8.10

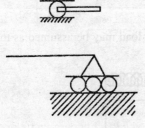

Fig 8.9

Fig. 8.10

2. Hinged (Pin) Support: At a hinged end, a beam cannot move in any direction support will not develop any resisting moment, but it can develop reaction in any direction.

In hinged support, there are two reaction is acting, one is vertical and another is horizontal. i.e., R_H and R_V

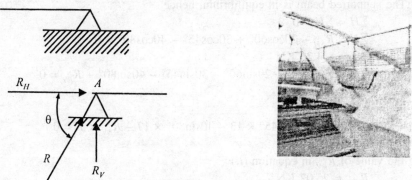

Fig 8.11

3. Fixed Support: At such support the beam end is not free to translate or rotate at fixed end there are three reaction a horizontal reaction (R_H), a vertical reaction(R_V), and moment(M)

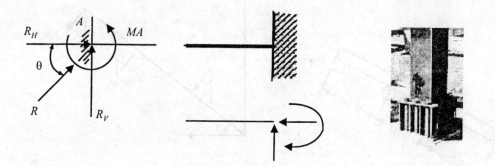

Fig 8.12

14.3.5 Rocker Support: Only one reaction i.e., R_H

Q. 3: Determine algebraically the reaction on the beam loaded as shown in fig 8.13. Neglect the thickness and mass of the beam.

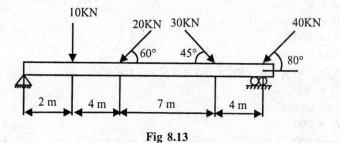

Fig 8.13

Sol.: Resolved all the forces in horizontal and vertical direction.

Let reaction at hinged i.e., point A is R_{AH} and R_{AV}, and reaction at roller support is R_{BV}
Let ΣH & ΣV is the sum of horizontal and vertical component of the
forces, The supported beam is in equilibrium, hence

$$\Sigma H = \Sigma H = 0$$
$$\Sigma H = R_{AH} - 20\cos 60° + 30\cos 45° - 40\cos 80° = 0$$
$$R_{AH} = -4.26 \text{ KN} \qquad ...(i)$$
$$\Sigma V = R_{AV} - 10 - 20\sin 60° - 30\sin 45° - 40\sin 80° + R_{BV} = 0$$
$$R_{AV} + R_{BV} = 87.92 \text{ KN} \qquad ...(ii)$$

Taking moment about point A
$$10 \times 2 + 20\sin 60° \times 6 + 30\sin 45° \times 13 - 40\sin 80° \times 17 - R_{BV} \times 17 = 0$$
$$R_{BV} = 62.9 \text{ KN} \qquad ...(iii)$$

Putting the value of R_{BV} in equation (ii)
$$R_{AV} = 25.02 \text{ KN}$$

Hence $R_{AH} = -4.26\text{KN}, R_{AV} = 25.02\text{KN}, R_{BV} = 62.9\text{KN}$ ....ANS

Q. 4: A light rod AD is supported by frictionless pegs at B and C and rests against a frictionless wall at A as shown in fig 8.14. A force of 100N is applied at end D. Determine the reaction at A, B and C.

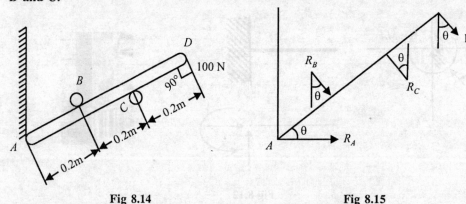

Fig 8.14   Fig 8.15

Sol.: Since roller support at point B, C, so only vertical reactions are there say R_B, R_C. At point A rod is in contact with the wall that is wall give a contact reaction to the rod say R_A.

Let rod is inclined at an angle of θ. Rod is in equilibrium position.
$$\Sigma V = 0$$
$$R_B \cos\theta - R_C \cos\theta + 100\cos\theta = 0$$
$$R_C - R_B = 100 \qquad ...(i)$$

Taking moment about point A:
$$\Sigma M_A = 100 \times 0.6 - RC \times 0.4 + RB \times 0.2 = 0$$
$$2R_C - R_B = 300 \qquad ...(ii)$$

Solving equation (i) and (ii)
$$R_C = 200 \qquad\qquad\qquad ....\text{ANS}$$
$$R_B = 100 \qquad\qquad\qquad ....\text{ANS}$$
$$\Sigma H = 0$$

$$R_A + R_B \sin\theta - RC\sin\theta + 100\sin\theta = 0$$
$$R_A + 100\sin\theta - 200\sin\theta + 100\sin\theta = 0$$
$$RA = 0 \qquad \text{.......ANS}$$

Q. 5: Find the reaction at the support as shown in fig 8.16.

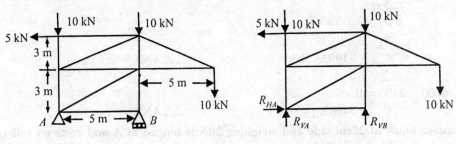

Fig 8.16   Fig 8.17

Sol.: First draw the *FBD* of the system as shown in fig 8.17.

Since hinged at point A and Roller at point B. let at point A R_{AH} and R_{AV} and at point B R_{BV} is the support reaction.

$$\Sigma H = 0$$
$$R_{AH} - 5 = 0$$
$$R_{AH} = \mathbf{5KN} \qquad \text{.......ANS}$$
$$\Sigma V = 0$$
$$R_{AV} + R_{BV} - 10 - 10 - 10 = 0$$
$$R_{AV} + R_{BV} = 30KN \qquad \qquad \qquad ...(i)$$

Taking moment about point A:
$$\Sigma M_A = 10 \times 5 + 10 \times 10 - 5 \times 6 - R_{VB} \times 5 = 0$$
$$R_{BV} = \mathbf{24KN} \qquad \text{.......ANS}$$

Putting the value of R_{BV} in equation (i)
$$R_{AV} = \mathbf{6KN} \qquad \text{.......ANS}$$

Q. 6: A fixed crane of 1000Kg mass is to lift 2400Kg crates. It is held in place by a pin at A and a rocker at B. the C.G. is located at G. Determine the components of reaction at A and B after drawing the free body diagram.

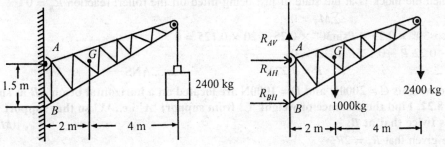

Fig 8.18   Fig 8.19

Sol.: Since two reaction (Vertical and Horizontal) at pin support i.e., R_{AH} and R_{AV}. And at rocker there will be only one Horizontal reaction i.e., R_{BH}.

First draw the *FBD* of the Jib crane as shown in fig 8.19. The whole system is in equilibrium. Take moment about point A

$$\Sigma M_A = - R_{BV} \times 1.5 + 1000 \times 2 + 2400 \times 6 = 0$$
$$R_{BH} = 10933.3 \text{Kg} \quad \text{.......ANS}$$
$$\Sigma H = 0$$
$$R_{AH} + R_{BH} = 0$$
i.e., $\quad R_{AH} = - R_{BH}$
$$R_{AH} = -10933.3 \text{Kg} \quad \text{.......ANS}$$
$$\Sigma V = 0$$
$$R_{AV} - 1000 - 2400 = 0$$
$$R_{AV} = 3400 \text{Kg} \quad \text{.......ANS}$$

Q. 7: A square block of 25cm side and weighing 20N is hinged at *A* and rests on rollers at *B* as shown in fig 8.20. It is pulled by a string attached at *C* and inclined at 300 with the horizontal. Make calculations for the force *P* to be applied so that the block gets just lifted off the roller.

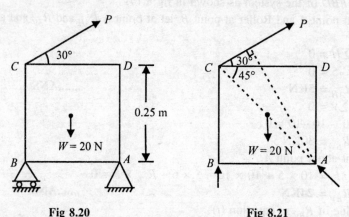

Fig 8.20   Fig 8.21

Sol.: From the Free body diagram the block is subjected to the following set of forces.
1. Force *P*
2. Weight of the block *W*
3. Reaction R_A at the hinged point
4. When the block is at the state of just being lifted off the roller, reaction $R_B = 0$

$$\Sigma M_A = 0$$
$$- P\cos 30° \times 0.25 - P\sin 30° \times 0.25 + 20 \times 0.125 = 0$$
$$- 0.22 P - 0.125P + 2.5 = 0$$
$$P = 7.27\text{N} \quad \text{.......ANS}$$

Q. 8: Two weights *C* = 2000N and *D* = 1000N are located on a horizontal beam *AB* as shown in the fig 8.22. Find the distance of weight '*C*' from support '*A*' i.e., '*X*' so that support reaction at *A* is twice that at *B*.   (*May–00–01*)

Sol.: Since given that $R_A = 2R_B$
$$\Sigma H = 0$$
$$R_{AH} = 0 \quad \quad \quad \quad \quad \quad \quad \quad \quad \quad ....(i)$$
$$\Sigma V = 0$$

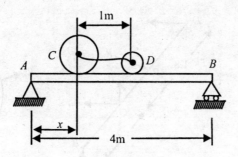

Fig. 8.22

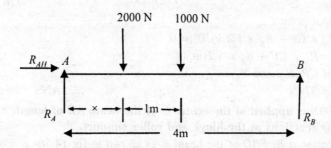

Fig. 8.23

$$R_A + R_B - 2000 - 1000 = 0$$
$$R_A + R_B = 3000N$$
$$\text{But } R_A = 2R_B$$

i.e.,
$$R_B = 1000N \qquad ...(ii)$$
$$R_A = 2000N \qquad ...(iii)$$

Taking moment about point A:
$$\Sigma M_A = 2000 \times x + 1000 \times (x + 1) - R_B \times 4 = 0$$
$$2000 \times x + 1000 \times (x + 1) - 1000 \times 4 = 0$$
$$2000x + 1000x + 1000 - 4000 = 0$$
$$3000x = 3000$$
$$x = 1m \qquad ....ANS$$

Q. 9: A 500N cylinder, 1 m in diameter is loaded between the cross pieces AE and BD which make an angle of 60° with each other and are pinned at C. Determine the tension in the horizontal rope DE assuming that the cross pieces rest on a smooth floor. *(Dec–01–02)*

Sol.: Consider the equilibrium of the entire system.

C is the pin joint, making the free body diagram of ball and rod separately.
$$2R_N \cos 60° = 500 \qquad ...(i)$$
$$R_N = 500KN$$
$$R_A + R_B = 500N \qquad ...(ii)$$
Due to symmetry $R_A = R_B = 250N$
$$CP = 0.5\cot 30° = 0.866m$$

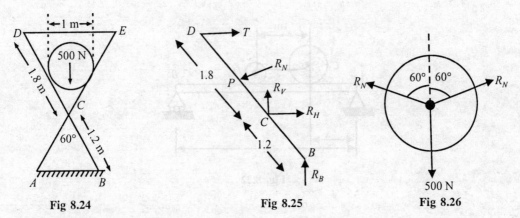

Fig 8.24   Fig 8.25   Fig 8.26

Taking moment about point C,
$T \times 1.8\cos30° - R_N \times CP - R_B \times 1.2\sin30° = 0$
$T \times 1.8\cos300 = R_N \times CP + R_B \times 1.2\sin30°$
Putting the value of CP, R_N, and R_B
$T = 374N$ ....ANS

Q. 10: A Force P = 5000N is applied at the centre C of the beam AB of length 5m as shown in the fig 8.27. Find the reactions at the hinge and roller support. *(May–01–02)*

Sol.: Hinged at A and Roller at B, FBD of the beam is as shown in fig 14.70

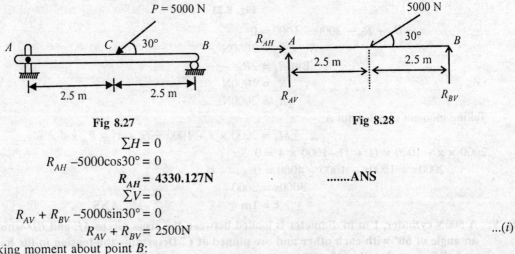

Fig 8.27   Fig 8.28

$\Sigma H = 0$
$R_{AH} - 5000\cos30° = 0$
$R_{AH} = 4330.127N$ ....ANS
$\Sigma V = 0$
$R_{AV} + R_{BV} - 5000\sin30° = 0$
$R_{AV} + R_{BV} = 2500N$ ...(i)

Taking moment about point B:
$\Sigma M_B = R_{AV} \times 5 - 5000\sin30° \times 2.5 = 0$
$R_{AV} = 1250N$ ....ANS

From equation (i)
$R_{BV} = 1250N$ ....ANS

Q. 11: The cross section of a block is an equilateral triangle. It is hinged at A and rests on a roller at B. It is pulled by means of a string attached at C. If the weight of the block is Mg and the string is horizontal, determine the force P which should be applied through string to just lift the block off the roller. *(Dec–02–03)*

Sol.: When block is just lifted off the roller the reaction at B i.e., R_B will be zero.

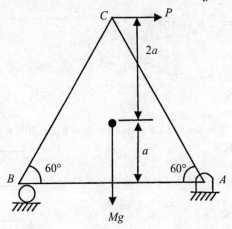

Fig. 8.29

For equilibrium, $R_A = R_B = Mg/2$
At this instance, taking moment about 'A'
$$P \times 3a = Mg.a\sqrt{3}$$
$$P = Mg/\sqrt{3} \qquad \text{........ANS}$$

Q. 12: A beam 8m long is hinged at A and supported on rollers over a smooth surface inclined at 300 to the horizontal at B. The beam is loaded as shown in fig 8.30. Determine the support reaction.

(*May–02–03*)

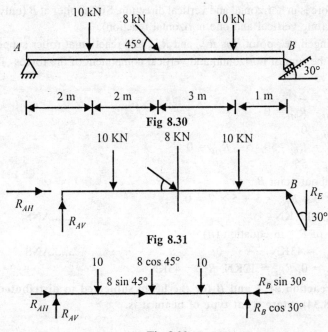

Fig 8.30

Fig 8.31

Fig 8.32

Sol.: F.B.D. is as shown in fig 8.32

$$\Sigma H = 0$$
$$R_{AH} + 8\cos 45° - R_B \sin 30° = 0$$
$$0.5R_B - R_{AH} = 5.66 \qquad \qquad \qquad ...(i)$$
$$\Sigma V = 0$$
$$R_{AV} - 10 - 8\cos 45° - 10 + R_B \cos 30° = 0$$
$$R_{AV} + 0.866 R_B = 25.66 \qquad \qquad \qquad ...(ii)$$

Taking moment about point A:
$$\Sigma M_A = 10 \times 2 + 8\cos 45° \times 4 + 10 \times 7 - R_B \cos 30° \times 8 = 0$$
$$R_B = 16.3 \text{KN} \qquad \qquad \qquad \text{.......ANS}$$

From equation (ii)
$$R_{AV} = 11.5 \text{KN} \qquad \qquad \qquad \text{.......ANS}$$

From equation (1)
$$R_{AH} = 2.5 \text{KN} \qquad \qquad \qquad \text{.......ANS}$$

Q. 13: Calculate the support reactions for the following. Fig(8.33).

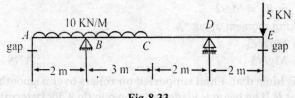

Fig 8.33

Sol.: First change *UDL* in to point load.

Resolved all the forces in horizontal and vertical direction. Since roller at B (only one vertical reaction) and hinged at point A (one vertical and one horizontal reaction).

Let reaction at hinged i.e., point B is R_{BH} and R_{BV}, and reaction at roller support i.e. point D is R_{DV}

Let ΣH & ΣV is the sum of horizontal and vertical component of the forces, The supported beam is in equilibrium, hence

$$\Sigma H = \Sigma V = 0$$
$$R_H = R_{BH} = 0$$
$$R_{BH} = 0 \qquad \qquad \qquad ...(i)$$
$$\Sigma V = R_{BV} - 50 - 5 - R_{DV} = 0$$
$$R_{BV} + R_{DV} = 55 \qquad \qquad \qquad ...(ii)$$

Taking moment about point B
$$50 \times 0.5 - R_{BV} \times 0 - R_{DV} \times 5 + 5 \times 7 = 0$$
$$R_{DV} = 12 \text{ KN} \qquad \qquad \qquad \text{.......ANS}$$

Putting the value of R_{BV} in equation (ii)
$$R_{BV} = 43 \text{KN} \qquad \qquad \qquad \text{.......ANS}$$

Hence $R_{BH} = 0$, $R_{DV} = 12 \text{KN}$, $R_{BV} = 43 \text{KN}$

Q. 14: Compute the reaction at A and B for the beam subjected to distributed and point loads as shown in fig (8.34). State what type of beam it is.

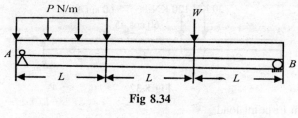

Fig 8.34

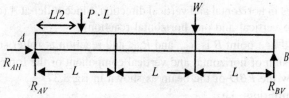

Fig 8.35

Sol.: First change *UDL* in to point load.

Resolved all the forces in horizontal and vertical direction. Since roller at B (only one vertical reaction) and hinged at point A (one vertical and one horizontal reaction).

Let reaction at hinged *i.e.*, point A is R_{AH} and R_{AV}, and reaction at roller support *i.e.*, point B is R_{BV}

Let ΣH & ΣV is the sum of horizontal and vertical component of the forces ,The supported beam is in equilibrium, hence Draw the *FBD* of the diagram as shown in fig 8.35

Since beam is in equilibrium, i.e.,

$$\Sigma H = 0;$$
$$R_{AH} = 0 \qquad \qquad \text{.......ANS}$$
$$\Sigma V = 0 ; \quad R_{AV} + R_{BV} - \text{P.L} - W = 0$$
$$R_{AV} + R_{BV} = \text{P.L} + W \qquad \qquad \text{...(i)}$$

Taking moment about point A,

$$\text{P.L} \times L/2 + W \times 2L - RBV \times 3L = 0 \qquad \qquad \text{...(ii)}$$
$$R_{BV} = \text{P.L}/6 + 2W/3 \qquad \qquad \text{.......ANS}$$

Put the value of R_{BV} in equation (*i*)

$$R_{AV} = 5\text{P.L}/6 + W/3 \qquad \qquad \text{.......ANS}$$

Q. 15: Find the reactions at supports A and B of the loaded beam shown in fig 8.36.

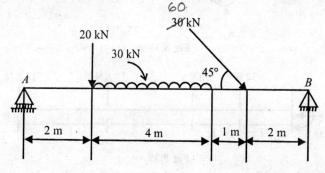

Fig 8.36

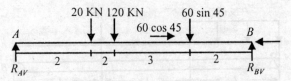

Fig 8.37

Sol.: First change *UDL* in to point load.

Resolved all the forces in horizontal and vertical direction. Since roller at *A* (only one vertical reaction) and hinged at point *B* (one vertical and one horizontal reaction).

Let reaction at hinged i.e., point *B* is R_{BH} and R_{BV}, and reaction at roller support i.e.. point *A* is R_{AV}

Let ΣH & ΣV is the sum of horizontal and vertical component of the forces, The supported beam is in equilibrium, hence Draw the *FBD* of the beam as shown in fig 8.37.

Since beam is in equilibrium, i.e.,
$$\Sigma H = 0;$$
$$R_{BH} - 60\cos 45° = 0$$
$$R_{BH} = 42.42 \text{KN} \quad \text{.......ANS}$$
$$\Sigma V = 0;$$
$$R_{AV} + R_{BV} - 20 - 120 - 42.4 = 0$$
$$R_{AV} + R_{BV} = 182.4 \text{KN} \quad \text{...(i)}$$

Taking moment about point *A*,
$$20 \times 2 + 120 \times 4 + 42.4 \times 7 - R_{BV} \times 9 = 0 \quad \text{...(ii)}$$
$$R_{BV} = 90.7 \text{KN} \quad \text{.......ANS}$$

Put the value of R_{BV} in equation (*i*)
$$R_{AV} = 91.6 \text{KN} \quad \text{.......ANS}$$

Hence reaction at support *A* i.e., $R_{AV} = 91.6$KN

reaction at support *B* i.e., $R_{BV} = 90.7$KN, $R_{BH} = 42.4$KN

Q. 16: The cantilever is shown in fig (8.38), Determine the reaction when it is loaded..

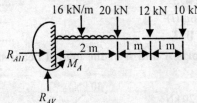

Fig 8.38

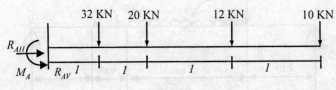

Fig 8.39

Sol.: In a cantilever at fixed end (Point *A*) there is three reaction i.e., R_{AH}, M_A, R_{AV}

First draw the *FBD* of the beam as shown in fig 8.39, Since beam is in equilibrium, i.e.,

$\Sigma H = 0$;
$R_{AH} = 0$
$R_{AH} = 0$ ....ANS
$\Sigma V = 0$;
$R_{AV} - 32 - 20 - 12 - 10 = 0$
$R_{AV} = 74KN$ ....ANS

Taking moment about point A,
$-M_A + 32 \times 1 + 20 \times 2 + 12 \times 3 + 10 \times 4 = 0$
$M_A = 148KN\text{–m}$ ....ANS

Hence reaction at support A i.e., $R_{VA} = 74KN$, $R_{HA} = 0KN$, $M_A = 148KN\text{–m}$

Q. 17: Determine the reactions at A and B of the overhanging beam as shown in fig (8.40).

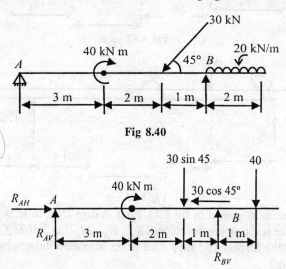

Fig 8.40

Fig 8.41

Sol.: First change *UDL* in to point load.

Resolved all the forces in horizontal and vertical direction. Since hinged at point A (one vertical and one horizontal reaction).

Let reaction at hinged i.e., point A is R_{AH} and R_{AV}, Let ΣH & ΣV is the sum of horizontal and vertical component of the forces, The supported beam is in equilibrium, hence Draw the *FBD* of the beam as shown in fig 8.42, Since beam is in equilibrium, i.e.,

$\Sigma H = 0$;
$R_{AH} = 30\cos45° = 21.2KN$
$R_{AH} = 21.21KN$ ....ANS
$\Sigma V = 0$;
$R_{AV} - 30\sin45 - 40 + R_{BV} = 0$
$R_{AV} + R_{BV} = 61.2KN$ ...(*i*)

Taking moment about point B,
$R_{AV} \times 6 + 40 - 30\sin45 \times 1 + 40 \times 1 = 0$

$R_{AV} = -9.8$ KN ....ANS

Putting the value of R_{AV} in equation (i), we get

$R_{BV} = 71$KN ....ANS

Hence reaction at support A i.e., $R_{AV} = -9.8$KN, $R_{AH} = 21.2$KN, $R_{BV} = 71$KN

Q. 18: Find out reactions at the grouted end of the cantilever beam shown in fig 8.42.

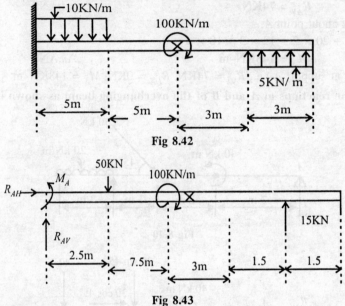

Fig 8.42

Fig 8.43

Sol.: Draw F.B.D. of the beam as shown in fig 8.43. First change *UDL* in to point load. Since Point A is fixed point i.e., there is three reaction are developed, R_{AH}, R_{AV}, M_A. Let ΣH & ΣV is the sum of horizontal and vertical component of the forces, The supported beam is in equilibrium, hence

$R = 0$,

$\Sigma H = ?V = 0$

$\Sigma H = 0; R_{AH} = 0$ ....ANS

$\Sigma V = 0; R_{AV} - 50 + 15 = 0, R_{AV} = 35$KN ....ANS

Now taking moment about point 'A'

$-M_A + 50 \times 2.5 + 100 - 15 \times 14.5 = 0$

$M_A = 7.5$ KN–m ....ANS

Q. 19: Find the support reaction at A and B in the beam as shown in fig 8.44.

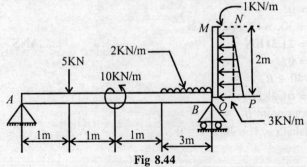

Fig 8.44

Force: Support Reaction / 181

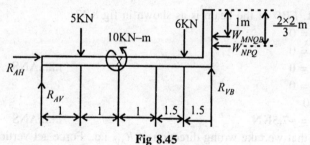

Fig. 8.45

Sol.: First draw the FBD of the beam as shown in fig 8.45
In the fig 8.46,
6KN is the point load of UDL
W_{MNQB} = Weight of MNQB
   = UDL × Distance(MB)
   = 1 × 2
   = 2KN, act at a point 1m vertically from point B
W_{NPQ} = Weight of Triangle NPQ
   = 1/2 × MB × (BP − BQ)
   = 1/2 × 2 × (3 − 1)
   = 2KN and will act at MB/3 = 2/3m from point B

Since hinged at point A and Roller at point B. let at point A R_{HA} and R_{VA} and at point B R_{VB} is the support reaction, Also beam is in equilibrium under action of coplanar non concurrent force system, therefore:

$$\Sigma H = 0$$
$$R_{AH} - W_{MNQB} - W_{NPQ} = 0$$
$$R_{AH} - 2 - 2 = 0$$
$$R_{AH} = 4KN \qquad \ldots\ldots\ldots\text{ANS}$$
$$\Sigma V = 0$$
$$R_{AV} + R_{BV} - 5 - 6 = 0$$
$$R_{AV} + R_{BV} = 11KN \qquad \ldots(i)$$

Taking moment about point A:
$$M_A = 5 \times 1 - 10 + 6 \times 4.5 - R_{BV} \times 6 - W_{NPQ} \times (2 - 4/3) - W_{MNQB} \times 1 = 0$$
$$5 \times 1 - 10 + 6 \times 4.5 - R_{BV} \times 6 - 2 \times (2 - 4/3) - 2 \times 1 = 0$$
$$R_{BV} = 3.11KN \qquad \ldots\ldots\ldots\text{ANS}$$

Putting the value of R_{BV} in equation (i)
$$R_{AV} = 7.99KN \qquad \ldots\ldots\ldots\text{ANS}$$

Q. 20: What force and moment is transmitted to the supporting wall at A in the given cantilever beam as shown in fig 8.46. *(May–02–03)*

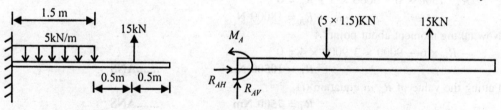

Fig 8.46   Fig 8.47

Sol.: Fixed support at A, *FBD* of the beam is as shown in fig 8.47

$$\Sigma H = 0$$
$$R_{AH} = 0$$
$$R_{AH} = 0 \qquad \text{......ANS}$$
$$\Sigma V = 0$$
$$R_{AV} - 7.5 + 15 = 0$$
$$R_{AV} = -7.5 \text{KN} \qquad \text{......ANS}$$

–ive sign indicate that we take wrong direction of R_{AV}, i.e., Force act vertically downwards.

Taking moment about point A:
$$\Sigma M = -M_A + 7.5 \times 0.75 - 15 \times 2 = 0$$
$$M_A = 7.5 \times 0.75 - 15 \times 2$$
$$\Rightarrow \qquad M_A = -24.357 \text{KN-m} \qquad \text{......ANS}$$

–ive sign indicate that we take wrong direction of moment, i.e., moment is clockwise.

Q. 21: Determine the reactions at supports of simply supported beam of 6m span carrying increasing load of 1500N/m to 4500N/m from one end to other end.

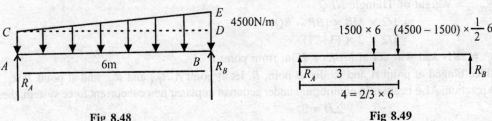

Fig 8.48   Fig 8.49

Sol.: Since Beam is simply supported i.e., at point A and point B only point load is acting. First change *UDL* and *UVL* in to point load. As shown in fig 8.49. Let ΣH & ΣV is the sum of horizontal and vertical component of the Resultant forces, The supported beam is in equilibrium, hence resultant force is zero.

Draw the *FBD* of the beam as shown in fig 8.49,
Divided the diagram *ACBE* in to two parts A triangle *CDE* and a rectangle *ABCE*.
Point load of Triangle $CDE = 1/2 \times CD \times DE = 1/2 \times 6 \times (4.5 - 1.5) = 9$KN
act at a distance 1/3 of *CD* (i.e., 2.0m) from point *D*
Point load of Rectangle $ABCD = AB \times AC = 6 \times 1.5 = 9$KN
act at a distance 1/2 of *AB* (i.e., 3m) from point *B*
Now apply condition of equilibrium:

$$\Sigma H = 0;$$
$$R_{AH} = 0 \qquad \text{......ANS}$$
$$\Sigma V = 0;$$
$$R_A - 1500 \times 6 - 3000 \times 3 + R_B = 0$$
$$R_A + R_B = 18000 \text{ N} \qquad \text{...(i)}$$

Now taking moment about point 'A'
$$-R_B \times 6 + 9000 \times 3 \; 9000 \times 4 = 0$$
$$R_B = 10500 \text{ Nm} \qquad \text{......ANS}$$

Putting the value of R_B in equation (*i*)
$$R_A = 7500 \text{ Nm} \qquad \text{......ANS}$$

Q. 22: Calculate the support reactions for the beam shown in fig (8.50).

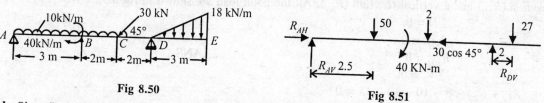

Fig 8.50

Fig 8.51

Sol.: Since Beam is overhang. At point A hinge support and point D Roller support is acting. First change UDL and UVL in to point load as shown in fig 8.51. Let ΣH & ΣV is the sum of horizontal and vertical component of the Resultant forces, the supported beam is in equilibrium, hence resultant force is zero. Convert UDL and UVL in point load and draw the FBD of the beam as shown in fig 8.51

$$\Sigma H = 0;$$
$$R_{AH} = 30\cos 45°$$
$$R_{AH} = 21.21 KN \quad \text{.......ANS}$$
$$\Sigma V = 0;$$
$$R_{AV} - 50 - 30\sin 450 + R_{DV} - 27 = 0$$
$$R_{AV} + R_{DV} = 98.21 \text{ KN} \quad ...(i)$$

Now taking moment about point 'A'
$$-R_{DV} \times 7 + 50 \times 2.5 + 40 + 30\sin 450 \times 5 + 27 \times 9 = 0$$
$$R_{DV} = 73.4 \text{ KN} \quad \text{.......ANS}$$

Putting the value of R_{DV} in equation (i).
$$R_{AV} = 24.7 KN \quad \text{.......ANS}$$

Q. 23: Determine the reactions at supports A and B of the loaded beam as shown in fig 8.52.

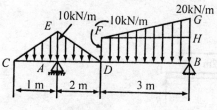

Fig 8.52

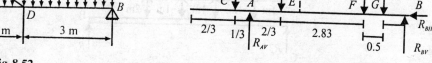

Fig 8.53

Sol.: First consider the FBD of the diagram 8.52. as shown in fig 8.53. In which Triangle CEA, AED and FHG shows point load and also rectangle FHDB shows point load.

Point load of Triangle $CEA = 1/2 \times AC \times AE = 1/2 \times 1 \times 10 = 5KN$,
act at a distance 1/3 of AC (i.e., 0.333m)from point A
Point load of Triangle $AED = 1/2 \times AD \times AE = 1/2 \times 2 \times 10 = 10KN$
act at a distance 1/3 of AD (i.e., 0.666m)from point A
Now divided the diagram DBGF in to two parts A triangle FHG and a rectangle FHDB.
Point load of Triangle $FHG = 1/2 \times FH \times HG = 1/2 \times 3 \times (20 - 10) = 15KN$
act at a distance 1/3 of FH (i.e., 1.0m)from point H
Point load of Rectangle $FHDB = DB \times BH = 3 \times 10 = 30KN$
act at a distance 1/2 of DB (i.e., 1.5m)from point D

At Point A roller support i.e., only vertical reaction (R_{AV}), and point B hinged support i.e., a horizontal reaction (R_{BH}) and a vertical reaction (R_{BV}). All the point load are shown in fig 8.53

$$\Sigma H = 0;$$
$$R_{BH} = 0$$
$$R_{BH} = 0 \quad \text{.......ANS}$$
$$\Sigma V = 0;$$
$$R_{AV} + R_{BV} - 5 - 10 - 30 - 15 = 0$$
$$R_{AV} + R_{BV} = 60 KN \quad ...(i)$$

Now taking moment about point 'A'
$$-5 \times 1/3 + 10 \times 0.66 + 30 \times 3.5 + 15 \times 4 - R_{BV} \times 5 = 0$$
$$R_{BV} = 34 \text{ KN} \quad \text{.......ANS}$$

Putting the value of RDV in equation (1)
$$R_{AV} = 26 KN \quad \text{.......ANS}$$

Q. 24: Determine the reactions at the support A, B, C, and D for the arrangement of compound beams shown in fig 8.54

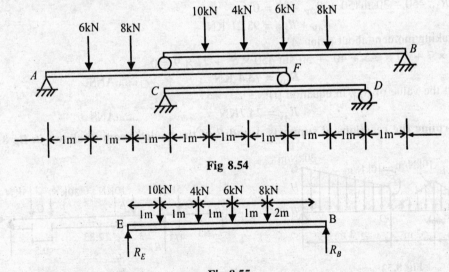

Fig 8.54

Fig 8.55

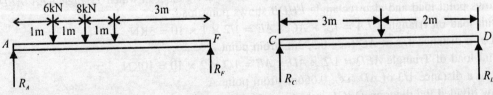

Fig 8.56 Fig 8.57

Sol.: This is the question of multiple beam (i.e., beam on a beam). In this type of question, first consider the top most beam, then second last beam as, In this problem on point E and F, there are roller support, and this support give reaction to both up and down beam. Consider FBD of top most beam EB as shown in fig 8.55

$\Sigma V = 0$;
$R_E + R_B - 10 - 4 - 6 - 8 = 0$
$R_E + R_B = 28$ KN ...(i)

Now taking moment about point 'E'
$10 \times 1 + 4 \times 2 + 6 \times 3 + 8 \times 4 - R_B \times 6 = 0$
$R_B = 11.33$ KN ....ANS

Putting the value of R_B in equation (i)
$R_E = 16.67$ KN ....ANS

Consider FBD of second beam AF as shown in fig 8.56:
$\Sigma V = 0$;
$R_A + R_F - 6 - 8 - R_E = 0$
$R_A + R_F = 30.67$ KN ...(ii)

Now taking moment about point 'A'
$6 \times 1 + 8 \times 2 + 16.67 \times 3 - RF \times 6 = 0$
$RF = 12$ KN ....ANS

Putting the value of R_F in equation (ii)
$R_A = 18.67$ KN ....ANS

Consider FBD of third beam CD as shown in fig 8.57:
$\Sigma V = 0$;
$R_C + R_D - R_F = 0$
$R_C + R_D = 12$ KN ...(iii)

Now taking moment about point 'C'
$- R_D \times 5 + 12 \times 3 = 0$
$R_D = 7.2$ KN ....ANS

Putting the value of R_D in equation (iii)
$R_C = 4.8$ KN ....ANS

Q. 25: Determine the reactions at A,B and D of system shown in fig 8.58. *(Dec-01-02)*

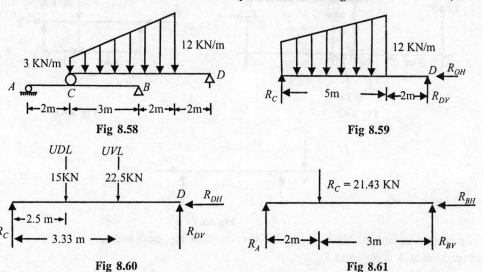

Fig 8.58

Fig 8.59

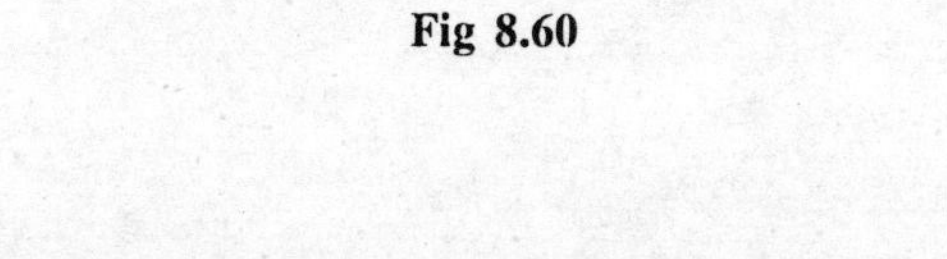

Fig 8.60

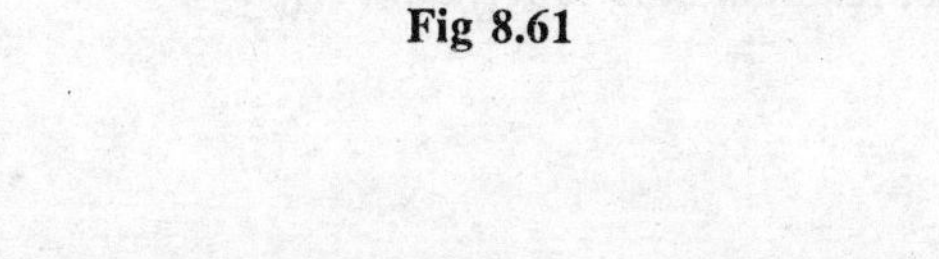

Fig 8.61

Solution: Consider FBD of top most beam as shown in fig 8.59 and 8.60
$$\sum H = 0$$
$$R_{DH} = 0 \quad \text{...(i)}$$
$$\sum V = 0$$
$$R_C + R_{DV} - 15 - 22.5 = 0$$
$$R_C + R_{DV} = 37.5 KN \quad \text{...(ii)}$$
Taking moment about point C:
$$\sum M_C = 15 \times 2.5 + 22.5 \times 3.33 - R_{DV} \times 7 = 0$$
$$R_{DV} = 16.07 KN \quad \text{.......ANS}$$
From equation (ii)
$$R_C = 21.43 KN \quad \text{.......ANS}$$
Consider FBD of bottom beam as shown in fig 8.61
$$\sum H = 0$$
$$R_{BH} = 0 \quad \text{...(iii)}$$
$$\sum V = 0$$
$$R_A + R_{BV} - R_C = 0$$
$$R_A + R_{BV} = 21.43 KN \quad \text{...(iv)}$$
Taking moment about point A:
$$\sum M_A = R_C \times 2 - R_{BV} \times 5 = 0$$
$$R_{BV} = 8.57 KN \quad \text{.......ANS}$$
From equation (iv)
$$R_A = 12.86 KN \quad \text{.......ANS}$$

Q. 26: Determine the reactions at supports A and D in the structure shown in fig–8.62

(Dec–(C.O)–03)

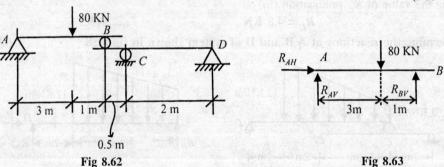

Fig 8.62   Fig 8.63

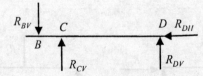

Fig 8.64

Sol.: Since there is composite beam, there fore first consider top most beam,
Let reaction at A is R_{AH} and R_{AV}

Reaction at B is R_{BV}
Reaction at C is R_{AV}
Reaction at D is R_{DH} and R_{DV}
Draw the *FBD* of Top beam as shown in fig 8.63,
$$\Sigma H = 0$$
$$R_{AH} = 0 \qquad \text{......ANS}$$
$$\Sigma V = 0$$
$$R_{AV} + R_{BV} - 80 = 0$$
$$R_{AV} + R_{BV} = 80 \text{KN} \qquad \text{...(i)}$$
Taking moment about point A:
$$\Sigma M_A = 80 \times 3 - R_{BV} \times 4 = 0$$
$$R_{BV} = 60 \text{KN} \qquad \text{......ANS}$$
From (i), $\qquad R_{AV} = 20 \text{KN} \qquad \text{......ANS}$
Consider the *FBD* of bottom beam as shown in fig 8.64,
$$\Sigma H = 0$$
$$R_{DH} = 0 \qquad \text{......ANS}$$
$$\Sigma V = 0$$
$$R_{CV} + R_{DV} - R_{BV} = 0$$
$$R_{CV} + R_{DV} = 60 \text{KN} \qquad \text{...(ii)}$$
Taking moment about point D:
$$\Sigma M_D = -60 \times 4.5 + R_{CV} \times 4 = 0$$
$$R_{CV} = 67.5 \text{KN} \qquad \text{......ANS}$$
From (ii), $\qquad R_{DV} = -7.5 \text{KN} \qquad \text{......ANS}$

Q. 27: Explain Jib crane Mechanism.

Sol.: Jib crane is used to raise heavy loads. A load W is lifted up by pulling chain through pulley D as shown in adjacent figure 8.65. Member CD is known as tie, and member AD is known as jib. Tie is in tension and jib is in compression. AC is vertical post. Forces in the tie and jib can be calculated. Very often chain BD and Tie CD are horizontal. Determination of forces for a given configuration and load is illustrated through numerical examples.

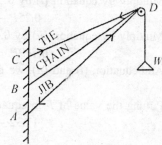

Q. 28: The frictionless pulley A is supported by two bars AB and AC which are hinged at B and C to a vertical wall. The flexible cable DG hinged at D goes over the pulley and supports a load of 20KN at G. The angle between the various members shown in fig 8.66. Determine the forces in AB and AC. Neglect the size of pulley. *(Dec–01–02)*

Sol.: Here the system is jib–crane. Hence Member CA is in compression and AB is in tension. As shown in fig 8.67.

Cable DG goes over the frictionless pulley, so
Tension in AD = Tension in AG
$\qquad = 20 \text{KN}$
FBD of the system is as shown in fig 8.67
$$\Sigma H = 0$$

$P\sin 30° - T\sin 60° - 20\sin 60° = 0$
$0.5P - 0.866T = 17.32 \text{KN}$ ...(i)

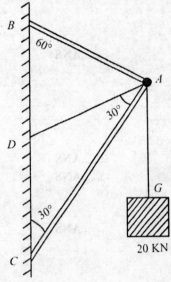

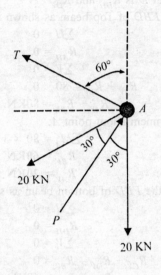

Fig 8.66   Fig 8.67

$\Sigma V = 0$
$P\cos 30° + T\cos 60° - 20 - 20\cos 60° = 0$
$0.866P + 0.5T = 30 \text{KN}$ ...(ii)

Multiply by equation (i) by 0.5 , we get
$0.25P - 0.433T = 8.66$ ...(iii)

Multiply by equation (ii) by 0.866 , we get
$0.749P + 0.433T = 25.98$ ...(iv)

Add equation (i) and (ii), we get
$P = 34.64 \text{KN}$ ....ANS

Putting the value of P in equation (i), we get
$17.32 - 0.866T = 17.32$
$T = 0$ ....ANS

Q. 29: **The lever ABC of a component of a machine is hinged at B, and is subjected to a system of coplanar forces. Neglecting friction, find the magnitude of the force P to keep the lever in equilibrium.**

Sol.: The lever ABC is in equilibrium under the action of the forces 200KN, 300KN, P and R_B, where R_B required reaction of the hinge B on the lever.

Hence the algebraic sum of the moments of above forces about any point in their plane is zero.
Moment of R_B and B is zero, because the line of action of R_B passes through B.
Taking moment about B, we get
$200 \times BE - 300 \times CE - P \times BF = 0$
since $CE = BD$,
$200 \times BE - 300 \times BD - P \times BF = 0$

$$200 \times BC\cos30° - 300 \times BC\sin30° - P \times AB\sin60° = 0$$
$$200 \times 12 \times \cos30° - 300 \times 12 \times \sin30° - P \times 10 \times \sin60° = 0$$

$$P = 32.10 \text{KN} \qquad \text{.......ANS}$$

Let

R_{BH} = Resolved part of R_B along a horizontal direction BE
R_{BV} = Resolved part of R_B along a horizontal direction BD
$\sum H$ = Algebraic sum of the Resolved parts of the forces along horizontal direction
$\sum v$ = Algebraic sum of the Resolved parts of the forces along vertical direction

$$\sum H = 300 + R_{BH} - P\cos20°$$
$$\sum H = 300 + R_{BH} - 32.1\cos20° \qquad ...(i)$$
$$\sum v = 200 + R_{BV} - P\sin20°$$
$$\sum v = 200 + R_{BV} - 32.1\sin20° \qquad ...(ii)$$

Since the lever ABC is in equilibrium

$$\sum H = R_V = 0, \text{ We get}$$
$$R_{BH} = -269.85 \text{KN}$$
$$R_{BV} = -189.021 \text{KN}$$
$$R_B = \{(RBH)2 + (RBV)2\}1/2$$
$$R_B = \{(-269.85)2 + (-189.02)2\}1/2$$
$$R_B = 329.45 \text{KN} \qquad \text{.......ANS}$$

Let θ = Angle made by the line of action of R_B with the horizontal
Then, $\tan\theta = R_{BV}/R_{BH} = -189.021/-269.835$

$$\theta = 35.01° \qquad \text{.......ANS}$$

Chapter 9

FRICTION

Q. 1: Define the term friction?
Sol.: When a body moves or tends to move over another body, a force opposing the motion develops at the contact surfaces. This force, which opposes the movement or the tendency of movement, is called frictional force or simply friction. Frictional force always acts parallel to the surface of contact, opposite to the moving direction and depends upon the roughness of surface.

A frictional force develops when there is a relative motion between a body and a surface on application of some external force.

A frictional force depends upon the coefficient of friction between the surface and the body which can be minimized up to a very low value equal to zero (theoretically) by proper polishing the surface.

Q. 2: Explain with the help of neat diagram, the concept of limiting friction.
Sol.: The maximum value of frictional force, which comes into play, when a body just begins to slide over the surface of the other body, is known as limiting friction. Consider a solid body placed on a horizontal plane surface.

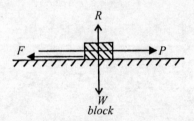

Fig 9.1

Let
W = Weight of the body acting through C.G. downwards.
R = Normal reaction of body acting through C.G. downwards.
P = Force acting on the body through C.G. and parallel to the horizontal surface.
F = Limiting force of friction

If 'P' is small, the body will not move as the force of friction acting on the body in the direction opposite to 'P' will be more than 'P'. But if the magnitude of 'P' goes on increasing a stage comes, when the solid body is on the point of motion. At this stage, the force of friction acting on the body is called **'LIMITING FORCE OF FRICTION (F)'**.
$R = W; F = P$

If the magnitude of 'P' is further increased the body will start moving. The force of friction, acting on the body is moving, is called KINETIC FRICTION.

Q. 3: Differentiate between;
   (a) Static and Kinetic Friction
   (b) Sliding and rolling Friction.

Sol.: (*a*) **Static Friction:** When the applied force is less than the limiting friction, the body remains at rest and such frictional force is called static friction and this law is known as law of static friction.

It is the friction experienced by a body, when it is at rest. Or when the body tends to move.

Kinetic (Dynamic) Friction

When the applied force exceeds the limiting friction the body starts moving over the other body and the friction of resistance experienced by the body while moving. This is known as law of Dynamic or kinetic friction.

Or

It is the friction experienced by a body when in motion. It is of two type;
1. Sliding Friction
2. Rolling Friction

(*b*) **Sliding Friction:** It is the friction experienced by a body, when it slides over another body.

Rolling Friction

It is the friction experienced by a body, when it rolls over the other.

Q. 4: Explain law of coulomb friction? What are the factor affecting the coefficient of friction and effort to minimize it.

Sol.: Coulomb in 1781 presented certain conclusions which are known as Coulomb's law of friction. These conclusions are based on experiments on block tending to move on flat surface without rotation. These laws are applicable at the condition of impending slippage or once slippage has begun. The laws are enunciated as follows:
1. The total force of friction that can be developed is independent of area of contact.
2. For low relative velocities between sliding bodies, total amount of frictional force is independent of the velocity. But the force required to start the motion is greater than that necessary to maintain the motion.
3. The total frictional force that can be developed is proportional to the normal reaction of the surface of contact.

So, coefficient of friction(μ) is defined as the ratio of the limiting force of friction (F) to the normal reaction (R) between two bodies.

Thus,

μ = Limiting force of friction/ Normal reaction

= F/R

or, $F = \mu.R$, Generally $\mu < 1$

The factor affecting the coefficient of friction are:
1. The material of the meeting bodies.
2. The roughness/smoothness of the meeting bodies.
3. The temperature of the environment.

Efforts to minimize it:
1. Use of proper lubrication can minimize the friction.
2. Proper polishing the surface can minimize it.

Q 5: Define the following terms;
   (a) Angle of friction
   (b) Angle of Repose
   (c) Cone of Friction

Sol.: (a) **Angle of Friction (θ)**

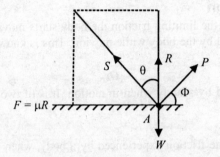

Fig 9.2

It is defined as the angle made by the resultant of the normal reaction (R) and the limiting force of friction (F) with the normal reaction (R).

Let, S = Resultant of the normal reaction (R) and limiting force of friction (F)
θ = Angle between S and R
Tan θ = F/R = μ
Note: The force of friction (F) is always equal to μR

(b) **Angle of Repose (α)**

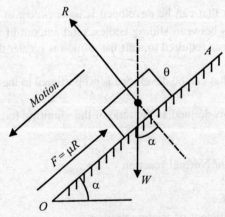

Fig 9.3

It is the max angle of inclined plane on which the body tends to move down the plane due to its own weight.

Consider the equilibrium of the body when body is just on the point of slide.

Resolving all the forces parallel and perpendicular to the plane, we have:
$$\mu R = W.\sin a$$
$$R = W.\cos a$$
Dividing 1 by 2 we get Tan a μ
But $\mu = \tan \theta$, θ = Angle of friction
i.e., $\theta = \alpha$

The value of angle of repose is the same as the value of limiting angle of friction.

(c) **Cone of Friction:** When a body is having impending motion in the direction of P, the frictional force will be the limiting friction and the resultant reaction R will make limiting friction angle θ with the normal. If the body is having impending motion in some other direction, again the resultant reaction makes limiting frictional angle θ with the normal in that direction. Thus, when the direction of force P is gradually changed through 360°, the resultant R generates a right circular cone with semi-central angle equal to θ.

If the resultant reaction is on the surface of this inverted right circular cone whose semi-central angle is limiting frictional angle (θ), the motion of body is impending. If the resultant is within this cone, the body is stationary. This inverted cone with semi-central angle, equal to limiting frictional angle θ, is called cone of friction.

It is defined as the right circular cone with vertex at the point of contact of the two bodies (or surfaces), axis in the direction of normal reaction (R) and semi-vertical angle equal to angle of friction (θ). Fig (9.4) shows the cone of friction in which,

O = Point of contact between two bodies.
R = Normal reaction and also axis of the cone of friction.
θ = Angle of friction

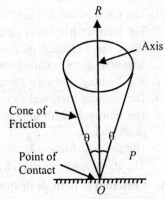

Fig 9.4

Q. 6: What are the types of Friction?
Sol.: There are mainly two types of friction,
  (i) Dry Friction   (ii) Fluid Friction
  (i) **Dry Friction:** Dry friction (also called coulomb friction manifests when the contact surfaces are dry and there is tendency for relative motion.
  Dry friction is further subdivided into:

Sliding Friction
Fiction between two surfaces when one surface slides over another.

Rolling Friction
Friction between two surfaces, which are separated by balls or rollers.

It may be pointed out that rolling friction is always less than sliding friction.

(ii) **Fluid Friction:** Fluid friction manifests when a lubricating fluid is introduced between the contact surfaces of two bodies.

If the thickness of the lubricant or oil between the mating surfaces is small, then the friction between the surfaces is called **GREASY OR NON-VISCOUS FRICTION.** The surfaces absorb the oil and the contact between them is no more a metal-to-metal contact. Instead the contact is through thin layer of oil and that ultimately results is less friction.

When a thick film of lubricant separates the two surfaces, metallic contact is entirely non-existent. The friction is due to viscosity of the oil, or the shear resistance between the layers of the oil rubbing against each other. Obviously then these occurs a great reduction in friction. This frictional force is known as **Viscous Or Fluid Friction.**

Q. 7: Explain the laws of solid friction?
Sol.: The friction that exists between two surfaces, which are not lubricated, is known as solid friction. The two Surfaces may be at rest or one of the surface is moving and other surface is at rest. The following are the laws of solid friction.
1. The force of friction acts in the opposite direction in which surface is having tendency to move.
2. The force of friction is equal to the force applied to the surfaces, so long as the surface is at rest.
3. When the surface is on the point of motion, the force of friction is maximum and this maximum frictional force is called the limiting friction force.
4. The limiting frictional force bears a constant ratio to the normal reaction between two surfaces.
5. The limiting frictional force does not depend upon the shape and areas of the surfaces in contact.
6. The ratio between limiting friction and normal reaction is slightly less when the two surfaces are in motion.
7. The force of friction is independent of the velocity of sliding.

The above laws of solid friction are also called laws of static and dynamic friction or law of friction.

Q. 8: "Friction is both desirable and undesirable" Explain with example
Sol.: Friction is Desirable: A friction is very much desirable to stop the body from its moving condition. If there is no friction between the contact surfaces, then a body can't be stopped without the application of external force. In the same time a person can't walk on the ground if there is no friction between the ground and our legs also no vehicle can run on the ground without the help of friction.

Friction is Undesirable: A friction is undesirable during ice skating or when a block is lifted or put down on the truck with the help of some inclined plane. If the friction is more between the block and inclined surface, then a large force is required to push the block on the plane.

Thus friction is desirable or undesirable depending upon the condition and types of work.

Q. 9: A body on contact with a surface is being pulled along it with force increasing from zero. How does the state of motion of a body change with force. Draw a graph and explain.
Sol.: When an external force is applied on a body and increases gradually then initially a static friction force acts on the body which is exactly equal to the applied force and the body will remain at rest. The graph is a straight line for this range of force shown by *OA* on the graph. When the applied force reaches to a value at which body just starts moving, then the value of this friction force is known as limiting friction. Further increase in force will cause the motion of the body and the friction in this case will be dynamic friction. This dynamic friction remains constant with further increase in force.

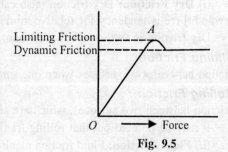

Fig. 9.5

Points to be Remembered
1. If applied force is not able to start motion; frictional force will be equal to applied force.
2. If applied force is able to start motion, and then applied force will be greater than frictional force.
3. The answer will never come in terms of normal reaction.
4. Assuming the body is in limiting equilibrium.

Solved Problems on Horizontal Plane

Q. 10: A body of weight 100N rests on a rough horizontal surface (μ = 0.3) and is acted upon by a force applied at an angle of 300 to the horizontal. What force is required to just cause the body to slide over the surface?

Sol.: In the limiting equilibrium, the forces are balanced. That is

$$\Sigma H = 0;$$
$$F = P\cos\theta$$
$$\Sigma V = 0;$$
$$R = W - P\sin\theta$$

Also
$$F = \mu R$$
$$P.\cos\theta = \mu(W - P.\sin\theta)$$
$$P.\cos\theta = \mu.W - \mu.P.\sin\theta$$
$$\mu.P.\sin\theta + P.\cos\theta = \mu.W$$
$$P(\mu.\sin\theta + \cos\theta) = \mu.W$$
$$P = \mu.W/(\cos\theta + \mu.\sin\theta)$$
$$= 0.3 \times 100 / (\cos 30° + 0.3\sin 30°)$$
$$= 29.53N \quad \text{.......Ans}$$

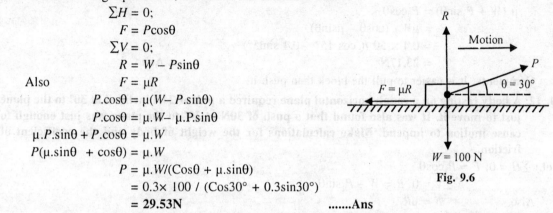

Fig. 9.6

Q. 11: A wooden block of weight 50N rests on a horizontal plane. Determine the force required which is acted at an angle of 150 to just (*a*) Pull it, and (*b*) Push it. Take coefficient friction = 0.4 between the mating surfaces. Comment on the result.

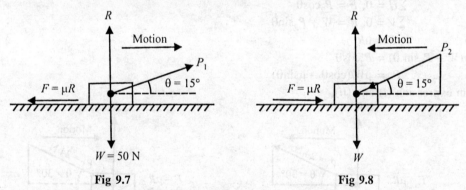

Fig 9.7   Fig 9.8

Sol.: Let P_1 be the force required to just pull the block. In the limiting equilibrium, the forces are balanced. That gives

$$\Sigma H = 0; \; F = P_1\cos\theta$$
$$\Sigma V = 0; \; R = W - P_1\sin\theta$$

Also
$$F = \mu R$$
$$\mu(W - P_1\sin\theta) = P_1\cos\theta$$

or
$$P_1 = \mu W / (\cos\theta + \mu\sin\theta)$$
$$= 0.4 \times 50 /(\cos 15° + 0.4 \sin 15°)$$
$$= 18.70N \quad \text{.......Ans}$$

(*b*) Let P_2 be the force required to just push the block. With reference to the free body diagram (Fig. 9.8),

Let us write the equations of equilibrium,
$$\Sigma H = 0; \quad F = P_2\cos\theta$$
$$\Sigma V = 0; \quad R = W + P_2\sin\theta$$
Also $\quad F = \mu R$
$$\mu(W + P_2\sin\theta) = P_2\cos\theta$$
or $\quad P_2 = \mu W / (\cos\theta - \mu\sin\theta)$
$$= 0.4 \times 50 /(\cos 15° - 0.4 \sin15°)$$
$$= 23.17 N \quad\quad\quad\quad\quad ....\text{Ans}$$

Comments. It is easier to pull the block than push it.

Q. 12: *A body resting on a rough horizontal plane required a pull of 24N inclined at 30° to the plane just to move it. It was also found that a push of 30N at 30° to the plane was just enough to cause motion to impend. Make calculations for the weight of body and the coefficient of friction.*

Sol.: $\Sigma H = 0; \quad F = P_1\cos\theta$
$$\Sigma V = 0; \quad R = W - P_1\sin\theta$$
Also $\quad F = \mu R$
$$\mu(W - P\sin\theta) = P_1\cos\theta$$
or $\quad P_1 = \mu W / (\cos\theta + \mu\sin\theta) \quad\quad\quad ...(i)$

With reference to the free body diagram (Fig (9.9) when push is applied)
$$\Sigma H = 0; \quad F = P_2\cos\theta$$
$$\Sigma V = 0; \quad R = W + P_2\sin\theta$$
Also $\quad F = \mu R$
$$\mu(W + P_2\sin\theta) = P_2\cos\theta$$
$$P_2 = \mu W/(\cos\theta - \mu\sin\theta) \quad\quad\quad ...(ii)$$

From expression (i) and (ii),

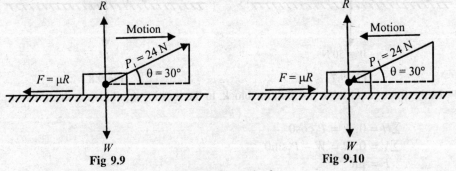

Fig 9.9   Fig 9.10

$$P_1/P_2 = (\cos\theta - \mu\sin\theta)/(\cos\theta + \mu\sin\theta)$$
$$24/30 = (\cos 30° - \mu\sin 30°)/(\cos 30° + \mu\sin 30°)$$
$$= (0.866 - 0.5\mu)/(0.866 + 0.5\mu)$$
$$0.6928 + 0.4\mu = 0.866 - 0.5\mu$$

On solving
$$\mu = 0.192 \quad\quad\quad\quad\quad ....\text{Ans}$$

Putting the value of μ in equation (i) we get the value of W
$$W = 120.25N \quad\quad\quad\quad\quad ....\text{Ans}$$

Q. 13: A block weighing 5KN is attached to a chord, which passes over a frictionless pulley, and supports a weight of 2KN. The coefficient of friction between the block and the floor is 0.35. Determine the value of force P if (*i*) The motion is impending to the right (*ii*) The motion is impending to the left.

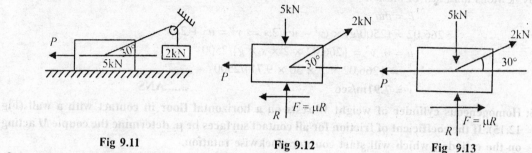

Fig 9.11   Fig 9.12   Fig 9.13

Sol.: Case-1
From the *FBD* of the block,
$$\Sigma V = 0 \rightarrow -5 + R + 2\sin 30° = 0$$
$$R = 4KN$$
$$\Sigma H = 0 \rightarrow -P + 2\cos 30° - 0.35N = 0$$
$$P = 2\cos 30° - 0.35 \times 4 = 0$$
$$\mathbf{P = 0.332KN} \qquad \text{.......Ans}$$

Case-2: Since the motion impends to the left, the friction force is directed to the right, from the *FBD* of the block:
$$\Sigma V = 0 \rightarrow -5 + R + 2\sin 30° = 0$$
$$R = 4KN$$
$$\Sigma H = 0 \rightarrow -P + 2\cos 30° + 0.35N = 0$$
$$P = 2\cos 30° + 0.35 \times 4 = 0$$
$$\mathbf{P = 3.132KN} \qquad \text{.......Ans}$$

Q. 14: A block of 2500N rest on a horizontal plane. The coefficient of friction between block and the plane is 0.3. The block is pulled by a force of 1000N acting at an angle 30° to the horizontal. Find the velocity of the block after it moves over a distance of 30m, starting from rest.

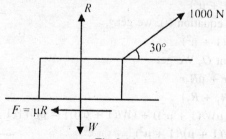

Fig 9.14

Sol.: Here $\Sigma V = 0$ but $\Sigma H \neq 0$, Because ΣH is converted into ma
$$\Sigma V = 0$$
$$R + 1000\sin 30° - W = 0, \ W = 2500N$$
$$R = 2000N$$
...(*i*)

$\Sigma H \neq 0$

$\Sigma H = \mu R - 1000\cos 30° = 266.02 N$ ...(ii)

Since $\Sigma H \neq 0$

By newtons third law of motion

$F = ma$

$266.02 = (2500/g) \times (v^2 - u^2)/2.s \Rightarrow v^2 = u^2 + 2as$

$u = 0$, $v^2 = \{266.02 \times 2 \times s \times g\}/2500$

$v^2 = \{266.02 \times 2 \times 30 \times 9.71\}/2500$

$v = 7.91 m/sec$ ....ANS

Q. 15: Homogeneous cylinder of weight W rests on a horizontal floor in contact with a wall (Fig 12.15). If the coefficient of friction for all contact surfaces be μ, determine the couple M acting on the cylinder, which will start counter clockwise rotation.

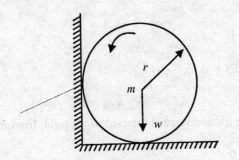

Fig 9.15   Fig 9.16

Sol.: $\Sigma H = 0 \Rightarrow R_1 - \mu R_2 = 0$

$R_1 = \mu R_2$ ...(i)

$\Sigma V = 0 \Rightarrow R_2 + \mu R_1 = W$ ...(ii)

Putting the value of R_1 in equation (ii), we get

$R_2 + \mu 2 R_2 = W$

$R_2 = W/(1 + \mu^2)$ ...(ii)

Putting the value of R_2 in equation (i), we get

$R_1 = \mu W/(1 + \mu^2)$ ...(iv)

Taking moment about point O, We get

$M_O = \mu R_1 r + \mu R_2 r$

$= \mu r\{R_1 + R_2\}$

$= \mu r\{(\mu W/(1 + \mu^2)) + (W/(1 + \mu^2))\} = \mu r W\{(1 + \mu)/(1 + \mu^2)\}$

$M_O = \mu r W(1 + \mu)/(1 + \mu^2)$ ....ANS

Q. 16: A metal box weighing 10KN is pulled along a level surface at uniform speed by applying a horizontal force of 3500N. If another box of 6KN is put on top of this box, determine the force required.

Friction / 199

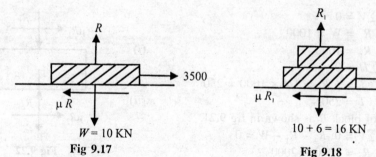

Fig 9.17   Fig 9.18

Sol.: In first case as shown in fig 12.17

$\Sigma H = 0$
$\mu R = 3500$ ...(i)
$\Sigma V = 0$
$R = W = 10KN = 10000$ ...(ii)

Putting the value of R in equation (i)
$\mu = 0.35$ ...(iii)

Now consider second case: as shown in fig 9.18
Now normal reaction is N_1,

$\Sigma H = 0$
$P - \mu R_1 = 0$
$P = \mu R_1$ ...(iv)
$\Sigma V = 0$
$R_1 = W = 10KN + 6KN = 16000$
$R_1 = 16000$ ...(v)

Putting the value of R_1 in equation (iv)
$P = 0.35 \times 16000$
$P = 5600N$ ....ANS

Q. 17 Block A weighing 1000N rests over block B which weights 2000N as shown in fig (9.19). Block A is tied to wall with a horizontal string. If the coefficient of friction between A and B is 1/4 and between B and floor is 1/3, what should be the value of P to move the block B. If (1) P is horizontal (2) P is at an angle of 300 with the horizontal.

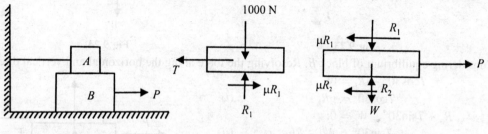

Fig 9.19   Fig 9.20   Fig 9.21

Sol.: (a) When P is horizontal
Consider FBD of block A as shown in fig 12.20.

$\Sigma V = 0$
$R_1 = W = 1000$
$R_1 = 1000$ ...(i)
$\Sigma H = 0$
$T = \mu_1 R_1 = 1/4 \times 1000 = 250$
$T = 250N$ ...(ii)

Consider FBD of block B as shown in fig 9.21.
$\Sigma V = 0; R_2 - R_1 - W = 0$
$R_2 = 1000 + 2000$
$R_2 = 3000 \, N$ ...(iii)
$\Sigma H = 0$
$P = \mu_1 R_1 + \mu_2 R_2 = 250 + 1/3 \times 3000$
$P = 1250N$ ....ANS

(2) When P is inclined at an angle of 30° Consider fig 9.22
$\Sigma H = 0$
$P\cos 30° = \mu_1 R_1 + \mu_2 R_2 = 250 + 1/3 \times R_2$
$R_2 = 3(P\cos 30° - 250)$ ...(iv)
$\Sigma V = 0$
$R_2 - R_1 - W + P\sin 30° = 0$
$R_2 + P\sin 30° = R_1 + W = 3000$ ...(v)

Putting the value of R_2 in equation (v)
$3(P\cos 30° - 250) + 0.5 \times P = 3000$
On solving $P = 1210.43N$ ....ANS

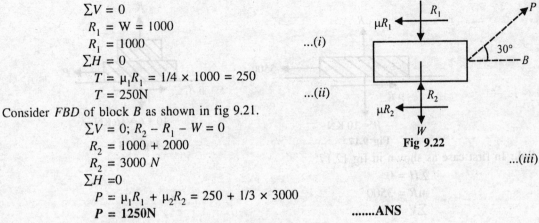

Fig 9.22

Q. 18: Two blocks A and B of weight 4KN and 2KN respectively are in equilibrium position as shown in fig (9.23). Coefficient of friction for both surfaces are same as 0.25, make calculations for the force P required to move the block A.

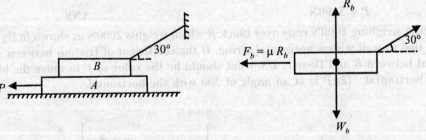

Fig 9.23 Fig 9.24

Sol.: Considering equilibrium of block B. Resolving the force along the horizontal and vertical directions:
$T\cos 30° - \mu R_b = 0;$
$T\cos 30° = \mu R_b$ ...(i)
$R_b + T\sin 30° - W_b = 0;$
$T\sin 30° = W_b - R_b$ ...(ii)

Dividing Equation (i) and (ii), We get
$\tan 30° = (W_b - R_b)/\mu R_b$

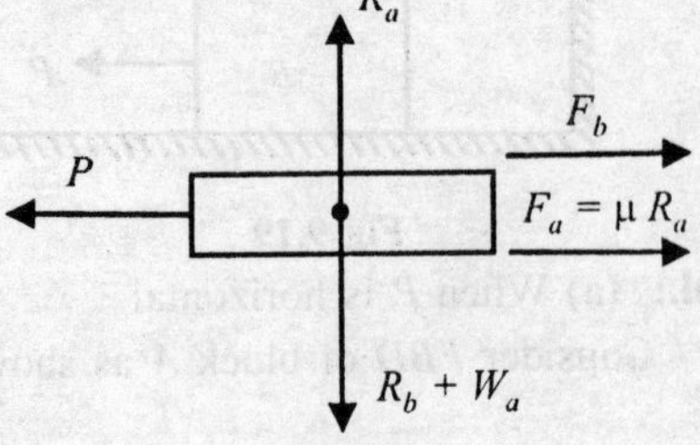

Fig. 9.25

$0.5773 = (2 - R_b)/0.25R_b$;
$0.1443R_b = 2 - R_b$
$R_b = 1.748N$
$F_b = \mu R_b = 0.25 \times 1.748 = 0.437N$

Considering the equilibrium of block A: Resolving the forces along the horizontal and vertical directions,
$F_b + \mu R_a - P = 0$; $P = F_b + \mu R_a$
$R_a - R_b - W_a = 0$; $R_a = R_b + W_a = 1.748 + 4 = 5.748$
$P = 0.437 + 0.25 \times 5.748$
$P = 1.874N$ ....Ans

Q. 19: Determine the force P required to impend the motion of the block B shown in fig (9.26). Take coefficient of friction = 0.3 for all contact surface.

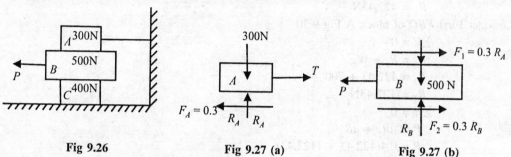

Fig 9.26   Fig 9.27 (a)   Fig 9.27 (b)

Sol.: Consider First *FBD* of block A Fig 12.27 (a)
$\Sigma V = 0 \rightarrow R_A = 300N$
$\Sigma H = 0 \rightarrow T = 0.3 N_A$
$T = 90N$

Consider *FBD* of Block B
$\Sigma V = 0 \rightarrow R_B = R_A + 500$
$R_B = 800N$
$\Sigma H = 0 \rightarrow P = 0.3 N_A + 0.3 R_B$
$= 0.3 (300 + 800)$
$P = 330N$ ....Ans

Q. 20: Block A of weight 520N rest on the horizontal top of block B having weight 700N as shown in fig (9.28). Block A is tied to a support C by a cable at 300 horizontally. Coefficient of friction is 0.4 for all contact surfaces. Determine the minimum value of the horizontal force P just to move the block B. How much is the tension in the cable then.

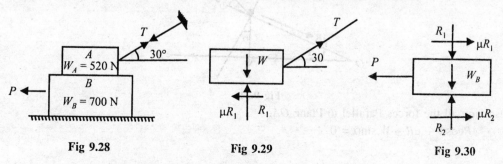

Fig 9.28   Fig 9.29   Fig 9.30

Sol.: Consider First *FBD* of block A Fig 9.29

$$\Sigma H = 0$$
$$\mu R_1 = T\cos 30°$$
$$0.4 R_1 = 0.866 T$$
$$R_1 = 2.165 T \qquad ...(i)$$
$$\Sigma V = 0$$
$$W = R_1 + T\sin 30°$$
$$520 = 2.165 T + 0.5 T$$
$$520 = 2.665 T$$
$$T = 195.12 N \qquad ...(ii)$$

Putting in (*i*) we get
$$R_1 = 422.43 N \qquad ...(iii)$$

Consider First *FBD* of block A Fig 9.30
$$\Sigma V = 0$$
$$R_2 = R_1 + W_B$$
$$R_2 = 422.43 + 700$$
$$R_2 = 1122.43 N \qquad ...(iv)$$
$$\Sigma H = 0$$
$$P = \mu R_1 + \mu R_2$$
$$P = 0.4(422.43 + 1122.43)$$
$$P = 617.9 N \qquad ....ANS$$

Q. 21: Explain the different cases of equilibrium of the body on rough inclined plane.

Sol.: If the inclination is less than the angle of friction, the body will remain in equilibrium without any external force. If the body is to be moved upwards or downwards in this condition an external force is required. But if the inclination of the plane is more than the angle of friction, the body will not remain in equilibrium. The body will move downward and an upward external force will be required to keep the body in equilibrium.

Such problems are solved by resolving the forces along the plane and perpendicular to the planes. The force of friction (*F*), which is always equal to $\mu.R$ is acting opposite to the direction of motion of the body

CASE -1: magnitude of minimum force '*p*' which is required to move the body up the plane. When '*p*' is acted with an angle of φ.

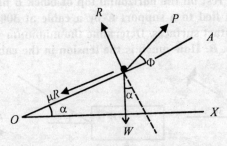

Fig 9.31

Resolving all the forces Parallel to Plane *OA*:
$$P\cos\Phi - \mu R - W.\sin\alpha = 0 \qquad ...(i)$$

Resolving all the forces Perpendicular to Plane OA:
$$R + P\sin F - W.\cos\alpha = 0 \quad ...(i)$$
Putting value of 'R' from (ii) in equation (i) we get
$$P = W.[(\mu.\cos\alpha + \sin\alpha)/(\mu.\sin\Phi + \cos\Phi)]$$
Now putting $\mu = \tan\theta$, on solving
$$P = W.[\sin(\alpha + \theta)/\cos(\Phi - \theta)] \quad ...(iii)$$
Now P is minimum at $\cos(\Phi - \theta)$ is max
i.e., $\cos(\Phi - \theta) = 1$ or $\Phi - \theta = 0$ i.e., $\Phi = \theta$
$$P_{min} = W.\sin(\theta + \Phi)$$

CASE-2: magnitude of force 'p' which is required to move the body down the plane. When 'p' is acted with an angle of φ.

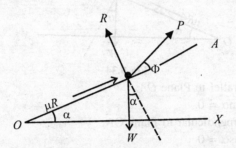

Fig 9.32

Resolving all the forces Parallel to Plane OA:
$$P\cos\Phi + \mu R - W.\sin\alpha = 0 \quad ...(i)$$
Resolving all the forces Perpendicular to Plane OA:
$$R + P\sin\Phi - W.\cos\alpha = 0 \quad ...(ii)$$
Putting value of 'R' from (ii) in equation (i) we get
$$P = W.[(\sin\alpha - \mu.\cos\alpha)/(\cos\Phi - \mu.\sin\Phi)]$$
Now putting $\mu = \tan\theta$, on solving
$$P = W.[\sin(\alpha - \Phi)/\cos(\Phi + \theta)]$$

CASE-3: magnitude of force 'p' which is required to move the body down the plane. When 'p' is acted horizontally

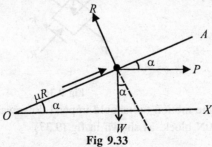

Fig 9.33

Resolving all the forces Parallel to Plane OA:
$$P\cos\alpha + \mu R - W.\sin\alpha = 0 \quad ...(i)$$
Resolving all the forces Perpendicular to Plane OA:

$R - P\sin\alpha - W.\cos\alpha = 0$ ...(ii)

Putting value of 'R' from (ii) in equation (i) we get

$P = W.[(\sin\alpha - \mu.\cos\alpha)/(\cos\alpha + \mu.\sin\alpha)]$

Now putting $\mu = \tan\theta$, on solving,

$P = W.\tan(\alpha - \theta)$

CASE-4: magnitude of force 'p' which is required to move the body up the plane. When 'p' is acted horizontally

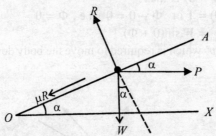

Fig 9.34

Resolving all the forces Parallel to Plane OA:

$P\cos\alpha - \mu R - W.\sin\alpha = 0$ ...(i)

Resolving all the forces Perpendicular to Plane OA:

$R - P\sin\alpha - W.\cos\alpha = 0$ ...(ii)

Putting value of 'R' from (ii) in equation (i) we get

$P = W.[(\sin\alpha + \mu.\cos\alpha)/(\cos\alpha - \mu.\sin\alpha)]$

Now putting $\mu = \tan\theta$, on solving,

$P = W.\tan(\alpha + \theta)$

Problems on Rough Inclined Plane

Q. 22: Determine the necessary force P acting parallel to the plane as shown in fig 9.35 to cause motion to impend. $\mu = 0.25$ and pulley to be smooth.

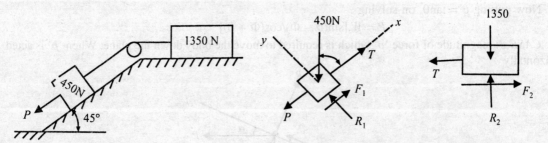

Fig 9.35   Fig 9.36   Fig 9.37

Sol.: Since P is acting downward; the motion too should impend downwards.

Consider first the *FBD* of 1350N block, as shown in fig (9.37)

$\Sigma V = 0$

$R_2 - W = 0$

$R_2 = 1350N$ ...(i)

$\Sigma H = 0$

$-T + \mu R_2 = 0$

Putting the value of R_2 and μ

$T = 0.25(1350)$
$= 337.5N$ ...(ii)

Now Consider the FBD of 450N block, as shown in fig (9.36)

$\Sigma V = 0$
$R_1 - 450\sin 45° = 0$
$R_1 = 318.2$ N ...(i)
$\Sigma H = 0$
$T - P + \mu R_1 - 450\sin 45° = 0$

Putting the value of R_1, μ and T we get

$P = T + \mu R_1 - 450\sin 45° = 0$
$= 337.5 + 0.25 \times 318.2 - 450\sin 45°$
$P = 98.85$ N ....ANS

Q. 23: Determine the least value of W in fig(9.38) to keep the system of connected bodies in equilibrium μ for surface of contact between plane AC and block = 0.28 and that between plane BC and block = 0.02

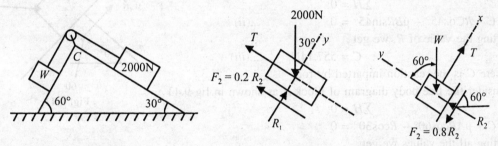

Fig 9.38   Fig 9.39   Fig 9.40

Sol.: For least value W, the motion of 2000N block should be impending downward.

From FBD of block 2000N as shown in fig 12.39

$\Sigma V = 0$
$R_1 - 2000\cos 30° = 0$
$R_1 = 1732.06N$ ...(i)
$\Sigma H = 0$
$T + \mu_1 R_1 - 2000\sin 30° = 0$
$T = 2000\sin 30° - 0.20 \times 1732.06, T = 653.6N$ ...(ii)

Now Consider the FBD of WN block, as shown in fig (9.40)

$\Sigma V = 0$
$R_2 = W\cos 60° = 0$
$R_2 = 0.5W$ N ...(iii)
$\Sigma H = 0$
$T - \mu_2 R_2 - W\sin 60° = 0$
$653.6 = W\sin 60° - 0.28 \times 0.5W$
$W_{LEAST} = 649.7N$ ....ANS

Q. 24: Block A and B connected by a rigid horizontally bar planed at each end are placed on inclined planes as shown in fig (9.41). The weight of the block B is 300N. Find the limiting values of the weight of the block A to just start motion of the system.

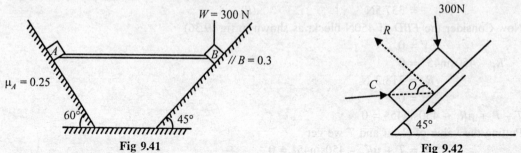

Fig 9.41   Fig 9.42

Sol.: Let W_a be the weight of block A. Consider the free body diagram of B. As shown in fig 12.42. And Assume A_B be the Axis of reference.

$$\Sigma V = 0;$$
$$R\sin45° - \mu BR\cos45° - 300 = 0$$
On solving, $R = 606.09N$ ...(i)
$$\Sigma H = 0;$$
$$C - R\cos45° - \mu BR\sin45° = 0$$ ...(ii)
Putting the value of R, we get
$$C = 557.14N$$ ...(iii)
Where C is the reaction imparted by rod.
Consider the free body diagram of block A as shown in fig 9.43
$$\Sigma H = 0;$$
$$C + \mu AR\cos60° - R\cos30° = 0$$ ...(iv)
Putting all the values we get
$$R = 751.85N$$ ...(v)
$$\Sigma V = 0;$$
$$\mu_A R\sin60° + R\sin60° - W = 0$$
On solving, $W = 538.7N$ ...(vi)
Hence weight of block   A = **538.7N**   ....ANS

Q. 25: What should be the value of the angle shown in fig 9.44 so that the motion of the 90N block impends down the plane? The coefficient of friction for the entire surface = 1/3.

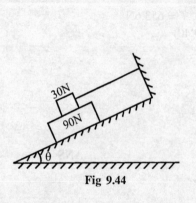

Fig 9.44

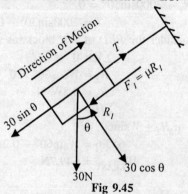

Fig 9.45

Sol.: Consider the equilibrium of block 30N
$$\Sigma V = 0;$$
$$R_1 - 30\cos\theta = 0,$$
$$R_1 = 30\cos\theta \qquad ...(i)$$
$$\Sigma H = 0;$$
$$T - \mu R_1 - 30\sin\theta = 0,$$
$$T = 10\cos\theta + 30\sin\theta \qquad ...(ii)$$
Consider the equilibrium of block 90N
$$\Sigma V = 0;$$
$$R_2 - R_1 - 90\cos\theta = 0$$
$$R_2 = 120\cos\theta \qquad ...(iii)$$
$$\Sigma H = 0;$$
$$90\sin\theta - \mu R_1 - \mu R_2 = 0,$$
$$90\sin\theta = 10\cos\theta + 40\sin\theta \qquad ...(vi)$$
$$\tan\theta = 5/9 \text{ i.e., } \theta = 29.050$$

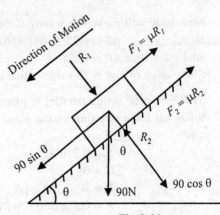

Fig 9.46

Q. 26: A block weighing 200N is in contact with an inclined plane (Inclination = 30°). Will the block move under its own weight. Determine the minimum force applied (1) parallel (2) perpendicular to the plane to prevent the motion down the plane. What force P will be required to just cause the motion up the plane, $\mu = 0.25$?

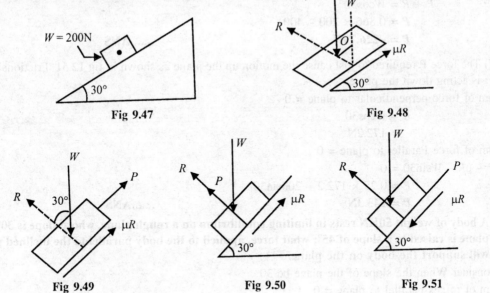

Fig 9.47   Fig 9.48

Fig 9.49   Fig 9.50   Fig 9.51

Sol.: Consider the *FBD* of block as shown in fig 9.48
From the equilibrium condition
Sum of forces perpendicular to plane = 0
$$R - W\cos 30° = 0;$$
$$R = W\cos 30° \qquad ...(i)$$
Sum of forces parallel to plane = 0
$$\mu R - W\sin 30° = 0 \qquad ...(ii)$$

Now body will move down only if the value of μR is less than Wsin30°
Now,   μR = 0.25 × W(0.866) = 0.2165W   ...(iii)
And   Wsin30° = 0.5W   ...(vi)
Since value of (iv) is less than value of (iii) So the body will move down.

(i) When Force acting parallel to plane as shown in fig 9.49
Frictional force is acting up the plane
$$\Sigma V = 0;$$
$$R = 0.216W$$ ...(v)
$$\Sigma H = 0;$$
$$P = W\sin 30° - \mu R$$
$$P = 0.5 \times 200 - 0.216 \times 200$$
$$P = 56.7N$$ ....ANS

(ii) When Force acting perpendicular to plane as shown in fig 12.50
Frictional force is acting up the plane
$$\Sigma H = 0;$$
$$W\sin 30° - \mu R = 0$$
$$R = 400$$ ...(vii)
$$\Sigma V = 0;$$
$$P + R = W\cos 30°$$
$$P = 0.866 \times 200 - 400$$
$$P = -226.79N$$ ....ANS

(iii) The force P required to just cause the motion up the plane as shown in fig 12.51. Frictional force is acting down the plane
Sum of force perpendicular to plane = 0
$$R = W\cos 30$$
$$= 172.2N$$ ...(viii)
Sum of force Parallel to plane = 0
$$P - \mu R - W\sin 30 = 0$$
$$P = 0.25 \times 172.2 - 200\sin 30$$
$$P = 143.3N$$ ....ANS

Q. 27: A body of weight 50KN rests in limiting equilibrium on a rough plane, whose slope is 30°. The plane is raised to a slope of 45°; what force, applied to the body parallel to the inclined plane, will support the body on the plane.

Sol.: Consider When the slope of the plane be 30°
Sum of forces parallel to plane = 0
$$\mu R - W\sin 30° = 0$$ ...(i)
Sum of forces perpendicular to plane = 0
$$R - W\cos 30° = 0$$
$$R = W\cos 30°$$ ...(ii)
Putting the value of R in equation (i) We get
$$\mu = \tan 30° = 0.577$$ ...(iii)

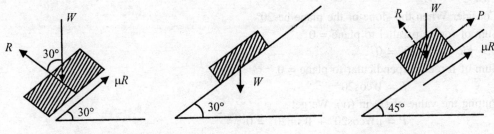

Fig. 9.52   Fig. 9.53   Fig. 9.54

Now consider the case when the slope is 45°, Let Force P required to support the body.
In this case
Sum of forces parallel to plane = 0
$P - \mu R - W\sin 45° = 0$ ...(iv)
Sum of forces perpendicular to plane = 0
$R - W\cos 45° = 0$
$R = W\cos 45°$ ...(v)
Putting the value of R in equation (iv) we get
$P = \mu W\cos 45° - W\sin 45°$
P = 15.20KN ....ANS

Q. 28: Force of 200N is required just to move a certain body up an inclined plane of angle 150, the force being parallel to plane. If angle of indication is made 20° the effort again required parallel to plane is found 250N. Determine the weight of body and coefficient of friction.

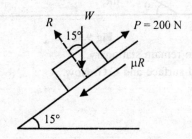 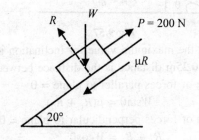

Fig 9.55   Fig 9.56

Sol.:
Case-1
Consider When the slope of the plane be 15°
Sum of forces parallel to plane = 0
$P - \mu R - W\sin 15° = 0$ ...(i)
Sum of forces perpendicular to plane = 0
$R = W\cos 15°$ ...(ii)
Putting the value of (ii) in (i), We get
$P = \mu W\cos 15° + W\sin 15° = 0$
$200 = 0.96\mu W + 0.25 W$ ...(iii)

Case-2
Consider When the slope of the plane be 20°
Sum of forces parallel to plane = 0
$$P - \mu R - W\sin 20° = 0 \qquad \ldots(iv)$$
Sum of forces perpendicular to plane = 0
$$R = W\cos 20° \qquad \ldots(v)$$
Putting the value of (v) in (iv), We get
$$P = \mu W\cos 20° + W\sin 20° = 0$$
$$250 = 0.939\mu W + 0.34W \qquad \ldots(vi)$$
Solved equation (iii) and (vi) we get
$$W = 623.6N \text{ and } \mu = 0.06 \qquad \ldots\text{ANS}$$

Q. 29: A four wheel drive can as shown in fig (9.57) has mass of 2000Kg with passengers. The roadway is inclined at an angle with the horizontal. If the coefficient of friction between the tyres and the road is 0.3, what is the maximum inclination that can climb?

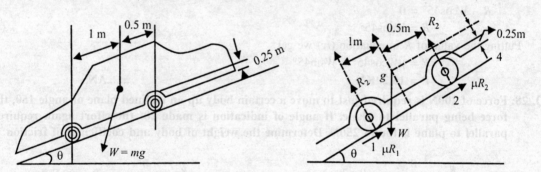

Fig 9.57 Fig 9.58

Sol.: Let the maximum value for inclination is θ for body to remain stationary.
Let 0.25m distance is the distance between the inclined surface and C.G. Now,
Sum of forces parallel to plane = 0
$$W\sin\theta = \mu(R_1 + R_2) \qquad \ldots(i)$$
Sum of forces perpendicular to plane = 0
$$R_1 + R_2 = W\cos\theta \qquad \ldots(ii)$$
Putting the value of (ii) in (i), We get
$$W\sin\theta = \mu(W\cos\theta)$$
Or
$$\mu = \tan\theta$$
$$\theta = \tan^{-1}(0.3)$$
$$\theta = 16.69° \qquad \ldots\text{ANS}$$

Q. 30: A weight 500N just starts moving down a rough inclined plane supported by force 200N acting parallel to the plane and it is at the point of moving up the plane when pulled by a force of 300N parallel to the plane. Find the inclination of the plane and the coefficient of friction between the inclined plane and the weight.

Sol.: In first case body is moving down the plane, so frictional force is acting up the plane
Let θ be the angle of inclination and μ be the coefficient of friction.

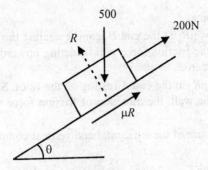
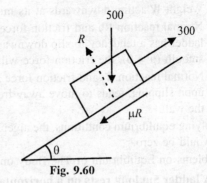

Fig. 9.59   Fig. 9.60

Sum of forces parallel to plane = 0
$$200 + \mu R = 500\sin\theta \qquad ...(i)$$
Sum of forces perpendicular to plane = 0
$$R = 500\cos\theta \qquad ...(ii)$$
Putting the value of (ii) in equation (i)
$$200 + 500\mu\cos\theta = 500\sin\theta \qquad ...(iii)$$

Now 300N is the force when applied to block, it move in upward direction. Hence in this case frictional force acts downward.

Sum of forces perpendicular to plane = 0
$$R = 500\cos\theta \qquad ...(vi)$$
Sum of forces parallel to plane = 0
$$300 = \mu R + 500\sin\theta \qquad ...(v)$$
Putting the value of (iv) in equation (v)
$$300 = 500\mu\cos\theta + 500\sin\theta \qquad ...(vi)$$
Adding equation (iii) and (vi), We get
$$\sin\theta = 1/2; \text{ or } \theta = 30° \qquad ....ANS$$
Putting the value in any equation we get
$$\mu = 0.115 \qquad ....ANS$$

Q. 31: What is ladder friction? How many forces are acting on a ladder?

Sol.: A ladder is an arrangement used for climbing on the walls It essentially consists of two long uprights of wood or iron and connected by a number of cross bars. These cross bars are called rungs and provide steps for climbing. Fig 9.61. shows a ladder AB with its end A resting on the ground and end B leaning against a wall. The ladder is acted upon by the following set of forces:

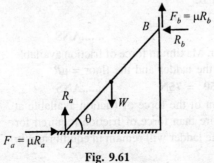

Fig. 9.61

(1) Weight W acting downwards at its mid point.
(2) Normal reaction Rh and friction force $F_h = \mu R_h$ at the end B leaning against the wall. Since the ladder has a tendency to slip downwards, the friction force will be acting upwards. If the wall is smooth ($\mu = 0$), the friction force will be zero.
(3) Normal reaction R_a and friction force $F_a = \mu R_a$ at the end A resting on the floor. Since the ladder, upon slipping, tends to move away from the wall, the direction of friction force will be towards the wall.

Applying equilibrium conditions, the algebraic sum of the horizontal and vertical component of forces would be zero.

Problems on Equilibrium of The Body on Ladder

Q. 32: A ladder 5m long rests on a horizontal ground and leans against a smooth vertical wall at an angle 70° with the horizontal. The weight of the ladder is 900N and acts at its middle. The ladder is at the point of sliding, when a man weighing 750N stands 1.5m from the bottom of the ladder. Calculate coefficient of friction between the ladder and the floor.

Sol.: Forces acting on the ladder is shown in fig 9.62
Resolving all the forces vertically,
$$RV = R - 900 - 750 = 0$$
$$R = 1650N \qquad ...(i)$$
Now taking moment about point B,
$$R \times 5\sin 20° - F_r \times 5\cos 20°$$
$$- 900 \times 2.5\sin 20° - 750 \times 3.5\sin 20° = 0$$
Since $F_r = \mu R$, and $R = 1650N$; $F_r = 1650\mu$
Putting, the value of R and F_r
$$\mu = 0.127 \qquad ....ANS$$

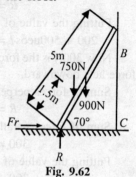

Fig. 9.62

Q. 33: A uniform ladder of length 13m and weighing 250N is placed against a smooth vertical wall with its lower end 5m from the wall. The coefficient of friction between the ladder and floor is 0.3. Show that the ladder will remain in equilibrium in this position. What is the frictional force acting on the ladder at the point of contact between the ladder and the floor?

Sol.: Since the ladder is placed against a smooth vertical wall, therefore there will be no friction at the point of contact between the ladder and wall. Resolving all the force horizontally and vertically.
$$\Sigma H = 0, F_r - R_2 = 0 \qquad ...(i)$$
$$\Sigma V = 0, R_1 - 250 \qquad ...(ii)$$
From the geometry of the figure, BC = 12m
Taking moment about point B,
$$R_1 \times 5 - F_r \times 12 - 250 \times 2.5 = 0$$
$$F_r = 52N \qquad ....ANS$$
For equilibrium of the ladder, Maximum force of friction available at the point of contact between the ladder and the floor = μR
$$= 0.3 \times 250 = 75N \qquad ....ANS$$

Thus we see that the amount of the force of friction available at the point of contact (75N) is more than force of friction required for equilibrium (52N). Therefore, the ladder will remain in equilibrium in this position.

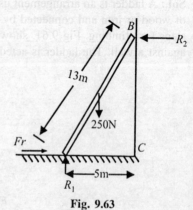

Fig. 9.63

Q. 34: A uniform ladder of 7m rests against a vertical wall with which it makes an angle of 45°, the coefficient of friction between the ladder and the wall is 0.4 and that between ladder and the floor is 0.5. If a man, whose weight is one half of that of the ladder, ascends it, how high will it be when the ladder slips?

Sol.: Let,

X = Distance between A and the man, when the ladder is at the point of slipping.
W = Weight of the ladder
Weight of man = $W/2 = 0.5W$

$$Fr_1 = 0.5R_1 \quad ...(i)$$
$$Fr_2 = 0.4R_2 \quad ...(ii)$$

Resolving the forces vertically
$$R_1 + Fr_2 - W - 0.5W = 0$$
$$R_1 + 0.4R_2 = 1.5W \quad ...(iii)$$

Resolving the forces Horizontally
$$R_2 - Fr_1 = 0; \quad R_2 = 0.5R_1 \quad ...(iv)$$

Solving equation (iii) and (iv), we get
$$R_2 = 0.625W, \quad Fr_2 = 0.25W$$

Now taking moment about point A,
$$W \times 3.5\cos 45° + 0.5W \times x\cos 45° - R_2 \times 7\sin 45° - Fr_2 \times 7\cos 45°$$

Putting the value of R_2 and F_{r2}, we get
$$X = 5.25 m \qquad ....ANS$$

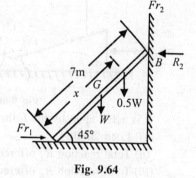

Fig. 9.64

WEDGE FRICTION

Q. 35: Explain how a wedge is used for raising heavy loads. Also gives principle.

Sol.: Principle of wedge : A wedge is small piece of material with two of their opposite faces not parallel. To lift a block of weight W, it is pushed by a horizontal force P which lifts the block by imparting a reaction on the block in a direction 1^r to meeting surface which is always greater than the total downward force applied by block W This will cause a resultant force acting in a upward direction on block and it moves up.

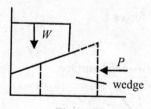

Fig 9.65

Q. 36: Two wedge blocks A and B are employed to raise a load of 2000 N resting on another block C by the application of force P as shown in Fig. 9.66. Neglecting weights of the wedge blocks and assuming co-efficient of friction $\mu = 0.25$ for all the surfaces, determine the value of P for impending upward motion of the block C.

Sol.: The block C, under the action of forces P on blocks A and B, tends to move upward. Hence the frictional forces will act downward. What holds good for block A, the same will hold good for block B.

$$\tan \varphi = \mu = 0.25 \text{ (given)},$$

where φ is the angle of friction
φ = 14° Refer Fig. 9.66
Consider the equilibrium of block C : Refer Fig. 9.67

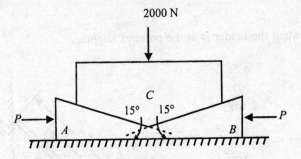

Fig 9.66

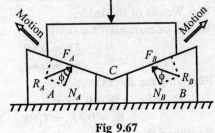

Fig 9.67

It is acted upon by the following forces :
(i) Load 2000 N,
(ii) Total reaction R_A offered by wedge block A, and
(iii) Total reaction R_B offered by wedge block B.

Using Lami's theorem, we get

$$\frac{2000}{\sin 50°} = \frac{R_A}{\sin(180° - 29°)} = \frac{R_B}{\sin(180° - 29°)}$$

$$\frac{2000}{\sin 58°} = \frac{R_A}{\sin 29°} = \frac{R_B}{\sin 29°}$$

$$R_A = R_B = \frac{2000 \times \sin 29°}{\sin 58°} = 1143 \text{ N}$$

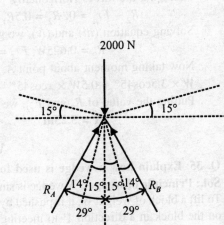

Fig 9.68

Refer Fig. 9.69
Consider equilibrium of block A:
It is acted upon by the following forces :
(i) Force P,
(ii) R_A (from block C), and
(iii) Total reaction R offered by horizontal surface.

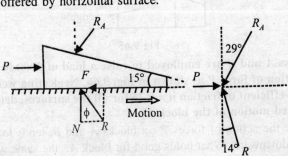

Fig 9.69

Using Lami's theorem, we have

$$\frac{P}{\sin[180° - (29° + 14°)]} = \frac{R_A}{\sin(90° + 14°)}$$

$$\frac{P}{\sin 137°} = \frac{R_A}{\sin 104°}$$

$$P = \frac{1143 \times \sin 137°}{\sin 104°} \quad (\because R_A = 1143\ N)$$

$$= \frac{1143 \times 0.682}{0.97} = 803\ N$$

Hence $P = 803\ N$ ....ANS

CHAPTER **10**

APPLICATION OF FRICTION: BELT FRICTION

Q. 1: What is belt? How many types of belt are used for power transmission?
Sol: The power or rotary motion from one shaft to another at a considerable distance is usually transmitted by means of flat belts, Vee belts or ropes, running over the pulley. But the pulleys contain some friction.

Types of Belts
Important types of belts are:
1. Flat belt
2. V- belt
3. Circular Belt

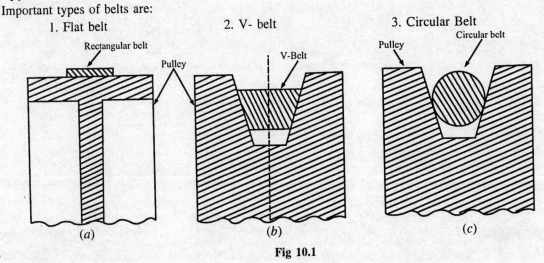

Fig 10.1

Flat Belt
The flat belt is mostly used in the factories and workshops. Where a moderate amount of power is to be transmitted, from one pulley to another, when the two pulleys are not more than 10m apart.

V-Belt
The V-belt is mostly used where a great amount of power is to be transmitted, from one pulley to another, when the two pulleys are very near to each other.

Circular Belt or Rope
The circular belt or rope is mostly used where a great amount of power is to be transmitted from one pulley to another, when the two pulleys are more than 5m apart.

Application of Friction: Belt Friction / 217

Q. 2: Explain how many types of belt drive used for power transmission? Also derive their velocity ratio.

Sol: There are three types of belt drive:
(1) Open belt drive
(2) Cross belt drive
(3) Compound belt drive

(1) Open Belt Drive

When the shafts are arranged in parallel and rotating in the same direction, open belt drive is obtained.

In the diagram 10.2, pulley 'A' is called as driver pulley because it is attached with the rotating shaft.

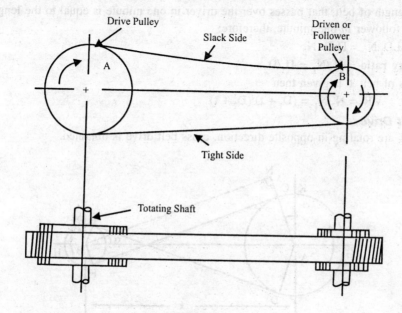

Fig 10.2

Velocity Ratio (V.R.) for Open Belt Drive

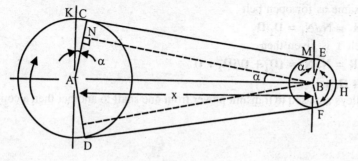

Fig 10.3

Consider a simple belt drive (i.e., one driver and one follower) as shown in fig 10.3.

Let
- D_1 = Diameter of the driver
- N_1 = Speed of the driver in R.P.M.
- D_2, N_2 = Corresponding values for the follower

Length of the belt,
that passes over the driver, in one minute = $\Pi.D_1.N_1$
Similarly,
Length of the belt,
That passes over the follower, in one minute = $\Pi.D_2.N_2$

Since the length of belt, that passes over the driver in one minute is equal to the length of belt that passes over the follower in one minute, therefore:

$$\Pi.D_1.N_1 = \Pi.D_2.N_2$$

Or, **velocity ratio** = $N_2/N_1 = D_1/D_2$

If thickness of belt 't' is given then

$$V.R = N_2/N_1 = (D_1 + t)/(D_2 + t)$$

(2) Cross Belt Drive
When the shafts are rotating in opposite direction, cross belt drive is obtained.

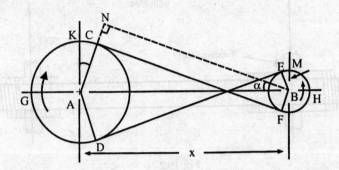

Fig 10.4

In the diagram 13.4, pulley 'A' is called as driver pulley because it is attached with the rotating shaft. Velocity ratio is same as for open belt

$$V.R. = N_2/N_1 = D_1/D_2$$

If thickness of belt 't' is given then

$$V.R = N_2/N_1 = (D_1 + t)/(D_2 + t)$$

(3) Compound Belt Drive
When a number of pulleys are used to transmit power from one shaft to another then a compound belt drive is obtained.

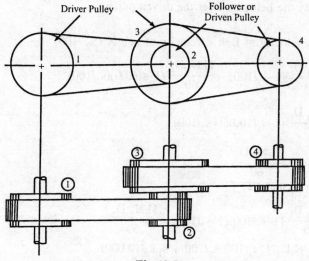

Fig 10.5

Velocity Ratio for Compound Belt Drive

$$\frac{\text{Speed of last follower}}{\text{Speed of first driver}} = \frac{\text{Product of diameter of driver(odd dia)}}{\text{Product of diameter of follower(even dia)}}$$

$$N_4/N_1 = (D_1.D_3)/(D_2.D_4)$$

Q. 3: What is slip of the belt? How slip of belt affect the velocity ratio?

Sol: When the driver pulley rotates, it carries the belt, due to a firm grip between its surface and the belt. The firm between the pulley and the belt is obtained by friction. This firm grip is known as frictional grip. But sometimes the frictional grip is not sufficient. This may cause some forward motion of the driver pulley without carrying the belt with it. This means that there is a relative motion between the driver pulley and the belt. The difference between the linear speeds of the pulley rim and the belt is a measure of slip. Generally, the slip is expressed as a percentage. In some cases, the belt moves faster in the forward direction, without carrying the driver pulley with it. Hence in case of driven pulley, the forward motion of the belt is more than that of driver pulley.

Slip of belt is generally expressed in percentage(%).

Let v = Velocity of belt, passing over the driver pulley/min

N_1 = Speed in R.P.M. of driver

N_2 = Speed in R.P.M. of follower

S_1 = Slip between driver and belt in percentage

S_2 = Slip between follower and belt in percentage

The peripheral velocity of the driver pulley

$$= \omega_1.r_1 = \frac{2\Pi N_1}{60} \times (D_1/2) = \frac{\Pi.D_1.N_1}{60} \qquad ...(i)$$

Now due to Slip between the driver pulley and the belt, the velocity of belt passing over the driver pulley will decrease

Velocity of belt $= \dfrac{\Pi.N_1.D_1}{60} \dfrac{(\Pi.D_1.N_1)}{60} \times \dfrac{s_1}{100} = \dfrac{\Pi.N_1.D_1}{60}(1-s_1/100) \qquad ...(ii)$

Now with this velocity the belt pass over the driven pulley,
Now
Velocity of Driven = Velocity of Belt - Velocity of belt X $(S_2/100)$

$$\frac{\Pi.N_1.D_1}{60}(1-s_1/100) - \frac{\Pi.N_1.D_1}{60}(1-s_1/100)(s_2/100)$$

$$\frac{\Pi.N_1.D_1}{60}(1-s_1/100)(1-s_2/100) \qquad \ldots(iii)$$

But velocity of driven = $\dfrac{\Pi.N_2.D_2}{60}$ ...(iv)

Equate the equation (iii) and (iv)

$$\frac{\Pi.N_1.D_1}{60}(1-s_1/100)(1-s_2/100) = \frac{\Pi.N_2.D_2}{60}$$

$N_2D_2 = N_1D_1(1-s_1/100-s_2/100 + s_1.s_2/10,000)$
    $= N_1D_1[1-(s_1+s_2)/100]$, Neglecting $s_1.s_2/10,000$, since very small

If $s_1 + s_2 = S$ = Total slip in %

$N_2/N_1 = D_1/D_2[1-S/100]$

This formula is used when total slip in % is given in the problem

NOTE: If Slip and thickness both are given then, Velocity ratio is,

$$V.R = N_2/N_1 = \frac{(D_1+t)}{(D_2+t)}[1-s/100]$$

Q. 4: Write down different relations used in belt drive.

Sol: Let:
 D_1 = Diameter of the driver
 N_1 = Speed of the driver in R.P.M.
 D_2 = Diameter of the driven or Follower
 N_2 = Speed of the driven or follower in R.P.M.
 R_1 = Radius of the driver
 R_2 = Radius of the driven or Follower
 t = Belt thickness (if given)
 X = Distance between the centers of two pulleys
 α = Angle of lap (Generally less than 10°)
 θ = Angle of contact (Generally greater than 150°)
   (always express in radian.)
 μ = Coefficient of friction
 s = Total slip in percentage(%)
 L = Total length of belt

Formula For	Open Belt Drive
V.R.	$V.R = N_2/N_1$
Thickness is considered	$V.R = N_2/N_1 = \dfrac{(D_1+t)}{(D_2+t)}$

Slip is considered	$V.R = N_2/N_1 = \dfrac{D_1}{D_2}[1-s/100]$
Slip and thickness both are considered	$V.R = N_2/N_1 = \dfrac{(D_1+t)}{(D_2+t)}[1-s/100]$
Angle of contact	$\theta = \Pi - 2\alpha$
Angle of lap	$\sin\alpha = (r_1-r_2)/X$
Length of belt	$L = \Pi(r_1+r_2) + \dfrac{(r_1-r_2)^2}{X} + 2X$

Q. 5: Prove that the ratio of belt tension is given by the $T_1/T_2 = e^{\mu\theta}$

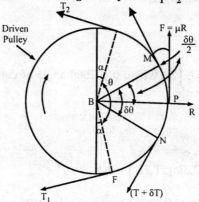

Fig 10.6

Let T_1 = Tension in the belt on the tight side
T_2 = Tension in the belt on the slack side
θ = Angle of contact
μ = Co-efficient of friction between the belt and pulley.
α = Angle of Lap

Consider a driven or follower pulley. Belt remains in contact with EBF. Let T_1 and T_2 are the tensions in the tight side and slack side.

Angle EBF called as angle of contact = $\Pi - 2\alpha$

Consider a driven or follower pulley.

Belt remains in contact with NPM. Let T_1 and T_2 are the tensions in the tight side and slack side.

Let T be the tension at point M & $(T + \delta T)$ be the tension at point N. Let $d\theta$ be the angle of contact of the element MN. Consider equilibrium in horizontal Reaction be 'R' and vertical reaction be μR.

Since the whole system is in equilibrium, i.e.,

$$\Sigma V = 0;$$
$$T\sin(90 - \delta\theta/2) + \mu R - (T + \delta T)\sin(90 - \delta\theta/2) = 0$$
$$T\cos(\delta\theta/2) + \mu R = (T + \delta T)\cos(\delta\theta/2)$$
$$T\cos(\delta\theta/2) + \mu R = T\cos(\delta\theta/2) + \delta T\cos(\delta\theta/2)$$
$$\mu R = \delta T\cos(\delta\theta/2)$$

Since $\delta\theta/2$ is very small & $\cos 0° = 1$, So $\cos(\delta\theta/2) = 1$
$$\mu R = \delta T \qquad \ldots(i)$$
$$\Sigma H = 0;$$
$$R - T\cos(90 - \delta\theta/2) - (T + \delta T)\cos(90 - \delta\theta/2) = 0$$
$$R = T\sin(\delta\theta/2) + (T + \delta T)\sin(\delta\theta/2)$$
Since $\delta\theta/2$ is very small So $\sin(\delta\theta/2) = \delta\theta/2$
$$R = T(\delta\theta/2) + T(\delta\theta/2) + \delta T(\delta\theta/2)$$
$$R = T.\delta\theta + \delta T(\delta\theta/2)$$
Since $\delta T(\delta\theta/2)$ is very small So $\delta T(\delta\theta/2) = 0$
$$R = T.\delta\theta \qquad \ldots(ii)$$
Putting the value of (ii) in equation (i)
$$\mu.T.\delta\theta = \delta T$$
or, $\qquad \delta T/T = \mu.\delta\theta$

Integrating both side: $\int_{T_2}^{T_1} \delta T/T = \mu \int_0^\theta \delta\theta$, Where θ = Total angle of contact

$$\ln(T_1/T_2) = \mu.\theta$$
or, $\qquad T_1/T_2 = e^{\mu.\theta}$

Ratio of belt tension = $T_1/T_2 = e^{\mu\theta}$
Belt ratio is also represent as $2.3\log(T_1/T_2) = \mu.\theta$
Note that θ is in radian
In this formula the main important thing is Angle of contact(θ)

For Open belt drive:
Angle of contact (θ) for larger pulley = $\Pi + 2\alpha$
Angle of contact (θ) for smaller pulley = $\Pi - 2\alpha$

For cross belt drive:
Angle of contact (θ) for larger pulley = $\Pi + 2\alpha$
Angle of contact (θ) for smaller pulley = $\Pi + 2\alpha$
(i.e. for both the pulley, it is same)
But for solving the problems, We always take the Angle of contact (θ) for smaller pulley
Hence,
Angle of contact (θ) = $\Pi - 2\alpha$ – for open belt
Angle of contact (θ) = $\Pi + 2\alpha$ – for cross belt

Q. 6: Explain how you evaluate power transmitted by the belt.
Sol: Let T_1 = Tension in the tight side of the belt
$\qquad T_2$ = Tension in the slack side of the belt
$\qquad V$ = Velocity of the belt in m/sec.
$\qquad\quad$ = $\pi DN/60$ m/sec, D is in meter and N is in RPM
$\qquad P$ = Maximum power transmitted by belt drive

The effective tension or force acting at the circumference of the driven pulley is the difference between the two tensions (i.e., $T_1 - T_2$)

Effective driving force = (T_1-T_2)
Work done per second = Force X Velocity
$= F \times V$ N.m
$= (T_1-T_2) \times V$ N.m
Power Transmitted $= (T_1-T_2).V/1000$ Kw
(Here T_1 & T_2 are in newton and V is in m/sec)
Note:
1. Torque exerted on the driving pulley = $(T_1-T_2).R_1$
   Where R_1 = radius of driving pulley
2. Torque exerted on the driven pulley = $(T_1-T_2).R_2$
   Where R_2 = radius of driven pulley

Q. 7: What is initial tension in the belt?
Sol: The tension in the belt which is passing over the two pulleys (i.e driver and follower) when the pulleys are stationary is known as initial tension in the belt.

When power is transmitted from one shaft to another shaft with the help of the belt, passing over the two pulleys, which are keyed, to the driver and driven shafts, there should be firm grip between the pulleys and belt. When the pulleys are stationary, this firm grip is increased, by tightening the two ends of the belt. Hence the belt is subjected to some tension. This tension is known as initial tension in the belt.

Let To = initial tension in the belt
T_1 = Tension in the tight side
T_2 = Tension in the slack side
To = $(T_1 + T_2)/2$

Q. 8: With the help of a belt an engine running at 200rpm drives a line shaft. The Diameter of the pulley on the engine is 80cm and the diameter of the pulley on the line shaft is 40cm. A 100cm diameter pulley on the line shaft drives a 20cm diameter pulley keyed to a dynamo shaft. Find the speed of the dynamo shaft when: (1) There is no slip (2) There is a slip of 2.5% at each drive.

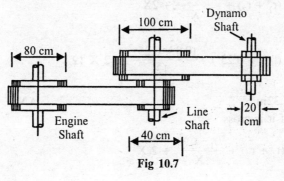

Fig 10.7

Sol:
Dia. of driver pulley (D_1) = 80cm
Dia. of follower pulley (D_2) = 40cm
Dia. of driver pulley (D_3) = 100cm
Dia. of follower pulley (D_4) = 20cm
Slip on each drive, $s_1 = s_2 = 2.5$

Let N_4 = Speed of the dynamo shaft
(i) When there is no slip
Using equation

$$N_4/N_1 = (D_1.D_3)/(D_2.D_4)$$
$$N_4 = N_1 \times (D_1.D_3)/(D_2.D_4)$$
$$= [(80 \times 100) \times 200]/(40 \times 20)$$
$$N_4 = 2000 \text{RPM} \qquad \text{......ANS}$$

(ii) When there is a slip of 2.5% at each drive
In this case we will have the equation of:

$$N_4/N_1 = [(D_1.D_3)/(D_2.D_4)][1-s_1/100][1-s_2/100]$$

Putting all the values, we get

$$N_4 = N_1 \times [(D_1.D_3)/(D_2.D_4)][1- s_1/100][1- s_2/100]$$
$$N_4 = 200 \times [(80 \times 100)/(40 \times 20)][1 - 2.5/100][1- 2.5/100]$$
$$N_4 = 1901.25 \text{R.P.M.} \qquad \text{......ANS}$$

Q. 9: Find the length of belt necessary to drive a pulley of 500mm diameter running parallel at a distance of 12m from the driving pulley of diameter 1600m.

Sol: Given Data

Dia. of driven pulley $(D_2) = 500\text{mm} = 0.5\text{m}$
Radius of driven pulley $(r_2) = 0.25\text{m}$
Centre distance $(X) = 12\text{m}$
Dia. of driver pulley $(D_1) = 1600\text{mm} = 1.6\text{m}$
Radius of driver pulley $(r_1) = 0.8\text{m}$

Since there is no mention about type of belt(Open or cross type)
So we find out for both the cases.
(i) Length of the belt if it is open

WE know that: $L = \Pi (r_1 + r_2) + \dfrac{(r_1-r_2)^2}{X} + 2X$

Putting all the value

$$L = \Pi (0.8 + 0.25) + \dfrac{(0.8 - 0.25)^2}{12} + 2 \times 12$$

$$L = 27.32\text{m} \qquad \text{......ANS}$$

(ii) Length of the belt if it is cross

WE know that: $L = \Pi (r_1 + r_2) + \dfrac{(r_1-r_2)^2}{X} + 2X$

Putting all the value

$$L = \Pi (0.8 + 0.25) + \dfrac{(0.8 - 0.25)^2}{12} + 2 \times 12$$

$$L = 27.39\text{m} \qquad \text{......ANS}$$

Q. 10: Find the speed of shaft driven with the belt by an engine running at 600RPM. The thickness of belt is 2cm, diameter of engine pulley is 100cm and that of shaft is 62cm.

Sol: Given that
· Speed of driven shaft (N_2) = ?
Thickness of belt (t) = 2cm
Diameter of driver shaft (D_2) = 100cm
Diameter of driven shaft (D_1) = 62cm
Speed of driver shaft (N_1) = 600rpm
Since we know that,

$$V.R = N_2/N_1 = \frac{(D_1 + t)}{(D_2 + t)}$$

$$N_2 = N_1 \times [(D_1 + t)/(D_2 + t)]$$

Putting all the value,
$$N_2 = 600 \times [(62 + 2)/(100 + 2)]$$
$$N_2 = 376.47 \text{RPM} \qquad \text{.......ANS}$$

Q. 11: A belt drives a pulley of 200mm diameter such that the ratio of tensions in the tight side and slack side is 1.2. If the maximum tension in the belt is not to exceed 240KN. Find the safe power transmitted by the pulley at a speed of 60rpm.

Sol: Given that,
D_1 = Diameter of the driver = 200mm = 0.2m
$T_1/T_2 = 1.2$
Since between T_1 and T_2, T_1 is always greater than T_2,
Hence T_1 = 240KN
N_1 = Speed of the driver in R.P.M. = 60PRM
P = ?
We know that
$T_1/T_2 = 1.2$
$T_2 = T_1/1.2 = 240/1.2 = 200$KN ...(i)
V = Velocity of the belt in m/sec.
= $\pi DN/60$ m/sec, D is in meter and N is in RPM
= (3.14 × 0.2 × 60)/60 = 0.628 m/sec ...(ii)
P = $(T_1 - T_2) \times V$
P = (240 - 200) × 0.628
P = 25.13KW ....ANS

Q. 12: Find the power transmitted by cross type belt drive connecting two pulley of 45.0cm and 20.0cm diameter, which are 1.95m apart. The maximum permissible tension in the belt is 1KN, coefficient of friction is 0.20 and speed of larger pulley is 100rpm.

Sol: Given that
D_1 = Diameter of the driver = 45cm = 0.45m
R_1 = Radius of the driver = 0.225m
D_2 = Diameter of the driven = 20cm = 0.2m
R_2 = Radius of the driven = 0.1m
X = Distance between the centers of two pulleys = 1.95m
T_1 = Maximum permissible tension = 1000N
μ = Coefficient of friction = 0.20
N_1 = Speed of the driver(Larger pulley) in R.P.M. = 100RPM

Since we know that,
Power Transmitted = $(T_1-T_2).V/1000$ Kw ...(i)
Tension is in KN and V is in m/sec
First ve find the velocity of the belt,
V = Velocity of the belt in m/sec.
Here we take diameter and RPM of larger pulley
= $\pi DN/60$ m/sec, D is in meter and N is in RPM
$\quad\quad\quad$ = (3.14 X0.45 X 100) /60
$\quad\quad\quad$ = 2.36m/sec ...(ii)
Now Ratio of belt tension, $T_1/T_2 = e^{\mu.\theta}$ ...(iii)
Here we don't know the value of θ. For θ, first find the value of α, by the formula,
Angle of Lap for cross belt $\alpha = \sin^{-1}(r_1 + r_2)/X$
$\quad\quad\quad$ = $\sin^{-1}(0.225 + 0.1)/1.95$
$\quad\quad\quad$ = 9.59° ...(iv)
Now Angle of contact (θ) = $\Pi + 2\alpha$ ----- for cross belt
$\quad\quad\theta = \Pi + 2 \text{ X } 9.59°$
$\quad\quad\quad$ = 199.19°
$\quad\quad\quad$ = 199.19°($\Pi/180°$) = 3.47rad ...(v)
Now putting all the value in equation (iii)
We get
$\quad\quad 1000/T_2 = e^{(0.2)(3.47)}$
$\quad\quad T_2 = 498.9$ N ...(vi)
Using equation (i), we get
$\quad\quad P = [(1000 - 498.9) \text{ X } 2.36]/1000$
$\quad\quad \mathbf{P = 1.18KW}$ ....ANS

Q. 13: A flat belt is used to transmit a torque from pulley A to pulley B as shown in fig 7.8. The radius of each pulley is 50mm and the coefficient of friction is 0.3. Determine the largest torque that can be transmitted if the allowable belt tension is 3KN.

Sol: Radius of each pulley = 50mm,
$\quad R_1 = R_2 = 50$mm
$\quad R_1$ = Radius of the driver = 50mm
$\quad R_2$ = Radius of the driven = 50mm
$\quad \theta$ = Angle of contact(In radian) = 1800 = p,
$\quad \mu$ = Coefficient of friction = 0.3

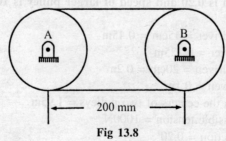

Fig 13.8

T_1 = Allowable tension = 3KN,

T_1 always greater than T_2
Using the relation $T_1/T_2 = e^{\mu\theta}$
Putting all the value,
$3/T_2 = e^{(0.3)(\pi)}$
On solving $T_2 = 1.169$ KN ....ANS
Since Radius of both pulley is same;
So, Torque exerted on both pulley is same and
$$= (T_1-T_2).R_1 = (T_1-T_2).R_2$$
Putting all the value we get,
$(3 - 1.169) \times 50 = 91.55$ KN-mm ....ANS

Q. 14: An open belt drive connects two pulleys 120cm and 50cm diameter on parallel shafts 4m apart. The maximum tension in the belt is 1855.3N. The coefficient of friction is 0.3. The driver pulley of diameter 120cm runs at 200rpm. Calculate (*i*) The power transmitted (*ii*) Torque on each of the two shafts.

Sol: Given data:
  D_1 = Diameter of the driver = 120cm = 1.2m
  R_1 = Radius of the driver = 0.6m
  N_1 = Speed of the driver in R.P.M. = 200RPM
  D_2 = Diameter of the driven or Follower = 50cm = 0.5m
  R_2 = Radius of the driven or Follower = 0.25m
  X = Distance between the centers of two pulleys = 4m
  μ = Coefficient of friction = 0.3
  T_1 = Tension in the tight side of the belt = 1855.3N
Calculation for power transmitting:
Let
  P = Maximum power transmitted by belt drive
  $= (T_1-T_2).V/1000$ KW ...(*i*)
Where,
  T_2 = Tension in the slack side of the belt
  V = Velocity of the belt in m/sec.
  $= \pi DN/60$ m/sec, D is in meter and N is in RPM ...(*ii*)
For T_2,
We use the relation Ratio of belt tension $= T_1/T_2 = e^{\mu\theta}$ ...(*iii*)
But angle of contact is not given,
let
  θ = Angle of contact and, θ = Angle of lap
for open belt, Angle of contact $(\theta) = \Pi - 2\alpha$ ...(*iv*)
$\sin\alpha \qquad = (r_1 - r_2)/X = (0.6 - 0.25)/4$
  $\alpha = 5.02°$ ...(*v*)
Using the relation (*iii*), $\theta = \Pi - 2\alpha = 180 - 2 \times 5.02 = 169.96°$
  $= 169.96° \times \Pi/180 = 2.97$ rad ...(*iv*)
Now using the relation (*iii*)

$1855.3/T_2 = e^{(0.3)(2.967)}$
$T_2 = 761.8N$ ...(vii)

For finding the velocity, using the relation (ii)
$V = (3.14 \times 1.2 \times 200)/60 = 12.56$ m/sec ...(viii)

For finding the Power, using the relation (i)
$P = (1855.3 - 761.8) \times 12.56$
P = 13.73 KW ....ANS

We know that,
1. Torque exerted on the driving pulley = $(T_1 - T_2).R_1$
   = $(1855.3 - 761.8) \times 0.6$
   = **656.1Nm** ....ANS
2. Torque exerted on the driven pulley = $(T_1 - T_2).R_2$
   = $(1855.3 - 761.8) \times 0.25$
   = **273.41Nm** ....ANS

Q. 15: Find the power transmitted by a belt running over a pulley of 600mm diameter at 200r.p.m. The coefficient of friction between the pulleys is 0.25; angle of lap 160° and maximum tension in the belt is 2.5KN.

Sol: Given data
D_1 = Diameter of the driver = 600mm = 0.6m
N_1 = Speed of the driver in R.P.M. = 200RPM
μ = Coefficient of friction = 0.25
θ = Angle of contact = 160°
= 1600 × (π/180) = 2.79rad

(Angle of lap is always less than 10°, so it is angle of contact which is always greater than 150°, always in radian)

T_1 = Maximum Tension = 2.5KN
Let
T_2 = Tension in the slack side of the belt
V = Velocity of the belt in m/sec.
= πDN/60 m/sec, D is in meter and N is in RPM
P = Power transmitted by belt drive

We know that
Power Transmitted = $(T_1 - T_2).V$ KW, T_1 & T_2 in KN
Here T_2 and V is unknown

Calculation for V
V = πDN/60 m/sec, D is in meter and N is in RPM
Putting all the value,
V = $(3.14 \times 0.6 \times 200)/60 = 6.28$m/sec ...(i)

Calculation for T_2
We also know that,
Ratio of belt tension, $T_1/T_2 = e^{\mu\theta}$
Putting all the value,

$$2.5 / T_2 = e^{(0.25 \times 2.79)}$$
$$T_2 = 1.24 \text{ KN}$$
Now, $\quad P = (2.5 - 1.24) \times 6.28$ ...(ii)
$\quad P = 7.92 \text{ KW}$ ....ANS

Q. 16: An open belt runs between two pulleys 400mm and 150mm diameter and their centers are 1000mm apart. If coefficient of friction for larger pulley is 0.3, then what should be the value of coefficient of friction for smaller pulley, so that the slipping is about to take place at both the pulley at the same time?

Sol: Given data
$D_1 = 400 \text{mm}, R_1 = 200 \text{mm}$
$D_2 = 150 \text{mm}, R_2 = 775 \text{mm}$
$X = 1000 \text{mm}$
$\mu_1 = 0.3$
$\mu_2 = ?$
$\sin\alpha = (r_1 - r_2)/X = (200 - 75)/1000$
$\alpha = 7.18° = 7.18° \times \Pi/180°$
$\alpha = 0.1256 \text{ rad}$ ...(i)

We know that
For Open belt drive:
Angle of contact (θ) for larger pulley $= \Pi + 2\alpha$
Angle of contact (θ) for smaller pulley $= \Pi - 2\alpha$
Since, Ratio of belt tension $= T_1/T_2 = e^{\mu\theta}$
It is equal for both the pulley, i.e.,
(T_1/T_2)larger pulley $= (T_1/T_2)$smaller pulley
or, $e^{\mu_1\theta_1} = e^{\mu_2\theta_2}$, or $\mu_1\theta_1 = \mu_2\theta_2$
putting all the value, we get,
$(0.3)(\Pi + 2\alpha) = (\mu_2)(\Pi - 2\alpha)$
$(0.3)(\Pi + 2 \times 0.1256) = (\mu_2)(\Pi - 2 \times 0.1256)$
on solving, $\mu_2 = 0.352$ ....ANS

Q. 17: A belt supports two weights W_1 and W_2 over a pulley as shown in fig 7.9. If $W_1 = 1000N$, find the minimum weight W_2 to keep W_1 in equilibrium. Assume that the pulley is locked and $\mu = 0.25$.

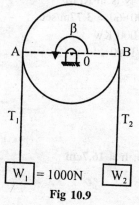

Fig 10.9

Sol : Let the tensions in the belt be T_1 and T_2 as shown, since the weight W_2 just checks the tendency of weight W_1 to move down, tension on the side of W_1 is larger.

That is, $T_1 > T_2$

$$\mu = 0.25, \theta = \Pi, W_1 = 1000N$$

Using the relation Ratio of belt tension

$$= T_1/T_2 = e^{\mu\theta}$$
$$W_1/T_2 = e^{(0.25)(\Pi)}$$

On solving, $T_2 = W_2 = 456N$ ....ANS

Q. 18: An open belt running over two pulleys 24cm and 60cm diameters. Connects two parallel shaft 3m apart and transmits 3.75KW from the smaller pulley that rotates at 300RPM, $\mu = 0.3$, and the safe working tension in 100N/cm width. Determine

(i) Minimum width of the belt.

(ii) Initial belt tension.

(iii) Length of the belt required.

Sol: Given that,

$$D_1 = 60cm$$
$$D_2 = 24cm$$
$$N_2 = 300rpm$$
$$\mu = 0.3$$
$$X = 3m = 300cm$$
$$P = 3.75KW$$

Safe Tension = Maximum tension = 100N/cm width = 100b N·b = width of belt

$$T_{max} = 100b \qquad \qquad ...(i)$$

Let θ = Angle of contact

$$\sin\alpha = (r_1 - r_2)/X = (30 - 12)/300 \ ; \ \alpha = 3.45°, \qquad ...(ii)$$
$$\theta = \Pi - 2\alpha = (180 - 2 \times 3.45) = 173.1°$$
$$= (173.1°) \times \Pi/180 = 3.02 rad \qquad ...(iii)$$

Now,

Using the relation, Ratio of belt tension = $T_1/T_2 = e^{\mu\theta} = e^{(0.3)(3.02)}$

$$T_1 = 2.474 T_2 \qquad \qquad ...(iv)$$

Now,

$V = \pi DN/60$ m/sec, D is in meter and N is in RPM

$$= 3.14 \times (0.24)(300)/60 = 3.77 m/sec \qquad ...(v)$$

Power Transmitted (P) = $(T_1 - T_2).v/1000$ Kw

$$3.75 = (T_1 - T_2) \times 3.77/1000$$
$$T_1 - T_2 = 994.7N \qquad \qquad ...(vi)$$

From relation (iv) and (v), we get:

$$T_1 = 1669.5N \qquad \qquad ...(vii)$$
$$T_2 = 674.8N \qquad \qquad ...(viii)$$

(i) For width of the belt

But $T_1 = T_{max} = 100b$; $1669.5 = 100b$; **b = 16.7cm** ....ANS

(ii) For initial tension in the belt

Let To = initial tension in the belt

$$T_o = (T_1 + T_2)/2$$
$$= (1669.5 + 674.8)/2$$
$$T_o = 1172.15 N \quad \text{.......ANS}$$

(iii) For length of belt

$$L = \Pi(r_1 + r_2) + \frac{(r_1 - r_2)^2}{X} + 2X$$

Putting all the value, we get
$$L = 7.33 m \quad \text{.......ANS}$$

Q. 19: Determine the minimum value of weight W required to cause motion of a block, which rests on a horizontal plane. The block weighs 300N and the coefficient of friction between the block and plane is 0.6. Angle of warp over the pulley is 90° and the coefficient of friction between the pulley and rope is 0.3.

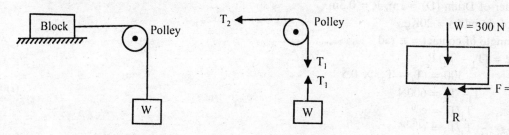

Fig 10.10   Fig 10.11   Fig 10.12

Sol: Since the weight W impend vertical motion in the downward direction, the tension in the two sides of the pulley will be as shown in fig 10.11

Given date:
$$T_1 = W, \mu = 0.3, \theta = 90° = \pi/2 \text{ rad}$$

Using the relation of Ratio of belt tension, $T_1/T_2 = e^{\mu.\theta}$
$$W/T_2 = e^{(0.3).(p/2)} = 1.6$$
$$W = 1.6 \times T_2 \quad ...(i)$$

Considering the equilibrium of block:
$$\Sigma V = 0$$
$$R = 300 N \quad ...(ii)$$
$$\Sigma H = 0$$
$$T_2 = \mu R = 0.3 \times 300 = 180 N \quad ...(iii)$$

Equating equation (i) and (iii), we get
$$W = 1.6 \times 180$$
$$W = 288 N \quad \text{.......ANS}$$

Q. 20: A horizontal drum of a belt drive carries the belt over a semicircle around it. It is rotated anti-clockwise to transmit a torque of 300N-m. If the coefficient of friction between the belt and rope is 0.3, calculate the tension in the limbs 1 and 2 of the belt shown in figure, and the reaction on the bearing. The drum has a mass of 20Kg and the belt is assumed to be mass less.

(May–01-02)

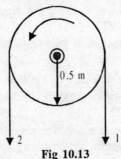

Fig 10.13

Sol: Given data:

Torque(t) = 300N-m
Coff. of friction(μ) = 0.3
Diameter of Drum (D) = 1m, R = 0.5m
Mass of drum(m) = 20Kg.
Since angle of contact = π rad
Torque = $(T_1 - T_2).R$
$300 = (T_1 - T_2) \times 0.5$
$T_1 - T_2 = 600N$ ...(i)

And, $T_1/T_2 = e^{\mu\theta}$
$T_1/T_2 = e^{(0.3)\pi}$
$T_1 = 2.566 T_2$ ...(ii)

Solving (i) and (ii)
We get,
$T_1 = 983.14N$ ....ANS
$T_2 = 383.14N$ ....ANS

Now reaction on bearing is opposite to the mass of the body, and it is equal to
$R = T_1 + T_2 + mg$
$R = 983.14 + 383.14 + 20 \times 9.81$
$R = 1562.484N$ ....ANS

Q. 21: A belt is stretched over two identical pulleys of diameter D meter. The initial tension in the belt throughout is 2.4KN when the pulleys are at rest. In using these pulleys and belt to transmit torque, it is found that the increase in tension on one side is equal to the decrease on the other side. Find the maximum torque that can be transmitted by the belt drive, given that the coefficient of friction between belt and pulley is 0.30.   *(Dec–02-03)*

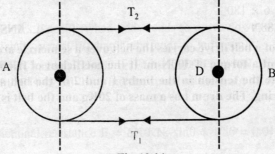

Fig 10.14

Sol: Given data:
Diameter of both pulley = D
Initial tension in belt (T_O) = 2.4KN
Torque = ?
Coefficient of friction (μ) = 0.3
Since dia of both pulley are same, i.e., Angle of contact = π
$$T_O = (T_1 + T_2)/2$$
$$T_1 + T_2 = 4.8\text{KN} \qquad \qquad ...(i)$$
Now, Ratio of belt tension = $T_1/T_2 = e^{\mu\theta}$
$$T_1/T_2 = e^{(0.3)\pi}$$
$$T_1 = 2.566 T_2 \qquad \qquad ...(ii)$$
Putting the value of (ii) in equation (i), We get
$$T_1 = \textbf{3.46KN} \qquad \qquad \text{.......ANS}$$
$$T_2 = \textbf{1.35KN} \qquad \qquad \text{.......ANS}$$
Now, Maximum torque transmitted by the pulley = $(T_1 - T_2)D/2$
(Since radius of both pulley are same)
Torque = $(3.46 - 1.35)D/2 = 1.055D$ KN–m
Torque = **1.055D KN-m** ....ANS

Q. 22: A belt is running over a pulley of 1.5m diameters at 250RPM. The angle of contact is 120° and the coefficient of friction is 0.30. If the maximum tension in the belt is 400N, find the power transmitted by the belt. *(Nov–03 C.O.)*

Sol: Given data
Diameter of pulley(D) = 1.5m
Speed of the driver(N) = 250RPM
Angle of contact(?) = 1200 = 1200 X (π/180°) = 2.09 rad
Coefficient of friction(μ) = 0.3
Maximum tension(Tmax) = 400N = T_1
Power (P) = ?
Since P = $(T_1 - T_2)$ X V Watt ...(i)
T1 is given, and for finding the value of T_2, using the formula
Ratio of belt tension = $T_1/T_2 = e^{\mu\theta}$
$$400/T_2 = e^{(0.3)(2.09)}$$
$$T_2 = 213.4\text{N} \qquad \qquad ...(ii)$$
Now We know that V = πDN/60 m/sec
$$V = [3.14 \text{ X } 1.5 \text{ X } 250]/60 = 19.64 \text{m/sec} \qquad ...(iii)$$
Now putting all the value in equation (i)
P = (400 – 213.4) X 19.64 watt
P = 3663.88Watt or 3.66KW ....ANS

Q. 23: Explain the concept of centrifugal tension in any belt drive. What are the main consideration for taking maximum tension?

Sol: We know that the belt continuously runs over both the pulleys. In the tight side and slack side of the belt tension is increased due to presence of centrifugal Tension in the belt. At lower speeds the centrifugal tension may be ignored but at higher speed its effect is considered.

The tension caused in the running belt by the centrifugal force is known as centrifugal tension. When ever a particle of mass 'm' is rotated in a circular path of radius 'r' at a uniform velocity 'v', a centrifugal force is acting radially outward and its magnitude is equal to $\dfrac{mv^2}{r}$.

i.e., $\qquad$ Fc = mv^2/r

The centrifugal tension in the belt can be calculated by considering the forces acting on an elemental length of the belt(i.e length MN) subtending an angle $\delta\theta$ at he center as shown in the fig 10.14.

Let
$\qquad$ v = Velocity of belt in m/s
$\qquad$ r = Radius of pulley over which belt run.
$\qquad$ M = Mass of elemental length of belt.
$\qquad$ m = Mass of the belt per meter length
$\qquad$ T_1 = Tight side tension
$\qquad$ T_c = Centrifugal tension acting at M and N tangentially
$\qquad$ F_c = Centrifugal force acting radially outwards

The centrifugal force R acting radially outwards is balanced by the components of Tc acting radially inwards. Now elemental length of belt
$$MN = r.\delta\theta$$
Mass of the belt MN = Mass per meter length X Length of MN
$$M = m \; X \; r \; X \; \delta\theta$$
Centrifugal force = F_c = M X v^2/r = $m.r.\delta\theta.v^2/r$
Now resolving the force horizontally, we get
$\qquad$ $T_c.\sin\delta\theta/2 + T_c.\sin\delta\theta/2 = F_c$
Or $\qquad$ $2T_c.\sin\delta\theta/2 = m.r.\delta\theta.v^2/r$
At the angle $\delta\theta$ is very small, hence = $\sin\delta\theta/2 = \delta\theta/2$
Then the above equation becomes as
$\qquad$ $2T_c.\delta\theta/2 = m.r.\delta\theta.v^2/r$
or $\qquad\qquad$ $T_c = m.v^2$

Important Consideration

1. From the above equation, it is clear that centrifugal tension is independent of T_1 and T_2. It depends upon the velocity of the belt. For lower belt speed (i.e., Belt speed less than 10m/s) the centrifugal tension is very small and may be neglected.

2. When centrifugal tension is to be taken into consideration then total tension on tight side and slack side of the belt is given by
$\qquad$ For tight side = T_1 + Tc
$\qquad$ For slack side = T_2 + Tc

3. Maximum tension(Tm) in the belt is equal to maximum safe stress in the belt multiplied by cross sectional area of the belt.
$$T_m = \sigma \; (b.t)$$
Where
$\qquad$ σ = Maximum safe stress in the belt
$\qquad$ b = Width of belt and

t = Thickness of belt
$T_m = T_1 + T_c$ ---- if centrifugal tension is to be considered
   $= T_1$ ------- if centrifugal tension is to be neglected

Q. 24: Derive the formula for maximum power transmitted by a belt when centrifugal tension in to account.

Sol: Let T_1 = Tension on tight side
T_2 = Tension on slack side
v = Linear velocity of belt
Then the power transmitted is given by the equation
$$P = (T_1 - T_2) \cdot V \qquad ...(i)$$
But we know that $T_1/T_2 = e^{\mu\theta}$
Or we can say that $T_2 = T_1 / e^{\mu\theta}$
Putting the value of T_2 in equation (i)
$$P = (T_1 - T_1/e^{\mu\theta}) \cdot v = T_1(1 - 1/e^{\mu\theta}) \cdot V \qquad ...(ii)$$
Let $(1 - 1/e^{\mu\theta}) = K$, K = any constant
Then the above equation is $P = T_1 \cdot K \cdot V$ or $KT_1 V$ ...(iii)
Let T_{max} = Maximum tension in the belt
T_c = Centrifugal tension which is equal to $m \cdot v^2$
Then $T_{max} = T_1 + T_c$
$T_1 = T_{max} - T_c$
Putting this value in the equation (iii)
$$P = K(T_{max} - T_c) \cdot V$$
$$= K(T_{max} - m \cdot V^2) \cdot V$$
$$= K(T_{max} \cdot v - m \cdot V^3)$$
The power transmitted will be maximum if $d(P)/dv = 0$
Hence differentiating equation w.r.t. V and equating to zero for maximum power, we get
$$d(P)/dv = K(T_{max} - 3 \cdot m \cdot V^2) = 0$$
$$T_{max} - 3mV^2 = 0$$
$$T_{max} = 3mV^2$$
$$V = (T_{max}/3m)^{1/2} \qquad ...(iv)$$
Equation (iv) gives the velocity of the belt at which maximum power is transmitted.
From equation (iv) $T_{max} = 3T_c$ ...(v)
Hence when the power transmitted is maximum, centrifugal tension would be 1/3rd of the maximum tension.
We also know that $Tmax = T_1 + T_c$
$$= T_1 + T_{max}/3 \qquad ...(vi)$$
$$T_1 = T_{max} - T_{max}/3$$
$$= 2/3 \cdot T_{max} \qquad ...(vii)$$
Hence condition for the transmission of maximum power are:
$$T_c = 1/3\ T_{max}, \text{ and } T_1 = 2/3 T_{max} \qquad ...(viii)$$
NOTE: Net driving tension in the belt = $(T_1 - T_2)$

STEPS FOR SOLVING THE PROBLEM FOR FINDING THE POWER
1. Use the formula stress (σ) = force (Maximum Tension)/Area
   Where; Area = b.t i.e., Tmax = σ.b.t
2. Unit mass (m) = ρ.b.t.L
   Where;
   ρ = Density of a material
   b = Width of Belt
   t = Belt thickness
   L = Unit length
   Take L = 1m, if b and t are in meter
   Take L = 100cm, if b and t are in cm
   Take L = 1000mm, if b and t are in mm
3. Calculate V using V = πDN/60 m/sec (if not given)
4. T_C = mV^2, For finding T_C
5. T_{max} = T_1 + T_c, for finding T_1
6. For T_2, Using relation Ratio of belt tension = T_1/T_2 = $e^{\mu\theta}$
7. Power Transmitted = $(T_1 - T_2)$.V/1000 Kw

Steps for Solving the Problem for Finding the Maximum Power
1. Use the formula stress (σ) = force (Maximum Tension)/Area
   Where; Area = b.t i.e. Tmax = σ.b.t
2. Unit mass (m) = ρ.b.t.L
   Where
   ρ = Density of a material
   b = Width of Belt
   t = Belt thickness
   L = Unit length
   Take L = 1m, if b and t are in meter
   Take L = 100cm, if b and t are in cm
   Take L = 1000mm, if b and t are in mm
3. T_C = 1/3 Tmax = mV^2, For finding T_C and velocity (If not given)
   We don't Calculate Velocity using V = πDN/60 m/sec (if not given)
5. T_{max} = T_1 + T_c, for finding T_1
6. For T_2, Using relation Ratio of belt tension = T_1/T_2 = $e^{\mu\theta}$
7. Maximum Power Transmitted = $(T_1 - T_2)$.v/1000 Kw

Initial Tension in The Belt
Let To = initial tension in the belt
T_1 = Tension in the tight side
T_2 = Tension in the slack side
T_C = Centrifugal Tension in the belt
T_o = $(T_1 + T_2)/2 + T_C$

Application of Friction: Belt Friction / 237

Q. 25: A belt 100mm wide and 8.0mm thick are transmitting power at a belt speed of 160m/minute. The angle of lap for smaller pulley is 165° and coefficient of friction is 0.3. The maximum permissible stress in belt is 2MN/m² and mass of the belt is 0.9Kg/m. find the power transmitted and the initial tension in the belt.

Sol.: Given data

Width of belt(b) = 100mm
Thickness of belt(t) = 8mm
Velocity of belt(V) = 160m/min = 2.66m/sec
Angle of contact(?) = 165° = 165° × Π/180 = 2.88rad
Coefficient of friction(μ) = 0.3
Maximum permissible stress(f) = 2 × 10⁶ N/m² = 2N/mm²
Mass of the belt material(m) = 0.9 Kg/m
Power = ?
Initial tension (To) = ?

We know that, $T_{max} = \sigma.b.t$
   = 2 × 100 × 8 = 1600N ...(i)

Since m and velocity (V) is given, then
Using the formula, $T_C = mV^2$, For finding T_C
   = 0.9(2.66)²
   = 6.4 N ...(ii)

Using the formula, $T_{max} = T_1 + T_c$, for finding T_1
   1600 = T_1 + 6.4
   T_1 = 1593.6N ...(iii)

Now, For T_2, Using relation Ratio of belt tension = $T_1/T_2 = e^{\mu\theta}$
   1593.6/T_2 = $e^{(0.3)(2.88)}$
   T_2 = 671.69 N ...(iv)

Now Power Transmitted = $(T_1-T_2).v/1000$ Kw
   P = (1593.6 – 671.69).2.66/1000 Kw
   P = 2.45KW ....ANS

Let To = initial tension in the belt
   $T_o = (T_1 + T_2)/2 + T_C$
   T_o = (1593.6 + 671.69)/2 + 6.4
   T_o = 1139.045N ....ANS

Q. 26: A belt embraces the shorter pulley by an angle of 165° and runs at a speed of 1700 m/min, Dimensions of the belt are Width = 20cm and thickness = 8mm. Its density is 1gm/cm³. Determine the maximum power that can be transmitted at the above speed, if the maximum permissible stress in the belt is not to exceed 250N/cm² and μ = 0.25.

Sol: Given date:

Angle of contact(θ) =165° = 165° × Π/180 = 2.88rad
Velocity of belt(V) = 1700m/min = 28.33m/sec
Width of belt(b) = 20cm
Thickness of belt(t) = 8mm 0.8cm

density of belt = 1gm/cm³
Maximum permissible stress(f) = 250 N/cm²
Coefficient of friction(μ) = 0.25
Maximum Power = ?
We know that, T_{max} = σ.b.t
$\quad\quad$ = 250 X 20 X 0.8 = 4000N $\quad\quad\quad\quad\quad\quad\quad\quad\quad\quad$ ...(i)
Since Unit mass (m) = ρ.b.t.L
$\quad\quad$ = 1/1000 X 20 X 0.8 X 100 = 1.6Kg $\quad\quad\quad\quad\quad\quad\quad\quad\quad$ ..(ii)
Since velocity(V) is given, So we don't find the velocity using formula T_C =1/3 Tmax = mV², then
Using the formula, T_C = mV², For finding T_C
$\quad\quad$ = 1.6(28.33)²
$\quad\quad$ = 1284 N $\quad\quad\quad\quad\quad\quad\quad\quad\quad\quad\quad\quad\quad\quad\quad\quad$ ...(iii)
Using the formula, T_{max} = T_1 + T_c, for finding T_1
$\quad\quad$ 4000 = T_1 + 1284
$\quad\quad\quad T_1$ = 2716N $\quad\quad\quad\quad\quad\quad\quad\quad\quad\quad\quad\quad\quad\quad\quad$ ...(iv)
Now, For T_2, Using relation Ratio of belt tension = T_1/T_2 = $e^{\mu\theta}$
$\quad\quad$ 2716/T_2 = $e^{(0.25)(2.88)}$
$\quad\quad\quad T_2$ = 1321 N $\quad\quad\quad\quad\quad\quad\quad\quad\quad\quad\quad\quad\quad\quad\quad$ ...(v)
Now Maximum Power Transmitted = (T_1–T_2).V/1000 KW
$\quad\quad$ P = (2716 – 1321) X 28.33/1000 KW
$\quad\quad$ **P = 39.52KW** $\quad\quad\quad\quad\quad\quad\quad\quad\quad\quad\quad\quad$ ....ANS

Q. 27: A belt of density 1gm/cm³ has a maximum permissible stress of 250N/cm². Determine the maximum power that can be transmitted by a belt of 20cm X 1.2cm if the ratio of the tight side to slack side tension is 2.

Sol: Given date
Density of belt = 1gm/cm³ = 1/1000 Kg/cm³
Maximum permissible stress(f) = 250 N/cm²
Width of belt(b) = 20cm
Thickness of belt(t) = 8mm 0.8cm
Ratio of tension (T_1/T_2) = 2
Maximum Power = ?
We know that, Tmax = σ.b.t
$\quad\quad$ = 250 X 20 X 1.2 = 6000N $\quad\quad\quad\quad\quad\quad\quad\quad\quad\quad$ ...(i)
Since Unit mass (m) = σ.b.t.L
$\quad\quad$ = 1/1000 X 20 X 1.2 X 100 = 2.4Kg $\quad\quad\quad\quad\quad\quad\quad\quad$ ...(ii)
Since velocity(V) is not given, So we find the velocity using formula T_C =1/3 Tmax = mV², for maximum power
Using the formula, 1/3 T_{max} = mV²
$\quad\quad$ V = (T_{max}/3m)^{1/2}
$\quad\quad$ V = (6000/3 X 2.4)^{1/2}
$\quad\quad$ V = 28.86 m/sec $\quad\quad\quad\quad\quad\quad\quad\quad\quad\quad\quad\quad\quad$ ...(iii)
Using the formula, T_C = mV², For finding T_C

$= 2.4(28.86)^2$
$= 1998.96 N$ ...(iv)

Using the formula, $T_{max} = T_1 + T_c$, for finding T_1
$6000 = T_1 + 1998.96$
$T_1 = 4001 N$ ...(v)

Now, For T_2, Using relation Ratio of belt tension $= T_1/T_2 = e^{\mu\theta} = 2$
$4001/T_2 = 2$
$T_2 = 2000.5 N$ ...(vi)

Now Maximum Power Transmitted $= (T_1 - T_2).V/1000$ KW
$P = (4001 - 2000.5) \times 28.86/1000$ KW
$P = 57.73$ KW ....ANS

Q. 28: What is V-belt. Drive the expression of Ratio in belt tension for V-belt

Sol: The power from one shaft to another shaft is also transmitted with the help of V-belt drive and rope drive. Fig shows a V-belt with a grooved pulley.

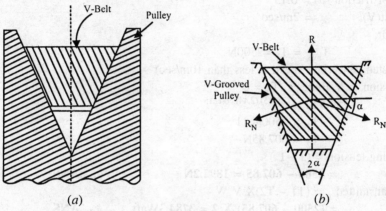

Fig 10.15

Sol: Let
R_N = Normal reaction between belt and sides with a grooved pulley.
2α = Angle of groove
μ = Co-efficient of friction between belt and pulley.
R = Total reaction in the plane of groove.

Resolving the forces vertically, we get
$R = R_N \sin \alpha + R_N \sin \alpha$
$= 2 R_N \sin \alpha$
$R_N = (R/2) \csc \alpha$ ...(i)

Frictional resistance $= \mu R_N + \mu R_N = 2\mu R_N = 2\mu(R/2)\csc \alpha$
$= \mu R \csc \alpha = R \cdot \mu \csc \alpha$

Since in flat belt frictional resistance is equal to μR, and in case of V-belt $\mu \csc \alpha \times R$
So,

Ratio of Tension in V-Belt:: $T_1/T_2 = e^{\mu.\theta.\csc \alpha}$

Q. 29: What do you mean by rope drive.

Sol: The ropes are generally circular in section. Rope-drive is mostly used when the distance between the driving shaft and driven shaft is large. Frictional grip in rope-drive is more than that in V-belt drive.

The ratio of tensions in this case will also be same as in case of V-belt. Hence ratio of tension will be as:

Ratio of Tension in Rope Drive:: $T_1/T_2 = e^{\mu.\theta.cosec\alpha}$

Q. 30: The maximum allowable tension, in a V-belt of groove angle of 30°, is 2500N. The angle of lap is 140° and the coefficient of friction between the belt and the material of the pulley is 0.15. If the belt is running at 2m/sec, Determine:

(i) Net driving tension (ii) Power transmitted by the pulley, Neglect effect of centrifugal tension.

Sol: Given data

Angle of groove(2α) = 30°, α = 15°
Max. Tension(T_{max}) = 2500N
Angle of lap(contact) (θ) = 140° = 140° X (Π/180°) = 2.44 rad
Coefficient of friction (μ) = 0.15
Speed of belt(V) = 2m/sec

We know that,

$$T_{max} = T_1 = 2500N$$

(T_C is neglected, since belt speed is less than 10m/sec)

Ratio of Tension in V-Belt:: $T_1/T_2 = e^{\mu.\theta.cosec\,\alpha}$

$$2500/T_2 = e^{(0.15).(2.44).cosec15}$$
$$T_2 = 2500/4.11$$
$$T_2 = 607.85N \qquad ...(i)$$

(i) Net driving tension = (T_1-T_2)
= 2500 – 607.85 = 1892.2N ....ANS

(iii) Power transmitted = (T1 – T2) X V W
= (2500 – 607.85) X 2 = 3784.3 Watt ....ANS

Q. 31: A pulley used to transmit power by means of ropes, has a diameter of 3.6m and has 15 groove of 45° angle. The angle of contact is 170° and the coefficient of friction between the ropes and the groove side is 0.28. The maximum possible tension in the ropes is 960N and the mass of the rope is 1.5Kg per m length. What is the speed of the pulley in rpm and the power transmitted if the condition of maximum power prevails?

Sol: Given data

Dia. Of pulley(D) = 3.6m
Number of groove(or ropes) = 15
Angle of groove(2a) = 45°, α = 22.50°
Angle of contact(θ) = 170° = 1700 X (Π/180°) = 2.97 rad
Coefficient of friction(μ) = 0.28
Max. Tension(Tmax) = 960N
Mass of rope(m) = 1.5Kg per m length

For maximum power:

$$T_c = 1/3\,Tm$$
$$= 1/3 \times 960 = 320N \qquad ...(i)$$

$$T_m = T_1 + T_C$$
$$960 = T_1 + 320$$
$$T_1 = 640N \qquad \qquad \ldots(ii)$$

Now
$$T_c = (1/3)T_m = mV^2$$
$$V = (T_m/3m)^{1/2}$$
$$= [960/(3 \times 1.5)]^{1/2}$$
$$= 14.6 \text{m/sec} \qquad \qquad \ldots(iii)$$

Since $V = \pi DN/60 = 14.6$,

N = 77.45 R.P.M. ....ANS

Now, Ratio of Tension in V-Belt:: $T_1/T_2 = e^{\mu.\theta.\cosec\,\alpha}$
$$640/T_2 = e^{(0.28).(2.97).\cosec 22.5}$$
$$T_2 = 73.08N \qquad \qquad \ldots(iv)$$

Maximum power transmitted$(P) = (T_1-T_2).v/1000$ Kw
$$P = [(640 - 73.08) \times 14.6]/1000 \text{ KW}$$
$$P = 8.277 \text{KW}$$

Total maximum power transmitted = Power of one rope X No. of rope

P = 8.277 X 15 = 124.16KW ....ANS

Chapter 11

LAWS OF MOTION

Q. 1 : Define Kinetics. What is plane motion?
Sol : Kinetics of that branch of mechanics, which deals with the force system, which produces acceleration, and resulting motion of bodies.
PLANE MOTION: The motion of rigid body, in which all particles of the body remain at a constant distance from a fixed reference plane, is known as plane motion.

Q. 2 : Define the following terms: Matter, Particle, Body, Rigid body, Mass, Weight and Momentum?
Sol : Matter: Matter is any thing that occupies space, possesses mass offers resistance to any stress, example Iron, stone, air, Water.
  Particle: A body of negligible dimension is called a particle. But a particle has mass.
  Body: A body consists of a No. of particle, It has definite shape.
  Rigid body: A rigid body may be defined as the combination of a large no. of particles, Which occupy fixed position with respect to another, both before and after applying a load.
  A rigid body may be defined as a body, which can retain its shape and size even if subjected to some external forces. In actual practice, no body is perfectly rigid. But for the shake of simplicity, we take the bodies as rigid bodies.
  Mass: The properties of matter by which the action of one body can be compared with that of another is defined as mass.

$$m = \rho.v$$

Where,

ρ = Density of body
V = Volume of the body

  Weight: Weight of a body is the force with which the body is attracted towards the center of the earth.
  Momentum : It is the total motion possessed by a body. It is a vector quantity. It can be expressed as,
  Momentum(M) = mass of the body(m) × Velocity(V) **Kg-m/sec**

Q. 3 : Define different Newton's law of motion.
Sol.: The entire system of Dynamics is based on three laws of motion, which are the basis assumptions, and were formulated by Newton.

First Law
A particle remains at rest (if originally at rest) or continues to move in a straight line (If originally in motion) with a constant speed. If the resultant force acting on it is Zero.

It is also called the **law of inertia**, and consists of the following two parts:
A body at rest has a tendency to remain at rest. It is called inertia of rest.
A body in motion has a tendency to preserve its motion. It is called inertia of motion.

Second Law

The rate of change of momentum is directly proportional to the external force applied on the body and take place, in the same direction in which the force acts.

Let a body of mass 'm' is moving with a velocity 'u' along a straight line. It is acted upon a force 'F' and the velocity of the body becomes 'v' in time 't' then.

Initial momentum = m.u
Initial momentum = m.v
Change in momentum = m(v-u)
Rate of change of momentum = change of momentum / Time
$\qquad$ = m(v-u)/t
but $\qquad$ v = u + a.t
$\qquad$ a = (v-u)/t
i.e Rate of change of momentum = m.a
But according to second law   F proportional to m.a
i.e. $\qquad$ F = k.m.a   Where K = constant.
Unit of force
$\qquad$ 1N = 1 kg-m/sec^2 = 10^5 dyne = 1 grm.cm/sec^2

Third Law

The force of action and reaction between interacting bodies are equal in magnitude, opposite in direction and have the same line of action.

Q. 4 : A car of mass 400kg is moving with a velocity of 20m/sec. A force of 200N acts on it for 2 minutes. Find the velocity of the vehicle:

(1) When the force acts in the direction of motion.

(2) When the force acts in the opposite direction of the motion.

Sol :
$\qquad$ m = 400Kg, u = 20m/sec, F = 200N, t = 2min = 120sec, v =?
Since $\qquad$ F = ma
$\qquad$ 200 = 400 X a
$\qquad$ a = 0.5m/sec^2 $\qquad$ ...(i)

(1) Velocity of car after 120sec, When the force acts in the direction of motion.
$\qquad$ v = u + at
$\qquad$ = 20 + 0.5 X 120
$\qquad$ **v = 80m/sec** $\qquad$ ....ANS

(2) Velocity of car after 120sec, When the force acts in the opposite direction of motion.
$\qquad$ v = u - at
$\qquad$ = 20 - 0.5 X 120
$\qquad$ **v = -40m/sec** $\qquad$ ....ANS
-ve sign indicate that the body is moving in the reverse direction

Q. 5 : A body of mass 25kg falls on the ground from a height of 19.6m. The body penetrates into the ground. Find the distance through which the body will penetrates into the ground, if the resistance by the ground to penetrate is constant and equal to 4998N. Take $g = 9.8 m/sec^2$.

Sol : Given that:

$m = 25 Kg, h = 19.6m, s = ?, F_r = 4998N, g = 9.8 m/sec^2$

Let us first consider the motion of the body from a height of 19.6m to the ground surface,

Initial velocity = u = 0,

Let final velocity of the body when it reaches to the ground = v,

Using the equation, $v^2 = u^2 + 2gh$

$$v^2 = (0)^2 + 2 \times 9.8 \times 19.6$$

$$v = 19.6 m/sec \qquad \qquad ...(i)$$

When the body is penetrating in to the ground, the resistance to penetration is acting in the upward direction. (Resistance is always acting in the opposite direction of motion of body.) But the weight of the body is acting in the downward direction.

Weight of the body = $mg = 25 \times 9.8 = 245N$ ...(ii)

Upward resistance to penetrate = 4998N

Net force acting in the upward direction = F

$$F = F_r - mg$$

$$= 4998 - 245 = 4753 N \qquad \qquad ...(iii)$$

Using F = ma, $4753 = 25 \times a$

$$a = 190.12 \, m/sec^2 \qquad \qquad ...(iv)$$

Now, calculation for distance to penetrate

Consider the motion of the body from the ground to the point of penetration in to ground.

Let the distance of penetration = s,

Final velocity = v,

Initial velocity = u = 19.6 m/sec,

Retardation a = $190.12 m/sec^2$

Using the relation, $v^2 = u^2 - 2as$

$$(0)^2 = (19.6)^2 - 2 \times 190.12 \times S$$

$$S = 1.01 m \qquad \qquad ....ANS$$

Q. 6 : A man of mass 637N dives vertically downwards into a swimming pool from a tower of height 19.6m. He was found to go down in water by 2m and then started rising. Find the average resistance of the water. Neglect the resistance of air.

Sol: Given that:

$W = 637N, h = 19.6m, S = 2m, g = 9.8 m/sec^2$

Let, F_r = Average resistance

Initial velocity of man u = 0,

$$V^2 = u^2 + 2gh$$

$$= 0 + 2 \times 9.8 \times 19.6$$

$$V = 19.6 \, m/sec \qquad \qquad ...(i)$$

Now distance traveled in water = 2m, v = 0, u = 19.6 m/sec now apply

$$V^2 = u^2 - 2as$$

$$0 = 19.6^2 - 2a \times 2$$

$$a = 96.04 \text{ m/sec}^2 \qquad ...(ii)$$

Since, net force acting on the man in the upward direction $= F_r - W$
But the net force acting on the man must be equal to the product of mass and retardation.

$$F_r - W = ma$$
$$F_r - 637 = (637/g) \times 96.04$$
$$F_r = 6879.6 \text{ N} \qquad ......\text{ANS}$$

Q. 7 : A bullet of mass 81gm and moving with a velocity of 300m/sec is fired into a log of wood and it penetrates to a depth of 10cm. If the bullet moving with the same velocity were fired into a similar piece of wood 5cm thick, with what velocity would it emerge? Find also the force of resistance assuming it to be uniform.

Sol: Given that
$$m = 81 \text{gm} = 0.081 \text{Kg}, u = 300 \text{m/sec}, s = 10 \text{cm} = 0.1 \text{m}, v = 0$$

As the force of resistance is acting in the opposite direction of motion of bullet, hence force of resistance will produce retardation on the bullet, Apply, $V^2 = u^2 - 2as$

$$0 = 300^2 - 2a(0.1)$$
$$a = 450000 \text{ m/sec}^2 \qquad ...(i)$$

Let F is the force of resistance offered by wood to the bullet.
Using equation, $F = ma$,
$$F = 0.081 \times 450000$$
$$F = 36450 \text{ N} \qquad ......\text{ANS}$$

Let v = velocity of bullet with which the bullet emerges from the piece of wood of 5cm thick,
$$U = 300 \text{m/sec}, a = 450000 \text{m/sec}^2, s = 0.05 \text{m}$$
Using equation, $V^2 = u^2 - 2as$
$$V^2 = 300^2 - 2 \times 450000 \times 0.05$$
$$V = 212.132 \text{ m/sec} \qquad ......\text{ANS}$$

Q. 8 : A particle of mass 1kg moves in a straight line under the influence of a force, which increases linearly with the time at the rate of 60N per sec. At time t = 0 the initial force may be taken as 50N. Determine the acceleration and velocity of the particle 4sec after it started from rest at the origin.

Sol: As the force varies linearly with time,
$$F = mt + C$$
Differentiate the equation with time,
$$dF/dt = m = 60 \text{(given)}$$
i.e., $\qquad m = 60 \qquad ...(i)$
Given that, at $t = 0$, $F = 50 \text{N}$,
$$50 = 60 \times 0 + C$$
$$C = 50 \qquad ...(ii)$$
Now the equation becomes,
$$F = 60t + 50 \qquad ...(iii)$$
Since, $\qquad F = ma, m = 1 \text{Kg}$
$$F = ma = 1.a = 60t + 50$$
At $t = 4$ sec,
$$a = 60 \times 4 + 50 = 290$$

$\qquad$ a = 290m/sec² ....ANS

also, $\qquad$ a = dv/dt
$\qquad$ a = dv/dt = 60t + 50

Integration both side for the interval of time 0 to 4sec.

$$V = \int_0^4 (60t + 50)dt$$

$\qquad$ V = (60t² + 50t), limit are 0 to 4
$\qquad$ V = 30(4)² + 50 X 4
$\qquad$ **V = 680m/sec** ....ANS

Q. 9 : Determine the acceleration of a railway wagon moving on a railway track if fraction force exerted by wagon weighing 50KN is 2000N and the frictional resistance is 5N per KN of wagon's weight.

Sol: Let a be the acceleration of the wagon
$\qquad$ Mass (m) = W/g = (50 X 1000/9.81)
$\qquad$ Friction force F_r = 5 X 50 = 250N
$\qquad$ Net force = F - F_r = ma
$\qquad$ 2000 – 250 = (50 X 1000/9.81)a
$\qquad$ **a = 0.3438m/sec²** ....ANS

Q.10: A straight link AB 40cm long has, at a given instant, its end B moving along line OX at 0.8m/s and acceleration at 4m/sec² and the other end A moving along OY, as shown in fig 11.1. Find the velocity and acceleration of the end A and of mid point C of the link when inclined at 30⁰ with OX.

Sol: Let the length of link is L = 40cm and AD = Y, OB = X
$\qquad$ X² + Y² = 1 $\qquad$ ...(i)

Diff with respect to time, and -ive sign is taken for down word motion of A, when B is moving in +ive direction, we get
$\qquad$ 2Xdx/dt – 2Ydy/dt = 0XV_B – YV_A = 0 $\qquad$ ...(ii)
$\qquad$ V_A = (X/Y)V_B = (Lcosθ/Lsinθ)V_B = V_B/tanθV_A = 0.8/tan30⁰ = 1.38m/sec ...(iii)
$\qquad$ **V_A = 1.38m/sec** ....ANS

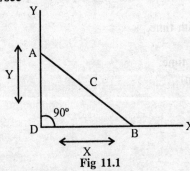

Fig 11.1

Again differentiating equation (2), we get
$\qquad$ Xd²x/dt² + (dx/dt)² – yd²y/dt² – (dy/dt)² = 0
$\qquad$ X.a_B + (V_B)² – Y.a_A – (V_A)² = 0
$\qquad$ 0.4cos30⁰ X 0.4 – (0.8)² – 0.4sin30⁰ X a_A – (1.38)² = 0
$\qquad$ 1.38 + 0.64 – 0.2a_A – 1.9 = 0
$\qquad$ **a_A = 0.6.6m/sec²** ....ANS

Q.11 : A 20KN automobile is moving at a speed of 70Kmph when the brakes are fully applied causing all four wheels to skid. Determine the time required to stop the automobile.

(1) on concrete road for which $\mu = 0.75$

(2) On ice for which $\mu = 0.08$

Sol: Given data: W = 20KN, u = 70Kmphr = 19.44m/sec, v = 0, t = ?

Consider FBD of the car as shown in fig 11.2

$\Sigma V = 0$, R = W ...(i)

$\Sigma H = 0$, $F_r = 0$ $F_r = \mu R$ ...(ii)

Here net force is the frictional force

i.e. $F = F_r$ $ma = \mu R = \mu mga = \mu g$ ...(iii)

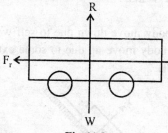

Fig 11.2

(1) on concrete road for which $\mu = 0.75$

$a = \mu g = 0.75 \times 9.81 = 7.3575$

$a = 7.35$ m/sec^2 ...(iv)

Using the relation $v = u - at$

$0 = 19.44 - 7.35t$

t = 2.64 seconds ....ANS

(1) On ice for which $\mu = 0.08$

$a = \mu g = 0.08 \times 9.81 = 0.7848$

$a = 0.7848$ m/sec^2 ...(v)

Using the relation $v = u - at$

$0 = 19.44 - 0.7848t$

t = 24.77 seconds ....ANS

Q. 12: Write different equation of motion on inclined plane for the following cases.

(a) Motion on inclined plane when surface is smooth.

(b) Motion on inclined plane when surface is rough.

Sol: CASE: 1 WHEN SURFACE SMOOTH

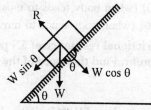

Fig 11.3

Fig 11.3 shows a body of weight W, sliding down on a smooth inclined plane.
Let,

θ = Angle made by inclined plane with horizontal
a = Acceleration of the body
m = Mass of the body = W/g

Since surface is smooth i.e. frictional force is zero. Hence the force acting on the body are its own weight W and reaction R of the plane.

The resolved part of W perpendicular to the plane is Wcos θ, which is balanced by R, while the resolved part parallel to the plane is Wsin θ, which produced the acceleration down the plane.

Net force acting on the body down the plane

F = W.sin θ, but F = m.a
m.a = m.g.sinθ

i.e.   a = g.sin θ (For body move down due to self weight.)
and,   a = -g.sin θ (For body move up due to some external force)

CASE: 2 WHEN ROUGH SURFACE

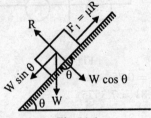

Fig 11.4

Fig 11.4 shows a body of weight W, sliding down on a rough inclined plane.
Let,

θ = Angle made by inclined plane with horizontal
a = Acceleration of the body
m = Mass of the body = W/g
μ = Co-efficient of friction
F_r = Force of friction

when body tends to move down:

R = w.cosθ
F_r = μ.R = μ.W.cosθ

Net force acting on the body F = W.sinθ - μ.W.cosθ
i.e.   m.a = W.sinθ - μ.W.cosθ
Put m = W/g we get

a = g.[sinθ - μ.cosθ] (when body tends to move down)
a = –g.[sinθ - μ.cosθ] (when body tends to move up)

Q. 13 : A train of mass 200KN has a frictional resistance of 5N per KN. Speed of the train, at the top of an inclined of 1 in 80 is 45 Km/hr. Find the speed of the train after running down the incline for 1Km.

Sol: Given data,

Mass m = 200KN, Frictional resistance F_r = 5N/KN, sinθ = 1/80 = 0.0125,

Initial velocity u = 45Km/hr = 12.5m/sec, s = 1km = 1000m
Total frictional resistance = 5 X 200 = 1000N = 1KN ...(i)
Force responsible for sliding = $W\sin\theta$ = 200 X 0.0125 = 2.5KN
Now, Net force, $F = F - F_r = ma$
$$2.5 - 1 = (200/9.81)a$$
$$a = 0.0735 m/sec^2$$ ...(ii)
Apply the equation, $v^2 = u^2 + 2as$
$$v^2 = 0 + 2 \times 0.0735 \times 1000$$
$$v = 12.1 \text{ m/sec}$$ ....ANS

Q.13: A train of wagons is first pulled on a level track from A to B and then up a 5% upgrade as shown in fig (11.5). At some point C, the least wagon gets detached from the train, when it was traveling with a velocity of 36Km.p.h. If the detached wagon has a mass of 5KN and the track resistance is 10N per KN, find the distance through which the wagon will travel before coming to rest. Take g = 9.8m/sec².

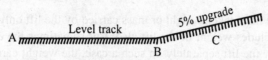

Fig 11.5

Sol: Given that, Grade = 5% or $\sin\theta$ = 5% = 0.05, u = 36Km.p.h. = 10m/sec,
W = 5KN, V = 0, F_r = 10N/KN
Let s = Distance traveled by wagon before coming to rest
Total track resistance F_r = 10 X 5 = 50N ...(i)
Resistance due to upgrade = $m\sin\theta$ = 5 X 0.05 = 0.25KN = 250N ...(ii)
Total resistance to wagon = Net force = 50 + 250 = 300N
But, F = ma, 300 = (5000/9.81)a
$$a = 0.588 m/sec^2$$ ...(iii)
Apply the equation, $v^2 = u^2 - 2as$
$$0 = (10)^2 - 2 \times 0.588 \times s$$
$$s = 85 \text{ m}$$ ....ANS

Q.14: Write equation of motion of lift when move up and when move down.

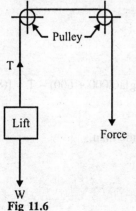

Fig 11.6

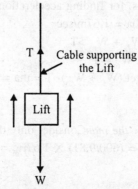

Fig 11.7 Lift is moving upward

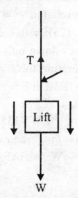

Fig 11.8 Lift is moving downward

Let,
W = Weight carried by the lift
m = Mass carried by lift = W/g
a = Uniform acceleration
T = Tension in cable supporting the lift, also called Reaction of the lift
For UP MOTION
Net force in upward direction = T-W
Also Net Force = m.a
i.e.   T-W = m.a   ...(i)
FOR DOWN MOTION
   Net force = W - T
Also Net Force = m.a
i.e.
      W - T = m.a   ...(ii)

Note: In the above cases, we have taken weight or mass carried by the lift only. We have assumed that the weight carried by the lift includes weight of the lift also. But sometimes the example contains weight of the lift and weight carried by the lift separately. In such a case, the weight carried by the lift or weight of the operator etc, will exert a pressure on the floor of the lift. Whereas tension in the cable will be given by the algebraic sum of the weight of the lift and weight carried by the lift.

Q.15: An elevator cage of a mineshaft, weighing 8KN when, is lifted or lowered by means of a wire rope. Once a man weighing 600N, entered it and lowered with uniform acceleration such that when a distance of 187.5m was covered, the velocity of cage was 25m/sec. Determine the tension in the rope and the force exerted by the man on the floor of the cage.

Sol: Given data;
Weight of empty lift W_L = 8KN = 8000N
Weight of man W_m = 600N
Distance covered by lift s = 187.5m
Velocity of lift after 187.5m v = 25m/sec
Tension in rope T = ?
Force exerted on the man F_m =?
Apply the relation $v^2 = u^2 + 2as$, for finding acceleration
   $(25)^2 = 0 + 2a(187.5)$ a = 1.67m/sec^2 ...(i)
Cage moves down only when $W_L + W_m$ >T
Net accelerating force = $(W_L + W_m)$- T
Using the relation F = ma, we get $(W_L + W_m)$- T = ma = [$(W_L + W_m)$/g]a(8000 + 600) – T = [(8000 + 600)/9.81] X 1.67
         T = 7135.98N   **.......ANS**

*Calculation for force exerted by the man*Consider only the weight of the man,
   $F_m - W_m$ = maF_m – 600 = (600/9.81) X 1.67F_m = 714.37N

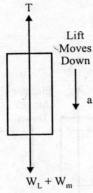

Fig 11.9

Since Newton's third law i.e The force of action and reaction between interacting bodies are equal in magnitude, opposite in direction and have the same line of action.

i.e., Force exerted by the man = F = 714.37N   ....ANS

Q.16: An elevator weight 2500N and is moving vertically downward with a constant acceleration.

(1) Write the equation for the elevator cable tension.

(2) Starting from rest it travels a distance of 35m during an interval of 10 sec. Find the cable tension during this time.

(3) Neglect all other resistance to motion. What are the limits of cable tension.

Sol: Given data;

Weight of elevator W_E = 2500N

Initial velocity u = 0

Distance traveled s = 35m

Time t = 10sec

(1) Since elevator is moving down

Net acceleration force in the down ward direction

$$= W_E - T = (2500 - T)N \qquad \ldots(i)$$

The net accelerating force produces acceleration 'a' in the down ward direction.

Using the relation, F = ma

$$2500 - T = (2500/9.81)a$$

$$T = 2500 - (2500/9.81)a \qquad \ldots\text{ANS}$$

Hence the above equation represents the general equation for the elevator cable tension when the elevator is moving downward.

(2) Using relation,

$$s = ut + \frac{1}{2}at^2 = 35 = 0 \times 10 + 1/2 \times a \,(10)^2 \qquad \ldots(ii)$$

∴ a = 0.7 m/sec²

Substituting this value of a in the equation of cable tension

$$T = 2500 - (2500/9.81) \times 0.7 T = 2321.61N \qquad \ldots\text{ANS}$$

(3) $T = 2500 - (2500/9.81)a$

Limit of cable tension is depends upon the value of a, which varies from 0 to g i.e. 9.81m/sec²

At   a = 0, T = 2500

i.e elevator freely down
At   a = 9.81, T = 0
i.e elevator is at the top and stationary.
Hence Limits are 0 to 2500N   ....ANS

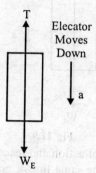

Fig 11.10

Q. 17: A vertical lift of total mass 500Kg acquires an upward velocity of 2m/sec over a distance of 3m of motion with constant acceleration, starting from rest. Calculate the tension in the cable supporting the lift. If the lift while stopping moves with a constant deceleration and comes to rest in 2sec, calculate the force transmitted by a man of mass 75kg on the floor of the lift during the interval.

Sol: Given data;
Mass of lift M_L = 500Kg
Final Velocity v = 2m/sec
Distance covered s = 3m
Initial velocity u = 0
Cable tension T = ?
Apply the relation $v^2 = u^2 + 2as$
$$2^2 = 0 + 2a \times 3 \quad a = 2/3 \text{ m/sec}^2 \quad ...(i)$$
Since lift moves up, T > $M_L \times g$ Net accelerating force = T – $M_L g$, and it is equal to, T – $M_L g$ = maT – 500 X 9.81 = 500 X 2/3
        T = 5238.5N   ....ANS
Let force transmitted by man of mass of 75Kg, is FF – mg = ma For finding the acceleration, using the relation v = u + at0 = 2 + a X 2

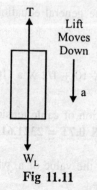

Fig 11.11

$$a = -1 \text{ m/sec}^2$$
...(ii)

Putting the value in equation, $F - mg = ma$
$$F - 75 \times 9.81 = 75(-1)$$
$$F = 660.75N \qquad \text{.......ANS}$$

Q. 18: An elevator weight 5000N is ascending with an acceleration of 3m/sec². During this ascent its operator whose weight is 700N is standing on the scale placed on the floor. What is the scale reading? What will be the total tension in the cable of the elevator during this motion?

Sol: Given data, $W_E = 5000N$, $a = 3\text{m/sec}^2$, $W_O = 700N$,

Let R = Reaction offered by floor on operator. This is also equal to the reading of scale.
T = total tension in the cable

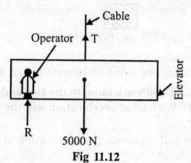

Fig 11.12

Net upward force on operator
$$= \text{Reaction offered by floor on operator} - \text{Weight of operator}$$
$$= R - 700$$

But, Net force = ma
$$R - 700 = (700/9.81) \times 3$$
$$R = 914.28N \qquad \text{.......ANS}$$

Now for finding the total tension in the cable, Total weight of elevator is considered.

Net upward force on elevator and operator
$$= \text{Total tension in the cable} - \text{Total weight of elevator and operator}$$
$$= T - 5700$$

But net force = mass × acceleration
$$T - 5700 = (5700/9.81) \times 3$$
$$T = 7445N \qquad \text{.......ANS}$$

Q.19: Analyse the motion of connected bodies, which is connected by a pulleys.

Sol: Fig 11.13 shows a light and inextensible string passing over a smooth and weightless pulley. Two bodies of weights W_1 and W_2 are attached to the two ends of the string.

Let $W_1 > W_2$, the weight W_1 will move downwards, whereas smaller weight W_2 will move upwards. For an inextensible string, the upward acceleration of the weight W_2 will be equal to the downward acceleration of the weight W_1.

As the string is light and inextensible and passing over a smooth pulley, the tension of the string will be same on both sides of the pulley.

Consider the Motion of weight W_1 (Down motion)
$$W_1 - T = m_1.a \qquad \text{...(i)}$$

Consider the Motion of weight W_2 (up motion)

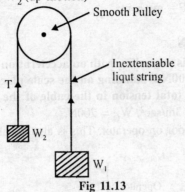

Fig 11.13

$T - W_2 = m_2 \cdot a$ ...(ii)

Solved both the equation for finding the value of Tension (T) or acceleration (a)

Q.20: Two bodies weighing 300N and 450N are hung to the two ends of a rope passing over an ideal pulley as shown in fig (11.14). With what acceleration will the heavier body come down? What is the tension in the string?

Sol: Since string is light, inextensible and frictionless, so the tension in the string on both side is equal to T, let acceleration of both the block is 'a'.

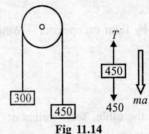

Fig 11.14

Let 450N block moves down,
Consider the motion of 450N block,
Apply the equation, $F = ma$
$450 - T = (450/9.81) a$
$450 - T = 45.87a$ ...(i)
Consider the motion of 300N block,
Apply the equation, $F = ma$
$T - 300 = (300/9.81) a$
$T - 300 = 30.58a$ ...(ii)
Add equation (1) and (2)
$150 = 76.45a$
$a = 1.962 m/sec^2$ ....ANS
Putting the value of a in equation (i), we get
$T = 360N$ ....ANS

Q.21: Find the tension in the string and accelerations of blocks A and B weighing 200N and 50N respectively, connected by a string and frictionless and weightless pulleys as shown in fig 11.15.

Sol: Given Data,
Weight of block A = 200N
Weight of block B = 50N
As the pulley is smooth, the tension in the string will be same throughout
Let, T = Tension in the string a = Acceleration of block B
Then acceleration of block A will be equal to half the acceleration of block B.
Acceleration of block
$$A = a/2 \qquad \ldots(i)$$
As the weight of block is more than the weight of block B, the block A will move downwards whereas the block B will move upwards.

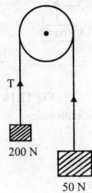

Fig 11.15

Consider the motion of block B,
Net force = T − 50   ...(ii)
Since Net force, F = ma
    T − 50 = (50/9.81) a
    T − 50 = 5.1a   ...(iii)
Consider the motion of block A,
Net force = 200− 2T   ...(iv)
Since Net force, F = ma
    200− 2T = (200/9.81)(a/2)
    200− 2T = 10.19a
    100 − T = 5.1a   ...(v)
Add equation (3) and (5)
    50 = 10.19a
    a = 4.9m/sec²   ....ANS
Putting the value of a in equation (5) we get
    T = 75N   ....ANS

Q.22: The system of particles shown in fig 11.16 is initially at rest. Find the value of force F that should be applied so that the system acquires a velocity of 6m/sec after moving 5m.
(Nov−03(C.O.))

Sol: Given data,
Initial velocity u = 0

Final velocity v = 6m/sec
Distance traveled s = 5m
For finding acceleration, using the relation, $v^2 = 4^2 + 2as$
∴   a = 3.6 m/sec² $6^2 = 0 + 2a \times 5$ ...(i)
Apply the relation F = ma,

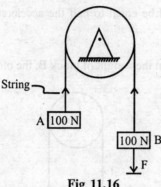

Fig 11.16

Let T = Tension in the string, same for both side
Using the relation F = ma, for block A
    T − 100 = ma
    T − 100 = (100/9.81) X 3.6   ...(ii)
Using the relation F = ma, for block B
    100 + F − T = ma
    100 + F − T = (100/9.81) X 3.6   ...(iii)
Add equation (2) and (3), we get
    F = 2[(100/9.81) X 3.6]
    F = 73.5N   ....ANS

Q.23: A system of weight connected by string passing over pulleys A and B is shown in fig. Find the acceleration of the three weights. Assume weightless string and ideal condition for pulleys.

Sol: As the strings are weightless and ideal conditions prevail, hence the tensions in the string passing over pulley A will be same. The tensions in the string passing over pulley B will also be same. But the tensions in the strings passing over pulley A and over pulley B will be different as shown in fig 11.17.

Let T_1 = Tension in the string passing over pulley A
T_2 = Tension in the string passing over pulley B

One end of the string passing over pulley A is connected to a weight 15N, and the other end is connected to pulley B. As the weight 15N is more than the weights (6 + 4 = 10N), hence weight 15N will move downwards, whereas pulley B will move upwards. The acceleration of the weight 15N and of the pulley B will be same.

Let, a = Acceleration of block 15N in downward directiona_1 = Acceleration of 6N downward with respect to pulley B.

Then acceleration of weight of 4N with respect to pulley B = a_1 in the upward direction.

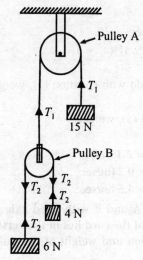

Fig 11.17

Absolute acceleration of weight 4N,

= Acceleration of 4N w.r.t. pulley B + Acceleration of pulley B.

= $a_1 + a$ (upward)

(as both acceleration are in upward direction, total acceleration will be sum of the two accelerations)

Absolute acceleration of weight 6N,

= Acceleration of 6 w.r.t. pulley B + Acceleration of pulley B.

= $a_1 - a$ (downward)

(As a_1 is acting downward whereas a is acting upward. Hence total acceleration in the downward direction)

Consider the motion of weight 15N

Net downward force = $15 - T_1$

Using F = ma,

$$15 - T_1 = (15/9.81)a \qquad \qquad \ldots(1)$$

Consider the motion of weight 4N

Net downward force = $T_2 - 4$

Using F = ma,

$$T_2 - 4 = (4/9.81)(a + a_1) \qquad \qquad \ldots(2)$$

Consider the motion of weight 6N

Net downward force = $6 - T_2$

Using F = ma,

$$6 - T_2 = (6/9.81)(a_1 - a) \qquad \qquad \ldots(3)$$

Consider the motion of pulley B,

$$T_1 = 2T_2 \qquad \qquad \ldots(4)$$

Adding equation (2) and (3)

$$2 = (4/9.81)(a + a_1) + (6/9.81)(a_1 - a)$$

$$9.81 = 5a_1 - a \qquad \qquad \ldots(5)$$

Multiply equation (2) by 2 and put the value of equation (4), we get

$T_1 - 8 = (8/9.81)(a_1 + a)$ ...(6)

Adding equation (1) and (6), we get

$15 - 8 = (15/9.81)a + (8/9.81)(a_1 + a)$

$23a + 8a_1 = 7 \times 9.81$ ...(7)

Multiply equation (5) by 23 and add with equation (7), we get

$a_1 = 2.39 m/sec^2$ ....ANS

Putting the value of a_1 in equation (5), we get

$a = 2.15 m/sec^2$ ....ANS

Acceleration of weight 15N = a = $2.15 m/sec^2$ ....ANS

Acceleration of weight 6N = a = $0.24 m/sec^2$ ....ANS

Acceleration of weight 4N = a = $4.54 m/sec^2$ ....ANS

Q.24: A cord runs over two pulleys A and B with fixed axles, and carries a movable pulleys 'c' if $P = 40N$, $P_1 = 20N$, $P_2 = 30N$ and the cord lies in the vertical plane. Determine the acceleration of pulley 'C. Neglect the friction and weight of the pulley.

Sol: $a = a_1 + a_2$ ...(1)

For pulley A, Apply F = ma,

$T - 20 = (20/10) a_1$, take g = $10 m/sec^2$

$T - 20 = 2a_1$ ...(2)

For pulley C, $40 - 2T = (40/10)a 40 - 2T = 4a$ ...(3)

For pulley B, $T - 30 = (30/10) a_2$ $T - 30 = 3a_2$ ...(4)

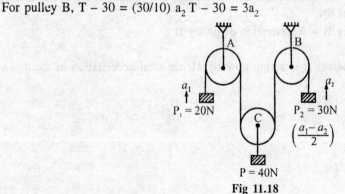

Fig 11.18

From equation (2) and (4)

$2a_1 - 3a_2 = 10$ ...(5)

Equation (3) can be rewritten as

$40 - 2T = 4(a_1 + a_2)$ ...(6)

Now (6) + 2 (4)

$40 - 2T + 2T - 2 \times 30 = 4(a_1 + a_2) + 6a_2$

$-20 = 4a_1 + 10a_2$ ...(7)

Solving equation (5) and (7), we get

$a_1 = 5/4 \, m/sec^2$ ....ANS

$a_2 = -5/2 m/sec^2$ ....ANS

Acceleration of 'C' = $a = a_1 + a_2$

= $5/4 - 5/2 = -1.25 m/sec^2$ **(downward)** ....ANS

Q.25: Analyse the motion of two bodies connected by a string when one body is lying on a horizontal surface and other is hanging free for the following cases.

1. The horizontal surface is smooth and the string is passing over a smooth pulley.
2. The horizontal surface is rough and string is passing over a smooth pulley.
3. The horizontal surface is rough and string is passing over a rough pulley.

Sol: CASE-1: THE HORIZONTAL SURFACE IS SMOOTH AND THE STRING IS PASSING OVER A SMOOTH PULLEY:

Fig shows the two weights W_1 and W_2 connected by a light inextensible string, passing over a smooth pulley. The weight W_2 is placed on a smooth horizontal surface, whereas the weight W_1 is hanging free. The weight W_1 is moving downwards, whereas the weight W_2 is moving on smooth horizontal surface. The velocity and acceleration of W_1 will be same as that of W_2. As the string is light and inextensible and passing over a smooth pulley, the tensions of the string will be same on both sides of the pulley.

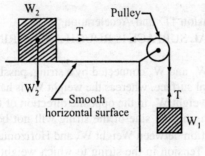

Fig 11.19

For W_1 block: Move down
$$W_1 - T = (W_1/g).a \qquad \ldots(1)$$
For W_2 block
$$T = (W_2/g).a \qquad \ldots(2)$$
(Since W act vertically and T act Horizontally & w.cos90 = 0)
Solve both the equation for the value of 'T' and 'a'.

CASE-2: THE HORIZONTAL SURFACE IS ROUGH AND STRING IS PASSING OVER A SMOOTH PULLEY.

Fig shows the two weights W_1 and W_2 connected by a light inextensible string, passing over a smooth pulley. The weight W_2 is placed on a rough horizontal surface, whereas the weight W_1 is hanging free. Hence in this case force of friction will be acting on the weight W_2 in the opposite direction of the motion of weight W_2.

Let, μ = Coefficient of friction between weight W_2 and horizontal surface. Force of friction = μR_2 = μW_2

Motion of W_1 (Down Motion)
$$W_1 - T = (W_1/g).a \qquad \ldots(1)$$

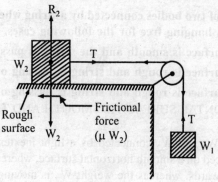

Fig 11.20

Motion of W_2

$$T - \mu.W_2 = (W_2/g).a \qquad ...(2)$$

Solve the equations for Tension 'T' and Acceleration 'a'

CASE-3: THE HORIZONTAL SURFACE IS ROUGH AND STRING IS PASSING OVER A ROUGH PULLEY.

Fig shows the two weights W_1 and W_2 connected by a string, passing over a rough pulley. The weight W_2 is placed on a rough horizontal surface, whereas the weight W_1 is hanging free. Hence in this case force of friction will be acting on the weight W_2 in the opposite direction of the motion. As the string is passing over a rough pulley. The tension on both side of the string will not be same.

Let, μ_1 = Coefficient of Friction between Weight W_2 and Horizontal plane μ_2 = Coefficient of Friction between String and pulley T_1 = Tension in the string to which weight W_1 is attached

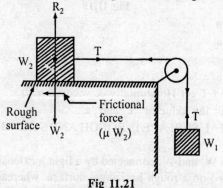

Fig 11.21

T_2 = Tension in the string to which weight W_2 is attached

Force of friction = $\mu_1 R_2 = \mu_1 W_2$

Consider block W_1

$$W_1 - T_1 = (W_1/g).a \qquad ...(1)$$

Consider block W_2

$$T_2 - \mu_2.W_2 = (W_2/g).a \qquad ...(2)$$

Another equation is, $T_1/T_2 = e^{\mu.\theta}$ ...(3)

Solve all three equation for the value of 'a', 'T_1' and 'T_2'

Q.26: Two bodies of weight 10N and 1.5N are connected to the two ends of a light inextensible String, passing over a smooth pulley. The weight 10N is placed on a rough horizontal surface while the weight of 1.5N is hanging vertically in air. Initially the friction between the weight

10N and the table is just sufficient to prevent motion. If an additional weight of 0.5N is added to the weight 1.5N, determine

(i) The acceleration of the two weight.

(ii) Tension in the string after adding additional weight of 0.5N to the weight 1.5N

Sol: Initially when $W_1 = 1.5N$, then the body is in equilibrium. i.e. both in rest or a = 0,
Then consider block W_1
$$R_V = 0; T = W_1 = 1.5N \qquad ...(1)$$
Consider block W_2
$$R_V = 0; R = W_2 = \qquad ...(2)$$
$$F_r - T = 0; F_r = T = 1.5N \qquad ...(3)$$
But, $\quad Fr = \mu R = \mu W_2; \mu W_2 = 1.5; \mu \times 10 = 1.5, \mu = 0.15 \qquad ...(4)$

Now when Weight $W_1 = 2.0N$, body moves down Now the tension on both side be T_1
Consider block $W_1 W_1 - T_1 = ma2 - T_1 = (2/g)a \qquad ...(5)$
Consider block W_2

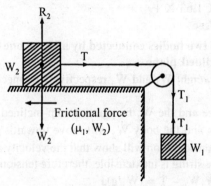

Fig 11.22

$$T_1 - F_r = ma$$
$$T_1 - \mu W_2 = (10/g)a$$
$$T_1 - 1.5 = (10/g)a \qquad ...(6)$$

Solve the equation (5) and (6) for T_1 and a, we get
$$T_1 = 1.916N, a = 0.408 m/sec^2 \qquad ....ANS$$

Q.27: Two blocks shown in fig 11.23, have masses A = 20N and B = 10N and the coefficient of friction between the block A and the horizontal plane, $\mu = 0.25$. If the system is released from rest, and the block B falls through a vertical distance of 1m, what is the velocity acquired by it? Neglect the friction in the pulley and the extension of the string.

Sol: Let T = Tension on both sides of the string.
a = Acceleration of the blocks
$\mu = 0.25$ Consider the motion of block B,
$$W_B - T = ma \qquad ...(1)$$
$$10 - T = \left(\frac{10}{2}\right) \cdot a$$

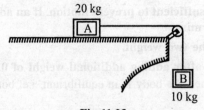

Fig 11.23

Consider the motion of block A,

$T - \mu W_A = ma$

$T - 0.25 \times 20 = (20/g)a$ ...(2)

Add equation (1) and (2)

$10 - 5 = (30/g)a$

$a = 1.63 \text{m/sec}^2$ ...(3)

Now using the relation, $v^2 = u^2 + 2as$

$v^2 = 0 + 2 \times 1.63 \times 1$

$v = 1.81 \text{m/sec}$ ....ANS

Q.28: Analyse the motion of two bodies connected by a string one of which is hanging free and other lying on a smooth inclined plane.

Sol.: Consider two bodies of weight W_1 and W_2 respectively connected by a light inextensible string as shown in fig 11.24

Let the body W_1 hang free and the W_2 be places on an inclined smooth plane.

W_1 will move downwards and the body W_2 will move upwards along the

inclined surface. A little consideration will show that the velocity and acceleration of the body W_1 will be same as that of W_2. Since the string is inextensible, therefore tension in both the string will also be equal.

Consider the motion of $W_1 W_1 - T = (W_1/g)a$ ...(1)

Consider the motion of W_1

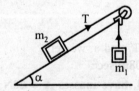

Fig 11.24

$T - W_2 \sin \pm = (W_1/g)a$ ...(2)

Solve the equations for 'T' and 'a'

Q.29: Analyse the motion of two bodies connected by a string one of which is hanging free and other lying on a rough inclined plane.

Sol.: Consider two bodies of weight W_1 and W_2 respectively connected by a light inextensible string as shown in fig 11.25.

Let the body W_1 hang free and the W_2 be places on an inclined rough plane. W_1 will move downwards and the body W_2 will move upwards along the inclined surface.

Consider the motion of $W_1 W_1 - T = (W_1/g)a$ ...(1)

Consider the motion of $W_1 T - W_2 \sin \alpha - \mu W_1 \cos \alpha = (W_1/g)a$ ...(2)

Solve the equations for 'T' and 'a'.

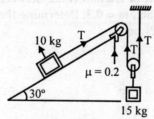

Fig 11.25

Q.30: Determine the resulting motion of the body A, assuming the pulleys to be smooth and weightless as shown in fig 11.26. If the system starts from rest, determine the velocity of the body A after 10 seconds.

Sol.: Given data:

Mass of Block A = 10Kg
Mass of Block B = 15Kg
Angle of inclination $\alpha = 30^0$
Co-efficient of friction m = 0.2
Consider the motion of block B,
The acceleration of block B will be half the acceleration of the block A i.e. a/2,
$$M_1g - 2T = m_1(a/2)$$

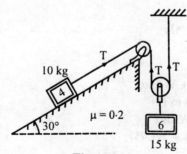

Fig 11.26

$$15 \times 9.81 - 2T = 15 (a/2)$$
$$147.15 - 2T = 7.5a \qquad ...(1)$$

Consider the motion of block B,
$$T - W_2 \sin \alpha - \mu W_1 \cos \alpha = (W_1/g)a$$
$$T - m_2 g \sin \alpha - 0.2 m_2 g \cos \alpha = m_2 a$$
$$T - 10 \times 9.81 \sin 30^0 - 0.2 \times 10 \times 9.81 \cos 30^0 = 10a$$
$$T - 66.04 = 10a \qquad ...(2)$$

Adding equation (1) with 2 X equation (2)
$$147.15 - 2T + 2T - 132.08 = 7.5a + 20a$$
$$a = 0.54 \text{ m/sec}^2 \qquad ....\text{ANS}$$

Now velocity of the block after 10 sec,
Apply v = u + at
$$V = 0 + 0.54 \times 10$$
$$V = 5.4 \text{m/sec} \qquad ....\text{ANS}$$

Q.31: In the fig 11.27, the coefficient of friction is 0.2 between the rope and the fixed pulley, and between other surface of contact, m = 0.3. Determine the minimum weight W to prevent the downward motion of the 100N body.

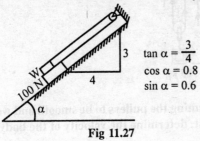

$\tan\alpha = \frac{3}{4}$
$\cos\alpha = 0.8$
$\sin\alpha = 0.6$

Fig 11.27

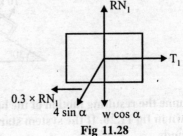

Fig 11.28

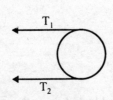

Fig 11.29

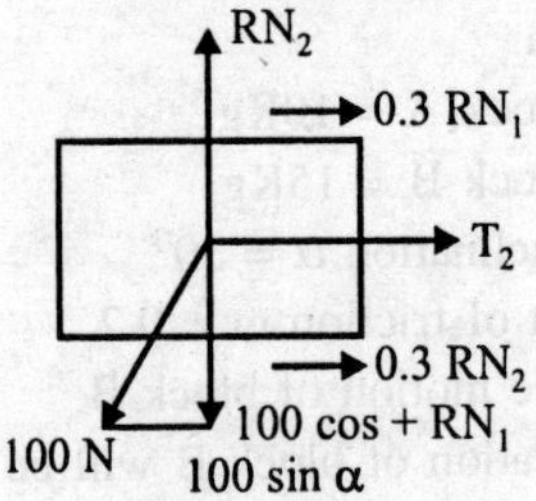

Fig 11.30

Sol.: From the given fig $\tan\alpha = 3/4$,
$\cos\alpha = 4/5$ & $\sin\alpha = 3/5$,
Consider equilibrium of block W
$$R_V = 0; \quad R_2 = W\cos\alpha \qquad \ldots(1)$$
$$R_H = 0; \quad T_1 = \mu R_2 + W\sin\alpha \qquad \ldots(2)$$
Putting the value of equation(1) in (2)
$$T_1 = \mu W\cos\alpha + W\sin\alpha$$
$$= 0.3 \times W(4/5) + W(3/5)$$
$$T_1 = 0.84W \qquad \ldots(3)$$
For pulley; $T_2/T_1 = e^{\mu_1 \theta}$
$$T_2 = T_1 \times e^{\mu_1 \theta}$$
$$= 0.84We^{(0.2 \times \pi)}$$
$$T_2 = 1.574W \qquad \ldots(4)$$
Consider equilibrium of block 100N
$$R_V = 0; \quad R_1 = 100\cos\alpha + R_2 \qquad \ldots(5)$$
$$R_1 = 100\cos\alpha + W\cos\alpha$$
$$= 100(4/5) + W(4/5)$$
$$R_1 = 80 + 0.8W \qquad \ldots(6)$$
$$R_H = 0;$$
$$T_2 = 100\sin\alpha - \mu R_1 - \mu R_2$$
$$T_2 = 100(3/5) - 0.3[(80 + 0.8W) - W(4/5)]$$
$$1.574W = 60 - 24 - 0.24W - 0.24W$$
$$W = 17.53N \qquad \text{......ANS}$$

Chapter 12

BEAM

Q.1: How you define a Beam, and about Shear force & bending moment diagrams?
Sol.: A beam is a structural member whose longitudinal dimensions (width) is large compared to the transverse dimension (depth). The beam is supported along its length and is acted by a system of loads at right angles to its axis. Due to external loads and couples, shear force and bending moment develop at ant section of the beams. For the design of beam, information about the shear force and bending moment is desired.

Shear Force (S.F.)
The algebraic sum of all the vertical forces at any section of a beam to the right or left of the section is known as shear force.

Bending Moment (B.M.)
The algebraic sum of all the moment of all the forces acting to the right or left of the section is known as bending Moment.

Shear Force (S.F.) and Bending Moment (B.M.) Diagrams
A S.F. diagram is one, which shows the variation of the shear force along the length of the beam. And a bending moment diagram is one, which shows the variation of the bending moment along the length of the beam.

Before drawing the shear force and bending moment diagrams, we must know the different types of beam, load and support.

Q.2: How many types of load are acting on a beam?
A beam is normally horizontal and the loads acting on the beams are generally vertical. The following are the important types of load acting on a beam.

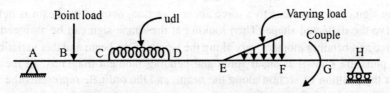

Fig 12.1 Various type of load acting on beam

Concentrated or Point Load

A concentrated load is one, which is considered to act at a point, although in practical it must really be distributed over a small area.

Uniformly Distributed Load (UDL)

A UDL is one which is spread over a beam in such a manner that rate of loading 'w' is uniform along the length (i.e. each unit length is loaded to the same rate). The rate of loading is expressed as w N/m run. For solving problems, the total UDL is converted into a point load, acting at the center of UDL.

Uniformly Varying Load (UVL)

A UVL is one which is spread over a beam in such a manner that rate of loading varies from point to point along the beam, in which load is zero at one end and increase uniformly to the other end. Such load is known as triangular load. For solving problems the total load is equal to the area of the triangle and this total load is assumed to be acting at the C.G. of the triangle i.e. at a distance of 2/3rd of total length of beam from left end.

Q.3: What sign convention is used for solving the problems of beam?

Although different sign conventions many be used, most of the engineers use the following sign conventions for shear forces and bending moment.

(i) The shear force that tends to move left portion upward relative to the right portion shall be called as positive shear force.

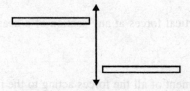

Fig 12.2

(ii) The bending moment that is trying to sag (Concave upward) the beam shall be taken as positive bending moment. If left portion is considered positive bending moment comes out to be clockwise moment.

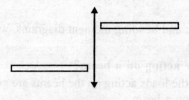

Fig 12.3

To decide the sign of moment due to a force about a section, assume the beam is held tightly at that section and observe the deflected shape. Then looking at the shape sign can be assigned.

The shear force and bending moment vary along the length of the beam and this variation is represented graphically. The plots are known as shear force and bending moment diagrams. In these diagrams, the abscissa indicates the position of section along the beam, and the ordinate represents the value of SF and BM respectively. These plots help to determine the maximum value of each of these quantities.

Fig 12.4

Q. 4: What is the relation between load intensity, shear force and bending moment?

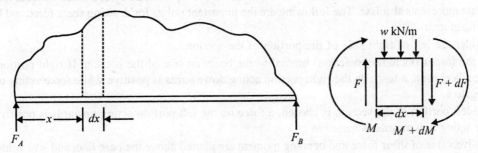

Fig 12.5

Sol.: Consider a beam subjected to any type of transverse load of the general form shown in fig 12.5. Isolate from the beam an element of length dx at a distance x from left end and draw its free body diagram as shown in fig 12.5. Since the element is of extremely small length, the loading over the beam can be considered to be uniform and equal to w KN/m. The element is subject to shear force F on its left hand side. Further, the bending moment M acts on the left side of the element and it changes to (M + dM) on the right side.

Taking moment about point C on the right side,
$$\Sigma M_C = 0$$
$$M - (M + dM) + F \times dx - (W \times dx) \times dx/2 = 0$$

The UDL is considered to be acting at its C.G.
$$dM = Fdx - [W(dx)^2]/2 = 0$$

The last term consists of the product of two differentials and can be neglected
$$DM = Fdx, \text{ or}$$
$$F = dM/dx$$

Thus the shear force is equal to the rate of change of bending moment with respect to x.

Apply the condition $\Sigma V = 0$ for equilibrium, we obtain
$$F - Wdx - (F + dF) = 0$$
Or
$$W = dF/dx$$

That is the intensity of loading is equal to rate of change of bending moment with respect to x.
$$F = dM/dx$$
and
$$W = dF/dx = dM^2/dx^2$$

Q.5: Define the nature of shear force and bending moment under load variation.

Sol.: The nature of SF and BM variation under two-load region is given in the table below

BETWEEN TWO POINTS, IF	S.F.D	B.M.D
No load	Constant	Linear
UDL	Inclined Linear	Parabolic
UVL	Parabolic	Cubic

Q.6: Define point of contraflexure or point of inflexion. Also define the point of zero shear force?
Sol.: The points (other than the extreme ends of a beam) in a beam at which B.M. is zero, are called points of contraflexure or inflexion.

The point at which we get zero shear force, we get the maximum bending moment of that section/beam at that point.

Q.7: How can you draw a shear force and bending moment diagram.
Sol.: In these diagrams, the shear force or bending moment are represented by ordinates whereas the length of the beam represents abscissa. The following are the important points for drawing shear force and bending moment diagrams:

1. Consider the left or right side of the portion of the section.
2. Add the forces (including reaction) normal to the beam on one of the portion. If right portion of the section is chosen, a force on the right portion acting downwards is positive while force acting upwards is negative.
3. If the left portion of the section is chosen, a force on the left portion acting upwards is positive while force acting downwards is negative.
4. The +ive value of shear force and bending moment are plotted above the base line, and -ive value below the base line.
5. The S.F. diagram will increase or decrease suddenly i.e. by a vertical straight line at a section where there is a vertical point load.
6. In drawing S.F. and B.M. diagrams no scale is to be chosen, but diagrams should be proportionate sketches.
7. For drawing S.F. and B.M. diagrams, the reaction of the right end support of a beam need not be determined. If however, reactions are wanted specifically, both the reactions are to be determined.
8. The Shear force between any two vertical loads will remain constant. Hence the S.F. diagram will be horizontal. The B.M. diagram will be inclined between these two loads.
9. For UDL S.F. diagram will be inclined straight line and the B.M. diagram will be curve.
10. The bending moment at the two supports of a simply supported beam and at the free end of a cantilever will be zero.
11. The B.M. is maximum at the section where S.F. changes its sign.
12. In case of overhanging beam, the maximum B.M. will be least possible when +ive max. B.M. is equal to the -ive max. B.M.
13. If not otherwise mentioned specifically, self-weight of the beam is to be neglected.
14. Section line is draw between that points on which load acts.

Numerical Problems Based on Simply supported beam

Q.8: Draw the SF and BM diagram for the simply supported beam loaded as shown in fig 12.6.

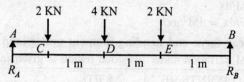

Fig 12.6

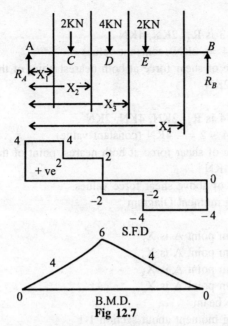

Fig 12.7

Sol.: Let reaction at support A and B be, R_A and R_B First find the support reaction
For that,
$$\Sigma V = 0$$
$$R_A + R_B - 2 - 4 - 2 = 0, \quad R_A + R_B = 8 \quad \ldots(1)$$
Taking moment about point A,
$$\Sigma M_A = 0$$
$$2 \times 1 + 4 \times 2 + 2 \times 3 - R_B \times 4 = 0$$
$$R_B = 4KN \quad \ldots(2)$$
From equation (1), $R_A = 4KN$ \quad \ldots(3)

Calculation for the Shear force Diagram
Draw the section line, here total 4 section line, which break the
load R_A and 2KN(Between Point A and C),
2KN and 4KN(Between Point C and D),
4KN and 2KN (Between Point D and E) and
2KN and RB(Between Point E and B)
Consider left portion of the beam
Consider section 1-1
Force on left of section 1-1 is R_A
$$SF_{1-1} = 4KN \text{ (constant value)}$$
Constant value means value of shear force at both nearest point of the section is equal i.e.
$$SF_A = SF_C = 4KN \quad \ldots(4)$$
Consider section 2-2
Forces on left of section 2-2 is R_A & 2KN
$$SF_{2-2} = 4 - 2 = 2KN \text{ (constant value)}$$
Constant value means value of shear force at both nearest point of the section is equal i.e.
$$SF_C = SF_D = 2KN \quad \ldots(5)$$

Consider section 3-3
Forces on left of section 3-3 is R_A, 2KN, 4KN
$$SF_{3-3} = 4 - 2 - 4 = -2KN \text{ (constant value)}$$
Constant value means value of shear force at both nearest point of the section is equal i.e.
$$SF_D = SF_E = -2KN \qquad \ldots(6)$$
Consider section 4-4
Forces on left of section 4-4 is R_A, 2KN, 4KN, 2KN
$$SF_{4-4} = 4 - 2 - 4 - 2 = -4KN \text{ (constant value)}$$
Constant value means value of shear force at both nearest point of the section is equal i.e.
$$SF_E = SF_B = -4KN \qquad \ldots(7)$$
Plot the SFD with the help of above shear force values.
Calculation for the Bending moment Diagram
Let
Distance of section 1-1 from point A is X_1
Distance of section 2-2 from point A is X_2
Distance of section 3-3 from point A is X_3
Distance of section 4-4 from point A is X_4
Consider left portion of the beam
Consider section 1-1, taking moment about section 1-1
$$BM_{1-1} = 4.X_1$$
It is Equation of straight line (Y = mX + C), inclined linear.
Inclined linear means value of bending moment at both nearest point of the section is varies with $X_1 = 0$ to $X_1 = 1$
At $\quad X_1 = 0$
$\quad BM_A = 0 \qquad \ldots(8)$
At $\quad X_1 = 1$
$\quad BM_C = 4 \qquad \ldots(9)$
i.e. inclined line 0 to 4
Consider section 2-2, taking moment about section 2-2
$$BM_{2-2} = 4.X_2 - 2.(X_2 - 1)$$
$$= 2.X_2 + 2$$
It is Equation of straight line (Y = mX + C), inclined linear.
Inclined linear means value of Bending moment at both nearest point of the section is varies with $X_2 = 1$ to $X_2 = 2$
At $\quad X_2 = 1$
$\quad BM_C = 4 \qquad \ldots(10)$
At $\quad X_2 = 2$
$\quad BMD = 6 \qquad \ldots(11)$
i.e. inclined line 4 to 6
Consider section 3-3, taking moment about section 3-3
$$BM_{3-3} = 4.X_3 - 2.(X_3 - 1) - 4.(X_3 - 2)$$
$$= -2.X_3 + 10$$
It is Equation of straight line (Y = mX + C), inclined linear.

Inclined linear means value of Bending moment at both nearest point of the section is varies with $X_3 = 2$ to $X_3 = 3$

At $\quad X_3 = 2$
$\quad\quad BM_D = 6$ ...(12)
At $\quad X_3 = 3$
$\quad\quad BM_E = 4$ ...(13)

i.e. inclined line 6 to 4

Consider section 4-4, taking moment about section 4-4

$BM_{4-4} = 4.X_4 - 2.(X_4 - 1) - 4.(X_4 - 2) - 2.(X_4 - 3)$
$\quad\quad\quad = -4.X_4 + 16$

It is Equation of straight line $(Y = mX + C)$, inclined linear.

Inclined linear means value of Bending moment at both nearest point of the section is varies with $X_4 = 3$ to $X_4 = 4$

At $\quad X_4 = 3; BM_E = 4$ ...(14)
At $\quad X_4 = 4; BM_B = 0$ ...(15)

i.e. inclined line 4 to 0

Plot the BMD with the help of above bending moment values.

Q.9: Draw the SF and BM diagram for the simply supported beam loaded as shown in fig. 12.8.

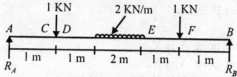

Fig 12.8

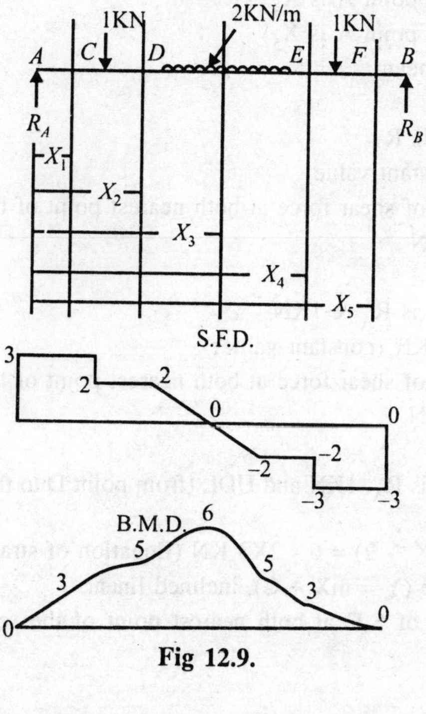

Fig 12.9.

Sol.: Let reaction at support A and B be, R_A and R_B

First find the support reaction.

For finding the support reaction, convert UDL in to point load and equal to $2 \times 2 = 4KN$, acting at mid point of UDL i.e. 3m from point A.

For that,
$$\Sigma V = 0$$
$$R_A + R_B - 1 - 4 - 1 = 0,$$
$$R_A + R_B = 6 \qquad \qquad ...(1)$$

Taking moment about point A,
$$\Sigma M_A = 0$$
$$1 \times 1 + 4 \times 3 + 1 \times 5 - R_B \times 6 = 0$$
$$R_B = 3KN \qquad \qquad ...(2)$$

From equation (1), $R_A = 3KN$ ...(3)

Calculation for the Shear force Diagram

Draw the section line, here total 5-section line, which break the
load R_A and 1KN (Between Point A and C),
1KN and starting of UDL (Between Point C and D),
end point of UDL and 1KN (Between Point E and F) and
1KN and R_B (Between Point F and B)

Let

Distance of section 1-1 from point A is X_1
Distance of section 2-2 from point A is X_2
Distance of section 3-3 from point A is X_3
Distance of section 4-4 from point A is X_4
Distance of section 5-5 from point A is X_5

Consider left portion of the beam

Consider section 1-1

Force on left of section 1-1 is R_A
$$SF_{1-1} = 3KN \text{ (constant value)}$$

Constant value means value of shear force at both nearest point of the section is equal i.e.
$$SF_A = SF_C = 3KN \qquad \qquad ...(4)$$

Consider section 2-2

Forces on left of section 2-2 is R_A & 1KN
$$SF_{2-2} = 3 - 1 = 2KN \text{ (constant value)}$$

Constant value means value of shear force at both nearest point of the section is equal i.e.
$$SF_C = SF_D = 2KN \qquad \qquad ...(5)$$

Consider section 3-3

Forces on left of section 3-3 is R_A, 1KN and UDL (from point D to the section line i.e. UDL on total distance of $(X_3 - 2)$
$$SF_{3-3} = 3 - 1 - 2(X_3 - 2) = 6 - 2X_3 \text{ KN (Equation of straight line)}$$

It is Equation of straight line $(Y = mX + C)$, inclined linear.

Inclined linear means value of S.F. at both nearest point of the section is varies with $X_3 = 2$ to $X_3 = 4$

At $X_3 = 2$
$SF_D = 2$ ...(6)
At $X_3 = 4$
$SF_E = -2$ ...(7)
i.e. inclined line 2 to -2

Since here shear force changes the sign so at any point shear force will be zero and at that point bending moment is maximum.

For finding the position of zero shear force equate the shear force equation to zero, i.e.
$6 - 2X_3 = 0$; $X_3 = 3m$, i.e. at 3m from point A bending moment is maximum.

Consider section 4-4

Forces on left of section 4-4 is R_A, 1KN, 4KN
$SF_{4-4} = 3 - 1 - 4 = -2KN$ (constant value)

Constant value means value of shear force at both nearest point of the section is equal i.e.
$SF_E = SF_F = -2KN$ ...(8)

Consider section 5-5

Forces on left of section 5-5 is RA, 1KN, 4KN, 1KN
$SF_{5-5} = 3 - 1 - 4 - 1 = -3KN$ (constant value)

Constant value means value of shear force at both nearest point of the section is equal i.e.
$SF_E = SF_B = -3KN$ ...(9)

Plot the SFD with the help of above shear force values.

Calculation for the Bending moment Diagram

Consider left portion of the beam

Consider section 1-1, taking moment about section 1-1
$BM_{1-1} = 3.X1$

It is Equation of straight line ($Y = mX + C$), inclined linear.

Inclined linear means value of bending moment at both nearest point of the section is varies with $X_1 = 0$ to $X_1 = 1$
At $X_1 = 0$
$BM_A = 0$ ...(10)
At $X_1 = 1$
$BM_C = 3$ ...(11)
i.e. inclined line 0 to 3

Consider section 2-2, taking moment about section 2-2
$BM_{2-2} = 3.X_2 - 1.(X_2 - 1)$
$= 2.X_2 + 1$

It is Equation of straight line ($Y = mX + C$), inclined linear.

Inclined linear means value of bending moment at both nearest point of the section is varies with $X_2 = 1$ to $X_2 = 2$
At $X_2 = 1$
$BM_C = 3$ ...(12)
At $X_2 = 2$
$BM_D = 5$ ...(13)

i.e. inclined line 3 to 5

Consider section 3-3, taking moment about section 3-3

$$BM_{3-3} = 3.X_3 - 1.(X_3 - 1) - 2.(X_3 - 2)[(X_3 - 2)/2]$$
$$= 2.X_3 + 1 - (X_3 - 2)^2$$

It is Equation of Parabola ($Y = mX^2 + C$),

Parabola means a parabolic curve is formed, value of bending moment at both nearest point of the section is varies with $X_3 = 2$ to $X_3 = 4$

At $\quad X_3 = 2$

$\quad BM_D = 5$ ...(14)

At $\quad X_3 = 4$

$\quad BM_E = 5$ ...(15)

But B.M. is maximum at $X_3 = 3$, which lies between $X_3 = 2$ to $X_3 = 4$

So we also find the value of BM at $X_3 = 3$

At $\quad X_3 = 3$

$\quad BM_{max} = 6$ ...(16)

i.e. curve makes with in 5 to 6 to 5 region.

Consider section 4-4, taking moment about section 4-4

$$BM_{4-4} = 3.X_4 - 1.(X_4 - 1) - 4.(X_4 - 3)$$
$$= -2.X_4 + 13$$

It is Equation of straight line ($Y = mX + C$), inclined linear.

Inclined linear means value of bending moment at both nearest point of the section is varies with $X_4 = 4$ to $X_4 = 5$

At $\quad X_4 = 4$

$\quad BM_E = 5$ ...(17)

At $\quad X_4 = 5$

$\quad BM_F = 3$ ...(18)

i.e. inclined line 5 to 3

Consider section 5-5, taking moment about section 5-5

$$BM_{5-5} = 3.X_5 - 1.(X_5 - 1) - 4.(X_5 - 3) - 1.(X_5 - 5)$$
$$= -3.X_5 + 18$$

It is Equation of straight line ($Y = mX + C$), inclined linear.

Inclined linear means value of bending moment at both nearest point of the section is varies with $X_5 = 5$ to $X_5 = 6$

At $\quad X_5 = 5$

$\quad BM_E = 3$ ...(19)

At $\quad X_4 = 6$

$\quad BM_F = 0$ ...(20)

i.e. inclined line 3 to 0

Plot the BM_D with the help of above bending moment values.

Q.10: Draw the SF and BM diagram for the simply supported beam loaded as shown in fig. 12.10

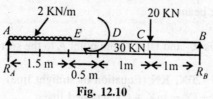

Fig. 12.10

Sol.: Let reaction at support A and B be, R_A and R_B First find the support reaction. For finding the support reaction, convert UDL in to point load and equal to $20 \times 1.5 = 30$ KN, acting at mid point of UDL i.e. 0.75m from point A.

For that,
$$\Sigma V = 0$$
$$R_A + R_B - 30 - 20 = 0, \quad R_A + R_B = 50 \qquad \ldots(1)$$

Taking moment about point A,
$$\Sigma M_A = 0$$
$$30 \times 0.75 + 30 + 20 \times 3 - R_B \times 4 = 0$$
$$R_B = 28.125 \text{ KN} \qquad \ldots(2)$$

From equation (1), $R_A = 21.875$ KN $\qquad \ldots(3)$

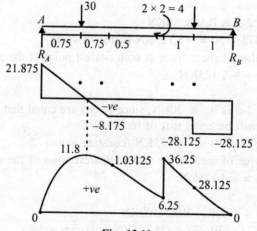

Fig. 12.11

Calculation for the Shear force Diagram
Draw the section line, here total 4-section line, which break the
load R_A and UDL (Between Point A and E),
30KN/m and 20KN (Between Point E and D),
30KN/M and 20KN (Between Point D and C) and
20KN and RB (Between Point C and B)
Let
Distance of section 1-1 from point A is X_1
Distance of section 2-2 from point A is X_2
Distance of section 3-3 from point A is X_3

Distance of section 4-4 from point A is X_4
Consider left portion of the beam
Consider section 1-1
Force on left of section 1-1 is R_A and UDL (from point A to the section line i.e. UDL on total distance of X_1

$$SF_{1-1} = 21.875 - 20X_1 \text{ KN (Equation of straight line)}$$

It is Equation of straight line (Y = mX + C), inclined linear.

Inclined linear means value of shear force at both nearest point of the section is varies with $X_1 = 0$ to $X_1 = 1.5$

At $\quad\quad X_1 = 0$

$\quad\quad\quad SF_A = 21.875$ ...(4)

At $\quad\quad X_1 = 1.5$

$\quad\quad\quad SF_E = -8.125$ ...(5)

i.e. inclined line 21.875 to − 8.125

Since here shear force changes the sign so at any point shear force will be zero and at that point bending moment is maximum.

For finding the position of zero shear force equate the shear force equation to zero, i.e.

$21.875 - 20X_1 = 0$; $X_1 = 1.09375$m, i.e. at 1.09375m from point A bending moment is maximum.

Consider section 2-2

Forces on left of section 2-2 is RA & 30KN

$$SF_{2-2} = 21.875 - 30 = -8.125 \text{KN (constant value)}$$

Constant value means value of shear force at both nearest point of the section is equal i.e.

$\quad\quad\quad SF_E = SF_D = -8.125 \text{KN}$ ...(6)

Consider section 3-3

Forces on left of section 3-3 is R_A & 30KN, since forces are equal that of section 2-2, so the value of shear force at section 3-3 will be equal that of section 2-2

$$SF_{3-3} = 21.875 - 30 = -8.125 \text{KN (constant value)}$$

Constant value means value of shear force at both nearest point of the section is equal i.e.

$\quad\quad\quad SF_D = SF_C = -8.125 \text{KN}$ ...(7)

Consider section 4-4

Forces on left of section 4-4 is R_A, 30KN, 20KN

$$SF_{4-4} = 21.875 - 30 - 20 = -28.125 \text{KN (constant value)}$$

Constant value means value of shear force at both nearest point of the section is equal i.e.

$\quad\quad\quad SF_C = SF_B = -28.125 \text{KN}$ ...(8)

Plot the SFD with the help of above shear force values.
Calculation for the Bending moment Diagram
Consider left portion of the beam
Consider section 1-1, taking moment about section 1-1

$$BM_{1-1} = 21.875 X_1 - 20 X_1 (X_1/2)$$

It is Equation of Parabola ($Y = mX^2 + C$),

Parabola means a parabolic curve is formed, value of bending moment at both nearest point of the section is varies with $X_1 = 0$ to $X_1 = 1.5$

At $\quad X_1 = 0$
$\quad\quad BM_A = 0$ ...(9)
At $\quad X_1 = 1.5$
$\quad\quad BM_C = 10.3125$ ...(10)

But B.M. is maximum at $X_1 = 1.09$, which lies between $X_1 = 0$ to $X_1 = 1.5$
So we also find the value of BM at $X_1 = 1.09$
At $\quad X_1 = 1.09$
$\quad\quad BM_{max} = 11.8$ ...(11)

i.e. curve makes with in 0 to 11.8 to 10.3125 region.
Consider section 2-2, taking moment about section 2-2
$$BM_{2-2} = 21.875X_2 - 30(X_2 - 0.75)$$
$$= -8.125.X_2 + 22.5$$

It is Equation of straight line $(Y = mX + C)$, inclined linear.
Inclined linear means value of bending moment at both nearest point of the section is varies with $X_2 = 1.5$ to $X_2 = 2$
At $\quad X_2 = 1.5$
$\quad\quad BM_E = 10.3125$ ...(12)
At $\quad X_2 = 2$
$\quad\quad BM_D = 6.25$ ...(13)

i.e. inclined line 10.3125 to 6.25
Consider section 3-3, taking moment about section 3-3
$$BM_{3-3} = 21.875X_3 - 30(X_3 - 0.75) + 30$$
$$= -8.125.X_2 + 52.5$$

It is Equation of straight line $(Y = mX + C)$, inclined linear.
Inclined linear means value of bending moment at both nearest point of the section is varies with $X_3 = 2$ to $X_3 = 3$
At $\quad X_3 = 2$
$\quad\quad BM_D = 36.25$ ...(14)
At $\quad X_3 = 3$
$\quad\quad BM_C = 28.125$ ...(15)

Consider section 4-4, taking moment about section 4-4
$$BM_{4-4} = 21.875X_4 - 30(X_4 - 0.75) + 30 - 20(X_4 - 3)$$
$$= -28.125.X_4 + 112.5$$

It is Equation of straight line $(Y = mX + C)$, inclined linear.
Inclined linear means value of bending moment at both nearest point of the section is varies with $X_4 = 3$ to $X_4 = 4$
At $\quad X_4 = 3$
$\quad\quad BM_C = 28.125$ ...(16)
At $\quad X_4 = 4$
$\quad\quad BM_B = 0$ ...(17)

i.e. inclined line 28.125 to 0
Plot the BM_D with the help of above bending moment values.

Q.11: Determine the SF and BM diagrams for the simply supported beam shown in fig 12.12. Also find the maximum bending moment.

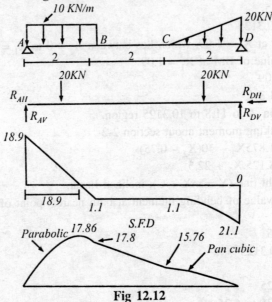

Fig 12.12

Sol.: Since hinged at point A and D, suppose reaction at support A and D be, R_{AH}, R_{AV} and R_{DH}, R_{DV} first find the support reaction. For finding the support reaction, convert UDL and UVL in to point load and,

Point load of UDL equal to 10 X 2 = 20KN, acting at mid point of UDL i.e. 1m from point A.

Point load of UVL equal to 1/2 X 20 X 2 = 20KN, acting at a distance 1/3 of total distance i.e. 1/3m from point D.

For that,

$$\Sigma V = 0$$

$R_{AV} + R_{DV} - 20 - 20 = 0$, $R_A + R_B = 40$ ...(1)

Taking moment about point A,

$$\Sigma M_A = 0$$

20 X 1 + 20 X 5.33 – R_{DV} X 6 = 0

R_{DV} = 21.1 KN ...(2)

From equation (1), R_{AV} = 18.9KN ...(3)

Calculation for the Shear force Diagram

Draw the section line, here total 3-section line, which break the load R_{AV} and UDL (Between Point A and B),

No load (Between Point B and C) and

UVL (Between Point C and D).

Let

Distance of section 1-1 from point A is X_1

Distance of section 2-2 from point A is X_2

Distance of section 3-3 from point A is X_3

Consider left portion of the beam

Consider section 1-1

Force on left of section 1-1 is R_{AV} and UDL (from point A to the section line i.e. UDL on total distance of X_1

$SF_{1-1} = 18.9 - 10X_1$ KN (Equation of straight line)

It is Equation of straight line (Y = mX + C), inclined linear.

Inclined linear means value of shear force at both nearest point of the section is varies with X1 = 0 to X_1 = 2

At $\quad X = 0$
$\quad SF_A = 18.9$
At $\quad X_1 = 2$ ...(4)
$\quad SF_B = -1.1$ ...(5)

i.e. inclined line 18.9 to -1.1

Since here shear force changes the sign so at any point shear force will be zero and at that point bending moment is maximum.

For finding the position of zero shear force equate the shear force equation to zero, i.e.
$18.9 - 10X_1 = 0$; $X_1 = 1.89$m, i.e. at 1.89m from point A bending moment is maximum.

Consider section 2-2

Forces on left of section 2-2 is R_{AV} & 20KN

$SF_{2-2} = 18.9 - 20 = -1.1$KN (constant value)

Constant value means value of shear force at both nearest point of the section is equal i.e.
$SF_B = SF_C = -1.1$KN ...(6)

Consider section 3-3

Forces on left of section 3-3 is R_{AV} & 20KN and UVL of 20KN/m over $(X_3 - 4)$ m length,

First calculate the total load of UVL over length of $(X_3 - 4)$

Consider triangle CDE and CGF

DE/GF = CD/CG

Since DE = 20

$20/GF = 2/(X_3 - 4)$

$GF = 10(X_3 - 4)$

Now load of triangle CGF = 1/2 X CG X GF = 1/2 X $(X_3 - 4)$ X $10(X_3 - 4)$

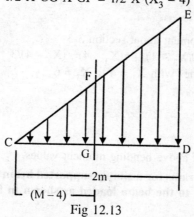

Fig 12.13

$= 5(X_3 - 4)^2$, at a distance of $(X_3 - 4)/3$ from G ...(7)

$SF_{3-3} = 18.9 - 20 - 5(X_3 - 4)^2 = -1.1 - 5(X_3 - 4)^2$ (Parabola)

Parabola means a parabolic curve is formed, value of bending moment at both nearest point of the section is varies with $X_3 = 4$ to $X_3 = 6$

At $X_3 = 4$
    $SF_C = -1.1 KN$ ...(8)
    $SF_D = -21.1 KN$ ...(9)

Calculation for the Bending moment Diagram
Consider left portion of the beam
Consider section 1-1, taking moment about section 1-1

$$BM_{1-1} = 18.9X_1 - 10X_1 \cdot X_1/2$$
$$= 18.9X_1 - 5 \cdot X_1^2$$

It is Equation of Parabola $(Y = mX^2 + C)$,

Parabola means a parabolic curve is formed, value of bending moment at both nearest point of the section is varies with $X_1 = 0$ to $X_1 = 2$

At $X_1 = 0$
    $BM_A = 0$ ...(10)
At $X_1 = 2$
    $BM_B = 17.8$ ...(11)

But B.M. is maximum at $X_1 = 1.89$, which lies between $X_1 = 0$ to $X_1 = 2$
So we also find the value of BM at $X_1 = 1.89$

At $X_1 = 1.89$
    $BM_{max} = 17.86$ ...(12)

i.e. curve makes with in 0 to 17.86 to 17.8 region.
Consider section 2-2, taking moment about section 2-2

$$BM_{2-2} = 18.9X_2 - 20(X_2 - 1)$$

It is Equation of straight line $(Y = mX + C)$, inclined linear.
Inclined linear means value of bending moment at both nearest point of the section is varies with $X_2 = 2$ to $X_2 = 4$

At $X_2 = 2$
    $BM_B = 17.8$ ...(13)
At $X_2 = 4$
    $BM_C = 15.76$ ...(14)

i.e. inclined line 17.8 to 15.76
Consider section 3-3, taking moment about section 3-3

$$BM_{3-3} = 18.9X_3 - 20(X_3 - 1) - 5(X_3 - 4)^2 \cdot (X_3 - 4)/3$$

It is cubic Equation which varies with $X_3 = 4$ to $X_3 = 6$

At $X_3 = 4$
    $BM_C = 15.76$ ...(15)
At $X_3 = 6$
    $BM_D = 0$ ...(16)

Plot the BM_D with the help of above bending moment values.

Q.12: Draw the SF and BM diagrams for a simply supported beam 5m long carrying a load of 200N through a bracket welded to the beam loaded as shown in fig 12.14

Sol.:

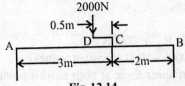

Fig 12.14

The diagram is of force couple system, let us apply at C two equal and opposite forces each equal and parallel to 2000N. Now the vertically upward load of 2000N at C and vertically downward load of 2000N at D forms an anticlockwise couple at C whose moment is 2000 X 0.5 = 1000Nm

And we are left with a vertically downward load of 2000N acting at C.

Let reaction at support A and B be, R_A and R_B first find the support reaction.

Taking moment about point A;

2000 X 3 – 1000 – R_B X 5 = 0

$\qquad R_B$ = 1000N ...(1)

$\qquad R_V$ = 0, $R_A + R_B$ – 2000 = 0

$\qquad R_A$ = 1000N ...(2)

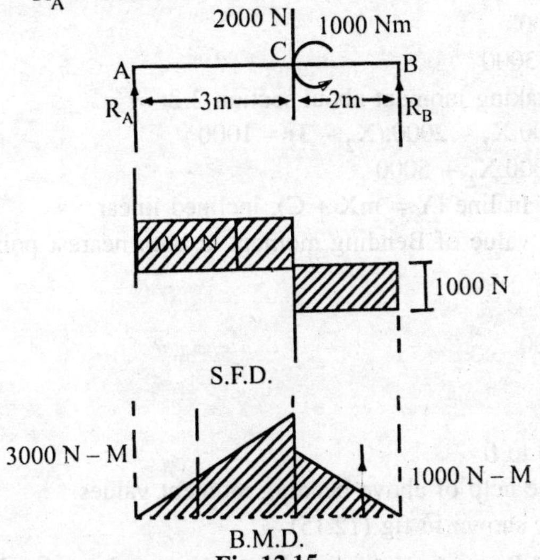

Fig 12.15

Calculation for the Shear force Diagram

Draw the section line, here total 2 section line, which break the load

R_A and 2000N (Between Point A and C),

2000N and R_B (Between Point C and B).

Let

Distance of section 1-1 from point A is X_1

Distance of section 2-2 from point A is X_2

Consider left portion of the beam

Consider section 1-1

Force on left of section 1-1 is R_A

$\qquad SF_{1-1}$ = 1000N (constant value)

Constant value means value of shear force at both nearest point of the section is equal i.e.

$SF_A = SF_C = 1000N$ ...(3)

Consider section 2-2

Forces on left of section 2-2 is R_A & 2000N

$SF_{2-2} = 1000 - 2000 = -1000$ (constant value)

Constant value means value of shear force at both nearest point of the section is equal i.e.

$SF_C = SF_B = -1000N$ ...(4)

Plot the SFD with the help of above shear force values.

Calculation for the bending moment Diagram

Consider section 1-1, taking moment about section 1-1

$BM_{1-1} = 1000.X_1$

It is Equation of straight line ($Y = mX + C$), inclined linear.

Inclined linear means value of bending moment at both nearest point of the section is varies with $X_1 = 0$ to $X_1 = 3$

At $X_1 = 0$

     $BM_A = 0$ ...(5)

At $X_1 = 3$

     $BM_C = 3000$ ...(6)

i.e. inclined line 0 to 3000

Consider section 2-2, taking moment about section 2-2

$BM_{2-2} = 1000.X_2 - 2000.(X_2 - 3) - 1000$

$= -1000.X_2 + 5000$

It is Equation of straight line ($Y = mX + C$), inclined linear.

Inclined linear means value of Bending moment at both nearest point of the section is varies with $X_2 = 3$ to $X_2 = 5$

At $X_2 = 3$

     $BM_C = 2000$ ...(7)

At $X_2 = 5$

     $BM_B = 0$ ...(8)

i.e. inclined line 2000 to 0

Plot the BMD with the help of above bending moment values.

The SFD and BMD is shown in fig (12.15).

Q.13: A simply supported beam 6m long is subjected to a triangular load of 6000N as shown in fig 12.16 below. Draw the S.F. and B.M. diagrams for the beam.

Sol.:

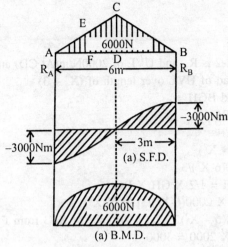

Fig 12.16

Let

Suppose reaction at support A and B be, R_A and R_B first find the support reaction.

Due to symmetry, $R_A = R_B = 6000/2 = 3000N$ ...(1)

Calculation for the Shear force Diagram

Draw the section line, here total 2-section line, which break the point A,D and Point D,B

Let

Distance of section 1-1 from point A is X_1

Distance of section 2-2 from point A is X_2

Consider left portion of the beam

Consider section 1-1

Forces on left of section 1-1 is R_A and UVL of 6000N/m over X_1 m length,

Since Total load = 6000 = 1/2 X AB X CD

  1/2 X 6 X CD = 6000, CD = 2000N ...(2)

First calculate the total load of UVL over length of X_1

Consider triangle ADC and AFE

  DC/EF = AD/AF

Since DC = 2000

  $2000/EF = 3/X_1$

  $EF = (2000X_1)/3$

Now load of triangle AEF = 1/2 X EF × AF

  = (1/2 X $2000X_1$)/3 × (X_1)

  = $\dfrac{1000 \cdot X_1^2}{3}$ a distance of $X_1/3$ from F ...(3)

SF1-1 = 3000 – (1000X12)/3 (Parabola)

Parabola means a parabolic curve is formed, value of bending moment at both nearest point of the section is varies with $X_1 = 0$ to $X_1 = 3$

At $X_1 = 0$

  $SF_A = 3000N$ ...(4)

At $\quad X_1 = 3$
$\quad SF_D = 0$ ...(5)

Consider section 2-2

Forces on left of section 2-2 is R_A and UVL of 2000N/m(At CD) and UVL over $(X_2 - 3)$ m length, First calculate the total load of UVL over length of $(X_2 - 3)$

Consider triangle CDB and BGH

$\quad DC/GH = DB/BG$

Since DC = 2000

$\quad 2000/GH = 3/(6 - X^2)$
$\quad GH = 2000(6-X^2)/3$

Now load of triangle BGH = 1/2 X GH X BG
$\quad = [1/2 \times 2000(6-X^2)/3] \times (6-X^2)$
$\quad = 1000(6 - X^2)^2/3$, at a distance of $X_1/3$ from F ...(6)

Load of CDB = 1/2 X 3 X 2000 = 3000

Now load of CDGH = load of CDB - load of BGH
$\quad = 3000 - 1000(6 - X_2)^2/3$ ...(7)

$SF_{2-2} = 3000 - 3000 - [3000 - 1000(6 - X^2)^2/3]$ (Parabola)

Parabola means a parabolic curve is formed, value of bending moment at both nearest point of the section is varies with $X_2 = 3$ to $X_2 = 6$

At $\quad X_2 = 3$
$\quad SF_A = 0$ ...(8)

At $\quad X_2 = 6$
$\quad SF_D = -3000N$ ...(9)

Plot the SFD with the help of above value as shown in fig.

Since SF change its sign at $X_2 = 3$, that means at a distance of 3m from point A bending moment is maximum.

Calculation for the Bending moment Diagram

Consider section 1-1

$\quad BM_{1-1} = 3000X_1 - [(1000X_1^2)/3]X_1/3$ (Cubic)

Cubic means a parabolic curve is formed, value of bending moment at both nearest point of the section is varies with $X_1 = 0$ to $X_1 = 3$

At $\quad X_1 = 0$
$\quad BM_A = 0$ ...(10)

At $\quad X_1 = 3$
$\quad BM_D = 6000$ ...(11)

Consider section 2-2

Point of CG of any trapezium is = h/3[(b + 2a)/(a + b)]

i.e. Distance of C.G of the trapezium CDGH is given by,
$\quad = 1/3 \times DG \times [(GH + 2CD)/(GH + CD)]$
$\quad = 1/3.(X2-3).\{[2000(6-X2)/3] + 2 \times 2000)\}/\{[2000(6-X2)/3]+[2000]\}$
$\quad = \{(X2 - 3)(12 - X2)\}/\{3(9 - X2)\}$ ...(12)

$BM_{2-2} = 3000X_2 - 3000(X_2-2) - [3000-1000(6 - X_2)^2/3]\{+ (X_2 - 3)(12 - X_2)\}/\{3(9 - X_2)\}$ (Equation of Parabola)

Parabola means a parabolic curve is formed, value of bending moment at both nearest point of the section is varies with $X_2 = 3$ to $X_2 = 6$

At $\quad X_2 = 3$
$\quad BM_A = 6000$ ...(13)

At $\quad X_2 = 6$
$\quad BM_D = 0N$ ...(14)
Plot the BMD with the help of above value.

Note: We also solve the problem by considering right hand side of the portion, example as given below.

Q.14: A simply supported beam carries distributed load varying uniformly from 125N/m at one end to 250N/m at the other. Draw the SF and BM diagram and determine the maximum B.M.

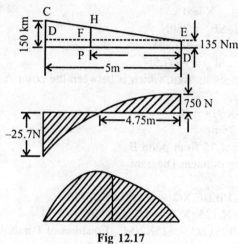

Fig 12.17

Sol.: Total load = Area of the load diagram ABEC
$\qquad$ = Rectangle ABED + Triangle DEC
$\qquad$ = (AB X BE) + (1/2 X DE X DC) = (9 X 125) + [1/2 X 9 X (250-125)]
$\qquad$ = 1125N + 562.5N ...(1)

Centroid of the load of 1125N(rectangular load) is at a distance of 9/2 = 4.5m from AD and the centroid of the load of 562.5N (Triangular load) is at a distance of 1/3 X DE = 1/3 X 9 = 3m from point A.

Let support reaction at A and B be R_A and R_B. For finding the support reaction,
Taking moment about point A,
$\qquad$ 1125 X 4.5 + 562.5 X 3 - R_B X 9 = 0
$\qquad\qquad R_B$ = 750N ...(2)
Now, $\qquad\qquad R_V$ = 0
$\qquad R_A + R_B$ = 1125 + 562.5 = 1687.5
$\qquad\qquad R_A$ = 937.5N ...(3)

Calculation for the Shear force Diagram
Draw the section line, here total 1-section line, which break the point A and B
Let
Distance of section 1-1 from point B is X
Consider right portion of the beam
Consider section 1-1
Forces on right of section 1-1 is R_B and Load of PBEF and Load of EFH

286 / *Problems and Solutions in Mechanical Engineering with Concept*

$$SF_{1-1} = RB - \text{load on the area PBEF} - \text{load on the area EFH}$$
$$= RB - X.125 - 1/2.X.FH$$

In the equiangular triangles DEC and FEH
$$DC/DE = FH/FE \text{ or, } 125/9 = FH/X$$
$$FH = 125X/9$$

S.F. between B and A = $750 - 125X - 125X^2/18$ (Equation of Parabola)

Parabola means a parabolic curve is formed, value of bending moment at both nearest point of the section is varies with X = 0 to X = 9

At   X = 0
$$SF_B = 750N \qquad \ldots(4)$$

At   X = 9
$$SF_A = -937.5N \qquad \ldots(5)$$

Since the value of SF changes its sign, which is between the point A and B we get max. BM
For the point of zero shear,
$$750 - 125X - 125X^2/18 = 0$$
On solving we get, X = 4.75m
That is BM is max. at X = 4.75 from point B
Calculation for the Bending moment Diagram
Consider section 1-1
$$BM_{1-1} = 750X - PB.BE.X/2 - 1/2.FE.FH.1/3.FE$$
$$= 750X - X.125.(X/2) - 1/2.X.(125X/9)(X/3)$$
$$= 750x - 125x2/2 - 125X2/54 \quad \text{(Equation of Parabola)}$$

Parabola means a parabolic curve is formed, value of bending moment at both nearest point of the section is varies with X = 0 to X = 9

At   X = 0
$$BM_B = 0 \qquad \ldots(6)$$

At   X = 4.75
$$BM_{max} = 1904 \text{N-m} \qquad \ldots(7)$$

At   X = 9
$$BM_A = 0 \qquad \ldots(8)$$

Numerical Problems Based on Cantilever Beam

Q.15: Draw the SF and BM diagram for the beam as shown in fig 12.18. Also indicate the principal values on the diagrams.

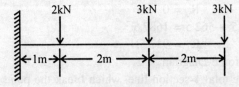

Fig 12.18

Sol.: Let reaction at support A be R_{AV}, R_{AH} and M(anti clock wise), First find the support reaction
For that,
$$\Sigma V = 0$$
$$R_{AV} - 2 - 3 - 3 = 0, R_{AV} = 8 \qquad \ldots(1)$$

Taking moment about point A,
$$\Sigma M_A = 0$$
$$-M + 2 \times 1 + 3 \times 3 + 3 \times 5 = 0$$
$$M = 26 KNm$$
$$\Sigma H = 0$$
$$R_{AH} = 0 \qquad \qquad ...(3)$$

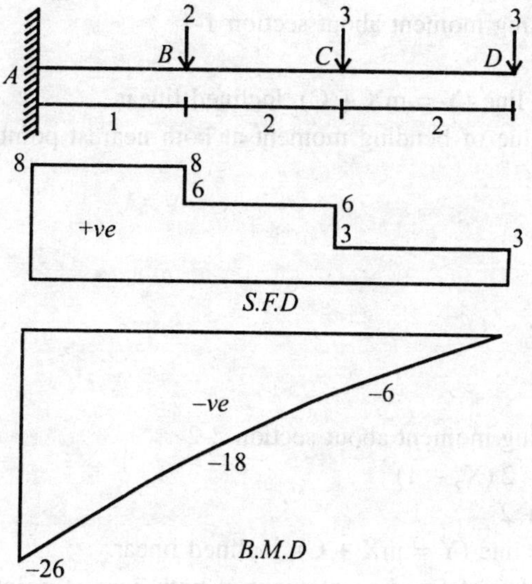

Fig 12.19

Calculation for the Shear force Diagram
Draw the section line, here total 3 section line, which break the
load R_{AV} and 2KN(Between Point A and B),
2KN and 3KN(Between Point B and C),
3KN and 3KN (Between Point C and D).
Consider left portion of the beam
Consider section 1-1
Force on left of section 1-1 is R_{AV}
$$SF_{1-1} = 8KN \text{ (constant value)}$$
Constant value means value of shear force at both nearest point of the section is equal i.e.
$$SF_A = SF_B = 8KN \qquad \qquad ...(4)$$
Consider section 2-2
Forces on left of section 2-2 is R_{AV} & 2KN
$$SF_{2-2} = 8 - 2 = 6KN \text{ (constant value)}$$
Constant value means value of shear force at both nearest point of the section is equal i.e.
$$SF_B = SF_C = 6KN \qquad \qquad ...(5)$$
Consider section 3-3
Forces on left of section 3-3 is R_A, 2KN, 3KN
$$SF_{3-3} = 8 - 2 - 3 = 3KN \text{ (constant value)}$$
Constant value means value of shear force at both nearest point of the section is equal i.e.

$SF_C = SF_D = 3KN$ ...(6)

Calculation for the Bending moment Diagram

Let

Distance of section 1-1 from point A is X_1

Distance of section 2-2 from point A is X_2

Distance of section 3-3 from point A is X_3

Consider section 1-1, taking moment about section 1-1

$$BM_{1-1} = 8.X_1$$

It is Equation of straight line (Y = mX + C), inclined linear.

Inclined linear means value of bending moment at both nearest point of the section is varies with $X_1 = 0$

to $X_1 = 1$

At $X_1 = 0$

  $BM_A = 0$ ...(8)

At $X_1 = 1$

  $BM_B = 8$ ...(9)

i.e. inclined line 0 to 8

Consider section 2-2, taking moment about section 2-2

$$BM_{2-2} = 8.X_2 - 2.(X_2 - 1)$$
$$= 6.X_2 + 2$$

It is Equation of straight line (Y = mX + C), inclined linear.

Inclined linear means value of Bending moment at both nearest point of the section is varies with $X_2 = 1$ to $X_2 = 3$

At $X_2 = 1$

  $BM_B = 8$ ...(10)

At $X_2 = 3$

  $BM_C = 20$ ...(11)

i.e. inclined line 8 to 20

Consider section 3-3, taking moment about section 3-3

$$BM_{3-3} = 8.X_3 - 2.(X_3 - 1) - 3.(X_3 - 3)$$
$$= 3.X_3 + 11$$

It is Equation of straight line (Y = mX + C), inclined linear.

Inclined linear means value of Bending moment at both nearest point of the section is varies with $X_3 = 3$ to $X_3 = 5$

At $X_3 = 3$

  $BM_C = 20$ ...(12)

At $X_3 = 5$

  $BM_D = 26$ ...(13)

i.e. inclined line 20 to 26

Plot the BM_D with the help of above bending moment values.

The SF_D and BM_D is shown in fig 12.19.

Q.16: A cantilever is shown in fig 12.20. Draw the BMD and SFD. What is the reaction at supports?
Sol.:

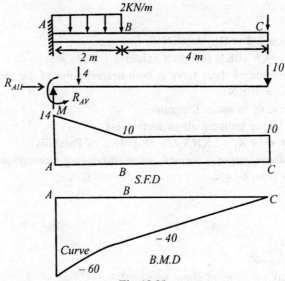

Fig 12.20

Let reaction at support A be R_{AV}, R_{AH} and M (anti clock wise), First find the support reaction
For that,

$$\Sigma V = 0$$
$$R_{AV} - 4 - 10 = 0, R_{AV} = 14 \qquad ...(1)$$

Taking moment about point A,
$$\Sigma M_A = 0$$
$$-M + 4 \times 1 + 10 \times 6 = 0$$
$$M = 64 \text{ KNm} \qquad ...(2)$$
$$\Sigma H = 0$$
$$R_{AH} = 0 \qquad ...(3)$$

Calculation for the Shear force Diagram

Draw the section line, here total 2 section line, which break the load R_{AV} and UDL (Between Point A and B), point B and 10KN(Between Point B and C).

Let

Distance of section 1-1 from point A is X_1
Distance of section 2-2 from point A is X_2
Consider left portion of the beam
Consider section 1-1
Force on left of section 1-1 is R_{AV} and UDL from point A to section line
$SF_{1-1} = 14 - 2X1$ KN (Equation of straight line)
It is Equation of straight line (Y = mX + C), inclined linear.
Inclined linear means value of shear force at both nearest point of the section is varies with X1 = 0 to $X_1 = 2$
    At $X_1 = 0$
                $SF_A = 14$ ...(4)

At $X_1 = 2$
$SF_B = 10$ ...(5)
i.e. inclined line 14 to 10
Consider section 2-2
Forces on left of section 2-2 is R_{AV} & 4KN
$SF_{2-2} = 14 - 4 = 10KN$ (constant value)
Constant value means value of shear force at both nearest point of the section is equal i.e.
$SF_B = SFC = 10KN$ ...(5)
Calculation for the Bending moment Diagram
Consider section 1-1, taking moment about section 1-1
$BM_{1-1} = -64 + 14.X_1 - 2.X1(X_1/2)$ (Equation of Parabola)
Parabola means a parabolic curve is formed, value of bending moment at both nearest point of the section is varies with $X_1 = 0$ to $X_1 = 2$
At $X_1 = 0$
$BM_A = -64$ ...(8)
At $X_1 = 2$
$BM_B = -40$ ...(9)
i.e. parabolic line -64 to -40
Consider section 2-2, taking moment about section 2-2
$BM_{2-2} = -64 + 14.X_2 - 4.(X_2 - 1)$
$= -60 + 10X_2$
It is Equation of straight line $(Y = mX + C)$, inclined linear.
Inclined linear means value of Bending moment at both nearest point of the section is varies with $X_2 = 2$ to $X_2 = 6$
At $X_2 = 2$
$BM_B = -40$ ...(10)
At $X_2 = 6$
$BM_C = 0$ ...(11)
i.e. inclined line -40 to 0
Plot the BMD with the help of above bending moment values.
The SFD and BMD is shown in fig (12.20).

Q.17: Fig 12.21 shows vertical forces 20KN, 40KN and UDL of 20KN/m in 3m lengths. Find the resultant force of the system and draw the shear force and B.M. diagram. *(Dec-01)*

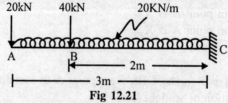

Fig 12.21

Sol.: Total force acting are 20KN, 40KN and 60KN (UDL),
Hence resultant of the system $= \sqrt{(\Sigma H)^2 + (\Sigma V)^2}$
$\Sigma H = 0$ and $\Sigma V = 20 + 40 + 60 = 120KN$
R = 120KN ....ANS

Here total two-section line, which cut AB, AC
Distance of section 1-1 from point A is X_1
Distance of section 2-2 from point A is X_2
Consider left portion of the beam
S.F. Calculations:
$S.F._{1-1} = -20 - 20.X_1$ (Equation of inclined line)
At $X_1 = 0$
     $SF_A = -20KN$
At $X_1 = 1$
     $SF_B = -40KN$
$S.F._{2-2} = -20 - 40 - 20X_2$
At $X_2 = 1$
     $SF_B = -80KN$
At $X_2 = 3$
     $SF_C = -120KN$
Plot the SFD with the help of above value

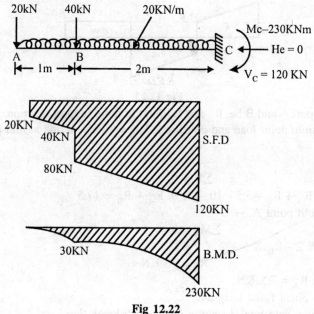

Fig 12.22

B.M. Calculations:
$B.M._{1-1} = -20X_1 - 20.X1(X_1/2)$ (Equation of Parabola)
At $X_1 = 0$
     $BM_A = 0$
At $X_1 = 1$
     $BM_B = -30KN\text{-}m$

$BM_{2-2} = -20X2 - 40(X_2-1) - 20X2(X_2/2)$
At $X_2 = 1$
  $BM_B = -30$KN-m
At $X_2 = 3$
  $SF_C = -230$KN-m

Plot the BMD with the help of above value

Numerical Problems Based on Overhanging Beam

Q.18: Draw the SF diagram for the simply supported beam loaded as shown in fig 12.23.

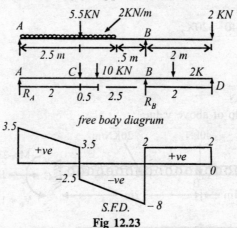

Fig 12.23

Sol.: Let reaction at support A and B be, R_A and R_B First find the support reaction. For finding the support reaction, convert UDL in to point load and equal to 2 X 5 = 10KN, acting at mid point of UDL i.e. 2.5m from point A.

For that,

$$\Sigma V = 0$$
$$R_A + R_B - 5.5 - 10 - 2 = 0, R_A + R_B = 17.5 \qquad ...(1)$$

Taking moment about point A,

$$\Sigma M_A = 0$$
$$10 \times 2.5 + 5.5 \times 2 - R_B \times 5 + 2 \times 7 = 0$$
$$R_B = 10 \text{ KN} \qquad ...(2)$$

From equation (1), $R_A = 7.5$ KN ...(3)

Calculation for the Shear force Diagram

Draw the section line, here total 3-section line, which break the
load R_A, 5.5KN (Between Point A and E),
5.5KN and UDL (Between Point E and B),
Point B and 2KN (Between Point B and C).

Let
Distance of section 1-1 from point A is X_1
Distance of section 2-2 from point A is X_2
Distance of section 3-3 from point A is X_3
Consider left portion of the beam
Consider section 1-1

Force on left of section 1-1 is R_A and UDL (from point A to the section line i.e. UDL on total distance of X_1
$SF_{1-1} = 7.5 - 2X_1$ KN (Equation of straight line)
It is Equation of straight line ($Y = mX + C$), inclined linear.
Inclined linear means value of shear force at both nearest point of the section is varies with $X_1 = 0$ to $X_1 = 2$

At $\quad X_1 = 0$
$\quad\quad SF_A = 7.5$
At $\quad X_1 = 2$ ...(4)
$\quad\quad SF_E = 3.5$ ...(5)
i.e. inclined line 7.5 to 3.5
Consider section 2-2
Forces on left of section 2-2 is RA, 5.5KN and UDL on X_2 length
$SF_{2-2} = 7.5 - 5.5 - 2X_2 = 2 - 2X_2$ (Equation of straight line)
It is Equation of straight line ($Y = mX + C$), inclined linear.
Inclined linear means value of shear force at both nearest point of the section is varies with $X_2 = 2$ to $X_2 = 5$

At $\quad X_2 = 2$
$\quad\quad SF_E = -2$
At $\quad X_2 = 5$ ...(4)
$\quad\quad SF_B = -8$ ...(5)
i.e. inclined line -2 to -8

Since here shear force changes the sign so at any point shear force will be zero and at that point bending moment is maximum.
For finding the position of zero shear force equate the shear force equation to zero, i.e.
$2 - 2X_2$; $X_2 = 1$m, i.e. at 1m from point A bending moment is maximum.
Consider section 3-3
Forces on left of section 3-3 is R_A, 5.5KN and 10KN and R_B
$SF_{3-3} = 7.5 - 5.5 - 10 + 10 = 2$KN (constant value)
Constant value means value of shear force at both nearest point of the section is equal i.e.
$\quad\quad SF_B = SF_C = 2$KN ...(7)
Plot the SFD with the help of above shear force values.

Q.19: Draw the shear force diagram of the beam shown in fig 12.24.
Sol.: First find the support reaction, for that
Convert UDL in to point load, Let reaction at C be R_{CH} and R_{CV}, and at point D be R_{DV}.
$\quad\quad R_V = 0$
$\quad\quad R_{CV} + R_{DV} = 1 \times 3 + 2 R_{CV} + R_{DV} = 5$KN ...(1)
Taking moment about point C,
$3 \times 0.5 + 2 \times 5 - R_{DV} \times 4 = 0$
$R_{DV} = 2.875$KN ...(2)
From equation (1)
$\quad\quad R_{CV} = 2.125$KN
Calculation for SFD
Here total 4 section line
$\quad\quad SF_{1-1} = 1 X_1$ (inclined line)

At $X_1 = 0$
       $SF_A = 0$
At $X_1 = 1; SF_C = 1$

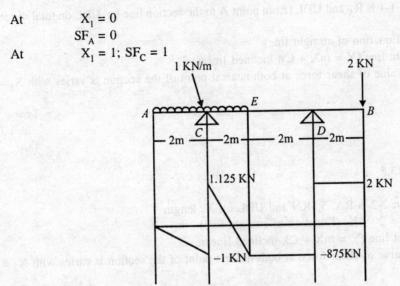

Fig 12.24

$SF_{2-2} = 1X_2 - R_{CV}$ (inclined line)
At $X_2 = 1$
       $SF_C = -1.125$
At $X_2 = 3$
       $SF_E = 0.875$
$SF_{3-3} = 3 - R_{CV}$ (Constant line)
At $X_3 = 3$
       $SF_C = 0.875$
At $X_3 = 5$
       $SF_D = 0.875$
$SF_{4-4} = 3 - R_{CV} - R_{DV}$ (Constant line)
At $X_4 = 5$
       $S_{FD} = -2$
At $X_4 = 7$
       $SF_B = -2$

Q.20: Find the value of X and draw the bending moment diagram for the beam shown below 12.25. Given that $R_A = 1000$ N & $R_B = 4000$ N.   *(May–01)*

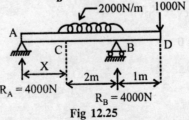

Fig 12.25

Sol.: For finding the Value of X, For that first draw the FBD, Taking moment about point A
UDL = 2000 X 2 = 4000 acting at a distance of (X + 1) from point A.

$M_A = 4000 \cdot (X + 1) - R_B \cdot (2 + X) + 1000 \cdot (X + 3) = 0$
$4000 + 4000X - 8000 - 4000X - 1000X - 3000 = 0$
$\qquad 1000X = 1000$
$\qquad\qquad X = 1m$

Calculation for Banding Moment diagram
Here total three-section line, which cut AC, CB and BD
Distance of section 1-1 from point A is X_1
Distance of section 2-2 from point A is X_2
Distance of section 3-3 from point A is X_3
Consider left portion of the beam
Consider section 1-1, taking moment about section 1-1
$\qquad BM_{1-1} = 1000.X_1$
It is Equation of straight line $(Y = mX + C)$, inclined linear.
Inclined linear means value of bending moment at both nearest point of the section is varies with $X_1 = 0$ to $X_1 = 1$
At $\qquad X_1 = 0$
$\qquad BM_A = 0$ ...(8)
At $\qquad X_1 = 1$
$\qquad BM_C = 1000$ ...(9)
i.e. inclined line 0 to 1000 (Inclined line)
Consider section 2-2, taking moment about section 2-2
$\qquad BM_{2-2} = 1000.X_2 - 2000(X_2 - 1)$
It is Equation of parabola $(Y = mX_2 + C)$,
Parabola means value of bending moment at both nearest point of the section is varies with $X_2 = 1$ to $X_2 = 3$ and make a curve
At $\qquad X_2 = 1$
$\qquad BM_C = 1000$ ...(8)
At $\qquad X_2 = 3$
$\qquad BM_B = -1000$ ...(9)
i.e. Curve between 1000 to -1000
Consider section 3-3, taking moment about section 3-3
$\qquad BM_{3-3} = 1000.X_3 - 4000(X_3 - 2) + 4000(X_3 - 3)$
It is Equation of straight line $(Y = mX + C)$, inclined linear.
Inclined linear means value of bending moment at both nearest point of the section is varies with $X_3 = 3$ to $X_3 = 4$
At $\qquad X_3 = 3$
$\qquad BM_B = -1000$ ...(8)
At $\qquad X_3 = 4$
$\qquad BM_B = 0$ ...(9)
i.e. Curve between -1000 to 0
Plot the BMD with the help of above value, BMD is show in fig 12.26.

296 / *Problems and Solutions in Mechanical Engineering with Concept*

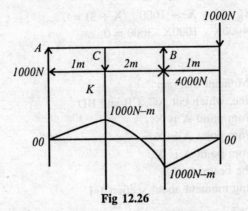

Fig 12.26

Q.21: Draw the SFD and BMD for the beam shown in the figure 12.27.

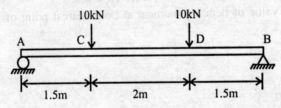

Fig 12.27

Sol.: Let Support reaction at A and B be Ra and Rb; and the diagram is symmetrical about y axis so the both reactions are equal; i.e.

$$R_a = R_b = 10 KN$$

S.F. Calculation

$$S.F._{1-1} = +10 \text{ KN}$$
$$S.F._A = S.F._C = 10$$
$$S.F._{2-2} = 10 - 10 = 0 \text{ KN}$$
$$S.F._C = S.F._D = 0$$

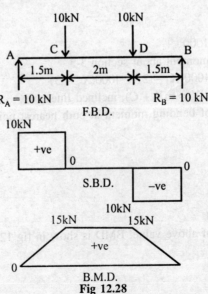

Fig 12.28

$S.F_{3-3} = 10 - 10 - 10 = -10$ KN
$S.F_D = S.F_B = -10$

B.M. Calculation
$B.M_{1-1} = 10.x1$ (Linear)
$B.M_A = 0$
$B.M_C = 15$ KN
$B.M_{2-2} = 10.x2 - 10(X_2 - 1.5)$ (Linear)
$B.M_C = 15$ KN
$B.M_D = 15$ KN
$B.M_{3-3} = 10.X_3 - 10(X_3 - 1.5) - 10(X_3 - 3.5)$ (Linear)
$B.M_D = 15$ KN
$B.M_B = 0$

Draw the SFD and BMD with the help of above values as shown in fig 12.28.
Note: The B.M. is zero at the point where shear force is zero. And the region where shear force is zero; the bending moment is constant as shown in fig.

Load Diagram and BM$_D$ from the Given SF$_D$

Q.22: The shear force diagram of simply supported beam is given below in the fig 12.29. Calculate the support reactions of the beam and also draw bending moment diagram of the beam. *(May–01(C.O.))*

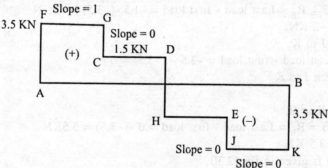

Fig 12.29

Sol.: For the given SF$_D$, First we draw the load diagram, and then with the help of load diagram we draw the BM$_D$.

As the slope in SFD is zero. So it indicates that the beam is only subjected to point loads. Let R_A and R_B be the support reaction at A and B and the load R_C, R_D and R_E in down ward direction at point C, D and E respectively.

Here the graph of SFD moves from A-F-G-C-D-H-E-J-K-B
Consider two points continuously,
Consider A-F
Load moves from A to F,
Load intensity at A = R_A = Last load - first load = 3.5 - 0 = 3.5KN
i.e. $R_A = 3.5$ KN ...(1)
Consider F-G

Load moves from F to G,
Load intensity = Last load - first load = 3.5 - 3.5 = 0
i.e. No load between F to G ...(2)
Consider G-C
Load moves from G to C,
Load intensity at C = R_C = Last load - first load = 3.5 – 1.5 = –2KN
i.e. R_C = –2 KN ...(3)
Consider C-D
Load moves from C to D,
Load intensity = Last load – first load = 1.5 – 1.5 = 0
i.e. No load between C to D ...(4)
Consider D-H
Load moves from D to H,
Load intensity at D = R_D = Last load - first load = –1.5 – 1.5 = –3KN
i.e. R_D = –3 KN ...(5)
Load moves from H to E,
Load intensity = Last load - first load = -1.5 -(-1.5) = 0
i.e. No load between H to G ...(6)
Consider E-J
Load moves from E to J,
Load intensity at E = R_E = Last load - first load = –1.5 -(–3.5) = –2KN
i.e. R_E = –2 KN ...(7)
Load moves from J to K,
Load intensity = Last load - first load = –3.5 – (–3.5) = 0
i.e. No load between J to K ...(8)
Consider K-B
Load moves from K to B,
Load intensity at B = R_B = Last load - first load = 0 –(–3.5) = 3.5KN
i.e. R_B = 3.5 KN ...(9)
Now load diagram is given in fig 12.30

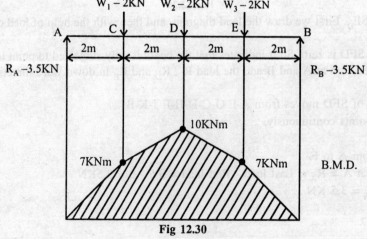

Fig 12.30

Now Calculation for BMD

Taking moment about any point gives the value of BM at that point.

Consider left portion of the beam

Taking moment about point A i.e. $M_A = BM_A = 0$

Taking moment about point C, $M_C = BM_C = 3.5 \times 2 = 7$ KN-m

Taking moment about point D, $M_D = BM_D = 3.5 \times 4 - 2 \times 2 = 10$ KN-m

Taking moment about point E, $M_E = BM_E = 3.5 \times 6 - 2 \times 4 - 3 \times 2$
$$= 7 \text{ KN-m}$$

Taking moment about point B, $M_B = BM_B = 3.5 \times 8 - 2 \times 6 - 3 \times 4 - 2 \times 2$
$$= 0 \text{ KN-m}$$

Draw the BMD with the help of above value.

Q.23: The shear force diagram of simply supported beam is given below in the fig. Calculate the support reactions of the beam and also draw bending moment diagram of the beam.

(Dec–01)

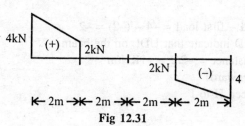

Fig 12.31

Sol.: For the given SFD, First we draw the load diagram, and then with the help of load diagram we draw the BMD.

Let R_A and R_B be the support reaction at A and B

Here the graph of SFD moves from A-F-G-C-D-E-H-J-B

Consider two points continuously,

Consider A-F

Load moves from A to F,

Load intensity at A = R_A = Last load - first load = $4 - 0 = 4$KN

i.e. $R_A = 4$ KN   ...(1)

Consider F-G

Load moves from F to G,

Load intensity = Last load − first load = $2 - 4 = -2$KN

Since inclined line in BMD indicate that UDL on the beam

Udl = Total Load/Total distance = $-2/2 = -1$KN/m

(-ive means UDL act downward)

i.e. UDL of 1KN/m between F to G   ...(2)

Consider G-C

Load moves from G to C,

Load intensity at C = R_C = Last load − first load = $0 - 2 = -2$KN

i.e. $R_C = -2$ KN   ...(3)

Consider C-D

Load moves from C to D,
Load intensity = Last load – first load = 0 – 0 = 0
i.e. No load between C to D ...(4)
Consider D-E
Load moves from D to E,
Load intensity at D = Last load – first load = 0 – 0 = 0
i.e. No load between D to E ...(8)
Load moves from E to H,
Load intensity = Last load - first load = –2 –0 = -2KN
i.e. R_E = –2KN ...(6)
Consider H-J
Load moves from H to J,
Load intensity = Last load – first load = –1.5 –(–3.5) = –2KN
i.e. R_E = –2 KN ...(7)
Load moves from J to K,
Load intensity = Last load – first load = –4 – (–2) = –2
Since inclined line in BMD indicate that UDL on the beam
Udl = Total Load/Total distance = –2/2 = –1KN/m
(-ive means UDL act downward)
i.e. UDL of 1KN/m between H to I ...(2)
Consider J-B
Load moves from J to B,
Load intensity at B = R_B = Last load - first load = 0 – (–4) = 4KN
i.e. R_B = 4KN ...(9)
Now load diagram is given in fig
Now Calculation for BMD
Here total three-section line, which cut AC, CD, DB
Distance of section 1-1 from point A is X_1
Distance of section 2-2 from point A is X_2
Distance of section 3-3 from point A is X_3
Consider left portion of the beam

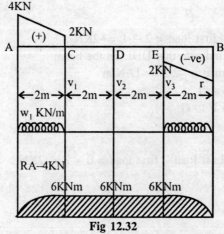

Fig 12.32

B.M. Calculations:
B.M. at A = 0KN.m
B.M. at C = 4 X 2 – 1 X 2 = 6KN.m
B.M. at D = 4 X 4 – 1 X 2 X (2/2 + 2) – 2 X 2 = 6KN.m
B.M. at E = 4 X 6 – 1 X 2 X (2/2 + 4) – 2 X 4 = 6KN.m
B.M. at B = 4 X 8 – 1 X 2 X (2/2 + 6) – 2 X 6 – 2 X 2 – 1 X 2 X (2/2) = 0KN.m

Loading Giagram and SF$_D$ from the given BM$_D$

Q.24: The bending moment diagram (BM$_D$) of a simple supported beam is given as shown in fig 12.33. Calculate the support reactions of the beam. *(Dec–00)*

Sol.:

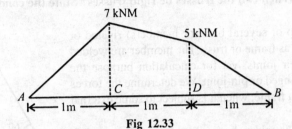

Fig 12.33

Linear variation of bending moment in the section AC, CD and DB indicate that there is no load on the beam in these sections. Change in the slope of the bending moment at point C and D is indicate that there must be concentrated vertical loads at these points.

Let point load acting at A, B, C, D are RA, RB, P, Q respectively.

Consider three section line of the beam which cut the line AC, CD and DB respectively. Since the value of moment at all the section is the last value of the BM at that section.

Consider Section 1-1, Taking moment from point C,

$M_C = 7 = RA \times 1$

$R_A = 7KN$ ...(1)

Consider Section 2-2, Taking moment from point D,

$M_D = 5 = R_A \times 2 - P \times 1$

$5 = 7 \times 2 - P$

$P = 9KN$ ...(2)

Consider Section 3-3, Taking moment from point B,

$M_B = 0 = R_A \times 3 - P \times 2 - Q \times 1$

$0 = 7 \times 3 - 9 \times 2 - Q$

$Q = 3KN$ ...(3)

NOW $R_A + R_B = P + Q$

$R_B = 5KN$

$R_A = 7KN$ and $R_B = 5KN$ ....ANS

Chapter 13

TRUSS

Q. 1: What are truss? When can the trusses be rigid trusses? State the condition followed by simple truss?

Sol.: A structure made up of several bars (or members) riveted or welded together is known as frame or truss. The member are welded or riveted together at their joints, yet for calculation purpose the joints are assumed to be hinged or pin-joint. We determine the forces in the members of a perfect frame, when it is subject to some external load.

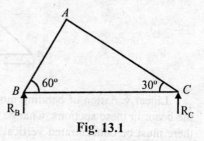

Fig. 13.1

Rigid Truss: A truss is said to be rigid in nature when there is no deformation on application of any external force.

Condition followed by simple truss: The truss which follows the law $n = 2j - 3$. is known as simple truss. Where n = Number of link or member j = Number of joints A triangular frame is the simplest truss.

Q. 2: Define and explain the term: (a) Perfect frame (b) Imperfect frame (c) Deficient frame (d) Redundant frame.

Perfect Frame

The frame, which is composed of such members, which are just sufficient to keep the frame in equilibrium, when the frame is supporting an external load, is known as perfect frame. Hence for a perfect frame, the number of joints and number of members are given as:

$$n = 2j - 3$$

Imperfect Fram

An Imperfect frame is one which does not satisfies the relation between the numbers of members and number of joints given by the equation $n = 2j - 3$.

This means that number of member in an imperfect frame will be either more or less than (2j-3)

It may be a deficient frame or a redundant frame.

Deficient Frame

If the numbers of member in a frame are less than (2j-3), then the frame is known as *deficient frame*.

Redundant Frame

If the numbers of member in a frame are more than $(2j-3)$, then the frame is known as *redundant frame*.

Q. 3: What are the assumptions made in the analysis of a simple truss?
Sol.: The assumptions made in finding out the forces in a frame are,
   (1) The frame is a perfect frame.
   (2) The frame carries load at the joints.
   (3) All the members are pin-joint. It means members will have only axial force and there will be no moment due to pin, because at a pin moment becomes zero.
   (4) Load is applied at joints only.
   (5) Each joint of the truss is in equilibrium, hence the whole frame or truss is also in equilibrium.
   (6) The weight of the members of the truss is negligible.
   (7) There is no deflection in the members on application of load.
   (8) Stresses induced on application of force in the members is negligible.

Q. 4: How can you evaluate the reaction of support of a frame?
Sol.: The frames are generally supported on a roller support or on a hinged support. The reactions at the supports of a frame are determined by the conditions of equilibrium (i.e. sum of horizontal forces and vertical forces is zero). The external load on the frame and the reactions at the supports must form a system of equilibrium.

There are three conditions of equilibrium.
1. $\Sigma V = 0$ (i.e. Algebraic sum of all the forces in a vertical direction must be equal to zero.)
2. $\Sigma H = 0$ (i.e. Algebraic sum of all the forces in a horizontal direction must be equal to zero.)
3. $\Sigma M = 0$ (i.e. Algebraic sum of moment of all the forces about a point must be equal to zero.)

Q. 5: How can you define the nature of force in a member of truss?
Sol.: We know that whenever force is applied on a cross section or beam along its axis, it either tries to compress it or elongate it. If applied force tries to compress the member force is known as compressive force as shown in fig (13.2). If force applied on member tries to elongate it, force is known as tensile force shown in fig. 13.3.

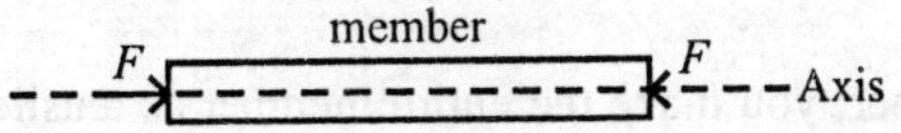

Fig. 13.2

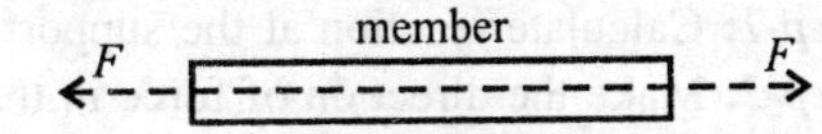

Fig. 13.3

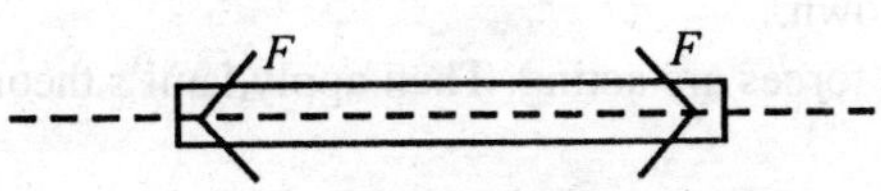

Fig. 13.4

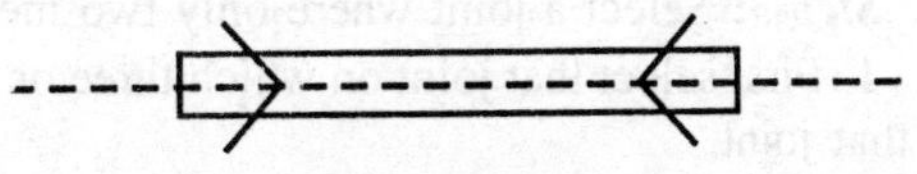

Fig. 13.5

If a compressive force is applied on the member as in fig. 13.2, the member will always try to resist this force & a force equal in magnitude but opposite to direction of applied force will be induced in it as shown in fig. 13.4, Similarly induced force in member shown in fig. 13.3 will be as shown in fig. 13.5

From above we can conclude that if induced force in a member of loaded truss is like the fig. 13.4 we will say nature of applied force on the member is compressive. If Nature of induced force in a member of truss like shown in fig. 13.5, then we can say that Nature of force applied on the member is tensile.

Q. 6: Explain, why roller support are used in case of steel trusses of bridges?

Sol.: In bridges most of time only external force perpendicular to links acts and the roller support gives the reaction to link, hence it is quite suitable to use roller support in case of steel trusses of bridges

Q. 7: Where do you find trusses in use? What are the various methods of analysis of trusses? What is basically found when analysis of a system is done?

Sol.: The main use of truss are:

1. The trusses are used to support slopping roofs.
2. Brick trusses are used in bridges to support deck etc.

Analysis of a frame consists of,

(a) Determinations of the reactions at the supports.

(b) Determination of the forces in the members of the frame.

The forces in the members of the frame are determined by the condition that every joint should be in equilibrium. And so, the force acting at every joint should form a system in equilibrium. A frame is analyzed by the following methods,

1. Method of joint.
2. Method of section.
3. Graphical method.

When analysis is done, we are basically calculation the forces acting at each joint by which we can predict the nature of force acting at the link after solving our basic equation of equilibrium.

Q. 8: How you can find the force in the member of truss by using method of joint? What are the steps involved in method of joint ?

Sol.: In this method, after determining the reactions at the supports, the equilibrium of every joint is considered. This means the sum of all the vertical forces as well as horizontal forces acting on a joint is equal to zero. The joint should be selected in such a way that at any time there are only two members, in which the forces are unknown.

The force in the member will be compressive if the member pushes the joint to which it is connected whereas the force in the member will be tensile if the member pulls the joint to which it is connected.

Steps for Method of Joint

To find out force in member of the truss by this method, following three Steps are followed.

Step-1: Calculate reaction at the support.

Step-2: Make the direction of force in the entire member; you make the entire member as tensile. If on solving the problems, any value of force comes to negative that means the assumed direction is wrong, and that force is compressive.

Step-3: Select a joint where only two members is unknown.

1- First select that joint on which three or less then three forces are acting. Then apply lami's theorem on that joint.

Step-4: Draw free body diagram of selected joint since whole truss is in equilibrium therefore the selected joint will be in equilibrium and it must satisfy the equilibrium conditions of coplanar concurrent force system.

$$\Sigma V = 0 \text{ and } \Sigma H = 0$$

Step-5: Now select that joint on which four forces, five forces etc are acting. On that joint apply resolution of forces method.

Note: If three forces act at a joint and two of them are along the same straight line, then for the equilibrium of the joint, the third force should be equal to zero.

Q. 9: Find the forces in the members AB, BC, AC of the truss shown in fig 13.6. C.O. Dec -04-05

Sol.: First determine the reaction R_B and R_C. The line of action of load 20KN acting at A is vertical. This load is at a distance of $AB \cos 60°$, from the point B. Now let us find the distance AB, The triangle ABC is a right angle triangle with angle BAC = 90°. Hence AB will be equal to $CB \cos 60°$. $AB = 5 \times \cos 60° = 2.5$m. Now the distance of line of action of 20KN from B is $= AB \cos 60° = 1.25$m.

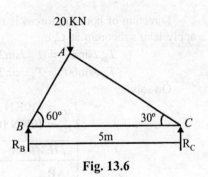

Fig. 13.6

Now, taking the moment about point B, we get
$R_C \times 5 - 20 \times 1.25 = 0$
$R_C = 5$KN ...(i)
$R_B = 15$KN ...(ii)

Let the forces in the member AC, AB and BC is in tension.
Now let us consider the equilibrium of the various joints.

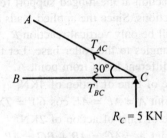

Fig 13.7(a)   Fig 13.7(b)

Joint B:
Consider FBD of joint B as shown in fig 13.7(a)
Let,
T_{AB} = Force in the member AB
T_{BC} = Force in the member BC
Direction of both the forces is taken away from point B. Since three forces are acting at joint B. So apply lami's theorem at B.

$T_{AB}/\sin 270° = T_{BC}/\sin 30° = R_B/\sin 60°$
$T_{AB}/\sin 270° = T_{BC}/\sin 30° = 15/\sin 60°$

On solving
$T_{AB} = -17.32$KN ...(iii)
$T_{AB} = 17.32$KN (Compressive) ....ANS
$T_{BC} = 8.66$KN ...(iv)
$T_{BC} = 8.66$KN (Tensile) ....ANS

Joint C Fig 13.7(b)
Consider FBD of joint C as shown in fig 13.7 (b)
Let,
T_{BC} = Force in the member BC
T_{AC} = Force in the member AC

Direction of both the forces is taken away from point C. Since three forces are acting at joint C. So apply lami's theorem at C.

$$T_{BC}/\sin60° = T_{AC}/\sin270° = R_C/\sin30°$$
$$T_{BC}/\sin60° = T_{AC}/\sin270° = 5/\sin30°$$

On solving

$$T_{AC} = -10KN$$ ...(v)
$$T_{AC} = 10KN \text{ (Compressive)}\quad ....ANS$$

MEMBER	FORCE	TYPE
AB	17.32KN	COMPRESSIVE
AC	8.66KN	TENSILE
BC	10KN	COMPRESSIVE

Q. 10: Determine the reaction and the forces in each member of a simple triangle truss supporting two loads as shown in fig 13.8.

Sol.: The reaction at the hinged support (end A) can have two components acting in the horizontal and vertical directions. Since the applied loads are vertical, the horizontal component of reaction at A is zero and there will be only vertical reaction R_A, Roller support (end C) is frictionless and provides a reaction R_C at right angles to the roller base. Let the forces in the entire member is tensile. First calculate the distance of different loads from point A.

Distance of Line of action of 4KN,
from point A = AF = AE cos 60° = 2X 0.5 = 1m
Distance of Line of action of 2KN,
from point A = AG = AB + BG = AB + BD cos 60°
= 2 + 2 x 0.5 = 3m

Taking moment about point A,

$$R_C \times 4 - 2 \times 3 + 4 \times 1 = 0 \quad R_C = 2.5KN \qquad ...(i)$$
$$R_A = 4 + 2 - 2.5 = 3.5 \text{ KN} \qquad ...(ii)$$

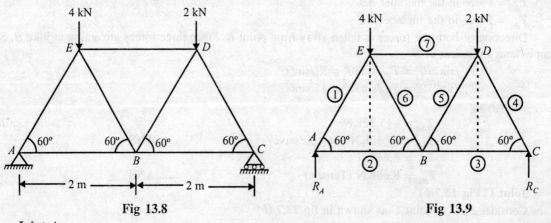

Fig 13.8   Fig 13.9

Joint A:

Consider FBD of joint A as shown in fig 13.10 Let, T_{AE} = Force in the member AE T_{AB} = Force in the member AB Direction of both the forces (T_{AE} & T_{AB}) is taken away from point A. Since three forces are

acting at joint A. So apply lami's theorem at B. $T_{AE}/\sin 270° = T_{AB}/\sin 30° = R_A/\sin 60°$
$T_{AE}/\sin 270° = T_{AB}/\sin 30° = 3.5/\sin 60°$
On solving

$T_{AE} = -4.04$ KN ...(iii)
$T_{AE} = 4.04$ KN (Compressive) ....ANS
$T_{AB} = 2.02$ KN ...(iv)
$T_{AB} = 2.02$ KN (Tensile) ....ANS

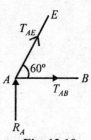

Fig. 13.10

Joint C:
Consider FBD of joint C as shown in fig 13.11 Let, T_{BC} = Force in the member BC T_{DC} = Force in the member DC Direction of both the forces(T_{BC} & T_{DC}) is taken away from point C. Since three forces are acting at joint C. So apply lami's theorem at C. $T_{BC}/\sin 30° = T_{DC}/\sin 270° = R_C/\sin 60°$ $T_{BC}/\sin 30° = T_{DC}/\sin 270° = 2.5/\sin 60°$
On solving

$T_{BC} = 1.44$ KN ...(v)
$T_{BC} = 1.44$ KN (Tensile) ....ANS
$T_{DC} = -2.88$ KN ...(iv)
$T_{DC} = 2.88$ KN (Compressive) ....ANS

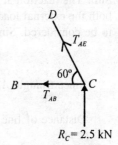

Fig. 13.11

Joint B:
Consider FBD of joint B as shown in fig 13.12 Since, $T_{AB} = 2.02$ KN(T) $T_{BC} = 1.44$ KN(T) Let, T_{BE} = Force in the member BE T_{DB} = Force in the member DB Direction of both the forces (T_{BE} & T_{DB}) is taken away from point B. Since four forces are acting at joint B. So apply resolution of forces as equilibrium at B.

$R_H = 0$
$-T_{AB} + T_{BC} - T_{BE} \cos 60° + T_{BD} \cos 60° = 0$
$-2.02 + 1.44 - 0.5 T_{BE} + 0.5 T_{BD} = 0$
$T_{BE} - T_{BD} = 1.16$ ...(vii)
$R_V = 0$
$T_{BE} \sin 60° + T_{BD} \sin 60° = 0$
$T_{BE} = -T_{BD}$ ...(viii)

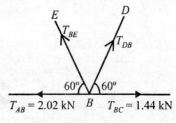

Fig. 13.12

Value of equation (viii) put in equation (vii), we get
i.e $-2 T_{BE} = 1.16$, or
$T_{BE} = -0.58$ KN ...(ix)
$T_{BE} = 0.58$ KN (Compression) ....ANS
$T_{BD} = 0.58$ KN (Tensile) ....ANS

Joint D:
Consider FBD of joint D as shown in fig 13.13 Since, $T_{CD} = -2.88$ KN(C) $T_{BD} = 0.58$ KN(T) Let, T_{ED} = Force in the member ED Direction of forces T_{CD}, T_{BD} & T_{ED} is taken away from point D. Since four forces are acting at joint D. So apply resolution of forces as equilibrium at D.

$R_H = 0$
$-T_{ED} - T_{BD} \cos 60° + T_{CD} \cos 60° = 0$
$-T_{ED} - 0.58 \times 0.5 + (-2.88) \times 0.5 = 0$
$T_{ED} = -1.73$ KN ....ANS

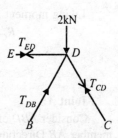

Fig. 13.13

Member	Force	Member	Force
AE	4.04KN(C)	BE	0.58KN(C)
AB	2.02KN(T)	BD	0.58KN(T)
BC	1.44KN(T)	DE	1.73KN(C)
CD	2.88KN(C)		

Q. 11: Determine the forces in all the members of the truss loaded and supported as shown in fig 13.14.

Sol.: The reaction at the supports can be determined by considering equilibrium of the entire truss. Since both the external loads are vertical, only the vertical component of the reaction at the hinged ends A need to be considered. Since the triangle AEC is a right angle triangle, with angle AEC = 90°. Then,

$$AE = AC \cos 60° = 5 \times 0.5 = 2.5m$$
$$CE = AC \sin 60° = 5 \times 0.866 = 4.33m$$

Since triangle ABE is an equilateral triangle and therefore,

$$AB = BC = AE = 2.5m$$

Distance of line of action of force 10KN from joint A,

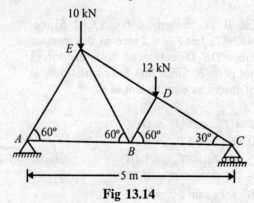

Fig 13.14

$$AF = AE \cos 60° = 2.5 \times 0.5 = 1.25m$$

Again, the triangle BDC is a right angle triangle with angle BDC = 90°.
Also,
$$BC = AC - AB = 5 - 2.5 = 2.5m$$
$$BD = BC \cos 60° = 2.5 \times 0.5 = 1.25m$$

Distance of line of action of force 12KN from joint A,

$$AG = AB + BG = AB + BD \cos 60° = 2.5 + 1.25 \times 0.5 = 3.125m$$

Taking moment about end A, We get

$$R_C \times 5 = 12 \times 3.125 + 10 \times 1.25 = 50$$
$$R_C = 10KN \qquad \qquad \ldots(i)$$
$$\Sigma V = 0, R_C + R_A = 10 + 12 = 22KN$$
$$R_A = 12KN \qquad \qquad \ldots(ii)$$

Joint A:
Consider FBD of joint A as shown in fig 13.15 Let, T_{AE} = Force in the member AE T_{AB} = Force in the member AB Direction of both the forces (T_{AE} & T_{AB}) is taken away from point A. Since three forces are

acting at joint A. So apply lami's theorem at A. $T_{AE}/\sin 270° = T_{AB}/\sin 30° = R_A/\sin 60°$ $T_{AE}/\sin 270° = T_{AB}/\sin 30° = 12/\sin 60°$

$T_{AE} = -13.85$ KN ...(iii)
$T_{AE} = 13.85$ KN (Compression) ....ANS
$T_{AB} = 6.92$ KN ...(iv)
$T_{AB} = 6.92$ KN (Tension) ....ANS

Fig. 13.15

Joint C:
Consider FBD of joint C as shown in fig 13.16 Let, T_{BC} = Force in the member BC T_{CD} = Force in the member CD Direction of both the forces (T_{BC} & T_{CD}) is taken away from point C. Since three forces are acting at joint C. So apply lami's theorem at C.

$T_{BC}/\sin 60° = T_{CD}/\sin 270° = R_C/\sin 30°$
$T_{BC}/\sin 60° = T_{CD}/\sin 270° = 10/\sin 30°$
$T_{BC} = 17.32$ KN ...(v)
$T_{BC} = 17.32$ KN (Tension) ....ANS
$T_{CD} = -20$ KN ...(vi)
$T_{CD} = 20$ KN (compression) ....ANS

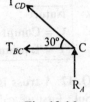

Fig. 13.16

Joint B:
Consider FBD of joint B as shown in fig 13.17
Since, $T_{AB} = 6.92$ KN
$T_{BC} = 17.32$ KN
Let, T_{BD} = Force in the member BD
T_{EB} = Force in the member EB
Direction of both the forces (T_{BD} & T_{EB}) is taken away from point B. Since four forces are acting at joint B. So apply resolution of forces at joint B.

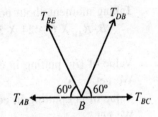

Fig. 13.17

$R_H = 0$
$-T_{AB} + T_{BC} - T_{EB} \cos 60° + T_{BD} \cos 60° = 0$
$- 6.92 + 17.32 - 0.5\, T_{EB} + 0.5\, T_{BD} = 0$
$T_{BD} - T_{EB} = -20.8$ KN ...(vii)
$R_V = 0$
$T_{EB} \sin 60° + T_{BD} \sin 60° = 0$
$T_{BD} = -T_{EB}$ ...(viii)
$T_{BD} = 10.4$ KN ...(ix)
$T_{BD} = 10.4$ KN (Tension) ....ANS
$T_{EB} = -10.4$ KN ...(x)
$T_{EB} = 10.4$ KN (compression) ....ANS

Joint D:
Consider FBD of joint D as shown in fig 13.18
Since, $T_{CD} = -20$ KN
Let, T_{ED} = Force in the member ED

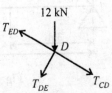

Fig. 13.18

Direction of the force (T_{ED}) is taken away from point D. Since four forces are acting at joint D. So apply resolution of forces at joint D.

Resolve all the forces along EDC, we get

$$T_{ED} + 12\cos 60° + T_{CD} = 0$$
$$T_{ED} + 6 - 20 = 0$$
$$T_{ED} = 14KN$$
$$T_{ED} = 14KN \text{ (Tension)} \quad \text{......ANS} \quad ...(xi)$$

Member	AE	AB	BC	CD	BD	BE	DE
Force in KN	13.85	6.92	17.32	20	10.4	10.4	14
Nature C = Compression T = Tension	C	T	T	C	T	C	T

Q. 12: A truss is as shown in fig 13.19. Find out force on each member and its nature.

Sol.: First we calculate the support reaction, Draw FBD as shown in fig 13.20

$$R_H = 0,$$
$$R_{AH} - R_{BV} \cos 60° = 0 \quad ...(i)$$
$$R_V = 0, R_{AV} + R_{BV} \sin 60° - 24 - 7 - 7 - 8 = 0 \quad ...(ii)$$

Taking moment about point

B, $R_{AV} \times 6 - 24 \times 3 - 7 \times 6 - 8 \times 3 = 0 \quad ...(iii)$
$$R_{AV} = 23KN \quad ...(iv)$$

Value of (iv) putting in equation (ii)

We get, $R_{BV} = 26.6KN \quad ...(v)$

Value of (v) putting in equation (i)

We get, $R_{AH} = 13.3KN \quad ...(vi)$

Joint E, Consider FBD as shown in fig 13.21 From article 8.8.2,

$$T_{ED} = 0, \quad ...(vii)$$
$$T_{ED} = 0 \quad \text{......ANS}$$

And,
$$T_{AE} = -7KN$$
$$T_{AE} = 7KN \text{ (compression)} \quad \text{......ANS} \quad ...(viii)$$

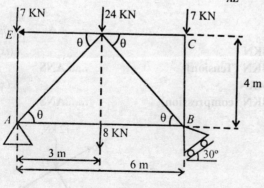

Fig 13.19

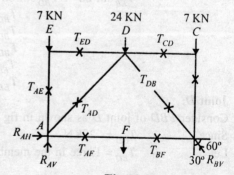

Fig 13.20

Joint C:
Consider FBD as shown in fig 13.22

$T_{CD} = 0$, ...(ix)
$T_{CD} = 0$ ....ANS

And, $T_{BC} = -7KN$ ...(x)
$T_{BC} = 7KN$ (compression) ....ANS

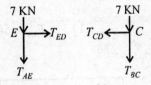

Fig. 13.21   Fig. 13.22

Note: Since for perfect frame the condition $n = 2j - 3$ is necessary to satisfied.
Here Point F is not a joint, if we take F as a joint then,
Number of joint $(j) = 6$ and No. of member $(n) = 7$
$$n = 2j - 3, \quad 7 = 2 \times 6 - 3$$
$$\neq 9 \text{ i.e}$$
i.e if F is not a joint, then $j = 5$
$$7 = 2 \times 5 - 3$$
$= 7$, i.e. F is not a joint. But at joint F a force of 8KN is acting. Which will effect on joint A and B, Since 8KN is acting at the middle point of AB, So half of its magnitude will equally effect on joint A and B. i.e. 4KN each acting on joint A and B downwards

Joint D:
Consider FBD of joint D as shown in fig 13.23
Since, $T_{CD} = T_{ED} = 0KN$
Let, T_{BD} = Force in the member BD
T_{AD} = Force in the member AD
Direction of the force (T_{AD}) & (T_{BD}) is taken away from point D. Since five forces are acting at joint D. So apply resolution of forces at joint D.
Resolve all the forces, we get

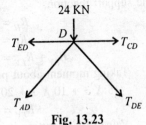

Fig. 13.23

$$R_H = -T_{AD} \cos\theta + \cos\theta = 0$$
$$T_{AD} = T_{BD}$$ ...(xi)
$$R_V = -T_{AD} \sin\theta - T_{BD} \sin\theta - 24 = 0$$
$$(T_{AD} + T_{BD}) = -24/\sin\theta$$
$$\sin\theta = 4/5$$
or, $$2T_{AD} = 2T_{BD} = -24/(4/5)$$
$$T_{AD} = T_{BD} = -15KN$$ ...(xii)
$$T_{AD} = 15KN \text{ (compression)} \quad ....ANS$$
$$T_{BD} = 15KN \text{ (compression)} \quad ....ANS$$

Joint A:
Consider FBD of joint A as shown in fig 13.24
Let, T_{AB} = Force in the member AB
Since, $T_{AD} = -15KN$ $T_{AE} = -7KN$
Direction of the force (T_{AD}) & (T_{AE}) & (T_{AB}) is taken away from point A. Since five forces are acting at joint A. So apply resolution of forces at joint A.
Resolve all the forces, we get
$$R_H = R_{AH} + T_{AB} + T_{AD} \cos\theta = 0$$

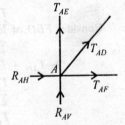

Fig. 13.24

$$= 13.3 + T_{AB} - 15 \cos \theta = 0$$
$$13.3 + T_{AB} - 15(3/5) = 0$$
$$T_{AB} = -4.3 \text{KN}$$
$$T_{AB} = 4.3 \text{KN (compression)} \quad \text{...ANS} \quad \text{...(xii)}$$

Member	AB	BC	CD	DE	EA	AD	DB
Force in KN	4.3	7	0	0	7	15	154
Nature C = Compression T = Tension	C	C	—	—	C	C	CT

Q. 13: A truss is shown in fig(13.25). Find forces in all the members of the truss and indicate whether it is tension or compression.

(Dec–00–01)

Sol.: Let the reaction at joint A and E are R_{AV} and R_{EV}. First we calculate the support reaction,

$$R_H = 0, R_{AH} = 0 \quad \text{...(i)}$$
$$R_V = 0, R_{AV} + R_{EV} - 10 - 15 - 20 - 10 = 0$$
$$R_{AV} + R_{EV} = 55 \quad \text{...(ii)}$$

Taking moment about point A and equating to zero; we get
$$15 \times 3 + 10 \times 3 + 20 \times 6 - R_{EV} \times 6 = 0 \quad \text{...(iii)}$$
$$R_{EV} = 32.5 \text{ KN} \quad \text{...(iv)}$$

Value of (iv) putting in equation (ii)
We get, $R_{BV} = 22.5$ KN ...(v)

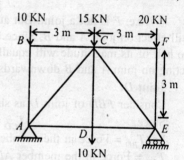

Fig. 13.25

Consider FBD of Joint B, as shown in fig 13.26
$$\Sigma H = 0; T_{BC} = 0$$
$$\Sigma V = 0; T_{BA} + 10 = 0; T_{BA} = -10 \text{KN(C)}$$

Consider FBD of Joint F, as shown in fig 13.27
$$\Sigma H = 0; T_{FC} = 0$$
$$\Sigma V = 0; T_{FE} + 20 = 0; T_{FE} = -20 \text{KN(C)}$$

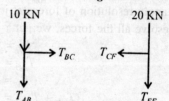

Fig. 13.26   Fig. 13.27

Consider FBD of Joint A, as shown in fig 13.28
$$\Sigma H = 0; T_{AD} + T_{AC} \cos 45 = 0$$
$$T_{AD} = -T_{AC} \cos 45 \quad \text{...(i)}$$
$$\Sigma V = 0; T_{CA} \sin 45 + 22.5 - 10 = 0$$
$$T_{CA} = -17.67 \text{KN(C)}$$

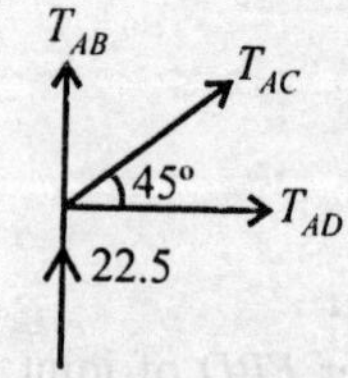

Putting this value in equation (i), we get
$$T_{AD} = 12.5 \text{KN(T)}$$

Consider FBD of Joint D, as shown in fig 8.29
$$\Sigma H = 0;$$
$$-T_{AD} + T_{DE} = 0$$
$$T_{AD} = T_{DE} = 12.5 \text{KN (T)}$$
$$\Sigma V = 0; T_{DC} - 10 = 0$$
$$T_{DC} = 10 \text{KN (T)}$$

Fig. 13.28

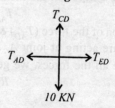

Fig. 13.29

Consider *FBD* of Joint *E*, as shown in fig 8.30

$\Sigma H = 0;$
$-T_{ED} - T_{ECn} \cos 45 = 0$
$-12.5 - T_{EC} \cos 45 = 0$
$T_{EC} = -17.67 KN(C)$
$\Sigma V = 0; \quad T_{FE} + T_{EC} \sin 45 + 32.5 = 0$
$T_{FE} + (-17.67) \sin 45 + 32.5 = 0$
$T_{FE} = -20 KN(C)$

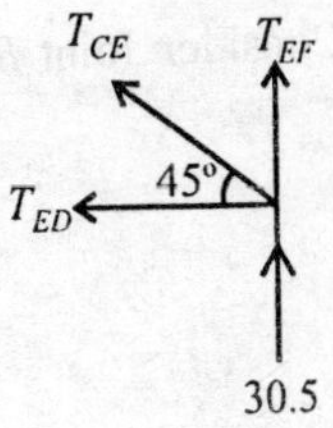

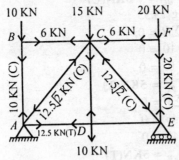

Fig. 13.30

Forces in all the members can be shown as in fig below.

Fig. 13.31

Q. 14: Find out the axial forces in all the members of a truss with loading as shown in fig 13.32.

(May–02 (C.O.))

Sol.: For Equilibrium

$\Sigma H = 0; \quad R_{AH} = 15 KN$
$\Sigma V = 0; \quad R_{AV} + R_{BV} = 0$
$\Sigma M_B = 0; \quad R_{AV} \times 4 + 10 \times 4 + 5 \times 8 = 0$
$R_{AV} = -20 \text{ KN}$

And
$R_{BV} = 20 KN$

Consider Joint A as shown in fig 13.33

$H = 0; \quad T_{AB} = 15 KN(T)$
$\Sigma V = 0; \quad T_{AF} = 20 KN(T)$

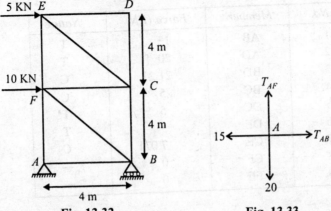

Fig. 13.32   Fig. 13.33

314 / *Problems and Solutions in Mechanical Engineering with Concept*

Consider Joint B as shown in fig 13.34

$$\Sigma H = 0;$$
$$-T_{AB} - T_{BF}\cos 45 = 0$$
$$T_{BF} = -15/\cos 45 = -21.21 KN$$
$$T_{BF} = -21.21 KN(C)$$
$$\Sigma V = 0;$$
$$T_{BC} + T_{BF}\sin 45 + 20 = 0$$
$$T_{BC} - 21.21 \sin 45 + 20 = 0$$
$$T_{BC} = -5 KN(C)$$

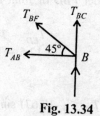

Fig. 13.34

Consider Joint F as shown in fig 13.35

$$\Sigma H = 0;$$
$$T_{FC} + T_{BF}\cos 45 + 10 = 0$$
$$T_{FC} - 21.21 \cos 45 + 10 = 0$$
$$T_{FC} = 5 KN(T)$$
$$\Sigma V = 0;$$
$$T_{FE} - T_{FA} - T_{BF}\sin 45 = 0$$
$$T_{FE} - 20 + 21.21 \sin 45 = 0$$
$$T_{FE} = 5 KN(T)$$

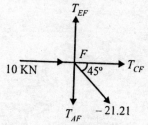

Fig. 13.35

Consider Joint C as shown in fig 13.36

$$\Sigma H = 0;$$
$$-T_{FC} - T_{CE}\cos 45 = 0$$
$$-5 - T_{CE}\cos 45 = 0$$
$$T_{CE} = -7.071 KN(C)$$
$$\Sigma V = 0;$$
$$T_{CD} - T_{CB} + T_{CE}\sin 45 = 0$$
$$T_{CD} + 5 - 7.071 \sin 45 + 20 = 0$$
$$T_{CD} = 0$$

Consider Joint D as shown in fig 13.37

$$\Sigma H = 0;$$
$$T_{ED} = 0$$

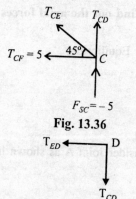

Fig. 13.36

Fig. 13.37

S.No.	Member	Force (KN)	Nature
1.	AB	15	T
2.	AD	20	T
3.	BD	21.21	C
4.	BC	5	C
5.	DC	5	T
6.	DE	5	T
7.	CE	7.071	C
8.	CF	0	—
9.	FE	0	—

Q. 15: Determine the magnitude and nature of forces in the various members of the truss shown in figure 13.38. *(C.O. August-05-06)*

Sol.: For Equilibrium

$\Sigma H = 0$; $R_{BH} = 0$
$\Sigma V = 0$; $R_{AV} + R_{BV} = 200$
$\Sigma M_B = 0$; $R_{AV} \times 6 - 50 \times 6 - 100 \times 3 = 0$
$R_{AV} = 100$ KN

And $R_{BV} = 100$ KN

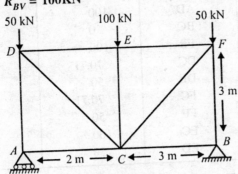

Fig. 13.38

Consider joint A; fig 13.39

$\Sigma H = 0$; $T_{AC} = 0$
$\Sigma V = 0$; $R_{AV} + T_{AD} = 0$ $T_{AD} = -100$ KN(C)

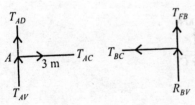 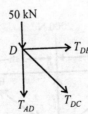 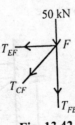 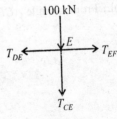

Fig. 13.39   Fig. 13.40   Fig. 13.41   Fig. 13.42   Fig. 13.43

Consider joint B; As shown in fig 13.40

$\Sigma H = 0$; $\mathbf{T_{BC} = 0}$
$\Sigma V = 0$; $R_{BV} + T_{FB} = 0$
$\mathbf{T_{FB} = -100}$ **KN(C)**

Consider joint D; As shown in fig 13.41

$\Sigma V = 0$; $-T_{AD} - T_{DC} \sin 45 - 50 = 0$
$T_{DC} = 70.71$ KN (T)
$\Sigma H = 0$; $T_{DE} + T_{DC} \cos 45 = 0$
$T_{DE} = -50$ KN (C)

Consider joint F; As shown in fig 13.42

$\Sigma V = 0$; $-T_{FB} - T_{FC} \sin 45 - 50 = 0$
$T_{FC} = 70.71$ KN (T)
$\Sigma H = 0$; $-T_{FE} - T_{FC} \cos 45 = 0$
$T_{FE} = -50$ KN (C)

Consider joint E; as shown in fig 13.43

$\Sigma V = 0; -T_{EC} - 100 = 0$
$T_{EC} = -100 KN$ (C)
$\Sigma H = 0; -T_{ED} + T_{EF} = 0$
$T_{ED} = -50 KN$ (C)

S.No.	Member	Force(KN)	Nature
1.	AC	0	—
2.	AD	100	C
3.	BC	0	—
4.	FB	100	C
5.	DC	70.71	T
6.	DE	50	C
7.	FC	70.71	T
8.	FE	50	C
9.	EC	100	C
10.	ED	50	C

Problems on Cantilever Truss

In case of cantilever trusses, it is not necessary to determine the support reactions. The forces in the members of cantilever truss can be obtained by starting the calculations from the free end of the cantilever.

Q. 16: Determine the forces in all the member of a cantilever truss shown in fig 13.44.

Sol.: From triangle ACE, we have

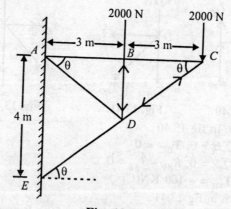

Fig. 13.44

$\tan \theta = AE/AC = 4/6 = 0.66$ ...(i)

Also,
$EC = \sqrt{4^2 + 6^2}$
$= 7.21 m$ ...(ii)
$\cos \theta = AC/EC = 6/7.21 = 0.8321$ ...(iii)
$\sin \theta = AE/CE = 4/7.21 = 0.5548$ ...(iv)

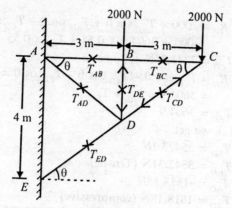

Fig 13.45

Joint C:

Consider FBD of joint C as shown in fig 13.46;
Since three forces are acting, so apply lami,s theorem at joint C.

$T_{BC}/\sin(90 - \theta) = T_{CD}/\sin 270 = 2000/\sin\theta$

$T_{BC}/\cos\theta = T_{CD}/\sin 270 = 2000/\sin\theta$

$T_{BC} = 2000/\tan\theta = 2000/0.66 = 3000.3 N$ ...(v)

$T_{BC} = 3000.3 N$ (Tensile) ....ANS

$T_{CD} = -2000/\sin\theta = 2000/0.55 = 3604.9 N$ ...(vi)

$T_{CD} = 3604.9 N$ (Compressive) ....ANS

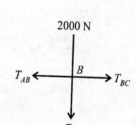

Fig. 13.46

Joint B:

Consider FBD of joint B as shown in fig 13.47
Since, $T_{BC} = 3000.3 N$
Let, T_{AB} = Force in the member AB
T_{DB} = Force in the member DB
Since four forces are acting at joint B, So apply resolution of forces at joint B

$R_H = T_{AB} - T_{BC} = 0, \; T_{AB} = T_{BC}$
$= 3000.03 = T_{AB}$
$T_{AB} = 3000.03$

$T_{AB} = 3000.03 N$ (Tensile) ...(vii) ....ANS

$R_V = -T_{DB} - 2000 = 0$
$T_{DB} = -2000 N$

$T_{DB} = 2000 N$ (compressive) ...(viii) ....ANS

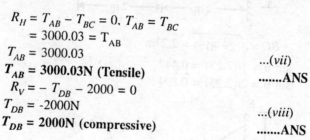

Fig. 13.47

Joint D:

Consider FBD of joint D as shown in fig 13.48
Since, $T_{DB} = -2000 N$
$T_{CD} = 3604.9 N$
Let, T_{AD} = Force in the member AD
T_{DE} = Force in the member DE
Since four forces are acting at joint D, So apply resolution of forces at joint D.

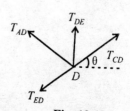

Fig. 13.48

$$R_V = 2000 + T_{CD} \sin\theta + T_{AD} \sin\theta - T_{ED} \sin\theta = 0$$
$$2000 + 3604.9 \times 0.55 + T_{AD} \times 0.55 - T_{ED} \times 0.55 = 0$$
...(ix)
$$T_{AD} - T_{ED} = 7241.26 N$$
$$R_H = T_{CD} \cos\theta - T_{AD} \cos\theta - T_{ED} \cos\theta = 0$$
$$= 3604.9 = T_{AD} + T_{ED}$$
...(x)
$$T_{AD} + T_{ED} = 3604.9$$

Solving equation (ix) and (x), we get
...(xi)
$$T_{ED} = 55423.1 N$$
$$T_{ED} = \mathbf{5542.31 N \text{ (Tensile)}} \quad \text{.......ANS}$$
...(xii)
$$T_{AD} = -1818.18 N$$
$$T_{AD} = \mathbf{1818.18 N \text{ (compressive)}} \quad \text{.......ANS}$$

Member	AB	BC	CD	DE	DB	AD
Force in N	3000.03	3000.03	3604.9	5542.31	2000	1818.18
Nature C = Compression T = Tension	T	T	C	T	C	C

Q. 17: Determine the forces in the various members of the cantilever truss loaded and supported as shown in fig. 13.49.

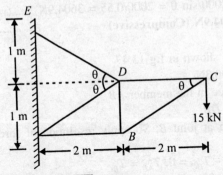

Sol.:

$$BC = \sqrt{(2^2 + 1^2)} = 2.23 m$$
$$\sin\theta = 1/(2.23) = 0.447$$
$$\sin\theta = 2/(2.23) = 0.894$$

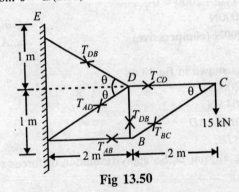

Fig 13.50

Let

T_{CD} = Force in the member CD
T_{CB} = Force in the member CB
T_{DB} = Force in the member DB
T_{AB} = Force in the member AB
T_{AD} = Force in the member AD
T_{BD} = Force in the member BD

Consider Joint C:
Consider FBD of joint C as shown in fig 13.51.
There are three forces are acting so apply lami's theorem at joint C

$T_{CD}/\sin(90 - \theta) = T_{BC}/\sin 270 = 15/\sin \theta$

$T_{CD} = 30 KN$ (Tensile) ....ANS
$T_{BC} = -33.56$ ...(i)
$T_{BC} = 33.56$ (Compressive) ....ANS

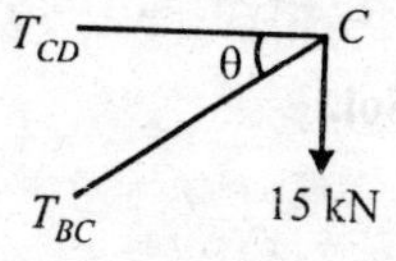

Fig. 13.51

Consider Joint B:
Consider FBD of joint B as shown in fig 13.52.
There are three forces are acting so apply lami's theorem at joint B

$T_{AB}/\sin(90 - \theta) = T_{BC}/\sin 90 = T_{DB}/\sin(180 + \theta)$

$T_4 = -30 KN$ ...(ii)
$T_{AB} = -30 KN$ (Compressive) ....ANS
$T_{DB} = 15$ ...(iii)
$T_{DB} = 15$ (Tensile) ....ANS

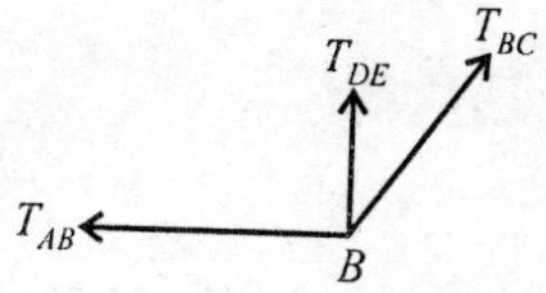

Fig. 13.52

Consider Joint D:
Consider FBD of joint D as shown in fig 13.53.
There are four forces are acting so apply resolution of forces at joint D

$R_H = 0$, $T_{CD} - T_{AD} \cos \theta - T_{ED} \cos \theta = 0$
$30 - (T_{AD} + T_{ED}) \cos \theta = 0$
$T_{AD} + T_{ED} = 30/\cos \theta = 30/0.89 = 33.56$ ...(iv)
$R_V = 0$
$T_{ED} \sin \theta - T_{AD} \sin \theta - T_{DB} = 0$
$(T_{ED} - T_{AD}) \sin \theta = 15$
$T_{ED} - T_{AD} = 15/\sin \theta$
$T_{ED} - T_{AD} = 33.56$ ...(v)

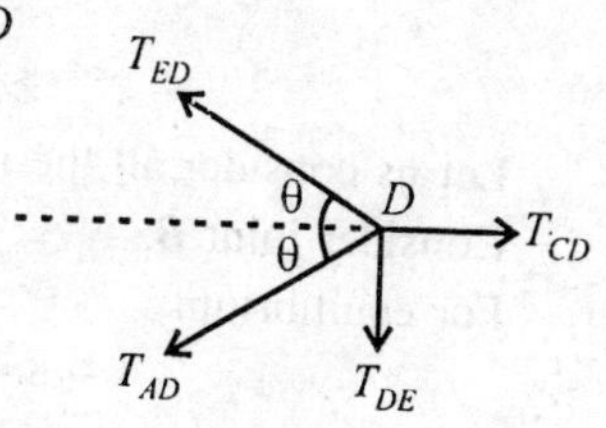

Fig. 13.53

Solve equation (iv) and (v), we get
$T_{AD} = 0$ ...(vi)
$T_{AD} = 0$ ....ANS
$T_{ED} = 33.56$ ...(vii)
$T_{ED} = 33.56$ (Tensile) ....ANS

Member	CD	BC	BD	BA	AD	DE
Force in kN	30	33.56	15	30	0	33.56
Nature C = Compression T = Tension	T	C	T	C	—	T

Q. 18: Find the axial forces in all the members of truss shown in fig- 13.54.   (Dec–02)

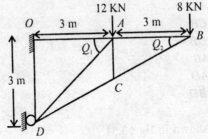

Fig. 1354

Sol.:

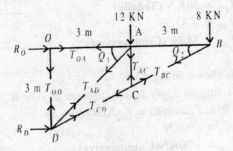

Fig 13.55

From the fig 13.55
$$\tan\theta = 3/6 = 1/2$$
$$\theta = 26.56°$$
$$\tan\theta_1 = 3/3 = 1$$
$$\theta_1 = 45°$$

Let us consider all the members subjected to tensile force as shown in fig 13.55.
Consider joint B:
For equilibrium $\Sigma V = 0;$
$$-8 - T_{BC}\cdot \sin 26.56° = 0$$
$$T_{BC} = -17.9 \text{ KN (Compressive)}$$
$$\Sigma H = 0;$$
$$-T_{AB} - T_{BC}\cos 26.56° = 0$$
$$T_{AB} = 17.9 \times \cos 26.56°$$
$$T_{AB} = 16 \text{ KN (Tensile)}$$

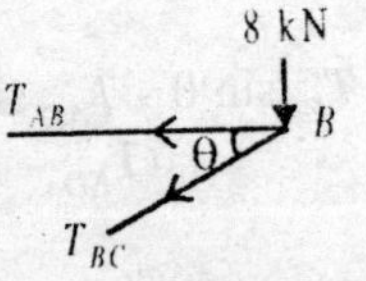

Consider joint C:
$$\Sigma H = 0;$$
$$T_{BC}\cdot \cos 26.56° - T_{CD}\cdot \cos 26.56° = 0$$
$$T_{BC} = T_{CD}$$
$$T_{BC} = -17.9 \text{ KN (Compressive)}$$
$$\Sigma V = 0;$$
$$T_{AC} + T_{BC}\cdot \sin 26.56° - T_{CD}\cdot \sin 26.56° = 0$$
$$T_{AC} = 0 \text{ (Because } T_{BC} = T_{CD}\text{)}$$

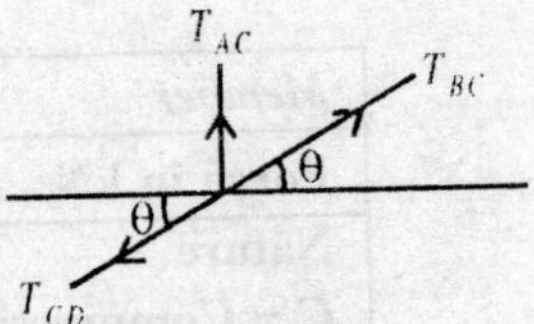

Fig. 13.56

Fig. 13.57

Consider joint A:

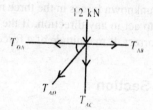

Fig. 13.58

$\Sigma V = 0$;
$-12 - T_{AC} - T_{AD} \cdot \sin 45° = 0$
$-12 - 0 - T_{AD} \cdot \sin 45° = 0$
$T_{AD} = -12/\sin 45°$
$T_{AD} = -16.97 KN$ (Compressive)
$\Sigma H = 0$;
$T_{AB} - T_{OA} - T_{AD} \cdot \cos 45° = 0$
$16 - T_{OA} + 16.97 \cdot \cos 45° = 0$
$T_{OA} = 28 KN$
$T_{OA} = 28 KN$ (Tensile)

Consider joint D:

$\Sigma V = 0$;
$T_{OD} + T_{AD} \cdot \cos \theta_1 + T_{CD} \cdot \sin \theta = 0$
$T_{OD} - 16.97 \cos 45° - 17.9 \cdot \sin 26.56° = 0$
$T_{OD} = 20 KN$
$T_{OD} = 20 KN$ (Tensile)

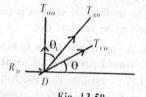

Fig. 13.59

Member	Force	Nature (T/C)
BC	17.9	C
AB	16	T
CD	17.9	C
AC	0	—
AD	16.97	C
AO	28	T
OD	20	T

Q. 19: Define method of section? How can you evaluate the problems with the help of method of section?

Sol.: This method is the powerful method of determining the forces in desired members directly, without determining the forces in the previous members. Thus this method is quick. Both the method, i.e. method of joint and method of sections can be applied for the analysis of truss simultaneously. For member near to supports can be analyzed with the method of joints and for members remote from supports can be quickly analyzed with the help of method of section.

In this method a section line is passed through the member, in which forces are to be determined in such a way that not more than three members are cut. Any of the cut part is then considered for equilibrium under the action of internal forces developed in the cut members and external forces on the cut part of the truss. The conditions of equilibrium are applied to the cut part of the truss under consideration. As three equations are available, therefore, three unknown forces in the three members can be determined. Unknown forces in the members can be assumed to act in any direction. If the magnitude of a force comes out to be positive then the assumed direction is correct. If magnitude of a force is negative than reverse the direction of that force.

Steps Involved for Method of Section

The various steps involved are:
(1) First find the support reaction using equilibrium conditions.
(2) The truss is split into two parts by passing an imaginary section.
(3) The imaginary section has to be such that it does not cut more than three members in which the forces are to be determined.
(4) Make the direction of forces only in the member which is cut by the section line.
(5) The condition of equilibrium are applied for the one part of the truss and the unknown force in the member is determined.
(6) While considering equilibrium, the nature of force in any member is chosen arbitrarily to be tensile or compressive.

If the magnitude of a particular force comes out positive, the assumption in respect of its direction is correct. However, if the magnitude of the forces comes out negative, the actual direction of the force is positive to that what has been assumed.

The method of section is particularly convenient when the forces in a few members of the frame is required to be worked out.

Q. 20: A cantilever truss is loaded and supported as shown in fig 13.60. Find the value of P, which would produce an axial force of magnitude 3KN in the member AC.

Sol.: Let us assume that the forces is find out in the member AC, DC and DF

Let T_1 = Force in the member AC
T_2 = Force in the member DC
T_3 = Force in the member DF

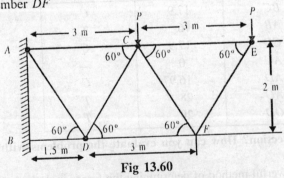

Fig 13.60

Draw a section line, which cut the member AC, DC, and DF.
Consider right portion of the truss, because Force P is in the right portion.
Taking moment about point D,

$$\Sigma M_D = 0$$

$$-T_1 \times AB + P \times (AC - AD) + P \times (AE - AD) = 0$$
$$-3 \times 2 + P \times 1.5 + P \times 4.5 = 0$$
$$P = 1KN \qquad \text{.......ANS}$$

Q. 21: Find the forces in members *BC*, *BE*, *FE* of the truss shown in fig 13.61, using method of section.

(May–04)

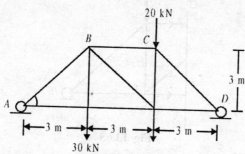

Fig 13.61

Sol.: First find the support reaction which can be determined by considering equilibrium of the truss.

$$\Sigma V = 0$$
$$R_A + R_D = 50 \qquad \ldots(i)$$

Taking moment about point A,

$$\Sigma M_A = 0$$
$$- R_D \times 9 + 20 \times 6 + 30 \times 3 = 0$$
$$R_D = 23.33 \text{KN} \qquad \ldots(ii)$$

Now, from equation (*i*); we get

$$R_A = 26.67 \text{KN} \qquad \ldots(iii)$$

Let draw a section line 1-1 which cut the member *BC*, *BE*, *F*E, and divides the truss in two parts *RHS* and *LHS* as shown in fig 13.62. Make the direction of forces only in those members which cut by the section line.

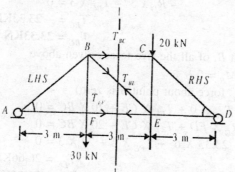

Fig 13.62

Choose any one part of them. Since both parts are separately in equilibrium. Let we choose right hand side portion (as shown in fig 13.63). And the Right hand parts of truss is in equilibrium under the action of following forces.

324 / *Problems and Solutions in Mechanical Engineering with Concept*

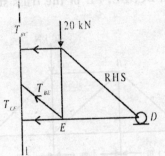

Fig 13.63

1. Reaction $R_D = 23.33$KN
2. 20KN load at joint C
3. Force T_{BC} in member BC (From C to B)
4. Force T_{BE} in member BE (From E to B)
5. Force T_{FE} in member FE (From E to F)

All three forces are assumed to be tensile.

Now we take moment of all these five forces only from any point of the truss for getting the answers quickly

Taking moment about point E, of all the five forces given above

$$\Sigma M_E = 0$$

(Moment of Force T_{BE}, T_{EF} and 20KN about point E is zero, since point E lies on the line of action of that forces)

$$- R_D \times ED + T_{EF} \times 0 + T_{BE} \times 0 - T_{BC} \times CE + 20 \times 0 = 0$$
$$- R_D \times 3 - T_{BC} \times 3 = 0$$
$$T_{BC} = - 23.33\text{KN} \quad \ldots (iii)$$
$$T_{BC} = 23.33\text{KN (Compressive)} \quad \ldots\ldots\text{ANS}$$

Taking moment about point B, of all the five forces given above

$$\Sigma M_B = 0$$

(Moment of Force T_{BE}, T_{BC} force about point B is zero)

$$- R_D \times FD + T_{BC} \times 0 + T_{BE} \times 0 + T_{FE} \times CE + 20 \times BC = 0$$
$$- R_D \times FD + T_{FE} \times CE + 20 \times BC = 0$$
$$-23.33 \times 6 + T_{FE} \times 3 + 20 \times 3 = 0$$
$$T_{FE} = 26.66\text{KN} \quad \ldots(iii)$$
$$T_{FE} = 26.66\text{KN (Tensile)} \quad \ldots\ldots\text{ANS}$$

Taking moment about point F, of all the five forces given above

$$\Sigma M_F = 0$$

(Moment of Force T_{FE} about point B is zero)

$$- R_D \times FD - T_{BC} \times EC - T_{BE} \cos 45° \times FE + T_{FE} \times 0 + 20 \times FE = 0$$
$$-23.33 \times 6 + 23.33 \times 3 - T_{BE} \cos 45° \times 3 + 20 \times 3 = 0$$
$$T_{BE} = 4.71\text{KN} \quad \ldots(iii)$$
$$T_{BE} = 4.71\text{KN (Tensile)} \quad \ldots\ldots\text{ANS}$$

Q. 22: Determine the support reaction and nature and magnitude of forces in members *BC* and *EF* of the diagonal truss shown in fig 13.64. (May–01, (C.O.))

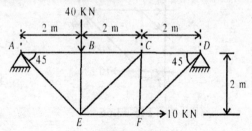

Fig 13.64

Sol.: First find the support reaction which can be determined by considering equilibrium of the truss.

Let R_{AH} & R_{AV} be the support reaction at hinged support A and R_{DV} be the support reaction at roller support D.

$$\Sigma H = 0$$
$$R_{AH} + 10 = 0$$
$$R_{AH} = -10 \text{ KN}$$
$$\Sigma V = 0$$
$$R_{AV} + R_{DV} = 40 \qquad ...(i)$$

Taking moment about point A,
$$\Sigma M_A = 0$$
$$40 \times 2 - 10 \times 2 - R_{DV} \times 6 = 0$$
$$R_{DV} = 10 \text{KN}$$

From equation (i); R_{AV} = **30KN**

Let draw a section line 1-1 which cut the member *BC*, *EC*, *FE*, and divides the truss in two parts *RHS* and *LHS* as shown in fig 13.65. Make the direction of forces only in those members which cut by the section line. i.e. in *BC*, *EF* and *EC*, Since the question ask the forces in the member *BC* and *EF*, but by draw a section line member *EC* is also cut by the section line, so we consider the force in the member *EC*.

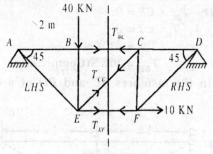

Fig 13.65

Choose any one part of them, Since both parts are separately in equilibrium. Let we choose right hand side portion (as shown in fig 13.66). And the Right hand parts of truss is in equilibrium under the action of following forces,

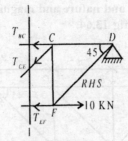

Fig 13.66

1. Reaction $R_{DV} = 10KN$ at the joint D
2. 10KN load at joint F
3. Force T_{BC} in member BC
4. Force T_{CE} in member CE
5. Force T_{FE} in member FE

All three forces are assumed to be tensile.

Now we take moment of all these five forces only from any point of the truss, for getting the answers quickly

Taking moment about point C, of all the five forces given above

$$\Sigma M_C = 0$$

(Moment of Force T_{BC}, T_{CE} about point C is zero, since point C lies on the line of action of that forces)

$$- R_D \times CD + T_{EF} \times CF - 10 \times CF = 0$$
$$-10 \times 2 + T_{EF} \times 2 - 10 \times 2 = 0$$
$$T_{EF} = 20KN \qquad \qquad ...(iii)$$
$$T_{EF} = 20KN \text{ (Tensile)} \qquad \qquad ....ANS$$

Taking moment about point E, of all the five forces given above

$$\Sigma M_E = 0$$

(Moment of Force T_{EF}, T_{EC} and 10KN about point E is zero, since point E lies on the line of action of that forces)

$$- R_D \times BD - T_{BC} \times CF = 0$$
$$-10 \times 4 - T_{BC} \times 2 = 0$$
$$T_{BC} = -20KN \qquad \qquad ...(iii)$$
$$T_{BC} = 20KN \text{(Compressive)} \qquad \qquad ....ANS$$

Q. 23: Determine the forces in the members BC and BD of a cantilever truss shown in the figure 13.67. *(May–04(C.O.))*

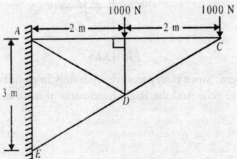

Fig 13.67

Sol.: In this problem; If we draw a section line which cut the member *BC, BD, AD, ED*, then the member *BC* and *BD* cut by this line, but this section line cut four members, so we don't use this section line. Since a section line cut maximum three members.

There is no single section line which cut the maximum three member and also cut the member *BC* and *BD*.

This problem is done in two steps
(1) Draw a section line which cut the member *BC* and *DC*. Select any one section and find the value of *BC*.
(2) Draw a new diagram, draw another section line which cut the member *AB, BD* and *CD*, and find the value of the member *BD*.

STEP–1

Draw a section line which cut the member *BC* and *CD*, as shown in fig 13.68.

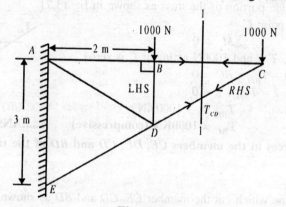

Fig 13.68

Consider Right hand side portion of the truss as shown in fig 13.69
Taking moment about point *D*

$$\Sigma M_D = 0$$
$$1000 \times 2 - T_{BC} \times BD = 0$$

Consider Triangle *CAE* and *CDB*, they are similar

$$BD/AE = BC/AC$$
$$BD/3 = 2/4$$
$$BD = 1.5m$$
$$1000 \times 2 - T_{BC} \times 1.5 = 0$$
$$T_{BC} = 1333.33 KN$$
$$T_{BC} = 1333.33 KN \text{ (Tensile)} \quad \dots(iii) \quad \dots\dots\text{ANS}$$

STEP–2:

Draw a section line which cut the member *AB, AD* and *BD*, as shown in fig 13.70.

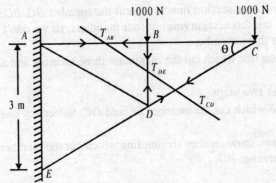

Fig 13.70

Consider Right hand side portion of the truss as shown in fig 13.71
Taking moment about point C

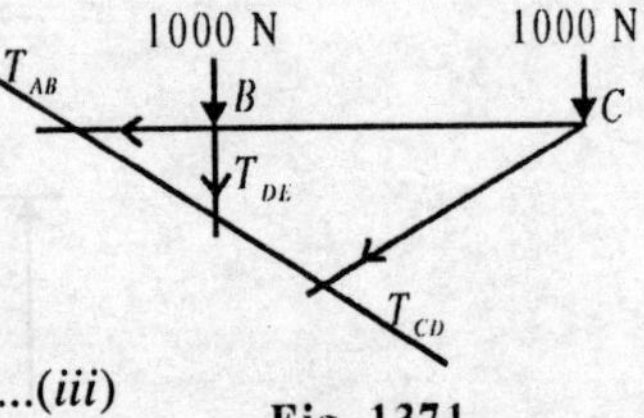

$$\Sigma M_C = 0$$

{Moment of force T_{AB}, T_{CD} and 1000N acting at C is zero}

$$-1000 \times BD - T_{BD} \times BC = 0$$
$$-1000 \times 2 - T_{BD} \times 2 = 0$$
$$T_{BD} = -1000 KN \quad ...(iii)$$
$$\mathbf{T_{BC} = 1000KN \text{ (compressive)}} \quad ....\mathbf{ANS}$$

Fig. 1371

Q. 24: Find the axial forces in the members CE, DE, CD and BD of the truss shown in fig 13.72.

(May–04(C.O.))

STEP–1:

First draw a section line which cut the member CE, CD and BD as shown in fig 13.73
Consider RHS portion of the truss as shown in fig 13.74.
Here Member $AB = BC = CD = BE$

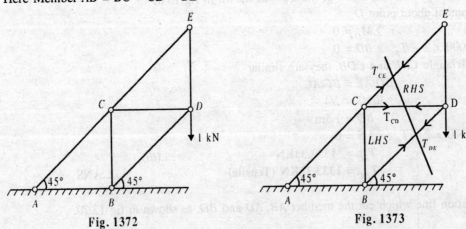

Fig. 1372 Fig. 1373

Taking moment about point C

$$\Sigma M_C = 0 \text{{Moment of force } } T_{CE}, T_{CD} \text{ is zero}\}$$
$$1 \times CD + T_{BD} \cos 45° \times BC = 0$$
$$T_{BD} = 1/\cos 45° = -1.414 KN$$
$$\mathbf{T_{BD} = 1.414 KN \text{ (compressive)}} \quad ....\mathbf{ANS}$$

Taking moment about point E

$\Sigma M_E = 0$

{Moment of force T_{CE}, 1KN is zero}

$T_{CD} \times ED + T_{DB} \cos 45° \times ED = 0$

$T_{CD} = T_{DB} \cos 45°$ $T_{CD} = -1KN$

$T_{CD} = 1KN$ (Tensile)   ....ANS

Taking moment about point B

$\Sigma M_B = 0$

{Moment of force T_{DB}, is zero}

$-T_{CD} \times CB + (1 + T_{CE} \cos 45°) \times CD - T_{CE} \sin 45° \times (ED + CB) = 0$

$CB = CD = ED$

$-1 + 1 + T_{CE} \cos 45° - 2 T_{CE} \sin 45° = 0$

$T_{CE} = 0$ ....ANS

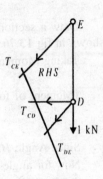

STEP–2:

Draw another section line which cut the member CE and ED Select RHS portion of the truss;

There are only two forces on the RHS portion

Taking moment about point C We get

$T_{ED} = 0$ ....ANS

Q. 25: A pin jointed cantilever frame is hinged to a vertical wall at A and E, and is loaded as shown in fig 13.75. Determine the forces in the member CD, CG and FG.

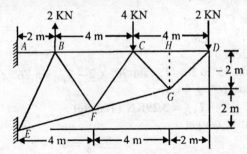

Fig 13.75

Sol.: First find the angle HDG

Let Angle $HDG = \theta$

$\tan \theta = GH/HD = 2/2 = 1$

$\theta = 45°$

Fig 13.76

Draw a section line which cut the member CD, CG and FG, Consider RHS portion of the truss as shown in fig 13.76.

Taking Moment about point G, we get
$$\sum M_G = 0$$
{Moment of force T_{FG} and T_{CG} is zero}
$$-T_{CD} \times 2 + 2 \times 2 = 0$$
$$T_{CD} = 2 \text{KN (Tensile)} \qquad \text{.......ANS}$$

Since Angle $HDG = DGH = HCG = HGC = 45°$
Now for angle GEK; $\tan_s = 2/8 = \frac{1}{4}$
Angle $GEK = 14°$
Angle $EGK = 76°$
Now resolved force T_{FG} and T_{CG} as

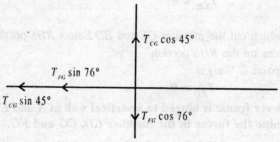

Taking Moment about point C, we get
$$\sum M_C = 0$$
{Moment of force T_{CD} is zero }
$$-T_{CG} \cos 45° \times 2 + T_{CG} \sin 45° \times 2 - T_{FG} \sin 76° \times 2 - T_{FG} \sin 76° \times 2 + 2 \times 4 = 0$$
$$-2 T_{FG} (\sin 76° + \cos 76°) + 8 = 0$$
$$T_{FG} = 3.29 \text{KN (Tensile)} \qquad \text{.......ANS}$$

Taking Moment about point E, we get
$$\sum M_E = 0$$
{Moment of force T_{FG} is zero}
$$-T_{CD} \times 4 + 2 \times 10 - T_{CG} \cos 45° \times 8 - T_{CG} \sin 45° \times 2 = 0$$
$$-2 \times 4 + 20 - 10 T_{CG} \cos 45° = 0$$
$$T_{CG} = 1.69 \text{KN (Tensile)} \qquad \text{.......ANS}$$

CHAPTER 14

SIMPLE STRESS AND STRAIN

Q. 1: Differentiate between strength of material and engineering mechanics.

Sol.: Three fundamental areas of mechanics of solids are statics, dynamics and strength of materials.

Strength of materials is basically a branch of `Solid Mechanics'. The other important branch of solid mechanics is Engineering Mechanics: statics and dynamics. Whereas `Engineering Mechanics' deals with mechanical behaviour of rigid (non-deformable) solids subjected to external loads, the 'Strength of Materials' deals with mechanical behaviour of non-rigid (deformable) solids under applied external loads. It is also known by other names such as Mechanics of Solids, Mechanics of Materials, and Mechanics of Deformable Solids. Summarily, the studies of solid mechanics can be grouped as follows.

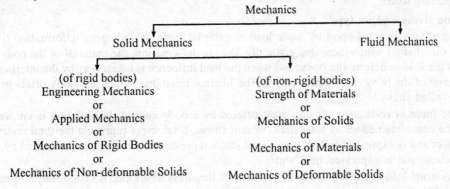

Fig. 14.1

Since none of the known materials are rigid, therefore the studies of Engineering Mechanics are based on theoretical aspects; but because all known materials are deformable, the studies of strength of materials are based on realistic concepts and practical footings. The study of Strength of Materials helps the design engineer to select a material of known strength at minimum expenditure.

Studies of Strength of Materials are applicable to almost all types of machine and structural components, all varieties of materials and all shapes and cross-sections of components. There are numerous variety of components, each behaving differently under different loading conditions. These components may be made of high strength steel, low strength plastic, ductile aluminium, brittle cast iron, flexiable copper strip, or stiff tungsten.

Q. 2: What is the scope of strength of materials?

Sol.: Strength of materials is the science which deals with the relations between externally applied loads and their internal effects on bodies.

The bodies are not assumed to be rigid, and the deformation, however small are of major interest.

Or, we can say that, When an external force act on a body. The body tends to undergoes some deformation. Due to cohesion between the molecules, the body resists deformation. This resistance by which material of the body oppose the deformation is known as strength of material, with in a certain limit (in the elastic stage). The resistance offered by the materials is proportional to the deformation brought out on the material by the external force.

So we conclude that the subject of strength of materials is basically a study of
(i) The behaviour of materials under various types of load and moment.
(ii) The action of forces and their effects on structural and machine elements such as angle iron, circular bars and beams etc.

Certain assumption are made for analysis the problems of strength of materials such as:
(i) The material of the body is homogeneous and isotropic,
(ii) There are no internal stresses present in the material before the application of loads.

Q. 3: What is load?
Sol. : A load may be defined as the combined effect of external forces acting on a body. The load is applied on the body whereas stress is induced in the material of the body. The loads may be classified as
1. Tensile load
2. Compressive load
3. Torsional load or Twisting load
4. Bending load
5. Shearing loads

Q. 4: Define stress and its type.
Sol. : When a body is acted upon by some load or external force, it undergoes deformation (i.e., change in shape or dimension) which increases gradually. During deformation, the material of the body resists the tendency of the load to deform the body, and when the load influence is taken over by the internal resistance of the material of the body, it becomes stable. The internal resistance which the body offers to meet with the load is called stress.

Or, The force of resistance per unit area, offered by a body against deformation is known as stress. Stress can be considered either as total stress or unit stress. Total stress represent the total resistance to an external effect and is expressed in N,KN etc. Unit stress represents the resistance developed by a unit area of cross section, and is expressed in KN/m^2.

If the external load is applied in one direction only, the stress developed is called simple stress Whereas If the external loads are applied in more than one direction, the stress developed is called compound stress.

Normal stress $(\sigma) = P/A$ N/m^2

$1\ Pascal(Pa) = 1\ N/m^2$
$1 KPa = 10^3\ N/m^2$
$1 MPa = 10^6\ N/m^2$
$1 GPa = 10^9\ N/m^2$

Generally stress are divided in to three group as:

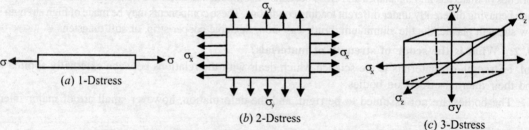

(a) 1-Dstress   (b) 2-Dstress   (c) 3-Dstress

Fig 14.2 *One, Two and Three dimensional stress*

But also the various types of stresses may be classified as:
1. Simple or direct stress (Tension, Compression, Shear)
2. Indirect stress (Bending, Torsion)
3. Combined Stress (Combination of 1 & 2)

(a) Tensile Stress
The stress induced in a body, when subjected to two equal and opposite pulls as shown in fig (14.3 (a)) as a result of which there is an increase in length, is known as tensile stress.

Let,
P = Pull (or force) acting on the body,
A = Cross - sectional area of the body,
σ = Stress induced in the body

Fig (a), shows a bar subjected to a tensile force P at its ends. Consider a section x-x, which divides the bar into two parts. The part left to the section x-x, will be in equilibrium if P = Resisting force (R). This is show in Fig (b), Similarly the part right to the section x-x, will be in equilibrium if P = Resisting force as shown in Fig (c), This resisting force per unit area is known as stress or intensity of stress. Tensile stress (σ) = Resisting force (R)/Cross sectional area

$\sigma_t = P/A \text{ N/m}^2$.

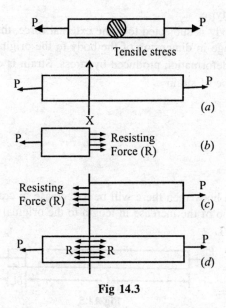

Fig 14.3

(b) Compressive Stress
The stress induced in a body, when subjected to two equal and opposite pushs as shown in fig (14.4 (a)) as a result of which there is an decrease in length, is known as tensile stress.

Let, an axial push P is acting on a body of cross sectional area A. Then compressive stress(σ_c) is given by;

σ_c = Resisting force (R)/Cross sectional area (A)
$\sigma_c = P/A \text{ N/m}^2$.

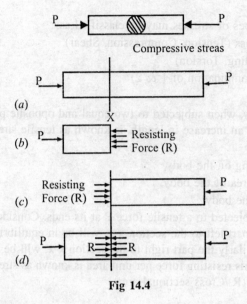

Fig 14.4

Q. 5: Define strain and its type.

Sol. : STRAIN(e) : When a body is subjected to some external force, there is some change of dimension of the body. The ratio of change in dimension of the body to the original dimension is known as strain.

Or, The strain (e) is the deformation produced by stress. Strain is dimensionless.

There are mainly four type of strain

1. tensile strain
2. Compressive strain
3. Volumetric strain
4. Shear strain

Tensile Strain

When a tensile load acts on a body then there will be a decrease in cross-sectional area and an increase in length of the body. The ratio of the increase in length to the original length is known as tensile strain.

$$e_t = \delta L/L$$

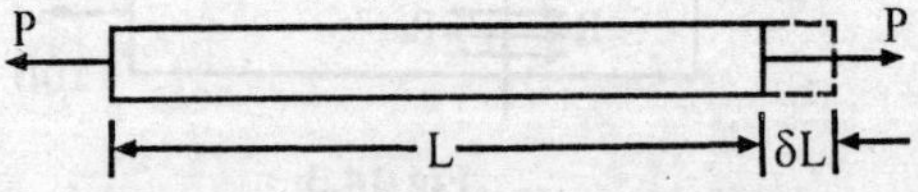

Fig 14.5

The above strain which is caused in the direction of application of load is called longitudinal strain. Another term lateral strain is strain in the direction perpendicular to the application of load i.e., $\delta D/D$

Compressive Strain

When a compressive load acts on a body then there will be an increase in cross-sectional area and decrease in length of the body. The ratio of the decrease in length to the original length is known as compressive strain.

$$e_c = \delta L/L$$

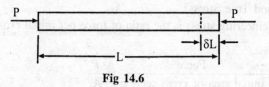

Fig 14.6

Q. 6: What do you mean by Elastic Limit?

Sol. : When an external force acts on a body, the body tends to undergo some deformation. If the external force is removed and the body comes back to its original shape and size (which means the deformation disappears completely), the body is known as elastic body. This property, by virtue of which certain materials return back to their original position after the removal of the external force, is called elasticity.

The body will regain its previous shape and size only when the deformation caused by the external force, is within a certain limit. Thus there is a limiting value of force upto and within which, the deformation completely disappears on the removal of the force. The value of stress corresponding to this limiting force is known as the elastic limit of the material.

If the external force is so large that the stress exceeds the elastic limit, the material loses to some extent its property of elasticity. If now the force is removed, the material will not return to its origin shape and size and there will be a residual deformation in the material.

Q. 7: State Hook's law.

Sol. : It states that when a material is loaded within elastic limit, the stress is proportional to the strain produced by the stress. This means the ratio of the stress to the corresponding strain is a constant within the elastic limit. This constant is known as Modulus of Elasticity or Modulus of Rigidity.

Stress /strain = constant

The constant is known as elastic constant

Normal stress/ Normal strain = Young's modulus or Modulus of elasticity (E)

Shear stress/ Shear strain = Shear modulus or Modulus of Rigidity (G)

Direct stress/ Volumetric strain = Bulk modulus (K)

Q. 8: What do you mean by Young's Modulus or Modulus of elasticity?

Sol. : It is the ratio between tensile stress and tensile strain or compressive stress and compressive strain. It is denoted by E. It is the same as modulus of elasticity

$$E = \sigma/e \ [\sigma_t/e_t \text{ or } \sigma_c/e_c]$$

S.No.	Material	Young's Modulus(E)
1	Mild steel	2.1×10^5 N/mm²
2	Cast Iron	1.3×10^5 N/mm²
3	Aluminium	0.7×10^5 N/mm²
4	Copper	1.0×10^5 N/mm²
5	Timber	0.1×10^5 N/mm²

The % error in calculation of Young's modulus is: $[(E_1 - E_2)/E_1] \times 100$

Q. 9: What is the difference between

(a) Nominal stress and true stress

(b) Nominal strain and true strain?

(a) Nominal Stress and True Stress

Nominal stress or engineering stress is the ratio of force per initial cross sectional area (original area of cross-section).

$$\text{Nominal stress} = \frac{\text{Force}}{\text{initial area of cross-section}} = \frac{P}{A_0}$$

True stress is the ratio of force per actual (instantaneous) cross-sectional area taking lateral strain into consideration.

$$\text{True stress} = \frac{\text{Force}}{\text{Actual area of cross-section}} = \frac{P}{A}$$

(b) Nominal Strain and True Strain

Nominal Strain is the ratio of change in length per initial length.

$$\text{Nominal strain} = \frac{\text{Change in length}}{\text{Initial length}} = \frac{\Delta L}{L}$$

True strain is the ratio of change in length per actual length (instantaneous length) taking longitudinal strain into consideration.

Q. 10: A load of 5 KN is to be raised with the help of a steel wire. Find the diameter of steel wire, if the maximum stress is not to exceed 100 MNm². *(UPTUQUESTION BANK)*

Sol.: Given data:

$$P = 5 \text{ KN} = 5000\text{N}$$
$$\sigma = 100 \text{MN/m}^2 = 100 \text{N/mm}^2$$

Let D be the diameter of the wire

We know that, $\sigma = P/A$

$$\sigma = P/(\Pi/4 \times D^2)$$
$$100 = 5000/(\Pi/4 \times D^2)$$
$$D = 7.28 \text{mm} \qquad \text{.......ANS}$$

Q. 11: A circular rod of diameter 20 m and 500 m long is subjected to tensile force of 45kN. The modulus of elasticity for steel may be taken as 200 kN/m². Find stress, strain and elongation of bar due to applied load. *(UPTUQUESTION BANK)*

Sol.: Given data:

$$D = 20\text{m}$$
$$L = 500\text{m}$$
$$P = 45\text{KN} = 45000\text{N}$$
$$E = 200 \text{KN/m}^2 = (200 \times 1000 \text{ N/mm}^2 = 200000 \text{ N/m}^2$$

Using the relation; $\sigma = P/A = P/(\Pi/4 \times D^2)$

$$\sigma = 45000/(\Pi/4 \times 20^2)$$
$$\sigma = 143.24 \text{ N/m}^2 \qquad \text{.......ANS}$$

$$E = \sigma/e$$
$$200000 = 143.24/e$$
$$e = 0.000716 \qquad \text{.......ANS}$$

Now, $e = dL_A/L$

$$0.000716 = dL_A/500$$
$$dL_A = 0.36 \qquad \text{.......ANS}$$

Q. 12: A rod 100 cm long and of 2 cm x 2 cm cross-section is subjected to a pull of 1000 kg force. If the modulus of elasticity of the materials 2.0 x 10⁶ kg/cm², determine the elongation of the rod. *(UPTUQUESTION BANK)*

Sol.: Given data:
$$A = 2 \times 2 = 4 \text{cm}$$
$$L = 100 \text{cm}$$
$$P = 1000 \text{kg} = 1000 \times 9.81 = 9810 \text{N}$$
$$E = 2.0 \times 10^6 \text{ kg/cm}^2 = 9.81 \times 2.0 \times 10^6 \text{ kg/cm}^2 = 19.62 \times 10^6 \text{ N/cm}^2$$

Using the relation; $\sigma = P/A$
$$\sigma = 9810/4$$
$$\sigma = 2452.5 \text{ N/cm2} \quad \text{........ANS}$$
$$E = \sigma/e$$
$$19.62 \times 10^6 = 2452.5/e$$
$$e = 0.000125 \quad \text{........ANS}$$

Now, $e = dL_A/L$
$$0.000125 = dL_A/100$$
$$dL = 0.0125 \quad \text{........ANS}$$

Q. 13: A hollow cast-iron cylinder 4 m long, 300 mm outer diameter, and thickness of metal 50 mm is subjected to a central load on the top when standing straight. The stress produced is 75000 kN/m². Assume Young's modulus for cast iron as 1.5 x 10⁸ KN/m² find

(*i*) magnitude of the load,

(*ii*) longitudinal strain produced and

(*iii*) total decrease in length.

Sol.: Outer diameter, D = 300 mm = 0.3 m Thickness, t = 50 mm = 0.05 m

Length, L = 4 m

Stress produced, σ = 75000 kN/m²
$$E = 1.5 \times 10^8 \text{ kN/m}^2$$

Here diameter of the cylinder, d = D − 2t = 0.3 − 2 × 0.05 = 0.2 m

(*i*) Magnitude of the load P:

Using the relation, $\sigma = P/A$

or $\quad P = \sigma \times A = 75000 \times \Pi/4 \, (D_2 - d_2)$
$$= 75000 \times \Pi/4 \, (0.3^2 - 0.2^2)$$

or $\quad P = 2945.2 \text{ kN} \quad \text{........ANS}$

(*ii*) Longitudinal strain produced, e :

Using the relation,

Strain, (e) = stress/E = 75000/1.5 × 10⁸ = 0.0005 ....ANS

(*iii*) Total decrease in length, dL:

Using the relation,

Strain = change in length/original length = dL_A/L
$$0.0005 = dL_A/4$$
$$dL_A = 0.0005 \times 4\text{m} = 0.002\text{m} = 2\text{mm}$$

Hence decrease in length = 2 mm ....ANS

Q. 14: A steel wire 2 m long and 3 mm in diameter is extended by 0.75 mm when a weight P is suspended from the wire. If the same weight is suspended from a brass wire, 2.5 m long and 2 mm in diameter, it is elongated by 4.64 mm. Determine the modulus of elasticity of brass if that of steel be 2.0×10^5 N/mm². *(UPTUQUESTION BANK)*

Sol.: Given: $L_S = 2$ m,

$\delta s = 3$ mm,

$\delta L_S = 0.75$ mm;

$E_s = 2.0 \times 10^5$ N/mm²;

$L_b = 2.5$ m; $d_b = 2$ mm;

$\delta L_b = 4.64$ m.

Modulus of elasticity of brass, E_b :

From Hooke's law, we know that;

$$E = \sigma/e$$
$$= (P/A)/(\delta L_A/L) = P.L/A. \delta L_A$$

or, $\qquad P = \delta L_A.A.E/L$

where,

δL = extension,

L = length,

A = cross-sectional area,

and E = modulus of elasticity.

Case I : For steel wire:

$$P = \delta L_s.A_s.E_s/L_s$$

or $\qquad P = [0.75 \times (\Pi/4 \times 3^2) \times 2.0 \times 10^5]/2000 \qquad \qquad ...(i)$

Case II : For bass wire

$$P = \delta L_b.A_b.E_b/L_b$$

or $\qquad P = [4.64 \times (\Pi/4 \times 2^2) \times E_b]/2500 \qquad \qquad ...(ii)$

Equating equation (*i*) and (*ii*), we get

$[0.75 \times (\Pi/4 \times 3^2) \times 2.0 \times 10^5 \times 2.0 \times 10^5]/2000 = P = [4.64 \times (\Pi/4 \times 2^2) \times E_b]/2500$

$\mathbf{E_b = 0.909 \times 10^5}$ **N/mm²** ....**ANS**

Q. 15: The wire working on a railway signal is 5mm in diameter and 300m long. If the movement at the signal end is to be 25cm, make calculations for the movement which must be given to the end of the wire at the signal box. Assume a pull of 2500N on the wire and take modulus of elasticity for the wire material as 2×10^5 N/mm².

Sol.: Given data:

$P = 2500$N

$D = 5$ mm

$L = 300$m $= 300 \times 1000$ mm

$D_m = 25$cm

$E = 2 \times 10^5$ N/mm².

We know that, $\sigma = P/A$

$\sigma = P/(\Pi/4 \times D^2)$

$\sigma = 2500/(\Pi/4 \times 5^2)$

$\sigma = 127.32$ N/mm².

$$e = \sigma/E = 127.32/2 \times 10^5$$
$$e = 0.0006366$$
Since $\quad e = \delta L/L$
$$\delta L = e.L = 0.0006366 \times 300 \times 1000 = 190.98 \text{mm} = 19.098 \text{ cm}$$
Total movement which need to be given at the signal box end = 25 + 19.098 = 44.098 cm ....ANS

Q. 16: Draw stress-strain diagrams, for structural steel and cast iron and briefly explain the various salient points on them. *(May–01, May–03)*

Or;

Draw a stress strain diagram for a ductile material and show the elastic limit, yield point and ultimate strength. Explain any one of these three. *(May–03(CO))*

Or;

Draw stress-strain diagram for a ductile material under tension. *(Dec–04)*

Or;

Draw the stress strain diagram for aluminium and cast iron. *(May–05)*

Or;

Explain the stress-strain diagram for a ductile and brittle material under tension on common axes single diagram. *(May–05(CO))*

Or

Define Ductile behaviour of a metal *(Dec–00)*

Sol.: The relation between stress and strain is generally shown by plotting a stress-strain (σ-e) diagram. Stress is plotted on ordinate (vertical axis) and strain on abscissa (horizontal axis). Such diagrams are most common in strength of materials for understanding the behaviour of materials. Stress-strain diagrams are drawn for different loadings. Therefore they are called
- Tensile stress-strain diagram
- Compressive stress-strain diagram
- Shear stress-strain. diagram

Stress-Strain Curves (Tension)

When a bar or specimen is subjected to a gradually increasing axial tensile load, the stresses and strains can be found out for number of loading conditions and a curve is plotted upto the point at which the specimen fails. giving what is known as stress-strain curve. Such curves differ in shape for various materials. Broadly speaking the curves can be divided into two categories.

(a) **Stress-strain carves for ductile materials :** A material is said to be ductile in nature, if it elongates appreciably before fracture. One such material is mild steel. The shape of stress-strain diagram for the mild steel is shown in Fig. 14.7.

A mild steel specimen of either circular cross-section (rod) or rectangular section (flat bar) is pulled until it breaks. The extensions of the bar are measured at every load increments. The stresses are calculated based on the original cross sectional area and strains by dividing the extensions by gauge length. When the specimen of a mild steel is loaded gradually in tension, increasing tensile load, in tension testing machine. The initial portion from O to A is linear where strain linearly varies with stress. The line is called line of proportionality and is known as proportionality limit. The stress corresponding to the point is called "Limit of Proportionality". Hook's law obeys in this part, the slope of the line gives, 'modulus of elasticity'.

Further increase in load increases extension rapidly and the stress-strain diagram becomes curved. At B, the material reaches its 'elastic limit' indicating the end of the elastic zone and entry into plastic zone. In most cases A and B coincide. if load is removed the material returns to its original dimensions.

Beyond the elastic limit, the material enters into the plastic zone and removal of load does not return the specimen to its original dimensions, thus subjecting the specimen to permanent deformation. On further loading the curve reaches the point `C' called the upper yield point at which sudden extension takes place which is known as ductile extension where the strain increases at constant stress. This is identified by the horizontal portion of the diagram. Point C gives 'yield stress'. beyond which the load decreases with increase in strain upto C' known as lower yield point.

After the lower yield point has been crossed, the stress again starts increasing, till the stress reaches the maximum value at point `D'. The increase in load causes non linear extension upto point D. The point D known as 'ultimate point' or 'maximum point'. This point gives the 'ultimate strength' or maximum load of the bar. The stress corresponding to this highest point `D' of the stress strain diagram is called the ultimate stress.

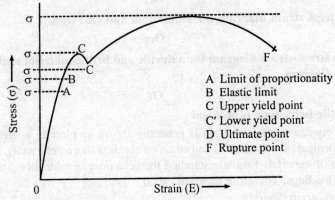

A Limit of proportionatity
B Elastic limit
C Upper yield point
C' Lower yield point
D Ultimate point
F Rupture point

Typical stress-stain diagram for a ductile material

Fig. 14.7

After reaching the point D, if the bar is strained further, a local reduction in the cross section occurs in the gauge length (i.e., formation of neck). At this neck stress increases with decrease in area at constant load, till failure take place. Point F is called 'rupture point.Note that all stresses are based on original area of cross section in drawing the curve of Fig 14.7.

1. Yield strength = $\dfrac{\text{Load at yield point}}{A_0}$

   where (original area) $A_0 = \dfrac{\pi}{A_0} D_0^2$

2. Ultimate strength = $\dfrac{\text{Ultimate load}}{A_0} = \dfrac{P_{max}}{A_0}$

3. % Elongation = $\dfrac{L_F - L_0}{L_0} \times 100$

where L_F = Final length of specimen
L_0 = Original length of specimen

4. % Reduction in area = $\dfrac{A_F - A_0}{A_0} \times 100$

    where A_F = Final area of cross section
    A_0 = Original area of cross section

5. Young's modulus of elasticity, $E = \dfrac{\text{Stress at any point with in elastic limit}}{\text{Strain at that point}}$

    From the figure clastic limit is upto point B.

(b) **Stress strain curves for brittle materials :** Materials which show very small elongation before they fracture are called brittle materials. The shape of curve for a high carbon steel is shown in Fig. 14.8 and is typical of many brittle materials such as G.I, concrete and high strength light alloys. For most brittle materials the permanent elongation (i.e., increase in length) is less than 10%.

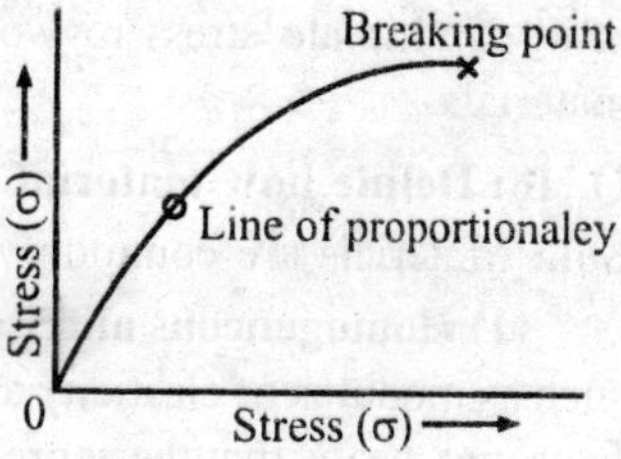

Stress-Strain Curves (Compression)

For ductile materials stress strain curves in compression are identical to those in tension at least upto the yield point for all practical purposes. Since tests in tension are simple to make, the results derived from tensile curves are relied upon for ductile materials in compression.

Fig. 14.8

Brittle materials have compression stress strain curves usually of the same form as the tension test but the stresses at various points (Limit of proportionality, ultimate etc) are generally considerably different.

Q. 17: Define the following terms:

   (1) limit of proportionality
   (2) yield stress and ultimate stress
   (3) working stress and factor of safety.

Sol.: (1) Limit of proportionality: Limit of proportionality is the stress at which the stress - strain diagram ceases to be a straight line i.e, that stress at which extension ceases to be proportional to the corresponding stresses.

(2) Yield stress and ultimate stress Yield stress : Yield stress is defined as the lowest stress at which extension of the test piece increases without increase in load. It is the stress corresponding to the yield point. For ductile material yield point is well defined whereas for brittle material it is obtained by offset method. It is also called yield strength.

Yield Stress = Lowest stress = Yield Point Load/ Cross sectional Area

Ultimate stress : Ultimate stress or Ultimate strength corresponds to the highest point of the stress-strain curve. It is the ratio of maximum load to the original area of cross-section. At this ultimate point, lateral strain gets localized resulting into the formation of neck.

Ultimate stress = Heighest value of stress = $\dfrac{\text{Maximum Load}}{\text{Original Cross sectional Area}}$

(3) Working stress and Factor of Safety Working Stress: Working stress is the safe stress taken within the elastic range of the material. For brittle materials, it is taken equal to the ultimate strength divided by suitable factor of safety. However, for materials possessing well defined yield point, it is equal to yield stress divided by a factor of safety. It is the stress which accounts all sorts of uncertainties.

$$\text{Working stress} = \frac{\text{Ultimate strength}}{\text{Factors of safety}} \text{ for brittle materials}$$

$$= \frac{\text{Yield strength}}{\text{Factors of safety}} \text{ for ductile materials}$$

It is also called allowable stress, permissible stress, actual stress and safe stress.

Factor of Safety : Factor of safety is a number used to determine the working stress. It is fixed based on the experimental works on the material. It accounts all uncertainties such as, material defects, unforeseen loads, manufacturing defects, unskilled workmanship, temperature effects etc. Factor of safety is a dimensionless number. It is fixed based on experimental works on each materials. It is defined as the ratio of ultimate stress to working stress for brittle materials or yield stress working stress for ductile materials.

Q. 18: Define how material can be classified?

Sol.: Materials are commonly classified as:

(1) Homogeneous and isotropic material: A homogeneous material implies that the elastic properties such as modulus of elasticity and Poisson's ratio of the material are same everywhere in the material system. Isotropic means that these properties are not directional characteristics, i.e., an isotropic material has same elastic properties in all directions at any one point of the body:

(2) Rigid and linearly elastic material: A rigid material is one which has no strain regardless of the applied stress. A linearly elastic material is one in which the strain is proportional to the stress.

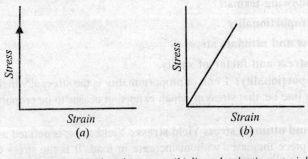

Fig. 14.9. (a) Rigid and   (b) linearly elastic material

(3) Plastic material and rigid-plastic material: For a plastic material, there is definite stress at which plastic deformation starts. A rigid-plastic material is one in which elastic and time-dependent deformations are neglected. The deformation remains even after release of stress (load).

Simple Stress and Strain / 343

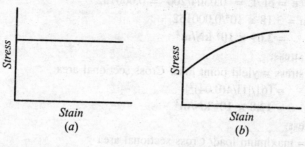

Fig. 14.10 (a) Plastic mid   (b) rigid-plastic material

(4) Ductile mid brittle material: A material which can undergo 'large permanent' deformation in tension, i.e., it can be drawn into wires is termed as ductile. A material which can be only slightly deformed without rupture is termed as brittle.

Ductility of a material is measured by the percentage elongation of the specimen or the percentage reduction in cross-sectional area of the specimen when failure occurs. If L is the original length and L' is the final length, then

$$\% \text{ increase in length} = \frac{L'-L}{L} \times 100$$

The length l' is measured by putting together two portions of the fractured specimen. Likewise if A is the original area of cross-section and A' is the minimum cross sectional area at fracture, then

$$\% \text{ age reduction in area} = \frac{A-A'}{A} \times 100$$

A brittle material like cast iron or concrete has very little elongation and very little reduction in cross-sectional area. A ductile material like steel or aluminium has large reduction in area and increase in elongation. An arbitrary percentage elongation of 5% is frequently taken as the dividing line between these two classes of material.

Q. 19: The following observations were made during a tensile test on a mild steel specimen 40 mm in diameter and 200 mm long. Elongation with 40 kN load (within limit of proportionality), δL = 0.0304 mm

Yield load =161 KN

Maximum load = 242 KN

Length of specimen at fracture = 249 mm

Determine:

(i) Young's modulus of elasticity

(ii) Yield point stress

(iii) Ultimate stress

(iv) Percentage elongation.

Sol.: (i) Young's modulus of elasticity E :

Stress, $\sigma = P/A$

$= 40/[\Pi/4(0.04)^2] = 3.18 \times 10^4$ kN/m²

Strain, $e = \delta L/L = 0.0304/200 = 0.000152$
E = stress/ strain = $3.18 \times 10^4/0.000152$
$\quad\quad\quad = 2.09 \times 10^8$ kN/m² ....ANS

(ii) Yield point stress:
Yield point stress = yield point load/ Cross sectional area
$\quad\quad\quad = 161/[\Pi/4(0.04)^2]$
$\quad\quad\quad = 12.8 \times 10^4$ kN/m² ....ANS

(iii) Ultimate stress:
Ultimate stress = maximum load/ Cross sectional area
$\quad\quad\quad = 242/[\Pi/4(0.04)2]$
$\quad\quad\quad = 19.2 \times 10^4$ kN/m² ....ANS

(iv) Percentage elongation:
Percentage elongation = (length of specimen at fracture - original length)/ Original length
$\quad\quad\quad = (249–200)/200$
$\quad\quad\quad = 0.245 = 24.5\%$ ....ANS

Q. 20: The following data was recorded during tensile test made on a standard tensile test specimen:

Original diameter and gauge length = 25 mm and 80 mm;

Minimum diameter at fracture = 15 mm;

Distance between gauge points at fracture = 95 mm;

Load at yield point and at fracture = 50 kN and 65 kN;

Maximum load that specimen could take = 86 kN.

Make calculations for

(a) Yield strength, ultimate tensile strength and breaking strength

(b) Percentage elongation and percentage reduction in area after fracture

(c) Nominal and true stress and fracture.

Sol.: Given data:
Original diameter = 25 mm
gauge length = 80 mm;
minimum diameter at fracture = 15 mm
distance between gauge points at fracture = 95 mm
load at yield point and at fracture = 50 kN
load at fracture = 65 kN;
maximum load that specimen could take = 86 kN.
Original Area $A_o = \Pi/4 (25)^2 = 490.87$ mm²
Final Area $A_f = \Pi/4 (15)^2 = 176.72$ mm²

(a) Yield Strength = Yield Load / Original Cross sectional Area
$\quad\quad\quad = (50 \times 10^3)/490.87 = $ **101.86 N/mm²** ....ANS

Ultimate tensile Strength Maximum Load / Original Cross sectional Area
$\quad\quad\quad = (86 \times 10^2)/490.87 = $ **175.2 N/mm²** ....ANS

Breaking Strength = fracture Load / Original Cross sectional Area
$\quad\quad\quad = (65 \times 103)/ 490.87 = $ **132.42 N/mm²** ....ANS

(b) Percentage elongation = (distance between gauge points at fracture - gauge length)/ gauge length
   = [(95 − 80)/80] × 100 = **18.75%**   ....ANS

percentage reduction in area after fracture = [(Original Area − Final Area)/ Original Area] × 100
   = [(490.87 − 176.72)/ 490.87] × 100 = **64%**   ....ANS

(c) Nominal Stress = Load at fracture / Original Area = (65 × 1000)/ 490.87
   = **132.42 N/mm^2**   ....ANS

True Stress = Load at fracture / Final Area = (65 × 1000)/ 176.72
   = **367.8 N/mm^2**   ....ANS

Q. 21: Find the change in length of circular bar of uniform taper.

Sol.: The stress at any cross section can be found by dividing the load by the area of cross section and extension can be found by integrating extensions of a small length over whole of the length of the bar. We shall consider the following cases of variable cross section:

Consider a circular bar that tapers uniformly from diameter d1 at the bigger end to diameter d2 at the smaller end, and subjected to axial tensile load P as shown in fig 14.11.

Let us consider a small strip of length dx at a distance x from the bigger end.

Diameter of the elementary strip:
$$d_x = d_1 - [(d_1 - d_2)x]/L$$
$$= d_1 - kx; \text{ where } k = (d_1 - d_2)/L$$

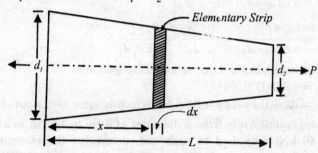

Fig 14.11

Cross-sectional area of the strip,
$$A_x = \frac{\pi}{4} d_x^2 = \frac{\pi}{4}(d_1 - kx)^2$$

Stress in the strip,
$$\sigma_x = \frac{P}{A_x} = \frac{P}{\frac{\pi}{4}(c_1 - kx)^2} = \frac{4P}{\pi(d_1 - kx)^2}$$

Strain in the strip
$$\varepsilon_x = \frac{\sigma_x}{E} = \frac{4P}{\pi(d_1 - kx)^2 E}$$

Elongation of the strip

$$\delta l_x = \varepsilon_x \, dx = \frac{4P\,dx}{\pi(d_1-kx)^2 E}$$

The total elongation of this tapering bar can be worked out by integrating the above expression between the limits $x = 0$ to $x = L$.

$$\delta l = \int_0^L \frac{4P\,dx}{\pi(d_1-kx)^2 E} = \frac{4P}{\pi E}\int_0^L \frac{dx}{(d_1-kx)^2}$$

$$= \frac{4P}{\pi E}\left[\frac{(d_1-kx)^{-1}}{(-1)\times(-k)}\right]_0^L = \frac{4P}{\pi EK}\left[\frac{1}{d_1-kx}\right]_0^L$$

Putting the value of $k = (d_1 - d_2)/l$ in the above expression, we obtain

$$\delta l = \frac{4PL}{\pi E(d_1-d_2)}\left[\frac{1}{d_1-\frac{(d_1-d_2)l}{l}} - \frac{1}{d_1}\right]$$

$$= \frac{4PL}{\pi E(d_1-d_2)}\left[\frac{1}{d_2} - \frac{1}{d_1}\right]$$

$$= \frac{4PL}{\pi E(d_1-d_2)} \times \frac{d_1-d_2}{d_1 d_2} = \frac{4PL}{\pi E d_1 d_2}$$

If the bar is of uniform diameter d throughout its length, then

$\delta L = 4.P.L/(\Pi.E.d^2)$

$= P.L/[(\Pi d^2/4).E] = P.l/A.E.$; Which is same as last article

Q. 22: A conical bar tapers uniformly from a diameter of 4 cm to 1.5 cm in a length of 40 cm. If an axial force of 80 kN is applied at each end, determine the elongation of the bar. Take $E = 2 \times 10^5$ N/mm² *(UPTU QUESTION BANK)*

Sol.: Given that: $P = 80 \times 10^3$ N,

$E = 200$GPa $= 2 \times 10^5$ N/mm2, dL = 40mm, d_2 = 15mm, L = 400mm

Since;

$$\delta L = \frac{4PL}{\pi E d_1 d_2}$$

Putting all the value, we get

$\delta L = [4 \times 80 \times 10^3 \times 400]/[\Pi(2 \times 10^5) \times (40 \times 15)]$

$= 0.3397$mm ....ANS

Q. 23: If the Tension test bar is found to taper from (D + a) cm diameter to (D – a)cm diameter, prove that the error involved in using the mean diameter to calculate Young's modulus is $(10a/D)^2$.

Sol.: Larger dia d_1 = D + a (mm)

Smaller dia d_2 = D – a (mm)

Let,

P = load applied on bar (tensile) N

L = length of bar (mm)
E_1 = Young's modulus using uniform cross section (N/mm²)
E_2 = Young's modulus using uniform cross section (N/mm²)
dL = Extension in length of bar (mm)
Using the relation of extension for tapering cross section, we have

$$\delta L = 4.P.L/(\Pi.E.d_1.d_2) = 4.P.L/[\Pi.E_1.(D + a).(D - a)] = 4.P.L/[\Pi.E_1.(D_2 - a_2)]$$
$$E_1 = 4.P.L/[\Pi. \, dL.(D_2 - a_2)] \qquad \ldots(i)$$

Now using the relation of extension for uniform cross - section; we have

$$\delta L = 4.P.L/(\Pi.E.d_2) = 4.P.L/(\Pi.E_2.D_2)$$
$$E_2 = 4.P.L/[\Pi. \, \delta L.D_2] \qquad \ldots(ii)$$

The % error in calculation of Young's modulus is: $[(E_1 - E_2)/E_1] \times 100$
Hence proved

Q. 24: A bar of length 25mm has varying cross section. It carries a load of 14KN. Find the extension if the cross section is given by $(6 + x^2/100)$ mm² where x is the distance from one end in cm. Take E = 200 GN/m². (Neglect weight of bar)

Sol.: Consider a small element of length dx at distance x from the small element. Due to tensile load applied at the ends, the element length dx elongates by a small amount Δx, and

$$\Delta x = Pdx/AE = Pdx/[(6 + x^2/100)]$$

The total elongation of the bar is then worked out by integrating the above identity between the limits $x = 0$ to $x = L$

$$\delta L = \int_0^L \frac{P}{\left(6 + \dfrac{x^2}{100}\right)E} = \frac{100P}{E} \int_0^{25} \frac{1}{600 + x^2} \, dx$$

Since, $\qquad \int \dfrac{dx}{a^2 + x^2} = \dfrac{1}{a} \tan^{-1}(x/a)$

Recalling ;
Here, $\qquad a = (600)^{1/2}$

$$\delta L = (100P/E) \times (1/(600)^{1/2})\tan^{-1}\left[x/(600)^{1/2}\right]_0^{25}$$

Putting $\qquad P = 14KN$ and $E = 2 \times 10^5 N/mm^2$; we get
We get;

$$\delta L = 0.227 \text{ mm} \qquad \ldots\ldots..ANS$$

Q. 25: A steel bar AB of uniform thickness 2 cm, tapers uniformly from 1.5 cm to 7.5 cm in a length of 50 cm. From first principles determine the elongation of plate; if an, axial tensile force of 100 kN is applied on it. [$E = 2 \times 10^5$ N/mm²]

Sol.: Consider a small element of length dx of the plate, at a distance x from the larger end. Then at this section,

$$\text{Width } w_x = 100 - (100 - 50)\frac{x}{400} = \left(100 - \frac{x}{8}\right) mm$$

Cross-section area A_x = thickness × width = $10\left(100-\dfrac{x}{8}\right)$ mm²

Stress $\sigma_x = \dfrac{P}{A_x} = \dfrac{50 \times 10^3}{10(100-x/8)} = \dfrac{5 \times 10^3}{(100-x/8)}$ N/mm²

Strain $\varepsilon_x = \dfrac{\sigma_x}{E} = \dfrac{5 \times 10^3}{(100-x/8) \times (2 \times 10^5)} = \dfrac{1}{40(100-x/8)}$

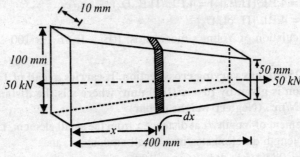

Fig 14.12

∴ Elongation of the elementary length,

$$\delta L_x = \varepsilon_x \times dx = \dfrac{dx}{40(100-x/8)}$$

The total change in length of the plate can be worked out by integrating the above identity between the limits $x = 0$ and $x = 400$ mm. That is:

$$\delta L = \int_0^{400} \dfrac{dx}{40(100-x/8)} = \dfrac{1}{40} \int_0^{400} \dfrac{dx}{(100-x/8)}$$

$$= \dfrac{1}{40}\left[\dfrac{\log_e(100-x/8)}{-1/8}\right]_0^{400} = -0.2\left[\log_e(100-x/8)\right]_0^{400}$$

$$= -0.2[\log_e 50 - \log_e 100] = 0.2(\log_e 100 - \log_e 50)$$

$$= 0.2 \log_e\left(\dfrac{100}{50}\right) = 0.2 \times 0.693 = 0.1386 \text{ mm}$$

Q. 26: Find ratio of upper end area to lower end area of a bar of uniform strength.

Sol: Figure 14.13 shows a bar subjected to an external tensile load P. If the bar had been of uniform cross-section, the tensile stress intensity at any section would be constant only if the self-weight of the member is ignored. If the weight of the member is also considered, the intensity of stress increases for sections at higher level. It is possible to maintain a uniform stress of all the sections by increasing the area from the lower end to the upper end. Let the areas of the upper and lower ends be A_1 and A_2 respectively. Let A be the area at a distance x and $A + dA$ at a distance $x + dx$ from the lower end. Let the weight per unit volume be w Making a force balance for the element ABCD.

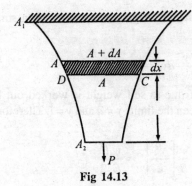

Fig 14.13

$$\sigma(A + dA) = \sigma A + w\,A\,dx$$

or,
$$\sigma\,dA = w\,A\,dx$$

or
$$\frac{dA}{A} = \frac{w}{\sigma} dx$$

Assuming w and s to be uniform,

$$\ln A = \frac{w}{\sigma} x + C_1$$

At $\quad x = 0, A = A_2 = C_1 = \ln A_2$

$$\ln \frac{A}{A_2} = \frac{wx}{\sigma}$$

$$A = A_2\, e^{ax/\sigma}$$

$x = L, A = A_1$

$$\frac{A_1}{A_2} = e^{wL/\sigma}$$

Q. 27: A vertical bar fixed at the upper end, and of uniform strength carries an axial load of 12KN. The bar is 2.4m long having a weight per unit volume of 0.0001N/mm^3. If the area of the bar at the lower end is 520mm^2, find the area of the bar at the upper end.

Sol.: The bar is of uniform strength, so the stress will remain the same everywhere.

$$\sigma = P/A_2; \quad = 12000/520 = 23.08 \text{N/mm}^2$$

Now $\quad \dfrac{A_1}{A_2} = e^{wL/\sigma}$

$$\frac{520 \times e^{-0.0001 \times 2400}}{23.08} = 520 e^{0.0104}$$

$$= 525.44 \text{mm}^2 \qquad \text{......ANS}$$

Q. 28: Find increase in length of a bar of uniform section due to self weight.

Sol.: Consider a bar of cross-sectional area A and length l hanging freely under its own weight (Fig. 14.14). Let attention be focused on a small element of length dy at distance x from the lower end. If 'ω' is the specific weight (weight per unit volume) of the bar material, then total tension at section m-n equals weight of the bar for length y and is given by

$$P = \omega A y$$

As a result of this load, the elemental length dx elongates by a small amount Δx, and

$$\Delta x = \frac{P\,dy}{AE} = \frac{wAy}{AE}\,dy = \frac{w}{E}$$

The total change in length of the bar due to self weight is worked out by integrating the above expression between the limits $y = a$ and $y = 1$. Therefore,

$$\delta x = \int_0^1 \frac{w}{E}\, y\,dy = \frac{w}{E}\left[\frac{y^2}{2}\right]_0^1 = \frac{w\,l^2}{E\,2}$$

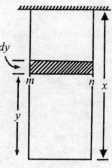

Fig 14.14

If W is the total weight of the bar ($W = wAl$), then $w = W/Al$. In that case, total extension of the bar

$$\delta x = \left(\frac{w}{Al}\right)\frac{l^2}{2E} = \frac{wl}{2\,AE}, \quad W = \text{Total weight}$$

Thus total extension of the bar due to self weight is equal to the extension that would be produced if one-half of the weight of the bar is applied at its end.

Q. 29: An aerial copper wire ($E = 1 \times 10^5$ N/mm^2) 40 m long has cross sectional area of 80 mm^2 and weighs 0.6 N per meter run. If the wire is suspended vertically, calculate

(a) the elongation of wire due to self weight,

(b) the total elongation when a weight of 200 N is attached to its lower end, and

(c) the maximum weight which this wire can support at its lower end if the limiting value of stress is 65 N/mm^2.

Sol.: (a) Weight of the wire $W = 0.6 \times 40 = 24$ N

The elongation due to self weight is,

$\delta L = \omega L^2/2\,E$; where ω is the specific weight (weight per unit volume)

In terms of total weight $W = \omega AL$

$\delta L = WL/2E = 24 \times (40 \times 10^3)/2 \times 80 \times (1 \times 10^5)$

$= 0.06$ mm ....ANS

(b) Extension due to weight P attached at the lower end,

$\delta L = PL/2E$

$= 200 \times (40 \times 10^3)/80 \times (1 \times 10^5)$

$= 1.0$ mm

Total elongation of the wire $= 0.06 + 1.0 =$ **1.06 mm**

(c) Maximum limiting stress $= 65\ N/$mm^2

Stress due to self weight equals that produced by a load of half its weight applied at the end. That is

Stress due to self weight $= (W/2)/A = (24/2)/80 = 0.15\ N/$ mm^2

Remaining stress $= 65 - 0.15 = 64.85$ N/mm2

Maximum weight which the wire can support $= 64.85 \times 80 =$ **5188 N**

Q. 30: A rectangular bar of uniform cross-section 4 cm × 2.5 cm and of length 2 m is hanging vertically from a rigid support. It is subjected to axial tensile loading of 10KN. Take the density of steel as 7850 kg/m³. And E = 200 GN/m². Find the maximum stress and the elongation of the bar. *(Dec–03)*

Sol.: Data given, cross sectional area $A = 4$ cm. $\times 2.5$ cm.

$$\text{length} \qquad L = 2m$$
$$\text{axial load}$$
$$P = 10KN$$
$$\text{density of steel } \rho = 7850 \text{ Kg/m}^3$$
$$E = 200 \text{ GN/m}^2$$
Since
$$\delta L = PL/AE + \rho L^2/2E$$
$$= (10 \times 10^3 \times 2)/(4 \times 2.5 \times 10^{-4} \times 200 \times 10^9)$$
$$+ (7850 \times 9.81 \times 2^2)/(2 \times 200 \times 10^9)$$
$$= 0.0001 \text{ m}$$
$$= 0.1 \text{ mm}$$
$$\text{stress }(\sigma_{max}) = E \times \text{strain}$$
$$= 200 \times 10^9 \times \delta L/L = 200 \times 10^9 \times (0.0001/2) = \mathbf{10.08 \text{ mpa}} \qquad \text{.......ANS}$$

Fig 14.15

Q. 31: Determine the elongation due to self weight of a conical bar.

Sol.: Consider a conical bar *ABC* of length *L* with its diameter d fixed rigidly at *AB*. Let attention be focussed on a small element of length dy at distance *y* from the lower end point *C*. Total tension P at section *DE* equals weight of the bar for length *y* and is given by

$$P = w \times \frac{1}{3}\left(\frac{\pi}{4} ds^2 y\right)$$

where ω is the specific weight of the bar material and *ds* = *DE* is the diameter of the elementary strip. From the similarity of triangles *ABC* and *DEC*,

$$\frac{AB}{DE} = \frac{L}{y} \text{ or } DE = AB\frac{y}{L} = d\frac{y}{L}$$

$$P = w \times \frac{1}{3}\left[\frac{\pi}{4}\frac{d^2 y^2}{L^2} y\right]$$

$$= w\frac{\pi}{L^2}\frac{d^2}{L^2} y^3$$

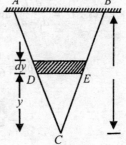

As a result of this load, the elemental length dy elongates by a small amount Dy and   **Fig 14.16**

$$\Delta y = \frac{P\,dy}{AE} = w\frac{\pi}{12}\frac{d^2}{L^2} y^3\, dy + \frac{\pi}{4}\frac{d^2 y^2}{L^2} E$$

$$= \frac{w}{3E} y\,dy$$

The total change in length of the conical bar due to self-weight is worked out by integrating the above expression between the limits $y = 0$ to $y = L$

$$\delta L = \frac{w}{3E}\int_0^L y\,dy = \frac{w}{3E}\left[\frac{y^2}{2}\right]_0^L = \frac{wL^2}{6E} = \frac{\rho g}{6E}L^2$$

Where ρ is the mass density of the bar material

Q. 32: Determine the elongation due to self weight of a tapering rod.

Sol.: Consider a tapering rod hung vertically and firmly fixed at the top position. The rod is of length L and it tapers uniformly from diameter d_1 to d_2, Let the sides AC and BD meet at point E when produced. Extension δL in the length L of the rod ABDE due to self weight is

δL = extension of conical rod AEB due to self weight – extension due to conical segment CEB due to self weight – extension in rod length L due to weight of segment CED

$$= \frac{wL'^2}{6E} = \frac{w(L'-L)^2}{6E} = \frac{4PL}{\pi E d_1 d_2}$$

where P is the weight of segment CED and w is the specific weight of bar material

$$P = w \frac{\pi}{4} \times \left[\frac{1}{3} d_2^2 \times (L' - L) \right]$$

Substituting this value of tensile load P in the above expression, we get

$$\delta l = \frac{wL^2}{6E} = \frac{w(L-L)^2}{6E} - \frac{w(L-L)L}{3E} \frac{d_2}{d_1}$$

From the geometrical configuration

$$\cot \theta = \frac{d_2/2}{(L-L)} = \frac{d_1/2}{L}$$

$$= \frac{d_2}{L-L} = \frac{d_1}{L}$$

This gives: $L' = \dfrac{d_1 L}{d_1 - d_2}$ and $L' - L = \dfrac{d_2 L}{d_1 - d_2}$

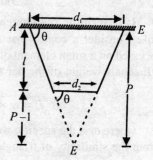

Fig. 14.17

$$\delta L = \frac{w}{6E} \left(\frac{d_1 L}{d_1 - d_2} \right) - \frac{w}{6E} \left(\frac{d_2 L}{d_1 - d_2} \right) - \frac{w}{6E} \left(\frac{d_2 L}{d_1 - d_2} \right) L \times \frac{d_2}{d_1}$$

$$= \frac{wL^2}{3E} \left[\frac{d_1^2}{2(d_1-d_2)^2} - \frac{d_2^2}{2(d_1-d_2)^2} - \frac{d_2^2}{2(d_1-d_2)d_1} \right]$$

$$= \frac{wL^2}{3E} \left[\frac{d_1^3 - d_1 d_2^2 - 2d_2^2(d_1-d_2)}{2d_1(d_1-d_2)^2} \right]$$

$$= \frac{wL^2}{6E} \left[\frac{d_1^3 + 2d_2^3 - 3d_1 d_2^2}{d_1(d_1-d_2)^2} \right]$$

If the rod is conical, *i.e.*, $d_2 = 0$

$$\delta L = \frac{wL^2}{6E}$$

which is the same expression as derived earlier

Q. 33: A vertical rod of 4 m long is rigidly fixed at upper end and carries an axial tensile load of 50kN force. Calculate total extension of the bar if the rod tapers uniformly from a diameter of 50 mm at top to 30 mm at bottom. Take density of material as 1×10^5 kg/m^3 and $E = 210$ GN/m^2. *(UPTU QUESTION BANK)*

Sol.: Extension in the rod due to external load

$$= \frac{4PL}{\pi E d_1 d_2}$$

$E = 2.1 \times 10^5$ N/mm^2
$= [4 \times (50 \times 1000) \times (4 \times 1000)]/[\Pi \times (2.1 \times 10^5) \times (50 \times 30)]$
$= 0.8088$ mm

Extension in the rod due to self weight

$$= \frac{wL^2}{6E} \left[\frac{d_1^3 + 2d_2^3 - 3d_1 d_2^2}{d_1 (d_1 - d_2)^2} \right]$$

where; $\omega = \rho.g$; $\rho = 1 \times 10^5$ Kg/m^3 = 1×10^{-4} Kg/m^3

$$= \frac{(1 \times 10^{-4} \times 9.81)(4 \times 1000)^2}{6 \times 2.1 \times 10^5} \left[\frac{50^3 + 2(30)^3 - 3 \times 50 \times 30^2}{50(50-30)^2} \right]$$

$= 0.0274$ mm

Total extension in the bar $= 0.8088 + 0.0274 = 0.8362$ mm   ....**ANS**

Q. 34: Explain the principle of superposition.

Sol.: A machine member is subjected to a number of forces acting on its outer edges as well as at some intermediate sections along its length. The forces are then split up and their effects are considered on individual sections. The resulting deformation is then given by the algebraic sum of the deformation of the individual sections. This is the principle of superposition which may be stated as

"The resultant elongation due to several loads acting on a body is the algebraic sum of the elongations caused by individual loads"

Or

"The total elongation in any stepped bar due to a load is the algebraic sum of elongations in individual parts of the bar".

Mathematically

$$\delta L = \sum_{i=L}^{i=n} \delta L_i$$

Q. 35: How you evaluate the elongation of a bar of varying cross section?

Sol.: Consider a bar made up of different lengths and having different cross-sections as shown in Fig. 14.18.

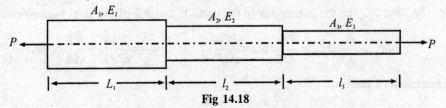

Fig 14.18

For such a bar, the following conditions apply:
(i) Each section is subjected to the same external pull or push
(ii) Total change in length is equal to the sum of changes of individual lengths
That is:

$$P_1 = P_2 = P_3 = P \text{ as } \Sigma H = 0 \text{ \& } \Sigma V = 0$$

and
$$\delta L = \delta L_1 + \delta L_2 + \delta L_3$$

$$= \frac{\sigma_1 L_1}{E_1} + \frac{\sigma_2 L_2}{E_2} + \frac{\sigma_3 L_3}{E_3}$$

$$= \frac{P_1 L_1}{A_1 E_1} + \frac{P_2 L_2}{A_2 E_2} + \frac{P_2 L_3}{A_3 E_3}$$

If the bar segments are made of same material, then In that case
$$E_1 = E_2 = E_3 = E.$$

$$\delta L = \frac{P}{E}\left[\frac{L_1}{A_1} + \frac{L_2}{A_2} + \frac{L_3}{A_3}\right]$$

Q. 36: A steel bar is 900 mm long; its two ends are 40 mm and 30 mm in diameter and the length of each rod is 200 mm. The middle portion of the bar is 15 mm in diameter and 500 mm long. If the bar is subjected to an axial tensile load of 15 kN, find its total extension. Take $E = 200$ GN/ m² (G stands for giga and $1G = 10^9$)

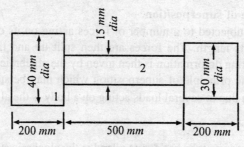

Fig 14.19

Sol.: Refer Fig. 14.19
  Load, P = 15kN
  Area, A_1 = (Π/4) × 40² = 1256.6 mm² = 0.001256 m²
  Area, A_2 = (Π/4) × 15² = 176.7 mm² = 0.0001767 m²
  Area, A_3 = (Π/4) × 30² = 706.8 mm² = 0.0007068 m²
  Lengths: L_1=200mm = 0.2m, L_2 = 500mm = 0.5m and L_3 =200mm = 0.2m
  Total extension of the bar:
  Let δL_1, δL_2 and δL_3, be the extensions in the parts 1, 2 and 3 of the steel bar respectively. Then,

$$\delta L_1 = \frac{PL_1}{A_1 E}, \delta L_2 = \frac{PL_2}{A_2 E}, \delta L_3 = \frac{PL_3}{A_3 E} \quad \left[\because E = \frac{\sigma}{e} = \frac{P/A}{\delta L/L} = \frac{P.L}{A.\delta} \text{ or } \delta L = \frac{PL}{AE}\right]$$

Total extension of the bar,

$$\delta L = \delta L_1 + \delta L_2 + \delta L_3$$
$$= \frac{PL_1}{A_1 E} + \frac{PL_2}{A_2 E} + \frac{PL_3}{A_3 E} = \frac{P}{E}\left[\frac{L_1}{A_1} + \frac{L_2}{A_2} + \frac{L_3}{A_3}\right]$$
$$= \frac{15 \times 10^3}{200 \times 10^9}\left[\frac{0.20}{0.001256} + \frac{0.50}{0.0001767} + \frac{0.20}{0.0007068}\right]$$
$$= 0.0002454 \text{ m} = 0.2454 \text{ mm}$$

Hence total extension of the steel bar = **0.2454 mm** ....ANS

Q. 37: A member *ABCD* is subjected to point loads P_1, P_2, P_3 and P_4 as shown in Fig. 14.20

Fig 14.20

Calculate the force P_3, necessary for equilibrium if $P_1 = 120\ kN$, $P_2 = 220\ kN$ and $P_4 = 160\ kN$. Determine also the net change in length of the member. Take $E = 200\ GN/m^2$.

(UPTU QUESTION BANK)

Sol.: Modulus of elasticity $E = 200\ GN/m^2 = 2 \times 10^5\ N/mm^2$.
Considering equilibrium of forces along the axis of the member.
$$P_1 + P_3 = P_2 + P_4;$$
$$120 + P_3 = 220 + 160$$
Force $\qquad P_3 = 220 + 160 - 120 = 260\ kN$

The forces acting on each segment of the member are shown in the free body diagrams shown below:

Let δL_1, δL_2 and δL_3, be the extensions in the parts 1, 2 and 3 of the steel bar respectively. Then,
$$\delta L_1 = \frac{pL_1}{A_1 E},\ \delta L_2 = \frac{PL_2}{A_2 E},\ \delta L_3 = \frac{PL_3}{A_3 E}$$

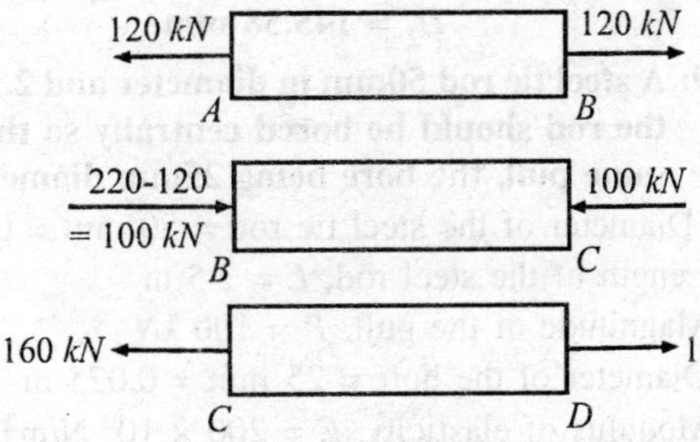

Since Tension in *AB* and *CD* but compression in *BC*, So,
Total extension of the bar,
$$\delta L = \delta L_1 - \delta L_2 + \delta L_3$$
$$\delta L = \frac{PL_1}{A_1 E} - \frac{PL_2}{A_2 E} - \frac{PL_3}{A_3 E}$$

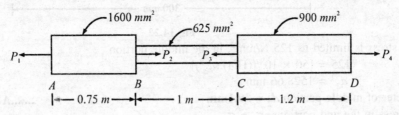

Fig 14.21

Extension of segment $AB = [(120 \times 10^3) \times (0.75 \times 10^3)]/[1600 \times (2 \times 10^5)] = 0.28125$ mm
Compression of segment $BC = [(100 \times 10^3) \times (1 \times 10^3)]/[625 \times (2 \times 10^5)] = 0.8$ mm
Extension of segment $CD = [(160 \times 10^3) \times (1.2 \times 10^3)]/[900 \times (2 \times 10^5)] = 1.0667$ mm
Net change in length of the member = $\delta l = 0.28125 - 0.8 + 1.0667 =$ **0.54795 mm (increase)** ....ANS

Q. 38: The bar shown in Fig. 14.22 is subjected to an axial pull of 150kN. Determine diameter of the middle portion if stress there is limited to 125N/mm². Proceed to determine the length of this middle portion if total extension of the bar is specified as 0.15 mm. Take modulus of elasticity of bar material $E = 2 \times 10^5$ N/mm².

Sol.: Each of the segment of this composite bar is subjected to axial pull $P = 150$ kN.

Axial Stress in the middle portion σ_2 = Axial pull/Area = $150 \times 10^3/[(\Pi/4).(d_2^2)]$

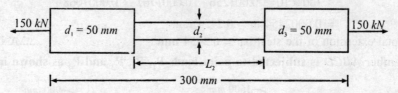

Fig 14.22

Since stress is limited to 125 N/mm², in the middle portion
$$125 = 150 \times 10^3/[(\Pi/4).(d_2^2)]$$
$$d_2^2 = 1528.66 \text{ mm}$$

Diameter of middle portion d_2 = **39.1mm** ....ANS

(ii) Stress in the end portions, $\sigma_1 = \sigma_3$
$$= 150 \times 10^3/[(\Pi/4).(50^2)] = 76.43 \text{ N/m}^2$$

Total change in length of the bar,
= change in length of end portions + change in length of mid portion
$$\delta L = \delta L_1 + \delta L_2 + \delta L_3$$
$$= \sigma_1 L_1/E + \sigma_2 L_2/E + \sigma_3 L_3/E; \text{ Since } E \text{ is same for all portions}$$
$$= \sigma_1(L_1 + L_3)/E + \sigma_2 L_2/E$$
$$L_1 + L_3 = 300 - L_2$$

Now putting all the values;
$$0.15 = [76.43(300 - L_2)]/2 \times 10^5 + 125 L_2/2 \times 10^5$$
$$L_2 = 145.58 \text{ mm} \quad \quad \text{........ANS}$$

Q. 39: A steel tie rod 50mm in diameter and 2.5m long is subjected to a pull of 100 KN. To what length the rod should be bored centrally so that the total extraction will increase by 15% under the same pull, the bore being 25mm diameter? For steel modulus of elasticity is 2×10^5 N/mm².

Sol.: Diameter of the steel tie rod = 50 mm = 0.05 m
Length of the steel rod, L = 2.5 m
Magnitude of the pull, P = 100 kN
Diameter of the bore = 25 mm = 0.025 m
Modulus of elasticity, $E = 200 \times 10^9$ N/m²

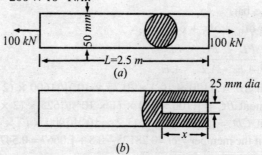

Fig. 14.23

Let length of the bore be 'x'.
Stress in the solid rod $\sigma = P/A$
$= \{(100 \times 1000)/[(\Pi/4)(0.05)2]\} = 50.92 \times 10^6$ N/m²
Elongation of the rod $\delta L = \sigma L/E$
$= (50.92 \times 10^6 \times 2.5) / (200 \times 10^9)$
$= 0.000636$m $= 0.636$mm
Elongation after the rod is bored $= 1.15 \times 0.636 = 0.731$mm
Area of the reduction section $= (\Pi/4) (0.05^2 - 0.025^2) = 0.001472$m²
Stress in the reduced section $\sigma_b = (100 \times 1000)/0.001472$m²
$= 67.93 \times 10^6$ N/m²
Elongation of the rod
$= \sigma(2.5 - x)/E + \sigma_b.x/E = 0.731 \times 10^{-3}$
$= [50.92 \times 10^6 (2.5 - x)]/(200 \times 10^9) + (67.93 \times 10^6.x)/(200 \times 10^9) = 0.731 \times 10^{-3}$
$x = 1.12$m

Hence length of the bore = 1.12m ....ANS

Q. 40: A square bar of 25 mm side is held between two rigid plates and loaded by an axial pull equal to 300kN as shown in Fig. 14.24. Determine the reactions at end A and C and elongation of the portion AB. Take $E = 2 \times 10^5$ N/mm².

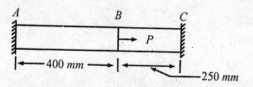

Fig 14.24

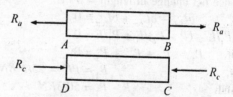

Fig 14.25

Sol.: Cross section area of the bar $A = 25 \times 25$ mm²
Since the bar is held between rigid support at the ends, the following observations need to be made:
(1) Portion AB will be subjected to tension and portion BC will be under compression
(2) Since each ends are fixed and rigid and therefore total Elongation;
$\delta L_{ab} - \delta L_{bc} = 0$; Elongation in portion AB equals shortening in portion BC. i.e., $\delta L_{ab} = \delta L_{bc}$
(3) Sum of reactions equals the applied axial pull i.e., $P = R_a + R_c$
Apply condition (2), we get
$[P_{ab} \times L_{ab}]/A_{ab}.E = [P_{bc} \times L_{bc}]/A_{bc}.E$
$(R_a \times 400)/(625 \times 2 \times 10^5) = (R_c \times 250)/(625 \times 2 \times 10^5)$
$R_c = 1.6 R_a$ ...(i)
Now apply condition (3) i.e., $P = R_a + R_c$
$300 \times 10^3 = R_a + 1.6 R_a$
$R_a = 1.154 \times 10^5$ N; $R_c = 1.846 \times 10^5$ N ....ANS

Q. 41: A rod $ABCD$ rigidly fixed at the ends A and D is subjected to two equal and opposite forces $P = 25$ kN at B and C as shown in the fig 14.26 given below: Make calculations for the axial stresses in each section of the rod.

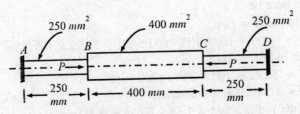

Fig 14.26

Sol.: The following observations need to be made.
(i) Due to symmetrical geometry and load, reaction at each of the fixed ends will be same both in magnitude and direction. That is $P_a = P_d = P_1$(say).
(ii) Segments AB and CD are in tension and the segment BC is in compression.
(iii) End supports are rigid and therefore total change in length of the rod is zero.

The forces acting on each segment will be as shown in Fig. 14.27

Using the relation $\delta L = PL/AE$

$\delta L_{ab} = (P_1 \times 250)/(250 \times E) = P_1/E$ ....Extension
$\delta L_{bc} = (P - P_1) \times 400)/(400 \times E) = (P - P_1)/E$ ....Compression
$\delta L_{cd} = (P_1 \times 250)/(250 \times E) = P_1/E$ ....Extension

Since net change in length = 0 i.e.,

$\delta L_{ab} - \delta L_{bc} + \delta L_{cd} = 0$
$P_1/E - (P - P_1)/E + P_1/E = 0$
Or, $P_1 - P + P_1 + P_1 = 0$;
or, $P_1 = P/3 = 25/3$ KN
And; $P - P_1 = 50/3$ KN

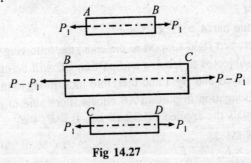

Fig 14.27

Now Stress in segment AB and CD
$= 25 \times 10^3/3 \times 250 = 33.33$ N/mm² (tensile)

Stress in segment
BC $= (50 \times 10^3)/(3 \times 400) = 41.67$ N/mm² (compressive)

Q. 42: A steel bar is subjected to loads as shown in fig. 14.28. Determine the change in length of the bar ABCD of 18 cm diameter. $E = 180$ kN/mm². *(May-05(C.O.))*

Sol.: Ref fig 14.28
Since $d = 180$mm
$E = 180 \times 10^3$ N/mm²

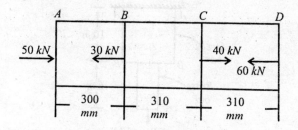

Fig. 14.28

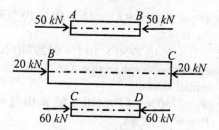

Fig. 14.29

L_{AB} = 300mm
L_{BC} = 310mm
L_{CD} = 310mm

From the fig 14.29
Load on portion $AB = P_{AB} = 50 \times 10^3$ N
Load on portion $BC = P_{BC} = 20 \times 10^3$ N
Load on portion $CD = P_{CD} = 60 \times 10^3$ N
Area of portion AB = Area of portion BC = Area of portion $CD = A = \Pi d^2/4$
        = $\Pi(180)^2/4$ = 25446.9 mm²
Using the relation $\delta l = Pl/AE$
$\delta L_{ab} = (50 \times 10^3 \times 300)/(25446.9 \times 180 \times 10^3) = 0.0033$ mm ---------------- Compression
$\delta L_{bc} = (20 \times 10^3 \times 310)/(25446.9 \times 180 \times 10^3) = 0.0012$ mm ---------------- Compression
$\delta L_{cd} = (60 \times 10^3 \times 310)/(25446.9 \times 180 \times 10^3) = 0.0041$ mm ---------------- Compression
Since net change in length = $-\delta L_{ab} - \delta L_{bc} - \delta L_{cd}$
        = $-0.0033 - 0.0012 - 0.0041$
        = -0.00856 mm
Decrease in length = 0.00856mm   ....ANS

Q. 43: Two prismatic bars are rigidly fastened together and support a vertical load of 45kN as shown in Fig. 14.30. The upper bar is steel having mass density 7750 kg/m³, length 10 m and cross-sectional area 65 cm². The lower bar is brass having mass density 9000 kg/m³, length 6 m and cross-sectional area 50 cm². Determine the maximum stress in each material. For steel $ES = 200$ GN/m^2 and for brass $Eb = 100$ GN/m^2.

Sol.: Refer Fig. 14.30. The maximum stress in the brass bar occurs at junction BB, and this stress is caused by the combined effect of 45kN load together with the weight of brass bar.

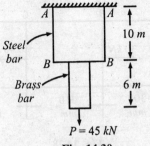

Fig. 14.30

Weight of brass bar, $W_b = \rho_b g V_b$ = 9000 × 9.81 × (6 × 50 × 10⁻⁴) = 2648.7 N
Stress at section BB, $\sigma_b = (P + W_b)/A_b$
$= (45000 + 4648.7)/50 × 10^{-4} = 9529740$ N/m² = 9.53 MN/m²

The maximum stress in the steel bar occurs at section A–A. Here the entire weight of steel and brass bars and 45kN load gives rise to normal stress.

Weight of steel bar, $W_S = \rho_s g V_s$ = 7750 × 9.81 × (10 × 65 × 10⁻⁴) = 4941.79 N
Stress at section AA ; $\sigma_s = (P + W_b + W_s)/A_s$
$= (45000 + 2648.7 + 4941.79)/ 65 × 10^{-4}$
$= 8090845 = 8.09$ MN/m² ....ANS

Q. 44: For the bar shown in Fig. 14.31, calculate the reaction produced by the lower support on the bar. Take $E = 200$ GN/m². Find also the stresses in the bars.

Sol.: Let R_1 = reaction at the upper support;
R_2 = reaction at the lower support when the bar touches it.
If the bar MN finally rests on the lower support,
we have
$$R_1 + R_2 = 55kN = 55000$$
N For bar LM, the total force = $R_1 = 55000 - R_2$ (tensile)
For bar MN, the total force = R_2 (compressive)
δL_1 = extension of $LM = [(55000 - R_2) × 1.2]/[(110 × 10^{-6}) × 200 × 10^9]$
δL_2 = contraction of MN = $[R_2 × 2.4]/[(220 × 10^{-6}) × 200 × 10^9]$
In order that N rests on the lower support, we have from compatibility equation
$$\delta L_1 - \delta L_2 = 1.2/1000 = 0.0012 \text{ m}$$
Or,
$[(55000 - R_2) × 1.2]/[(110 × 10^{-6}) × 200 × 10^9]$
$- [R_2 × 2.4]/[(220 × 10^{-6}) × 200 × 10^9] = 0.0012$
on solving;
$R_2 = 16500$N or, **16.5 KN**
.......ANS
$R_1 = 55 - 16.5 = $ **38.5 KN** ....ANS
Stress in LM = $R_1/A_1 = 38.5/110 × 10^{-6} = 0.350 × 10^6$ kN/m² = **350 MN/m²** ....ANS
Stress in MN = $R_2/A_2 = 16.5/220 × 10^{-6} = 0.075 × 10^6$ kN/m² = **75 MN/m²** ....ANS

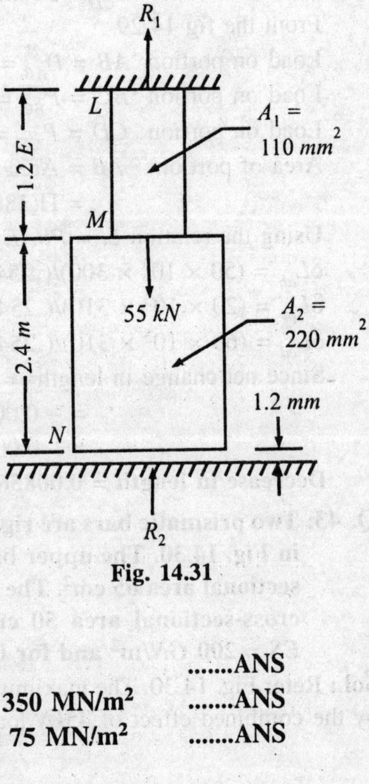

Fig. 14.31

Q. 45: A 700 mm length of aluminium alloy bar is suspended from the ceiling so as to provide a clearance of 0.3 mm between it and a 250 mm length of steel bar as shown in Fig. 14.32. A_{al} = 1250 mm², E_{al} = 70 GN/m^2, As = 2500 mm², Es = 210 GN/m^2. Determine the stress in the aluminium and in the steel due to a 300kN load applied 500 mm from the ceiling.

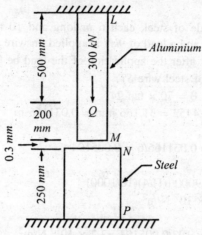

Fig. 14.32

Sol.: On application of load of 300kN at Q, the portion LQ will move forward and come in contact with N so that QM and NP will both be under compression. LQ will elongate, while QM and NP will contact and the net elongation will be equal to gap of 0.3 mm between M and N.

Let σ_1 = tensile stress in LQ

σ_2 = compressive stress in QM

σ_3 = compressive stress in NP

Elongation of $LQ = (\sigma_1 \times 0.5)/70 \times 10^9$

Contraction of $QM = (\sigma_2 \times 0.2) / 70 \times 10^9$

Contraction of $NP = (\sigma_3 \times 0.25)/ 210 \times 10^9$

But force in QM = force in NP

$$\sigma_2 \times 1250 \times 10^{-6} = \sigma_3 \times 2500 \times 10^{-6}$$

We get $\sigma_3 = \sigma_2/2$

So the Contraction of $NP = (\sigma_2 \times 0.25)/(2 \times 210 \times 10^9)$

Net elongation = $\delta L_{LQ} - \delta L_{QM} - \delta L_{NP}$ = 0.0003

$(\sigma_1 \times 0.5)/70 \times 10^9 - (\sigma_2 \times 0.2) / 70 \times 10^9 - (\sigma_2 \times 0.25)/(2 \times 210 \times 10^9) = 0.0003$

on solving we get;

$$3\sigma_1 - 1.45\sigma_2 = 2 \times 210 \times 10^9 \times 0.0003 \qquad ...(i)$$

Tensile force in LQ + compressive force in QM = 300000

$1250 \times 10^{-6} \times \sigma_1 + 1250 \times 10^{-6} \times \sigma_2 = 300000$

We get; $\sigma_1 + \sigma_2 = 2.4 \times 10^8$ N/m² ...(ii)

From equation (i) and (ii) we get

$\sigma_1 = 1.065 \times 10^8$ N/m² = 106.5 MN/m² (tensile) ....ANS

$\sigma_2 = 1.335 \times 10^8$ N/m² = 133.5 MN/m² (compressive) ....ANS

$\sigma_3 = \sigma_2/2 = 0.667 \times 10^8$ N/m² = 66.7 MN/m² (compressive) ....ANS

Q. 46: Two parallel steel wires 6 m long, 10 mm diameter are hung vertically 70 mm apart and support a horizontal bar at their lower ends. When a load of 9kN is attached to one of the wires, it is observed that the bar is 24° to the horizontal. Find 'E' for wire.

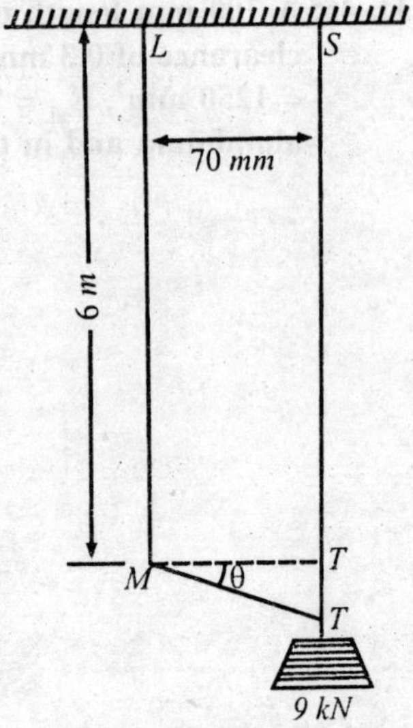

Sol.: Two wires LM and ST made of steel, each 6 m long and 10 mm diameter are fixed at the supports and a load of 9kN is applied on wire ST.

Let the inclination of the bar after the application of the load be θ.

The extension in the length of steel wire ST,

$$\delta L = 70 \cdot \tan \theta = 70 \times \tan 24°$$
$$= 70 \times 0.4452 = 31.166 \text{ mm} = 0.031166 \text{ m}$$

Strain in the wire,

$$e = \delta L/L = 0.031166/6 = 0.005194$$

and stress in the wire

$$\sigma = P/A = 9000/[(\Pi/4)(10/1000)^2]$$
$$= 11.46 \times 10^7 \text{ N/m}^2$$

Young's modulus $E = \sigma/e$

$$= 11.46 \times 10^7/0.005194 = 2.2 \times 10^{10} \text{ N/m}^2$$
$$= 22 \text{ GN/m}^2 \quad \text{.......ANS}$$

Fig. 14.33

Ques No-47: A rigid beam AB 2.4m. long is hinged at A and supported as shown in fig14.34. by two steel wires CD and EF. CD is 6m long and 12mm in diameter and EF is 3m long and 3mm in diameter. If a load of 2250N is applied at B find the stress in each wire $E = 2 \times 10^5$ N/mm².

Sol.: Let $CD = L_1 = 6$m
$EF = L_2 = 3$m
Tension in $CD = T_1$ N
Tension in $EF = T_2$ N
For the equilibrium of the beam, taking moment about the hinge A.

$$T_1 \times 0.60 + T_2 \times 1.80 = 2250 \times 24$$
$$T_1 + 3T_2 = 9000 \quad \text{...(i)}$$

Since the beam is rigid it will remain straight
Let extension of $CD = \delta_1 = DD_1$
Extension of $EF = \delta_2 = FF_1$
As the wires extend, the rigid beam takes the position AD_1FB_1

$$\delta_1/\delta_2 = AD/AF = 0.60/1.80 = 1/3$$
$$\delta_2 = 3\delta_1 \quad \text{...(ii)}$$

But $\delta_1 = T_1L_1/A_1E$
and $\delta_2 = T_2L_2/A_2E$
i.e.; $T_2L_2/A_2E = 3 \cdot T_1L_1/A_1E$
$T_1 = 1/3\{(A_1/A_2)(L_2/L_1)\}T_2$
$T_1 = 1/3\{(12/3)^2(3/6)\}T_2$
$T_1 = 8/3T_2 \quad \text{...(iii)}$

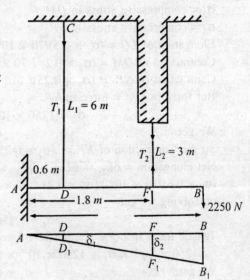

Fig. 14.34

Solve equation (i) and (iii) we get
$$T_1 = 4233N \text{ and } T_2 = 1589N \qquad \text{......ANS}$$
Now

Stress in the wire $CD = 4233/ (\Pi/4)12^2 = 37 \text{ N/mm}^2$

Stress in the wire $EF = 1589/ (\Pi/4)3^2 = 225 \text{ N/mm}^2$

Q. 48: How you determine the stress in composite bar? What is Modular Ratio?

Sol.: It becomes necessary to have a compound tie or strut (column) where two or more material elements are fastened together to prevent their uneven straining. The salient features of such a composite system are: System extends (or contracts) as one unit when subjected to tensile (or compressive) load. This implies that deformation (extension or contraction) of each element is same.o Strain, i.e., deformation per unit length of each element is same.o Total external load on the system equals the sum of loads carried by the different materials comprising the composite system. Fig 14.35

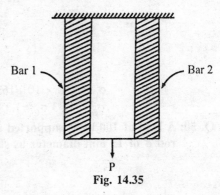

Fig. 14.35

Consider a composite bar subjected to load P and fixed at the top as shown in Fig. 14.35. Total load is shared by the two bars and as such

$$P = P_1 + P_2 = \sigma_1 A_1 + \sigma_2 A_2 \qquad ...(i)$$

Further elongation in two bars are same, i.e., $\delta L_1 = \delta L_2 = \delta L_3$ ...(ii)

If strain in the bar are equal $e_1 = e_2$; $\sigma_1/E_1 = \sigma_2/E_2$

or, $\quad \sigma_1/\sigma_2 = E_1/E_2$ (when length is same)

This ratio E_1/E_2 is called Modular ratio.

Modular Ratio: Modular ratio is the ratio of moduli of elasticity of two materials. It is denoted by μ.

$$\text{Modular ratio, } \mu = \frac{\text{Young's Modulas of Material 1}}{\text{Young's Modulas of Material 2}} = \frac{E_1}{E_2}$$

Q. 49: Two copper rods one steel rod lie in a vertical plane and together support a load of 50kN as shown in Fig. 14.36. Each rod is 25 mm in diameter, length of steel rod is 3 m and length of each copper rod is 2m. If modulus of elasticity of steel is twice that of copper, make calculations for the stress induced in each rod. It may be presumed that each rod deforms by the same amount.

Sol.: Each rod deforms by the same amount and accordingly i.e. $\delta L_S = \delta L_E$ or $\dfrac{\sigma_s \cdot L_s}{E_s} = \dfrac{\sigma_c \cdot L_c}{E_c}$

or $\quad \sigma_s = \dfrac{E_s}{E_c} \cdot \sigma_c \cdot \dfrac{L_c}{L_s}$

$\sigma_s = 2 \times (2/3)\, \sigma_c = 1.33\, \sigma_c$

Division of load between the steel and copper rods is as follows : total load = load carried by steel rod + rod carried by two copper rods

$50 \times 10^3 = \sigma_s A_s + 2\, \sigma_c A_c$

$= 1.33\, \sigma_c \times (\Pi/4)(25)^2 + 2\sigma_c \times (\Pi/4)(25)^2 = 1633.78 \sigma_c$

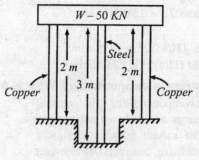

Fig. 14.36

$\sigma_C = (50 \times 10^3)/1633.78 = 30.60 \text{ N/mm}^2$ ....ANS

$\sigma_S = 1.33\, \sigma_C = 1.33 \times 30.60 = 40.7 \text{ N/mm}^2$ ....ANS

Q. 50: A load of 100 kg is supported upon the rods A and C each of 10 mm diameter and another rod B of 15 mm diameter as shown in figure 14.37. Find stresses in rods A, B and C.

(May-01)

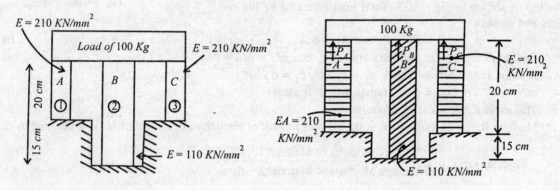

Fig. 14.37   Fig. 14.38

Sol.: Given data

m = 100Kg

Diameter of rod A and C = 10mm

Diameter of rod B = 15mm

Area of rod 'A' = Area of rod 'C'

$= A_A = \pi/4 . D^2 = \pi/4 . 10^2 = 78.54 \text{ mm}^2$

Area of rod 'B'

$= A_B = \pi/4 . D_B^2 = \pi/4 . 15^2 = 176.71 \text{ mm}^2$

Since; $P = P_A + P_B + P_C$

$100 \times 9.81 = \sigma_A \times A_A + \sigma_B \times A_B + \sigma_C \times A_C$

$981 = 78.54\, \sigma_A + 176.71 \sigma_B + 78.54\, \sigma_C$ ...(i)

The deflection or shorten in length in each rod will be same so,

$\delta L_A = \delta L_B = \delta L_C$

$e_A . L_A = e_B . L_B = e_C . L_C$

$$(\sigma_A/E_A) \times L_A = (\sigma_B/E_B) \times L_B = (\sigma_C/E_C) \times L_C$$
$$(\sigma_A/210) \times 200 = (\sigma_B/110) \times 350 = (\sigma_C/210) \times 200$$
$$\sigma_A = \sigma_C = 3.34\, \sigma_B \qquad \qquad ...(ii)$$

Putting the value of (ii) in equation (i); we get
$$981 = 78.54 \times 3.34\sigma_B + 176.71\sigma_B + 78.54 \times 3.34\, \sigma_B$$
$$981 = 701.5\sigma_B$$
$$\sigma_B = 1.34\ \text{N/mm}^2 \qquad \qquad \text{.......ANS}$$
$$\sigma_A = \sigma_C = 4.67\ \text{N/mm}^2 \qquad \qquad \text{.......ANS}$$

Q. 51: A beam weighing 50 N is held in horizontal position by three wires. The outer wires are of brass of 1.8 mm dia and attached to each end of the beam. The central wire is of steel of 0.9 mm diameter and attached to the middle of the beam. The beam is rigid and the wires are of the same length and unstressed before the beam is attached. Determine the stress induced in each of the wire. Take Young's modulus for brass as 80 GN/m^2 and for steel as 200 GN/m^2. Fig 14.39

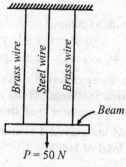

Fig. 14.39

Sol.: If P_b denotes load taken by each brass wire and P_s denotes load taken by steel wire, then Total load
$$P = 2P_b + P_s = 2\sigma_b A_b + \sigma_s A_s \qquad \qquad ...(i)$$
As the beam is horizontal, all the wire extend by the same amount. Further since each wire is of same length, the wires would experience the same amount of strain, thus
$$e_s = e_b$$
$$\sigma_s/E_s = \sigma_b/E_b$$
$$\sigma_s = (E_s \times \sigma_b)/E_b = (200 \times \sigma_b)/80 = 2.5\sigma_b \qquad \qquad ...(iii)$$
Putting the value of equation (ii) in equation (i)
$$50 = 2\sigma_b(\Pi/4)(1.8)^2 + 2.5\sigma_b\,(\Pi/4)(0.9)^2$$
$$50 = 6.678\ \sigma_b$$
$$\sigma_b = 7.49\ \text{N/mm}^2 \qquad \qquad \text{.......ANS}$$
$$\sigma_s = 2.5\sigma_b = 18.71\ \text{N/mm}^2 \qquad \qquad \text{.......ANS}$$

Q.52: A concrete column 300mm x 300mm in section, is reinforced by 10 longitudinal 20mm diameter round steel bars. The column carries a compressive load of 450KN. Find load carried and compressive stress produced in the steel bars and concrete. Take $E_s = 200 GN/m^2$ and $E_c = 15 GN/m^2$.

Sol.: Cross sectional Area of column = 300 × 300 = 90000mm²

Area of steel bars $A_s = 10 \times (\Pi/4)(20)^2 = 3141.59 \text{mm}^2$

Area of concrete $= 90000 - 3141.59 = 86858.4 \text{ mm}^2$

Each component (concrete and steel bars) shorten by the same amount under the compressive load, and therefore

Strain in concrete = strain in steel

$$\sigma_c/E_c = \sigma_s/E_s$$

Where σ_c, σ_s are stress induced in concrete and steel respectively

$$\sigma_s = (E_s \sigma_c)/E_c$$
$$= (200 \times 10^9 / 15 \times 10^9) \sigma_c = 13.33 \sigma_c$$

Further,

Total load on column = Load carried by steel + load carried by concrete

$$P = \sigma_s.A_s + \sigma_c.A_c$$
$$450 \times 10^3 = 13.33\sigma_c \times 3141.59 + \sigma_c. 86858.4$$
$$= 128735.8 \sigma_c$$
$$\sigma_c = 3.59 \text{ N/mm}^2 \quad \text{......ANS}$$
$$\sigma_s = 13.33 \times 3.59 = 46.95 \text{ N/mm}^2 \quad \text{......ANS}$$

Load carried by concrete $Pc = \sigma_{cAc} = 3.59 \times 86858.4 = 311821.656 \text{ N} = 311.82 \text{KN}$ ......ANS

Load carried by steel $P_s = \sigma_{sAs} = 46.95 \times 3141.59 = 147497.65 \text{ N} = 147.49 \text{ KN}$ ......ANS

Q. 53: A load of 300kN is applied on a short concrete column 250 mm × 250 mm. The column is reinforced by steel bars of total area 5600 mm². If the modulus of elasticity of steel is 15 times that of concrete, find the stresses in concrete and steel.

If the stress in the concrete should not exceed 4 N/mm², find the area of the steel required so that the column may support a load of 600 kN. [U.P.T.U. Feb., 2001]

Sol.: do your self.

Q. 54: A solid steel cylinder 500 mm long and 70 mm diameter is placed inside an aluminium cylinder having 75 mm inside diameter and 100 mm outside diameter. The aluminium cylinder is 0.16 mm. longer than the steel cylinder. An axial load of 500kN is applied to the bar and cylinder through rigid cover plates as shown in Fig. 14.40. Find the stresses developed in the steel cylinder and aluminium tube. Assume for steel, $E = 220 \text{ GN/m}^2$ and for Al $E = 70 \text{ GN/m}^2$

Sol.: Since the aluminium cylinder is 0.16 mm longer than the steel cylinder, the load required to compress this cylinder by 0.16 mm will be found as follows:

E = stress/ strain = $P.L/A.\delta L$

Or $\quad P = E.A.\delta L/L$
$\quad = 70 \times 10^9 \times \pi/4 \ (0.1^2 - 0.075^2) \times 0.00016/0.50016 = 76944 \text{ N}$

When the aluminium cylinder is compressed by its extra length 0.16 mm, the load then shared by both aluminium as well as steel cylinder will be,

$500000 - 76944 = 423056 \text{ N}$

Let e_s = strain in steel cylinder

e_a = strain in aluminium cylinder

σ_s = stress produced in steel cylinder

σ_a = stress produced in aluminium cylinder

$E_s = 220 \text{ GN/m}^2$

$E_a = 70 \text{ GN/m}^2$

As both the cylinders are of the same length and are compressed by the same amount

$$e_s = e_a$$
$$\sigma_s/E_s = \sigma_a/E_a$$
or; $\sigma_s = E_s/E_a \cdot \sigma_a$
$= (220 \times 10^9/70 \times 10^9).$
$\sigma_s = (22/7). \sigma_a$ Also $P_s + P_a = P$
or; $\sigma_s A_s + \sigma_a \cdot A_a = 423056$
$(22/7). \sigma_a A_s + \sigma_a A_a = 423056$ ...(i)
$A_s = \pi/4 (0.07^2) = 0.002199 \text{ m}^2$
$A_a = \pi/4 (0.12 - 0.075^2) = 0.003436 \text{ m}^2$

Putting the value of A_s and A_a in equation (i) we get
$\sigma_a = 27.24 \times 10^6 \text{ N/m}^2$
$= 27.24 \text{ MN/m}^2$ ....ANS
and $\sigma_s = 22/7 \times 27.24 = 85.61 \text{ MN/m}^2$

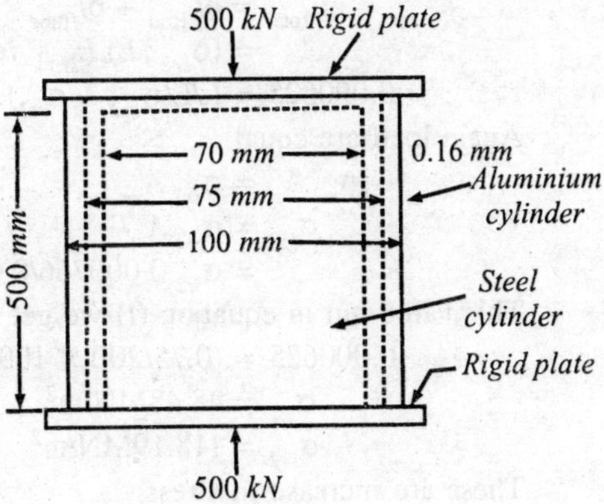

Stress in the aluminium cylinder due to load 76944 N
$= 76944/ \pi/4 (0.12 - 0.075^2)$
$= 23.39 \times 10^9 \text{ N/m}^2 = 22.39 \text{ MN/m}^2$

Total stress in aluminium cylinder
$= 27.24 + 22.39 = 49.63 \text{ MN/m}^2$ ....ANS
and stress in steel cylinder $= 85.61 \text{ MN/m}^2$ ....ANS

Fig 14.40

Ques No-55: A steel rod 20 mm diameter passes centrally through a steel tube 25 mm internal diameter and 40 mm external diameter. The tube is 750 mm long and is closed by rigid washers of negligible thickness which are fastened by nuts threaded on the rod. The nuts are tightened until the compressive load on the tube is 20kN. Calculate the stresses in the tube and the rod. Find the increase in these stresses when one nut is tightened by one quarter of a turn relative to the other. There are 0.4 threads per mm length. Take $E = 200 \text{ GN/m}^2$.

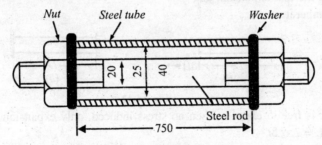

Fig 14.41

Sol.: Area of steel rod = $A_{sr} = \pi/4 (0.02^2) = 0.0003142 \text{ m}^2$
Area of steel tube = $A_{st} = \pi/4 (0.04^2 - 0.025^2) = 0.000766 \text{ m}^2$
Load is equally applied on both steel tube and steel rod.
i.e.; $\sigma_{st1}.A_{st} = \sigma_{sr1}.A_{sr} = 2000$
Since compression in steel tube and tension in steel rod.
$0.0003142 \; \sigma_{sr1} = 0.000766 \; \sigma_{st1} = 2000$
$\sigma_{sr1} = 63.6 \text{ MN/m}^2 (T)$
$\sigma_{st1} = 26.1 \text{ MN/m}^2 (C)$

Now when nut is tightened by one quarter, then let σ_{sr2} and σ_{st2} be the additional stresses produced in the rod and tube respectively.

Distance traveled by nut = ¼(1/0.4) = 0.625mm = 0.000625m = δL_{Total}

$\delta L_{Total} = \delta L_{Rod} + \delta L_{tube}$
$= (\sigma_{sr2}/E).L_{sr} + (\sigma_{st2}/E).L_{st}$; $L_{st} = L_{sr}$
$0.000625 = L/E(\sigma_{sr2} + \sigma_{st2})$ ...(i)

Again load are equal

$\sigma_{st2}.A_{st} = \sigma_{sr2}.A_{sr}$
$\sigma_{sr2} = \sigma_{st2}.A_{st}/A_{sr}$
$= \sigma_{st2}.0.000766/0.0003142 = 2.44\sigma_{st2}$ ...(ii)

This value put in equation (i) we get

$0.000625 = (0.75/200 \times 10^9)(2.44\ \sigma_{st2} + \sigma_{st2})$
$\sigma_{st2} = 48.48 MN/m^2$
$\sigma_{sr2} = 118.19 MN/m^2$

These are increase in stress

Q. 56: Explain the concept of temperature stress.

Sol.: When the temperature of a material changes, there will be corresponding changes in its dimensions. When a member is free to expand or contract due to the rise or fall of temperature, no stress will be induced in the member. But if the natural changes in length due to rise or fall of temperature be prevented, stress will be offered.

If prevented, then stress is induced, which offers strain, which is given by

$e = \delta L/L = \alpha.\delta t$
$\delta L = L.\alpha.\delta t$
$\sigma = e.E = \alpha.\delta t.E$

where
L is the length of the member,
α is coefficient of thermal expansion and
δt is change in temperature.

Fig 14.42

Case-1: If the bar is free to expand; Then no stress induced, only expansion in terms of δL

$\delta L = L.\alpha.\delta t$

Case-2: If the bar is rigidly fixed at both end to prevent expansion, or the grip do not yield or Expansion is prevented; then stress is induced

$\delta L = Ll.\alpha.\delta t$
$e = \delta L/L = \alpha.\delta t$
$\sigma = e.E = \alpha.\delta t.E$

Case-3: Some grip is provided for expansion (Yield), same as in railway track

$\delta L = L.\alpha.\delta t - yield$
$e = \delta L/L = (\alpha.\delta t.L - yield)/L$
$\sigma = e.E = \{(\alpha.\delta t.L - yield)/L\}E$

Q. 57: Two parallel walls 6 m apart, are stayed together by a steel rod 20 mm diameter, passing through metal plates and nuts at each end. The nuts are tightened, when the rod is at a temperature of 100°C. Determine the stress in the rod, when the temperature falls down to 20°C, if

(1) The ends do not yield.

(2) The ends yield by 1mm.

Take $E = 2.0 \times 10^5 N/mm^2$, $\alpha_s = 12 \times 10^{-6}/°C$. *(UPTU QB)*

Sol.: Given data;
$$L = 6m = 6000mm$$
$$d = 20mm$$
$$T_1 = 100°C$$
$$T_2 = 20°C$$
$$\alpha_s = 12 \times 10^{-6} /°C$$
$$E = 2 \times 10^5 N/mm^2$$

(1) When the ends do not yield

Thermal stress = $\sigma = \alpha E \Delta T = 12 \times 10^{-6} \times 2 \times 10^5 \times (100 - 20)$

$\sigma = 192\ N/mm^2$ ....ANS

(2) The ends yield by 1mm.

$\sigma = E(\alpha.L.\Delta T - \delta L)/L$

$= \{(12 \times 10^{-6} \times 80 \times 6000 - 1)/6000\} \times 2 \times 10^5 = 158.67\ N/mm^2$ ....ANS

Ques No-58: A copper rod 15 mm diameter, 0.8 m long is heated through 50°C. What is its expansion when free to expand? Suppose the expansion is prevented by gripping it at both ends, find the stress, its nature and the force applied by the grips, when:

(i) The grips do not yield.

(ii) One grip yields back by 0.5 mm.

Take $\alpha_c = 18.5 \times 10^{-6}/°C$ and $E_C = 1.25 \times 10^5\ N/mm^2$ *(Feb–01)*

Sol.: Given data;
$$L = 0.8m = 800mm$$
$$d = 15mm$$
$$\Delta T = 500C$$
$$\alpha_c = 18.5 \times 10^{-6}\ /°C$$
$$E_C = 1.25 \times 10^5 N/mm^2$$

(1) Expansion when free to expand

$\delta L = \alpha.\Delta T.L = 18.5 \times 10^{-6} \times 50 \times 800 = \mathbf{0.74mm}$ ....ANS

(2) When the ends do not yield

Thermal stress = $\sigma = \alpha E \Delta T = 18.5 \times 10^{-6} \times 1.25 \times 10^5 \times 50$

$\sigma = 115.63\ N/mm^2$ ....ANS

$P = \sigma A = 115.63 \times \pi/4 (15)^2 = \mathbf{20.43 KN(Compressive)}$ ....ANS

(Since gripping is provided so the force is compressive)

(3) The ends yield by 0.5mm.

$\sigma = (\alpha.L.\Delta T - \Delta L).E/L$

$= \{(18.5 \times 10^{-6} \times 800 \times 50 - 0.5)/800\} \times 1.25 \times 10^5 = \mathbf{37.5\ N/mm^2}$ ....ANS

$P = \sigma_A = 37.5 \times \pi/4(15)^2 = \mathbf{6.63 KN(Compressive)}$ ....ANS

Q. 59: A steam pipe is 30 m long at a temperature of 15°C. Steam at 180°C is passed through the pipe. Calculate the increase in length when the pipe is free to expand. What stress is induced in the material if the expansion is prevented ?

$(E = 200$ GN/m^2, $\alpha = 0.000012$ per °C) *(May-03 (C.O.))*

Sol.: Given data;
$$L = 30\text{m} = 30000\text{mm}$$
$$T_1 = 15°C$$
$$T_2 = 180°C$$
$$\alpha_s = 12 \times 10\text{-}6 \text{ /0C}$$
$$E = 200 \times \text{GN/m}^2 = 2 \times 10^5 \text{N/mm}^2$$

(1) When free to expand
$$\delta L = \Delta.L.\Delta T = 12 \times 10^{-6} \times 30000 \times (180 - 15)$$
$$\delta L = 59.4 \text{ mm} \quad\quad\quad\quad\text{.......ANS}$$

(2) When expansion is prevented
$$\sigma = E\alpha.\Delta T$$
$$= 2 \times 10^5 \times 12 \times 10^{-6} \times 165 = 396 \text{ N/mm}^2 \quad\quad\text{.......ANS}$$

Q. 60: A steel rod 2.5 m long is secured between two walls. If the load on the rod is zero at 20°C, compute the stress when the temperature drops to -20°C. The cross-sectional area of the rod is 1200 mm^2, $\alpha = 11.7$ μm/(m°C), and E = 200GPa, assuming

(a) that the walls are rigid and

(b) that the walls spring together a total distance of 0.5 mm as the temperature drops.

Sol.: (a) Let the rod be disconnected from the right wall. Temperature deformations can then freely occur. A temperature drop causes the contraction represented by δ_T in Fig. 14.43. To reattach the rod to the wall will evidently require a pull P to produce the load deformation δ_P so that $\delta_T = \delta_P$. Thus

$$\alpha \Delta t L = \frac{PL}{AE} = \frac{\sigma L}{E}$$

$$\sigma = E.\alpha.\Delta t = 200 \times 10^9 \times 11.7 \times 10^{-6} \times 40 = 93.6 \times 106 \text{ N/mm}^2$$
$$= 93.6 \text{ MPa} \quad\quad\quad\quad\text{.......ANS}$$

It may be noted that the stress is independent of the length of the rod.

(b) When the walls spring together, the free temperature contraction is equal to the sum of the load deformation and the yield of the walls.

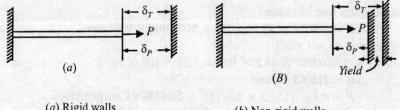

(a) Rigid walls (b) Non-rigid walls

Fig 14.43

Hence $\quad \delta_T = \delta_P + \text{yield}$

Now,
$$\alpha.L.\Delta t = \sigma.L/E + \text{yield}$$

$$11.7 \times 10^{-6} \times 2.5 \times 40 = (\sigma \times 2.5)/(200 \times 10^9) + 0.5 \times 10^{-3}$$
we get; $\sigma = 53.6$ MPa   ....ANS

Thus the yield of the walls reduces the stress considerably.

Q. 61: A circular bar of length 400 mm and tapering uniformly from 50 mm to 25 mm diameter is held between rigid supports at the ends. Calculate the maximum and minimum stress developed in the bar when the temperature is raised by 30°C. Take $E = 2 \times 10^5$ N/mm^2 and $\alpha = 1.2 \times 10^{-5}$ per°C.

Sol.: Increase in length due to temperature rise, $= L.\alpha.\Delta t = 400 \times (1.2 \times 10^{-5}) \times 30 = 0.144$ mm)

This elongation sets up a compressive reaction P at the supports and the corresponding shortening in length is given by;
$$= 4P.L/4.d_1.d_2.E = (4P \times 400)/(4 \times 50 \times 25 \times 2 \times 10^5) = 1.6 \times 10^{-6} P$$

From compatibility condition; $1.6 \times 10^{-6} P = 0.144$; $P = 0.09 \times 10^6$ N

Maximum stress $s_{max} = P/A_{min} = 0.09 \times 10^6/\{\pi/4 \, (25)^2\} = 183.44$ N/mm^2 ....ANS

Minimum stress $s_{min} = P/A_{max} = 0.09 \times 10^6/\{\pi/4 \, (50)^2\} = 45.86$ N/mm^2 ....ANS

Q. 62: Explain the effect of temperature change in a composite bar. What is compatibility condition?

Sol.: Consider temperature rise of a composite bar consisting of two members; one of steel and other of brass rigidly fastened to each other. If allowed to expand freely;

expansion of brass bar: $AB = L\alpha_b \Delta t$

expansion of steel bar: $AC = L\alpha_s \Delta t$

Since coefficient of thermal expansion of brass is greater than that of steel, expansion of brass will be more. But the bars are fastened together and accordingly both will expand to the same final position represented by DD with net expansion of composite system AD equal to dl. To attain this position, brass bar is pushed back and the steel bar is pulled.

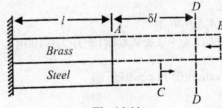

Fig 14.44

Obviously compressive stress will be induced in brass bar and tensile stress will be developed in steel bar. Under equilibrium state;

compressive force in brass = tensile force is steel
$$\sigma_c A_c = \sigma_s A_s$$

Corresponding to brass rod:

Reduction in elongation,
$$DB = AB - AD = L\alpha_b \sigma_t - \delta L$$

Strain $e_b = (L\alpha_b \Delta t - \delta L)/L = \alpha b \Delta t - e;$

Where $e = \delta L/L$, is the actual strain of the composite system

Corresponding to steel rod:

extra elongation, $CD = AD - AC = \delta L - L\alpha_s \Delta t$

Strain $e_s = (\delta L - L\alpha_s \Delta t)/L = e - \alpha_b \Delta t$

Adding e_b and e_s, we get
$$e_b + e_s = (\delta_b - \delta_s) \Delta t,$$
It is also called as compatibility condition.

It may be pointed out that the nature of the stresses in the bars will get reversed if there is reduction in the temperature of the composite system.

Q. 63: A steel tube with 2.4 cm external diameter and 1.8 cm internal diameter encloses a copper rod 1.5 cm diameter to which it is rigidly joined at each end. If at a temperature of 10°C, there is no longitudinal stress, calculate the stresses in the rod and tube when the temperature is raised to 200°C. $E_S = 210{,}000$ N/mm², $\alpha_s = 11 \times 10^{-6}$/°C, $E_C = 100{,}000$ N/mm², $\alpha_C = 18 \times 10^{-6}$/°C   [U.P.T.U. Feb–01]

Sol.: Given that:
$$E_S = 210{,}000 \text{ N/mm}^2,$$
$$E_C = 100{,}000 \text{ N/mm}^2$$
$$\alpha_s = 11 \times 10^{-6}/°C,$$
$$\alpha_C = 18 \times 10^{-6}/°C$$
$$\Delta T = 200°C$$

Apply compatibility condition:
$$e_C + e_S = (\alpha_C - \alpha_S) \Delta t,$$
$$e_C + e_S = (18 - 11) \times 10^{-6} \times 200$$
$$e_C + e_S = 0.0014 \qquad \qquad \ldots(i)$$

From the equilibrium condition;
Compressive force on copper = Tensile force on steel
$$P_B = P_C$$
$$e_C . A_C . E_C = e_S . A_S . E_S$$
$$e_C = e_S[(A_S/A_C)(E_S/E_C)]$$
$$= e_S[\{(\pi/4)(2.4^2 - 1.8^2)/(\pi/4)(1.5^2)\}(210/100)]$$
$$e_C = 2.35 e_S \qquad \qquad \ldots(ii)$$

Substituting the value of equation(ii) in equation (i)
$$2.35 e_S + e_S = 0.0014$$
$$e_S = 0.000418 \qquad \qquad \ldots(iii)$$
$$e_C = 0.000982 \qquad \qquad \ldots(iv)$$

Stress in steel tube $\delta_S = e_S . E_S$
$$= 0.000418 \times 210{,}000$$
$$\sigma_S = 87.7 \text{ N/mm}^2 \qquad \qquad \text{.......ANS}$$

Stress in copper tube $\sigma_C = e_C . E_C$
$$= 0.000982 \times 100{,}000$$
$$\sigma_C = 98.2 \text{ N/mm}^2 \qquad \qquad \text{.......ANS}$$

Q. 64: A steel bar is placed between two copper bars each having the same area and length as the steel bar at 15°C. At this stage they are rigidly connected together at both ends. When the temperature is raised to 315°C the length of the bar increases by 1.50 mm. Determine the original length and the final stresses in the bars. Take, $E_S = 2.1 \times 10^5$ MPa, $E_C = 1 \times 10^5$ N/mm², $\alpha_S = 0.000012$ per °C, $\alpha_{C_r} = 0.0000175$ per °C.   (*UPTU QUESTION BANK*)

Sol.: When the composite system is in equilibrium,
Tensile force in steel bar = compressive force in two copper bars
$$\sigma_s A_s = 2\sigma_c A_C; \quad \sigma_s = 2\sigma_c;$$
Since $\quad A_s = A_C$

Using the relation
$$e_C - e_S = (\alpha_C - \alpha_S)\Delta t$$
$$\sigma_C/E_C + \sigma_S/E_S = (1.75 \times 10^{-5} - 1.2 \times 10^{-5}) \times (250 - 20)$$
$$\sigma_C/E_C + \sigma_S/E_S = 0.001265$$

Substituting the value of
$$\sigma_s = 2\sigma_C; \text{ we get}$$
$$\sigma_C/1 \times 10^5 + 2\sigma_C/2 \times 10^5 = 0.001265$$
$$\sigma_C = 63.25 \text{ N/mm}^2 \quad \text{......ANS}$$
$$\sigma_S = 126.5 \text{ N/mm}^2 \quad \text{......ANS}$$

Further, $\quad e_C = \alpha_C.\Delta t - e;$
where $e = \delta L/L$, actual strain of the composite system
$$e = \delta L/L = \alpha_C.\Delta t - e_C = \alpha_C.\Delta_t - \sigma_C/E_C$$
$$= 1.75 \times 10^{-5} (250 - 20) - 63.25/1 \times 10^5 = 0.0033925$$

Original length of bar, $L = \delta L/0.0033925 = 1.25/0.0033925 = 368.46$ mm = **0.3685m** ......ANS

Q.65: A flat bar of aluminium alloy 25 mm wide and 5 mm thick is placed between two steel bars each 25 mm wide and 10 mm thick to form a composite bar 25 mm × 25 mm as shown in Fig. 14.45. The three bars are fastened together at their ends when the temperature is 15°C. Find the stress in each of the material when the temperature of the whole assembly is raised to 55°C. If at the new temperature a compressive load of 30kN is applied to the composite bar what are the final stresses in steel and alloy ?

Take $E_S = 200$ GN/m², $E_{al} = 200/3$ GN/m²

$\alpha_s = 1.2 \times 10^{-5}$ per°C, $\alpha_{al} = 2.3 \times 10^{-5}$ per°C.

Sol.: Refer Fig. 14.45.

Area of aluminium,
$$A_{al} = 25 \times 5 = 125 \text{ mm}^2 = 125 \times 10^{-6} \text{ m}^2$$

Area of steel,
$$A_S = 2 \times 25 \times 10 = 500 \text{ mm}^2 \text{ or } 500 \times 10^{-6} \text{ m}^2$$

(i) **Stresses due to rise of temperature:**

If the two members had been free to expand,

Free expansion of steel = $\alpha_s.\Delta t.L_s$

Free expansion of aluminium = $\sigma_{al}.\Delta t.L_{al}$

But since the members are fastened to each other at the ends, final expansion of each member would be the same.

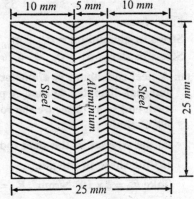

Fig 14.45

Let this expansion be δ. The free expansion of aluminium is greater than δ while the free expansion of steel is less than δ. Hence the steel is subjected to tensile stress while aluminium is subjected to compressive stress.

Let σS. and σ_{al} be the stresses in steel and aluminium respectively.

The whole system will be in equilibrium when

Total tension (pull) in steel = total compression (push) in aluminium

$$\sigma_S \cdot A_S = \sigma_{al} \cdot A_{al}$$
or $\quad \sigma_S \times 500 \times 10^{-6} = \sigma_{al} \times 125 \times 10^{-6}$

$$\sigma_s = \frac{\sigma_{al}}{4} = 0.25\ \sigma_{al}$$

Final increase in length of steel = final increase in length of aluminium

$$\sigma_s \cdot \Delta t \cdot L_s + \sigma_s \cdot L_S/E_S = \alpha_{al} \cdot \Delta t \cdot La_1 - s_{al} \cdot L_{al}/E_{al}$$
$$\alpha_s \cdot \Delta t + \sigma_S/E_S = \alpha_{al} \cdot \Delta t - \sigma_{al}/E_{al}\ ;\quad \text{Since } L_S = L_{AL}$$

But $\quad \Delta t = 55 - 15 = 40°C$

$$1.2 \times 10^{-5} \times 40 + 0.25\sigma_{al}/(200 \times 10^9) = 2.3 \times 10^{-5} \times 40 - \sigma_{al}/(200/3) \times 10^9$$
or; $\quad 1.2 \times 10^{-5} \times 40 \times 200 \times 10^9 + 0.25\sigma_{al} = 2.3 \times 10^{-5} \times 40 \times 200 \times 10^9 - 3\sigma al$

$$\sigma_{al} = 27.07\ \text{MN/m}^2\ \text{(Compressive)} \qquad \text{.......ANS}$$

(ii) Stresses due to external compressive load 30kN :

Let σ_{S1} and sal1 be the stresses due to external loading in steel and aluminium respectively.
Strain in steel = e_S
Strain in aluminium = e_{al}
$$\sigma_{S1}/E_S = \sigma_{al1}/E_{Al}$$
$$\sigma_{S1} = E_S \cdot \sigma_{al1}/E_{Al}$$
$$= (200 \cdot \sigma_{al1})/(200/3) = 3 \cdot \sigma_{al1}$$

But, load on steel + load on aluminium = total load
$$\sigma_{S1} \cdot A_S + \sigma_{al1} \cdot A_{al} = 30 \times 1000$$
or; $3 \cdot \sigma_{al1} \times 500 \times 10^{-6} + \sigma_{al1} \times 125 \times 10^{-6} = 3000$

$$\sigma_{al1} = 18.46\ \text{MN/m}^2\ \text{(compressive)}$$
$$\sigma S1 = 3 \cdot \sigma_{al1} = 55.38\ \text{MN/m}^2\ \text{(compressive)}$$

Final stress:
Stress in aluminium = $\sigma_{al} + \sigma_{al1}$ = 27.07 + 18.46 = 45.53 MN/m² (Compressive) ....ANS
Stress in steel= $\sigma_S + \sigma_{S1}$ = –6.76 + 55.38 = 48.62 MN/m² (Compressive) ....ANS

Q. 66: A steel rod 40mm in diameter is enclosed by a copper tube of external diameter 50mm and internal diameter 40 mm. A pin 25mm in diameter is fitted transversely to the assembly at each end so as to secure the rod and the tube. If the temperature of the assembly is raised by 60°C, find

(i) the stresses in the steel rod and the copper tube, and

(ii) the shear in the pin.

Take $E_S = 2 \times 10^5$ N/mm², $E_C = 1 \times 10^5$ N/mm², $\alpha_s = 1.2 \times 10^{-5}$ per °C and ac =1.6 × 10⁻⁵ per°C.

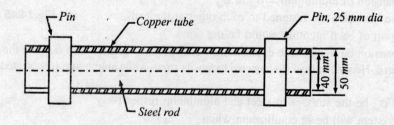

Fig 14.46

Sol.: Area of steel rod; $A_S = \pi/4 \times 40^2 = 400\pi$ mm^2
Area of copper tube $A_C = \pi/4 \times (50^2 - 40^2) = 225\pi$ mm^2
Let σ_S and σ_C be the stresses produced in steel and copper respectively due to change in temperature. It may be noted that steel is in tension and copper is in compression. For equilibrium of the assembly,
Total tension (pull) in steel = total compression (push) in copper

$$\sigma_S . A_S = \sigma_C . A_C$$

or $\quad\quad \sigma_S \times 400\pi = \sigma_C \times 225\pi$

$$\sigma S = (9/16) \sigma_C$$

Actual expansion of steel = Actual compression of copper

$$\alpha_s . \Delta t . L_s + \sigma_S . L_S / E_S = \alpha_C . \Delta t . L_C - \sigma_C . L_C / E_C$$
$$\alpha s . \Delta t + (9/16)\sigma_C / E_S = \alpha C . \Delta t - \sigma_C / E_C ;$$

Since $\quad\quad L_S = L_{AL}$
and $\quad\quad \sigma S = (9/16)\sigma_C$
But $\quad\quad \Delta t = 600 C$
$\quad 1.2 \times 10^{-5} \times 60 + (9/16)\sigma_C/(2 \times 10^5) = 1.6 \times 10^{-5} \times 60 - \sigma_C/1 \times 10^5$
$\quad 72 + (9/32) \sigma_C = 96 - \sigma s_C$
$\quad \sigma_C = 18.286$ N/m^2 **(Compressive)** ....**ANS**
$\quad \sigma_S = 9/16 \times 18.286 = 10.286$ N/m^2 **(tensile)** ....**ANS**

Q. 67: The composite bar consisting of steel and aluminium components as shown in Fig. 14.47 is connected to two grips at the ends at a temperature of 60°C. Find the stresses in the two rods when the temperature fall to 20°C

(i) if the ends do not yield,

(ii) if the ends yield by 0.25 mm.

Take $E_S = 2 \times 10^5$ and $E_a = 0.7 \times 10^5$ N/mm^2, $\alpha_s = 1.17 \times 10^{-5}$ and $\alpha_a = 2.34 \times 10^{-5}$ per°C. The areas of steel and aluminium bars are 250 mm^2 and 375 mm^2 respectively.

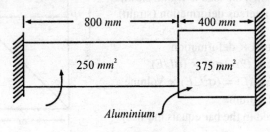

Fig 14.47

Sol.: $A_a/A_S = 375/250 = 1.5$

Free contraction of the composite bar

$= \alpha_s . \Delta t . L_s + \alpha a . \Delta t . La$
$= 1.17 \times 10^{-5} \times 40 \times 800 + 2.34 \times 10^{-5} \times 40 \times 400$
$= 0.7488$ mm

When the contraction is prevented, partially tensile stresses are induced in the rods.

$\quad A_S . \sigma_S = A_a . \sigma_a$
$\quad \sigma_S = (375/250).\sigma_a$
$\quad\quad = 1.5.\sigma_a$

(1) When the rod do not yield, contraction prevented in steel and aluminimu = 0.7488 mm.

$$\sigma_s/E_s \cdot L_s + \sigma_a/E_a \cdot L_a = 0.7488$$
$$(1.5.\sigma_a/2 \times 10^{-5}) \times 800 + (\sigma_a/0.70 \times 10^{-5}) \times 400 = 0.7488$$
$$11.7143 \times 10^{-3}\, \sigma_a = 0.7488$$
$$\sigma_a = 63.92 \text{ N/mm}^2 \quad \text{........ANS}$$
$$\sigma_s = 1.5 \times 63.92 = 95.88 \text{ N/mm}^2 \quad \text{........ANS}$$

(ii) When the ends yield by 0.25 mm
Contraction prevented = 0.7488 – 0.25 = 0.4988mm
$$11.7142 \times 10^{-3}.\sigma_a = 0.4988$$
$$\sigma_a = 42.58 \text{ N/mm}^2 \quad \text{........ANS}$$
$$\sigma_s = 1.5 \times 42.58 = 63.87 \text{ N/mm}^2 \quad \text{........ANS}$$

Q. 68: Explain strain energy and Resilience. *(Dec–01; May–05(C.O.), May–02)*

Sol.: When an external force acts on an elastic material and deforms it, internal resistance is developed in the material due to cohesion between the molecules comprising the material. The internal resistance does some work which is stored within the material as energy and this strain energy within elastic limit is known as resilience.

What-ever energy is absorbed during loading, same energy is recovered during unloading and the material springs back to its original dimension. Machine members like helical, spiral and leaf springs possess this property of resilience.

A body may be subjected to following types of loads:
(1) Gradually applied load
(2) Suddenly applied load
(3) Falling or impact loads

(A) Gradually Applied Loads:

Load applied to a bar starts from zero and increases linearly until the bar is fully loaded. When the load is within elastic limit, the plot of load (stress) versus deformation (strain) is linear (Fig. 14.48).

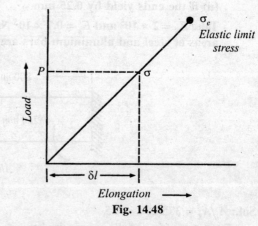

Fig. 14.48

Work done = average load × deformation
$$= (1/2)P.\delta L = (½) (\sigma A) \times (\sigma L/E)$$
$$= (½) (\sigma^2/E)(AL) = (\sigma^2/2E) \times \text{Volume}$$

Work done = $(\sigma^2/2E) \times$ Volume

The strain energy U stored in the bar equals the work done and therefore,
$$U = (\sigma^2/2E) \times \text{volume}$$
$$\sigma = P/A$$

The maximum strain energy absorbed by a body upto its elastic limit is termed as Proof Resilience and this proof resilience per unit volume is called Modulus of Resilience.

Proof resilience = $(\sigma_e^2/2E) \times$ volume
where se is the stress at elastic limit.
Modulus of resilience = $\sigma_e^2/2E$

(B) Suddenly Applied Load:
Load is applied suddenly and this remains constant throughout the process of deformation. Accordingly the plot between load and elongation will be parallel to x-axis.

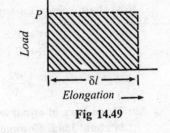

Fig 14.49

Work done = area of shaded portion = $P.\delta L = P(\sigma_{su}.L/E)$
The subscript su refers to suddenly applied load.
The work done equals the strain energy given by
$$(\sigma_{su}^2/2E)(AL)$$
$$(\sigma_{su}^2/2E)(AL) = p(\sigma_{su}.L/E)$$
or, $\sigma_{su} = 2P/A = 2 \times$ stress due to gradually applied load

Thus the instantaneous stress induced in a member due to suddenly applied load is twice that when the load is applied gradually.

(C) Impact Loads
Impact loading occurs when a weight is dropped on a member from some height. The kinetic energy of the falling weight is utilized in deforming the member. Refer Fig. 14.50, a rod of cross-sectional area A and length l is fixed at one end and has a collar at the outer end. A weight W, slides freely on the rod and is dropped on the collar through height. The falling weight causes impact load and that leads to extension δl_i, and tensile stress σL_i, (subscript i refers to impact load).

Now
external work done = energy stored in the rod
$$W(h + \delta L_i) = (\sigma_i^2/2E) \times \text{volume} = (\sigma_i^2/2E) \times AL$$
$$W[h + (\sigma_i.L)/L] = (\sigma_i^2/2E) \times AL$$
Or, $(AL/2E)\sigma_i^2 - (WL/E)\sigma_i - Wh = 0$
Solution of this quadratic equation gives

$$\frac{\dfrac{WL}{E} \pm \sqrt{\left(\dfrac{WL}{E}\right)^2 - 4 \times \left(\dfrac{AL}{2E}\right) \times (-Wh)}}{2 \times \dfrac{AL}{2E}} = \frac{W}{A} \pm \sqrt{\frac{W^2}{A^2} + \frac{2WhE}{AL}}$$

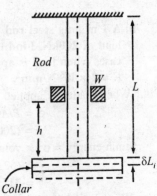

Fig 14.50

Negative sign is inadmissible as the stress cannot be compressive when the bar gets elongated.

$$s_i = \frac{W}{A} + \frac{W}{A}\sqrt{1 + \frac{2h\,AE}{WL}} = \frac{W}{A}\left[1 + \sqrt{1 + \frac{2h\,AE}{WL}}\right]$$

Following relations are worth noting:
(i) If δl_i is neglected as compared to h, (if $\delta L \times 1000 <<< h$), then
$$Wh = \frac{\sigma_i^2}{2E} \cdot AL \text{ and } \sigma_i = \sqrt{\frac{2WhE}{Al}}$$

(ii) If $h = 0$, then
$$W = \frac{\sigma \cdot L}{E} = \frac{\sigma_i^2}{2E} \cdot AL \text{ and } \sigma_i = \frac{2W}{A}$$

Note: Strain energy becomes smaller and smaller as cross sectional area of the bar is increased over more and more of its length.

Now strain energy for any types of load = $U = \frac{1}{2}\frac{\sigma^2}{E} \cdot AL$

where $\sigma = P/A$ for gradually applied load

$= 2 P/A$ for suddenly applied load

$= \frac{W}{A}\left[1 + \sqrt{1 + \frac{2hAE}{W \cdot L}}\right]$ for impact load

Q. 69: Three bars of equal length and having, cross-sectional areas in ratio 1:2:4, are all subjected to equal load. Compare their strain energy. *(Dec-03)*

Sol.: Let A, $2A$, $4A$ be the areas.

Loads are equal = P

Each have equal length = L

Now $\sigma_1 = P/A$; $\sigma_2 = P/2A$; $\sigma_3 = P/4A$;

Since strain energy $= U = \sigma_2 \cdot \text{Vol}/2E$

Strain energy in first bar $= U_1 = \sigma_1^2 \cdot \text{Vol}/2E = (1/2E)(P/A)^2 \cdot AL = P^2 \cdot L/2AE$

Strain energy in second bar $= U_2 = \sigma_2^2 \cdot \text{Vol}/2E = (1/2E)(P/2A)^2 \cdot 2AL = P^2 \cdot L/4AE$

Strain energy in third bar $= U_3 = \sigma_3^2 \cdot \text{Vol}/2E = (1/2E)(P/4A)^2 \cdot 4AL = P^2 \cdot L/8AE$

Now;

$U_1 : U_2 : U_3 = 1 : \frac{1}{2} : \frac{1}{4}$

Q. 70: A 1 m long steel rod of rectangular section 80 mm × 40 mm is subjected to an axial tensile load of 200kN. Find the strain energy and maximum stress produced in it for the following cases when load is applied gradually and when load falls through a height of 100 mm. Take $E = 2 \times 10^5$ N/mm². *(May-2005)*

Sol.: When gradually applied load

$\sigma = P/A$

$= (200 \times 1000)/(80 \times 40) = 62.5$ N/mm²

Strain energy $= \sigma^2 \times \text{volume} / 2E$

$= (62.5^2 \times 80 \times 40 \times 1000) / (2 \times 2 \times 10^5)$

$= 31250$ N-mm ....ANS

When load falls through a height of 100mm

$\sigma = \frac{W}{A} + \sqrt{\left(\frac{W}{A}\right)^2 + \frac{2EWh}{AL}}$

$= \frac{200 \times 1000}{80 \times 40} + \sqrt{\left(\frac{200 \times 1000}{80 \times 40}\right)^2 + \frac{2 \times 2 \times 10^5 \times 100 \times 200 \times 1000}{80 \times 40 \times 1000}}$

$= 62.5 + \sqrt{(625)^2 + 25 \times 10^5}$

$= 62.5 + \sqrt{3906.25 + 2500000} = 62.5 + 1582.37$

$= 1644.87$ N/mm² ....ANS

$U = \frac{1}{2} \times \frac{1644 \cdot 87^2}{2 \times 10^5} \times 80 \times 40 \times 1000$

$U = 21644778.5$ Nmm ....ANS

Q. 71: A bar of 1.2 cm diameter gets stretched by 0.3 cm under a steady load of 8KN. What stress would be produced in the bar by a weight of 0.8KN. Which falls through 8 cm before commencing the stretching of the rod, which is initially unstressed. Take $E = 200 GN/m^2$.
(UPTUQB)

Sol.: Cross-sectional area of the bar = $A = \pi/4.d^2 = \pi/4(1.2/1000)^2 = 0.0001131 \ m^2$
Steady load = 8 kN
Elongation under steady load, $\delta L = 0.3 \ cm = 0.003 \ m$
Falling load = 0.8 kN
Distance through which the weight falls, $h = 8 \ cm = 0.08 \ m$
Modulus of elasticity, $E = 200 \ GN/m^2$
Instantaneous stress produced due to the falling load, $\sigma_i = ?$
In order to first find length of the bar, using the following relation, we have
$\delta L = WL/AE$ or; $L = \delta L.A.E/W$

$$L = \frac{0.003 \times 0.0001131 \times 200 \times 10^9}{8 \times 1000} = 8.48 \ m$$

Now to calculate instantaneous stress si due to falling load 0.8 kN using the following relation, we have

$$\sigma_i = \frac{W}{A}\left[1 + \sqrt{1 + \frac{2hAE}{WL}}\right] = \frac{0.8 \times 1000}{0.0001131} \times \left[1 + \sqrt{\frac{2 \times 0.08 \times 0.0001131 \times 200 \times 10^9}{0.8 \times 1000 \times 8.84}}\right]$$

$= 7073386.3 \ (1 + 23.119)$
$\sigma_i = 170.6 \times 10^6 \ N/m^2$ or $170.6 \ MN/m^2$. ....ANS

Q. 72: A bar 3m long and 5 cm diameter hands vertically and has a collar securely attached at the lower end. Find the maximum stress induced when;

(i) a weight 2.5KN falls from 12 cm on the collar

(ii) a weight of 25KN falls from 1 cm on the collar Take $E = 2.0 \times 10^5 \ N/mm^2$. (UPTUQB)

Sol.: Cross-sectional area of the bar = $A = \pi/4.d^2 = \pi/4(50)^2 = 1962.5 \ mm^2$

(i) Instantaneous elongation of bar ;
$\delta L = WL/AE = (2500 \times 3000)/(1962.5 \times 2 \times 10^5) = 0.01911 mm$
This elongation is very small as compared to 120mm height of fall and as such can be neglected.
Accordingly stress induced in the bar can be worked out by using the relation. $\sigma_1 = \sqrt{\frac{2WhE}{A \cdot L}}$

$$= \sqrt{\frac{2 \times 2500 \times 120 \times 2 \times 10^5}{1962.5 \times 3000}}$$

$= 142.77 N/mm^2$ ....ANS

(ii) Instantaneous elongation
$\delta L = WL/AE = (25000 \times 3000)/(1962.5 \times 2 \times 10^5) = 0.1911 mm$
This elongation is comparable to 10 mm height of fall. Further the falling weight is large and hence extension of the bar cannot be neglected. Accordingly stress induced in the bar is worked out from the relation

$$\sigma_i = \frac{W}{A}\left[1 + \sqrt{1 + \frac{2hAE}{WL}}\right]$$

$$= \frac{25000}{1962.5} \times \left[1 + \sqrt{1 + \frac{2 \times 1962.5 \times 2 \times 10^5 \times 10}{25000 \times 3000}}\right]$$

$$= 143.68 \text{ N/mm}^2 \qquad \text{........ANS}$$

Q. 73: A load of 100 N falls by gravity a vertical distance of 300 cm When it is suddenly stopped by a collar at the end of a vertical rod of length 6m and diameter 2 cm. The top of the bar is rigidly fixed to a ceiling. calculate the maximum stress and the strain induced in the bar. Table $E = 1.96 \times 10^7 \text{ N/cm}^2$. *(UPTUQB)*

Sol.: Given data:

Weight of the object = 100N
Height of fall = 300 cm
Length of vertical rod = 6m = 600cm
Diameter of rod = 2cm
$E = 1.96 \times 10^7 \text{ N/cm}^2$
Cross-sectional area of the rod
$= A = \pi/4 . d^2 = \pi/4 . (2)^2 = 3.142 \text{ cm}^2$

Maximum stress induced in the bar;

$$\sigma_i = \frac{W}{A}\left[1 + \sqrt{1 + \frac{2hAE}{WL}}\right]$$

$$= \frac{100}{3.142} \times \left[1 + \sqrt{1 + \frac{2 \times 300 \times 3.142 \times 1.96 \times 10^7}{100 \times 600}}\right]$$

$$= 25007.945 \text{ N/cm}^2 \qquad \text{........ANS}$$

Strain induced in the bar due to impact e_i;
We know that

$$ei = \sigma_i/E = 25007.945/(1.96 \times 10^7) = 0.001276$$
$$ei = 0.001276 \qquad \text{........ANS}$$

Q. 74: A steel specimen 1.5cm² in cross section stretches 0.05mm over 5 cm gauge length under an axial load of 30 KN. Calculate the strain energy stored in the specimen at this point. If the load at the elastic limit for specimen is 50KN, Calculate the elongation at the elastic limit.

Sol.: Cross sectional area of specimen $A = 1.5 \text{ cm}^2 = 1.5 \times 10^{-4} \text{ m}^2$
Increase in length over 5cm gauge length $\delta L = 0.05\text{mm} = 0.05 \times 10^{-3}$ m
Axial load W = 30KN
Load at elastic limit = 50KN
Strain energy stored in the specimen

$$U = \sigma^2 A.L/2E = \frac{1}{2}. W. \delta L = \frac{1}{2} \times (30 \times 1000) \times 50 \times 10^{-3}$$
$$U = 0.75 \text{ Nm or } J \qquad \text{........ANS}$$

Also $\quad E = W.L/A.\delta L$
$$= \{(30 \times 1000) \times (5/100)\}/\{(1.5 \times 10^{-4}) \times (0.05 \times 10^{-3})\}$$
$$= 200 \times 10^9 = 200 \text{ GN/m}^2$$

Elongation at elastic limit, δL
$$\delta L = W.L/A.E = \{(50 \times 1000) \times (5/100)\}/\{(1.5 \times 10^{-4}) \times (200 \times 10^9)\}$$
$$\delta L = 0.0000833 \text{ m} = 0.0833 \text{ mm} \qquad \text{........ANS}$$

Q. 75: A wagon weighing 35KN is attached to a wire rope and moving down an incline plane at speed of 3.6Km/hr when the rope jams and the wagon is suddenly brought to rest. If the length of the rope is 60 meters at the time of sudden stoppage, calculate the maximum instantaneous stress and maximum instantaneous elongation produced. Diameter of rope = 30mm. $E = 200 GN/m^2$.

Sol.: Weight of the wagon $W = 35 KN$
Speed of the wagon, $v = 3.6$ km/hr = 1m/sec
Diameter of the rope, $d = 30$mm = 0.03m
Length of the rope at the time of sudden stoppage, $L = 60$m
Maximum instantaneous stress σ_i;
The kinetic energy of the wagon = $½.mv^2$ = Strain energy
$$= ½ . (35 \times 1000 /9.81) \times 1^2 = 1783 \text{ Nm or } J \qquad \ldots(i)$$
This energy is to be absorbed by the rope at a stress σ_i
Now, strain energy stored = $\sigma_i^2 . A . L / 2E$
$$= \{\sigma_i^2 \times \pi/4 \times (0.03^2) \times 60\}/(2 \times 200 \times 10^9) = 0.0106 \, \sigma_i^2/10^{11} \qquad \ldots(ii)$$
Equating equation (i) and (ii)
$$\sigma_i = 129.69 \times 10^6 \text{ N/m}^2 = 129.69 \text{ MN/m}^2 \qquad \text{......ANS}$$
Maximum instantaneous elongation of the rope, δL
Using the relation,
$$\delta L = \sigma_i . L / E = (129.69 \times 10^6 \times 60)/(200 \times 10^9)$$
$$\delta L = 389.07 \times 10^{-4} \text{ m} = 38.9 \text{ mm} \qquad \text{......ANS}$$

Q. 76: A steel wire 2.5 mm diameter is firmly held in clamp from which it hangs vertically. An anvil the weight of which may be neglected, is secured to the wire 1.8 m below clamp. The wire is to be tested allowing a weight bored to slide over the wire to drop freely from 1 m above the anvil. Calculate the weight required to stress the wire to 1000MN/m² assuming the wire to be elastic upto this stress. Take: $E = 210$ GN/m².

Sol.: Diameter of the steel wire, $d = 2.5$mm $= 2.5 \times 10^{-3}$ m
Height of fall, $h = 1$m
Length of the wire, $L = 1.8$m
Instantaneous stress produced $\sigma_i = 1000$ MN/m²
$E = 210$ GN/m²
Weight required to stress the wire, W;
Instantaneous extension,
$$\delta L = \sigma_i . L / E = (1000 \times 10^6 \times 1.8)/(210 \times 10^9) = 0.00857 \text{ m}$$
Equating the loss of potential energy to strain energy stored by the wire, we have
$$W(h + \delta L) = \sigma_i^2 . A . L / 2E$$
$$W (1 + 0.00857) = \{(1000 \times 10^6)^2 \times \pi/4 (2.5 \times 10^{-3})^2 \times 1.8\}/\{2 \times 210 \times 10^9\}$$
$$W = 20.85 N \qquad \text{......ANS}$$

Q. 77: Explain the concept of Complementary shear stress. (May–05(C.O.), Dec-05)

Sol.: It states that a set of shear stresses across a plain is always accompanied by a set of balancing shear stresses across the plane and normal to it.
Shear force on face $AB = \tau . AB$
Shear force on face $CD = \tau . CD$ \qquad ...(i)
These parallel and equal forces form a couple.
The moment of couple = $\tau . AB . AD$ or $\tau . CD . AD$ \qquad ...(ii)

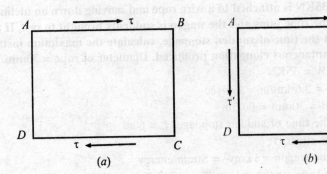

Fig. 14.51

For equilibrium they must be a restoring couple.
Shear force on face AD or $BC = \tau^1.AD$ or $\tau^1.BC$ ...(iii)
They also form a couple as
$$= \tau^1.AD.AB \text{ or } \tau^1.BC.AB \qquad ...(iv)$$
The moment of two couple must be equal;
$$\tau.AB.CD = \tau^1.AD.AB \text{ or } \tau = \tau^1$$
τ^1 is called complementary shear stress

Q. 78: Explain longitudinal strain and lateral strain. What is the relation between longitudinal and lateral strain.

Sol.: Longitudinal strain is the longitudinal deformation expressed as a dimensionless constant and is defined as the ratio of change in length to the initial length.
$$e = \delta L/L$$
Lateral strain is the lateral deformation expressed as a dimensionless constant and is defined as the ratio of change in lateral dimension to the initial lateral dimension

Lateral strain = Change in lateral dimension/Original lateral dimension
= Change in diameter/Original diameter; For circular bar
= Change in width or depth/Original width or depth; For rectangular bar

LATERAL STRAIN = POISSON'S RATIO X LONGITUDINAL STRAIN

Q. 79: What do you meant by shear stress and shear strain?

Sol.: Stress and strain produced by a force tangential to the surface of a body are known as shear stress and shear strain

Shear Stress

Shear stress exists between two parts of a body in contact, when the two parts exert equal and opposite force on each other laterally in a direction tangential to their surface of contact.

Figure 14.52 shows a section of rivet subjected to equal and opposite forces P causing sliding of the particles one over the other.

From figure it is clear that the resisting force of the rivet must be equal to P. Hence, shearing stress τ is given by
τ = total tangential force/(Surface Area)

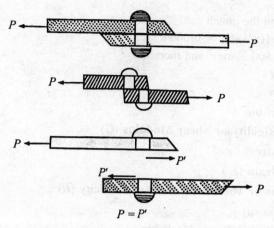

Fig. 14.52

$$= P/A \; ; \; A = \Pi.d.t$$

The tensile stress and compressive stress are also known as "direct stresses" and shearing stress as "tangential stress".

The common examples of a system involving shear stress are riveted and welded joint, towing device, punching operation etc.

Shear Strain

In case of a shearing load, a shear strain will be produced which is measured by the angle through which the body distorts. In Fig. 14.53 is shown a rectangular block LMNP fixed at one face and subjected to force F. After application of force, it distorts through an angle Φ and occupies new position LM'N'P. The shear strain (e_s) is given by

$e_s = NN'/NP = \tan\Phi$

$= \Phi$ (radians) ...... since Φ is very small.

The above result has been obtained by assuming NN' equal to arc (as NN' is small) drawn with centre P and radius PN.

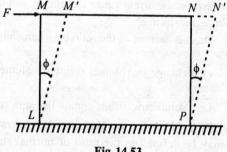

Fig 14.53

Q. 80: A steel punch can be worked to a compressive stress of 800 N/mm². Find the least diameter of hole which can be punched through a steel plate 10 mm thick if its ultimate shear strength is 350 N/mm².

(UPTU QUESTION BANK)

Sol.: Let d be the diameter of hole in mm

Area being sheared = $\pi.d.t = \pi d \times 10 = 10\pi d$ mm²

Force required to punch the hole = Ultimate shear strength x area sheared

$= 350 \times 10\pi d = 3500\pi d$ (N)

Cross sectional area of the hole = $(\pi/4)d^2$ mm²

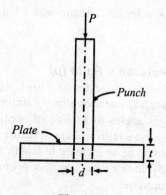

Fig. 14.54

Compressive stress on the punch

$\sigma_C = 3500\pi d / \{(\pi/4)d^2\} = 14000/d$

But σ_C is limited to 800 N/mm² and therefore

$800 = 14000/d$

$d = 17.5$ mm   ....ANS

Q. 81: Write short notes on:

(a) Modulus of Rigidity or Shear Modulus (G)

(b) Hydrostatic stress

(c) Volumetric strain (e_v)

(d) Bulk Modulus or Volume modulus of elasticity (K)

(e) Poisson's ratio (µ)

(a) Modulus of Rigidity or Shear Modulus

It is the ratio between shear stress(τ) and shear strain(e_s). It is denoted by G. It is the same as Shear modulus of elasticity

$G = \tau/e_s$

(b) Hydrostatic Stress

When a body is immersed in a fluid to a large depth, the body gets subjected to equal external pressure at all points of the body. This external pressure is compressive in nature and is called hydrostatic stress.

(c) Volumetric Strain

The hydrostatic stress cause change in volume of the body, and this change of volume per unit is called volumetric strain e_v.

Or, It is defined as the ratio between change is volume and original volume of the body, and is denoted by e_v,

e_v = change in volume/ original volume = $\delta V/V$

$e_v = e_x + e_y + e_z$

i.e., Volumetric strain equals the sum of the linear normal strain in x, y and z direction.

(d) Bulk Modulus or Volume Modulus of Elasticity

It may be defined as the ratio of normal stress(on each face of a solid cube) to volumetric strain. It is denoted by K. It is the same as Volume modulus of elasticity. K is a measure of the resistance of a material to change of volume without change of shape or form.

K = Hydrostatic pressure / Volumetric strain.

$$K = \frac{\sigma_n}{e_v}$$

(e) Poisson's Ratio (µ)

If a body is subjected to a load, its length changes; ratio of this change in length to the original length is known as linear or primary strain. Due to this load, the dimensions of the body change; in all directions at right angles to its line of application the strains thus produced are called lateral or secondary or transverse strains and are of nature opposite to that of primary strains. For example, if the load is tensile, there will be an increase in length and a corresponding decrease in cross-sectional area of the body (Fig. 14.55). In this case, linear or primary strain will be tensile and secondary or lateral or transverse strain compressive.

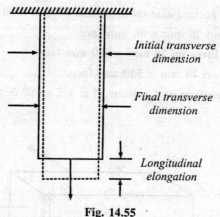

Fig. 14.55

Poisson's ratio is the ratio of lateral strain to the longitudinal strain. It is an elastic constant having the value always less than 1. It is denoted by 'µ' (1/m)

Poisson's Ratio (µ) = Lateral Strain / Longitudinal Strain; always less than 1.

Sl. No.	Material	Poisson's ratio
1.	Aluminium	0.330
2.	Brass	0.340
3.	Bronze	0.350
4.	Cast iron	0.270
5.	Concrete	0.200
6.	Copper	0.355
7.	Steel	0.288
8.	Stainless steel	0.305
9.	Wrought iron	0.278

Q. 82. A steel bar 2 m long, 20 mm wide and 10 mm thick is subjected to a pull of 20KN in the direction of its length. Find the changes in length, breadth and thickness. Take $E = 2 \times 10^5$ N/mm^2 and poisons ratio 0.30. *(UPTU QUESTION BANK)*

Sol.: Longitudinal strain = $\delta L/L$ = stress/ modulus of elasticity
$$= (P/A)/E = P/AE = 20 \times 10^3 /\{(20 \times 10) \times (2 \times 10^5)\} = 0.5 \times 10^{-3}$$
Change in length δL = longitudinal strain x original length
$$= (0.5 \times 10^{-3}) \times (2 \times 10^3) = 1.0 \text{ mm (increase)}$$
Lateral strain = Poisson's ratio × longitudinal strain $= 0.3 \times (0.5 \times 10^{-3}) = 0.15 \times 10^{-3}$
The lateral strain equals $\delta b/b$ and $\delta t/t$
Change in breadth $\delta b = b \times$ lateral strain $= 20 \times (0.15 \times 10^{-3})$
$$= 3 \times 10^{-3} \text{ mm (decrease)} \qquad \text{.......ANS}$$
Change in thickness $\delta t = t \times$ lateral strain $= 10 \times (0.15 \times 10^{-3})$
$$= 1.5 \times 10^{-3} \text{ mm (decrease)} \qquad \text{.......ANS}$$

Q. 83: A bar of steel 25 cm long, of rectangular cross-section 25 mm by 50 mm is subjected to a uniform tensile stress of 200 N/mm^2 along its length. Find the changes in dimensions. E = 205,000 N/mm^2 Poisson's ratio = 0.3. *(UPTU QUESTION BANK)*

Sol.: do your self

Q. 84: A 500 mm long bar has rectangular cross-section 20 mm × 40 mm. The bar is subjected to:
  (i) 40 kN tensile force on 20 mm × 40 mm face.
  (ii) 200kN compressive force on 20 mm × 500 mm face.
  (iii) 300kN tensile force on 40 mm × 500 mm face.

Find the change in dimensions and volume, if $E = 2 \times 10^5$ N/mm² and poisson ratio = 0.3.

[U.P.T.U. March–02]

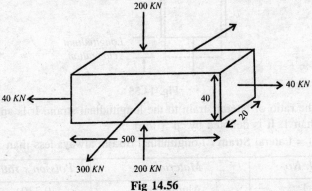

Fig 14.56

Sol.: $E = 2 \times 10^5$ N/mm², $\dfrac{1}{m} = \mu = \cdot 3$

$$\sigma_x = \frac{P_x}{A_x} = \frac{40 \times 10^3}{20 \times 40} = 50 \text{ N/mm}^2 \text{ (tensile stress)}$$

$$\sigma_y = \frac{P_y}{A_y} = \frac{200 \times 10^3}{20 \times 500} = 20 \text{ N/mm}^2 \text{ (compressive stress)}$$

$$= -20 \text{ N/mm}^2 \text{ (tensile stress)}$$

$$\sigma_z = \frac{P_z}{A_z} = \frac{300 \times 10^3}{40 \times 500} = 15 \text{ N/mm}^2 \text{ (tensile stress)}$$

$$e_x = \frac{\sigma_x}{E} - \frac{\sigma_y}{mE} - \frac{z}{mE}$$

$$= \frac{50}{2 \times 10^5} + \frac{20 \times 0.3}{2 \times 10^5} - \frac{15 \times 0.3}{2 \times 10^5}$$

$$e_x = 000257$$

$$e_y = \frac{\sigma_y}{E} - \frac{\sigma_x}{m \times E} - \frac{\sigma_z}{mE}$$

$$= \frac{-20}{2 \times 10^5} - \frac{50 \times 0.3}{2 \times 10^5}$$

$$e_y = -0001975$$

$$e_z = \frac{\sigma_z}{E} - \frac{\sigma_x}{mE} - \frac{\sigma_y}{mE}$$

$$= \frac{15}{2 \times 10^5} - \frac{50 \times 0.3}{2 \times 10^5} + \frac{20 \times 0.3}{2 \times 10^5}$$

$$e_z = \cdot 00003$$

Change in length in x direction $\Delta l = l \times e_x$
$$= 500 \times \cdot 000257$$
$$\Delta lx = \cdot 1285 \text{ mm}$$

in y direction
$$\Delta b = e_y \times b$$
$$= -\cdot 000197 \times 90$$
$$= -\cdot 0076 \text{ mm (decreases)}$$

in z direction
$$\Delta w = e_z \times w$$
$$= \cdot 00003 \times 20$$
$$ev = e_x + e_y + e_z$$
$$= \cdot 00009$$
$$\Delta v = v \times e_v = 20 \times 40 \times 500 \times \cdot 00009$$
$$= 36 \text{ mm}^3$$

Q. 85: A 2m long rectangular bar of 7.5 cm × 5 cm is subjected to an axial tensile load of 1000kN. Bar gets elongated by 2mm in length and decreases in width by 10×10^{-6} m. Determine the modulus of elasticity E and Poisson's ratio of the material of bar. *(Dec–2002)*

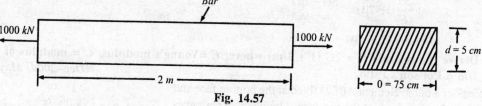

Fig. 14.57

Sol.: Given:
$$L = 2m;$$
$$B = 7.5\text{cm} = 0.075\text{m};$$
$$D = 5\text{cm} = 0.05\text{m}$$
$$P = 1000\text{kN}$$
$$\delta L = 2\text{mm} = 0.002\text{m}$$
$$\delta b = 10 \times 10^{-6}\text{m}.$$

Longitudinal strain $e_L = e_t = \delta L/L = 0.002/2 = 0.001$
Lateral strain $= \delta b/b = 10 \times 10^{-6}/0.075 = 0.000133$
Tensile stress (along the length) $\sigma_t = P/A = (1000 \times 1000)/(0.075 \times 0.05) = 0.267 \times 10^9$ N/m²
Modulus of elasticity, $E = \sigma_t/e_t = 0.267 \times 10^9/0.001 = \mathbf{267 \times 10^9}$ **N/m²** ....**ANS**
Poisson's ratio = Lateral strain/ Longitudinal strain = $(\delta b/b)/(\delta L/L) = 0.000133/0.001$

Q. 86: Prove that $E = 3K(1 - 2\mu)$.

Sol.: Consider a cubical element subjected to volumetric stress σ which acts simultaneously along the mutually perpendicular x, y and z-direction.

The resultant strains along the three directions can be worked out by taking the effect of individual stresses.

Strain in the x-direction,

e_x = strain in x-direction due to σ_x – strain in x-direction due to σ_Y – strain in x-direction due to

$$\sigma_Z = \sigma_x/E - \mu.\sigma_y/E - \mu.\sigma_z/E \quad ...(i)$$

But $\sigma_x = \sigma_y = \sigma_z = \sigma$

$$e_x = \frac{\sigma}{E} - \mu\frac{\sigma}{E} - \mu\frac{\sigma}{E} = \frac{\sigma}{E}(1-2\mu)$$

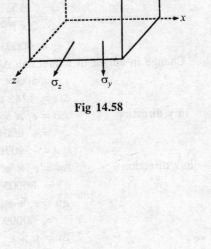

Fig 14.58

Likewise $e_y = \frac{\sigma}{E}(1-2\mu)$ and $e_z = \frac{\sigma}{E}(1-2\mu)$

Volumetric strain

$$e_x = e_x + e_y + e_z = \frac{3\sigma}{E}(1-2\mu)$$

Now, bulk modulus

$$K = \frac{\text{volumetric stress}}{\text{volumetric strain}}$$

$$= \frac{\sigma}{\frac{3\sigma}{E}(1-2\mu)} = \frac{E}{3(1-2\mu)} \text{ or, } E = 3K(1-2\mu)$$

$$E = 3K(1 - 2\mu) \quad ...(i)$$

Q. 87: Derive the relation $E = 2C(1 + 1/m)$ where; E = Young's modulus, C = modulus of rigidity $1/m$ = Poisson's ratio. *(Dec–2004, May–2005)*

Sol.: Consider a cubic element ABCD fixed at the bottom face and subjected to shearing force at the top face. The block experiences the following effects due to this shearing load:

- shearing stress t is induced at the faces DC and AB.
- complimentary shearing stress of the same magnitude is set up on the faces AD and BC.
- The block distorts to a new configuration ABC'D'.
- The diagonal AC elongates (tension) and diagonal BD shortens (compression).

Longitudinal strain in diagonal AC

$$= \frac{AC' - AC}{AC} = \frac{AC' - AE}{AC} = \frac{EC'}{AC} \quad ...(i)$$

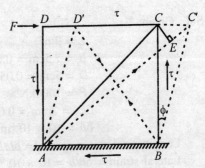

Fig 14.58

where CE is perpendicular from C onto AC'

Since extension CC' is small, $\angle ACB$ can be assumed to be equal $\angle ACB$ which is 45°. Therefore

$$EC' = CC' \cos 45° = \frac{CC'}{\sqrt{2}}$$

Longitudinal strain $= \dfrac{CC'}{\sqrt{2}\, AC} = \dfrac{CC'}{\sqrt{2} \times \sqrt{2}\, BC} = \dfrac{CC'}{2\, BC} = \dfrac{\tan \phi}{2} = \dfrac{\phi}{2}$ ...(ii)

Where, $\Phi = CC'/BC$ represents the shear strain
In terms of shear stress t and modulus of rigidity C, shear strain $= \tau/C$
longitudinal strain of diagonal $AC = \tau/2C$ ...(iii)
The strain in diagonal AC is also given by
= strain due to tensile stress in AC – strain due to compressive stress in BD

$$= \frac{\tau}{E} - \left(-\mu \frac{\tau}{E}\right) = \frac{\tau}{E}(1+\mu) \quad ...(v)$$

From equation (iv) and (v), we get

$$= \frac{\tau}{2C} = \frac{\tau}{E}(1+\mu)$$

or $\qquad E = 2C(1+\mu)$ ...(vi)

Q. 88: What is the relation between elastic constant E, C and K.?
Sol.: With reference to the relations (1) and (6) derived above,

$$E = 2C(1+\mu) = 3K(1-2\mu)$$

To eliminate μ from these two expressions for E, we have

$$\mu = \frac{E}{2C} - 1 \text{ and } E = 3k\left[1 - 2\left(\frac{E}{2C} - 1\right)\right]$$

or $\qquad E = 3K\left[1 - \left(\frac{E}{C} - 2\right)\right] = 3k\left[3 - \frac{E}{C}\right] = 9K - \frac{3KE}{C}$

or $\qquad E + \dfrac{3KE}{C} = 9K$; $E\left(\dfrac{C+3K}{C}\right) = 9K$

or $\qquad E = \dfrac{9KC}{C+3K}$

$$E = 2C(1+\mu) = 3K(1-2\mu) = \frac{9KC}{C+3K}$$

Q. 89: A circular rod of 100 mm diameter and 500 m long is subjected to a tensile force of 1000KN. Determine the modulus of rigidity, bulk modulus and change in volume if poisons ratio = 0.3 and Young's Modulus = 2×10^5 N/mm². *(UPTU QUESTION BANK)*

Sol.: Modulus of rigidity
$\qquad G = E/2.(1+\mu) = 2 \times 10^5/ 2(1+0.3) = \mathbf{0.769 \times 10^5}$ **N/mm²** ....ANS

Bulk modulus
$\qquad K = E/3(1-2\mu) = 2 \times 105/3(1 - 2 \times 0.3) = \mathbf{1.667 \times 10^5}$ **N/mm²** ....ANS

Normal stress $\sigma = P/A = 1000 \times 10^3/ \pi/4(100)^2 = 127.388$ N/mm²
Linear (Longitudinal) strain = $\delta L/L$ = Normal stress/ Young's modulus
$= 127.388/ 2 \times 10^5 = 0.000637$

Diametral (Lateral) strain
$= \delta d/d = \mu \cdot \delta L/L = 0.3 \times 0.000637 = 0.0001911$

Now volume of a circular rod
$= V = \pi/4 \cdot d^2 \cdot L$

Upon differentiation
$\delta V = \pi/4[\ 2.d.\delta d.L + d^2.\delta L]$

Volumetric strain
$\delta V/V = \pi/4[\ 2.d.\delta d.L + d^2.\delta L]/ \pi/4.d^2.L = 2\delta d/d + \delta L/L$

Substituting the value of $\delta d/d$ and $\delta L/L$ as calculated above, we have
$\delta V/V = 2\ (-0.0001911) + 0.000637 = 0.0002548$

The -ive sign with $\delta d/d$ stems from the fact that whereas the length increases with tensile force, there is decrease in diameter.

Change in volume
$\delta V = 0.0002548\ [\pi/4(100)^2 \times 500] = $ **1000.09 mm³**   ....**ANS**

Q. 90: What do you know about properties of a metal?

Sol.: Different materials posses different properties in varying degree and therefore behave in different ways under given conditions. These properties includes mechanical properties, electrical properties, thermal properties, chemical properties, magnetic properties and physical properties. We are basically interested in knowing as to how a particular material will behave under applied load i.e. in knowing the mechanical properties.

Q 91: What is mechanical properties of material? Define strength.

Sol.: Those characteristics of the materials which describe their behaviour under external loads are known as Mechanical Properties. The most important and useful mechanical properties are:

Strength:

It is the resistance offered by a material when subjected to external loading. So Stronger the material the greater the load it can withstand. Depending upon the type of load applied the strength can be tensile, compressive, shear or torsional.

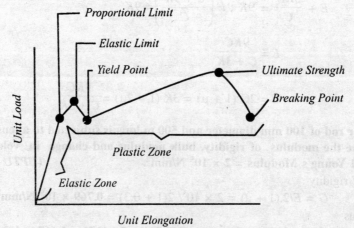

Fig 14.59 A Typical Stress-Strain Curve

The stress at the elastic limit is known as yield Strength.

And the maximum stress before the fracture is called ultimate strength. While in tension the ultimate strength of the material represents it tenacity.

Q. 92: Write short notes on:

Sol.: Elasticity, stiffness, Plasticity, Malleability, Ductility, Brittleness, Toughness

Elasticity: Elasticity of a material is its power of coming back to its original position after deformation when the stress or load is removed. Elasticity is a tensile property of its material.

Proportional Limit: It is the maximum stress under which a material will maintain a perfectly uniform rate of strain to stress.

Elastic Limit: The greatest stress that a material can endure without taking up some permanent set is called elastic limit.

Stiffness: It is the property of a material due to which it is capable of resisting deflection or elastic deformation under applied loads. also called rigidity.

The degree of stiffness of a material is indicated by the young's modulus. The steel beam is stiffer or more rigid than aluminium beam.

Plasticity: The plasticity of a material is its ability to change some degree of permanent deformation without failure. This property is widely used in several mechanical processes like forming, shaping, extruding, rolling etc. Due to this properties various metal can be transformed into different products of required shape and size. This conversion into desired shape and size is effected either by the application of pressure, heat or both. Plasticity increase with increase in temp.

Malleability: Malleability of a material is its ability to be flattened into their sheets without creaking by hot or cold working. Aluminum, copper, tin lead steel etc are malleable metals.

Ductility: Ductility is that property of a material, which enables it to draw out into thin wire. Mild steel is a ductile material. The percent elongation and the reduction in area in tension is often used as empirical measures of ductility.

Brittleness: The brittleness of a material is the property of breaking without much permanent distortion. There are many materials, which break or fail before much deformation take place. Such materials are brittle e.g. glass, cast iron. Therefore a non-ductile material is said to be a brittle material. Usually the tensile strength of brittle materials is only a fraction of their compressive strength. A brittle material should not be considered as lacking in strength. It only shows the lack of plasticity.

Toughness: Toughness is a measure of the amount of energy a material can absorb before actual fracture or failure takes place. The toughness of a material is its ability to withstand both plastic and elastic deformation. *"The work or energy a material absorbs is called modulus of toughness"*

For Ex: If a load is suddenly applied to a piece of mild steel and then to a piece of glass the mild steel will absorb much more energy before failure occurs. Thus mild steel is said to be much tougher than a glass.

Q. 93: Write short notes on: Hardness, Impact Strength

Sol.: Hardness: Hardness is defined in terms of the ability of a material to resist screeching, abrasion, cutting, indentation or penetration. Many methods are now in use for determining the hardness of a material. They are Brinell, Rockwell and Vickers.

Hardness of a metal does not directly related to the hardenability of the metal. Hardenability is indicative of the degree of hardness that the metal can acquire through the hardening process. i.e., heating or quenching.

Impact Strength: It can be defined as the resistance of the material to fracture under impact loading, i.e under quickly applied dynamic loads. Two standard tests are normally used to determine this property.

1. The IZOD impact test.
2. The CHARPY test.

Q. 94: What is fatigue; how it is related to creep?

Sol.: Fatigue : Failure of a material under repeated stress is known as fatigue and the maximum stress that a metal can withstand without failure for a specific large number of cycle of stress is called Fatigue limit.

Creep: The slow and progressive deformation of a material with time at constant stress is called creep. There are three stages of creep. In the first one, the material elongates rapidly but at a decreasing rate. In the second stage, the rate of elongation is constant. In the third stage, the rate of elongation increases rapidly until the material fails. The stress for a specifid rate of strain at a constant temperature is called creep strength.

Creep Curve and Creep Testing: Creep Test is carried out at high temp. A creep curve is a plot of elongation of a tensile specimen versus time, For a given temp. and under constant stress. Tests are carried out for a period of a few days to many years.

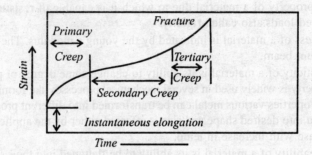

Fig 14.60

Creep curve shows four stages of elongation:
1. Instantaneous elongation on application of load.
2. Primary creep: Work hardening decreases and recovery is slow.
3. Secondary creep: Rate of work hardening and recovery processes are equal.
4. Tertiary creep: Grain boundary cracks. Necking reduces the cross sectional area of the test specimen.

Chapter 15

COMPOUND STRESS AND STRAIN

Q. 1: Define Compound Stress.
Sol.: Simple stresses mean only tensile stress or compressive stress or only shear stress. Tensile and compressive stresses act on a plane normal to the line of action of these stresses, and shear stress acts on a plane parallel to the line of action of this stress. But when a plane in a strained body is oblique to the applied external force, this plane may be subjected to tensile or compressive stress and shear stress, i.e.;

Such a system or a plane in which direct or normal stresses and shear stresses act simultaneously are called compound stress or complex stress.

Q. 2: Define the concept of plane stress.
Sol.: In a cubical element of a strained material is acted on by stresses acting on only two pairs of parallel planes and the third pair of parallel planes is free from any stress, it is said that the element is under the action of plane stresses. So, plane stress condition can be called two-dimensional stress condition. Let a cubical element ABCD taken from a strained body be subjected to normal stresses σ_1 and σ_2 and shear stress ^{1}.

Planes AD and BC are subjected to normal stress σ_1 and shear stress^1, and planes AB and CD are subjected to normal stress σ_2 and shear stress^1. But no stress acts on the third pair (front face and rear face of ABCD) of parallel planes. Hence it is said that the cubical element ABCD is under the action of plane stresses.

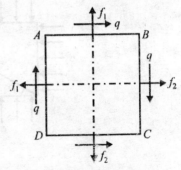

Fig. 15.1

Q. 3: Explain Principal Planes and Principal Stresses. *(Dec–00)*
Sol.: When an element in a strained body is under the action of plane stresses, it is found that there exist two mutually perpendicular planes of the element on which normal stresses are maximum and minimum and no shear stress acts on these planes. These planes are called *principal planes*.

From the figure 15.2; The principle planes can be determined by

$$\tan 2\theta = \frac{2\tau_{xy}}{\sigma_x - \sigma_y}$$

Since shear stress on these plane are zero, therefore they are also called the shear free plane.

The maximum and minimum principal stresses acting on principal planes are called *principal stress*.

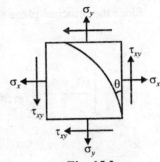

Fig. 15.2

The principal stress having maximum value is called "*major principal stress*" and the principal stress having minimum value is called "*minor principal stress*".

$$\sigma_{1,2} = \frac{1}{2}\left[(\sigma_x + \sigma_y) \pm \sqrt{(\sigma_x - \sigma_y)^2 (4\tau_{xy}^2)}\right]$$

Major principal stress $= \sigma_1 = \frac{1}{2}\left[(\sigma_x + \sigma_y) \pm \sqrt{(\sigma_x - \sigma_y)^2 (4\tau_{xy}^2)}\right]$

Minor principal stress $= \sigma_2 = \frac{1}{2}\left[(\sigma_x + \sigma_y) \pm \sqrt{(\sigma_x - \sigma_y)^2 (4\tau_{xy}^2)}\right]$

Where σ_x, σ_y be the direct stresses in x and y direction and σ_{xy} be the shear stress normal to plane *xy*.

Resultant stress is given by the equation

$$\sigma_r = \sqrt{\sigma_n^2 + \tau_t^2}$$

Q. 4: Derive the equation for principal stresses and principal planes for an element subjected to compound stresses.

Sol.: For the state of stress shown in fig. the normal stress and shear stress on any oblique plane inclined at an angle θ can be determined by,

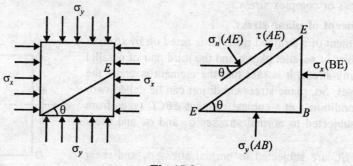

Fig 15.3

$$\sigma_n = \left(\frac{\sigma_x + \sigma_y}{2}\right) + \left(\frac{\sigma_x - \sigma_y}{2}\right)\cos 2\theta + \tau_{xy} \sin 2\theta$$

and

$$\tau = \left(\frac{\sigma_x - \sigma_y}{2}\right)\sin 2\theta - \tau_{xy} = \cos 2\theta$$

Since the principal plane should carry only normal stress, shear stress acting on it is to be zero.

i.e.,

$$\tau = \left(\frac{\sigma_x - \sigma_y}{2}\right)\sin 2\theta - \tau_{xy} = \cos 2\theta$$

$$\left[\frac{(\sigma_x - \sigma_y)}{2}\right]\sin 2\theta = \tau_{xy} = \cos 2\theta$$

or

$$\tan 2\theta = \left[\frac{2\tau_{xy}}{\sigma_x - \sigma_y}\right]$$

This gives one of the principal planes, other principal plane shall be perpendicular to this plane. Substituting the value of θ in σ_n expression by rearranging as below.

$$\sin 2\theta = \frac{\pm 2\tau_{xy}}{\sqrt{(\sigma_x - \sigma_y)^2 + 4\tau_{xy}^2}} \quad \text{and} \quad \cos 2\theta = \frac{\pm(\sigma_x - \sigma_y)}{\sqrt{(\sigma_x - \sigma_y)^2 + 4\tau_{xy}^2}}$$

$$\sigma_n = \frac{(\sigma_x - \sigma_y)}{2} + \frac{(\sigma_x - \sigma_y)}{2}\left[\frac{\pm(\sigma_x - \sigma_y)}{\sqrt{(\sigma_x^2 - \sigma_y^2) + 4\tau_{xy}^2}}\right] + \frac{2\tau_{xy}^2}{\sqrt{(\sigma_x - \sigma_y)^2 + 4\tau_{xy}^2}}$$

$$= \frac{(\sigma_x + \sigma_y)}{2} \pm \frac{(\sigma_x + \sigma_y)^2 \pm 4\tau_{xy}^2}{2 \times \sqrt{(\sigma_x + \sigma_y)^2 + 4\tau_{xy}^2}}$$

$$\sigma_{1,2} = \frac{1}{2}\left[(\sigma_x - \sigma_y) \pm \sqrt{(\sigma_x = \sigma_y)^2 + 4\tau_{xy}^2}\right]$$

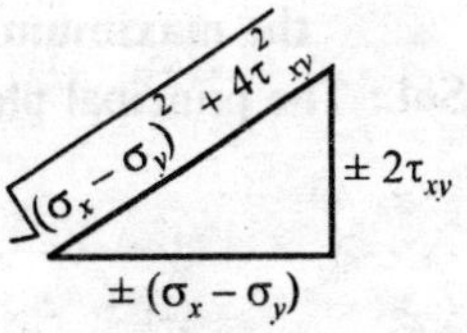

Fig. 15.4

It gives the values of principal stresses. The largest one is called major and smaller is called minor principal stress.

Q. 5: Obtain the expression for maximum shearing stress and maximum shearing planes.

Sol.: Condition for maximum shearing stress, :

$$\frac{d\tau}{d\theta} = 0$$

$$\frac{d}{d\theta}\left\{\left(\frac{\sigma_x - \sigma_y}{2}\right)\sin 2\theta - \tau_{xy}\cos 2\theta\right\} = 0$$

or

$$\frac{\sigma_x - \sigma_y}{2}\cos 2\theta = \tau_{xy}\sin 2\theta$$

or

$$\tan 2\theta = -\left(\frac{\sigma_x - \sigma_y}{2\tau_{xy}}\right)$$

This expression give the value of a maximum shearing plane. Another plane which also carries maximum shearing stress is normal to this plane. Substituting the value of θ in τ expression with the following rearrangement.

$$\sin 2\theta = \frac{\pm(\sigma_x - \sigma_y)}{\sqrt{(\sigma_x - \sigma_y)^2 + 4\tau_{xy}^2}}$$

$$\cos 2\theta = \frac{\mp 2\tau_{xy}}{\sqrt{(\sigma_x - \sigma_y)^2 + 4\tau_{xy}^2}}$$

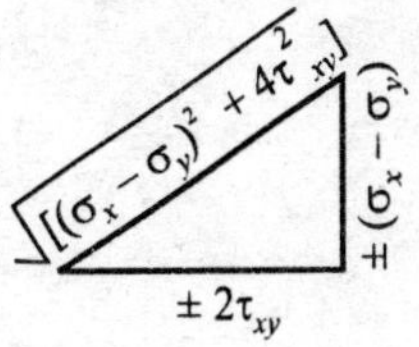

Fig. 15.5

$$\tau_{max} = \left(\frac{\sigma_x - \sigma_y}{2}\right)\sin 2\theta - \tau_{xy}\cos 2\theta$$

$$\tau_{max} = \left(\frac{\sigma_x - \sigma_y}{2}\right)\frac{\pm(\sigma_x - \sigma_y)}{\sqrt{(\sigma_x - \sigma_y)^2 + 4\tau_{xy}^2}} - \tau \cdot \frac{\mp 2\tau_{xy}}{\sqrt{(\sigma_x - \sigma_y)^2 + 4\tau_{xy}^2}}$$

$$= \pm \frac{(\sigma_x - \sigma_y)^2 + 4\tau_{xy}^2}{\sqrt{(\sigma_x - \sigma_y)^2 + 4\tau_{xy}^2}}$$

$$\tau_{max} = \pm \frac{1}{2}\sqrt{(\sigma_x - \sigma_y)^2 + 4\tau_{xy}^2}$$

or $$\tau_{max} = \pm \frac{1}{2}(\sigma_x - \sigma_y)$$

Q. 6: Prove that the plane inclined at 45° to the plane carrying the greatest normal stress carries the maximum shear stress.?

Sol.: The principal planes are determined by,

$$\tan 2\theta = \frac{2\tau_{xy}}{\sigma_x - \sigma_y}$$

and maximum shearing planes by,

$$\tan 2\theta' = \frac{\sigma_x - \sigma_y}{-2\tau_{xy}}$$

Multiply both expressions,

$$\tan 2\theta \cdot \tan 2\theta' = -1$$

When the products of slopes of two lines $= -1$; then the two line will be orthogonal,

$$2\theta' = 2\theta + 90°$$
$$\theta' = \theta + 45°$$

i.e., Maximum shearing planes are always inclined at 45° with principal planes.

Q. 7: Prove the when stress are unequal and alike

(*i*) Normal stress

$$\sigma_n = \left(\frac{\sigma_x + \sigma_y}{2}\right) + \left(\frac{\sigma_x - \sigma_y}{2}\right)\cos 2\theta$$

(*ii*) Shear stress

$$\tau = \frac{1}{2}(\sigma_x + \sigma_y)\sin 2\theta$$

Sol.: By drawing the F.B.D. of wedge *BCE*, considering unit thickness of element.

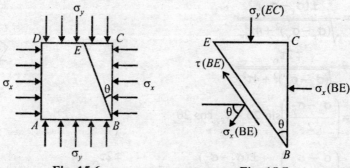

Fig. 15.6   Fig. 15.7

Let σ_n and τ be normal and shear stresses on the plane *BE*. Applying Equilibrium conditions to the wedge *BCE*:

$$\Sigma Fx = 0$$
$$\sigma_x (BC) - \sigma_n (BE) \cos \theta - \tau (BE) \cos (90 - \theta) = 0$$
$$\sigma_x (BC) - \sigma_n (BE) \cos \theta - \tau (BE) \sin \theta = 0 \qquad \text{...(i)}$$
$$\Sigma Fy = 0$$
$$\sigma_y (EC) - \sigma_n (BE) \sin \theta + \tau (BE) \sin (90 - \theta) = 0$$
$$\sigma_x EC - \sigma_n (BE) \sin \theta + \tau (BE) \cos \theta = 0 \qquad \text{...(ii)}$$

Dividing equations (i) and (ii) by BE and replacing
$$\frac{BC}{BE} = \cos \theta \quad \text{and} \quad \frac{EC}{BE} = \sin \theta$$
$$\sigma_x \cos \theta - \sigma_n \cos \theta - \tau \sin \theta = 0 \qquad \text{...(iii)}$$
$$\sigma_y \sin \theta - \sigma_n \sin \theta + \tau \cos \theta = 0 \qquad \text{...(iv)}$$

Multiplying equation (iii) by $\cos \theta$, (iv) by $\sin \theta$ and adding
$$\sigma_x \cos^2 \theta - \sigma_n \cos^2 \theta + \sigma_y \sin^2 \theta - \sigma_n \sin^2 \theta = 0$$
$$\sigma_n = \sigma_x \cos^2 \theta + \sigma_y \sin^2 \theta$$

Putting
$$\cos^2 \theta = \frac{1 + \cos \theta}{2} \quad \text{and} \quad \sin^2 \theta = \frac{1 - \cos 2\theta}{2}$$
$$\sigma_n = \sigma_x \left[\frac{1 + \cos 2\theta}{2}\right] + \sigma_y \left[\frac{1 - \cos 2\theta}{2}\right]$$
$$\sigma_n = \left(\frac{\sigma_x - \sigma_y}{2}\right) + \left(\frac{\sigma_s - \sigma_y}{2}\right) \cos 2\theta$$

Multiplying equation (iii) by $\sin \theta$ and (iv) by $\cos \theta$ and then subtracting.
$$\sigma_x \sin \theta \cos \theta - \tau \sin^2 \theta - \sigma_y \sin \theta \cos \theta - \tau \cos^2 \theta = 0 \qquad [\because (\sin^2 \theta + \cos^2 \theta) = \tau]$$
or $\qquad (\sigma_x - \sigma_y) \sin \theta \cos \theta - \tau = 0$
or $\qquad \tau = (\sigma_x - \sigma_y) \sin \theta \cos \theta$
$$\tau = \frac{1}{2}(\sigma_x - \sigma_y) \sin 2\theta .$$

Q. 8: Derive the expression for normal and shear stress on a plane AE inclined at an angle B with AB subjected to direct stresses of compressive nature (both) of σ_x and σ_y on two mutually perpendicular stresses as shown in fig 15.8.

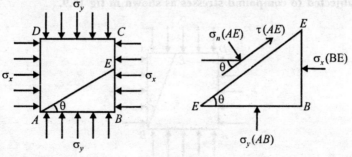

Fig 15.8

Sol.: Consider the unit thickness of the element. Applying equations of equilibrium to the free body diagram of wedge ABE.

Let σ_n and τ be the normal and shear stresses on the plane AE.

$$\Sigma Fx = 0$$
$$\sigma_x (BE) - \tau (AE) \cos\theta - \sigma_n (AE) \cos(90-\theta) = 0$$
$$\sigma_x (BE) + \tau (AE) \cos\theta + \sigma_n (AE) \sin\theta = 0 \qquad ...(i)$$
$$\Sigma Fy = 0$$
$$\sigma_y (AB) + \tau (AE) \sin\theta - \sigma_n (AE) \sin(90-\theta) = 0$$
$$\sigma_y (AB) + \tau (AE) \sin\theta - \sigma_n (AE) \cos\theta = 0 \qquad ...(ii)$$

Dividing equations (i) and (ii) by AE and replacing

$$\sin\theta = \frac{BE}{AE}$$

and
$$\cos\theta = \frac{AB}{AE}$$

$$-\sigma_x \sin\theta + \tau \cos\theta + \sigma_n \sin\theta = 0 \qquad ...(iii)$$
$$-\sigma_x \cos\theta + \tau \sin\theta - \sigma_n \cos\theta = 0 \qquad ...(iv)$$

Multiplying equation (iii) by $\sin\theta$, (iv) by $\cos\theta$ and subtracting
$$-\sigma_x \sin^2\theta + \sigma_n \sin^2\theta - \sigma_y \cos^2\theta + \sigma_n \cos^2\theta = 0$$
$$\sigma_n = \sigma_x \sin^2\theta + \sigma_y \cos^2\theta$$

Putting
$$\cos^2\theta = \frac{1+\cos 2\theta}{2}$$

and
$$\sin^2\theta = \frac{1-\cos 2\theta}{2}$$

$$\sigma_n = \left(\frac{\sigma_x + \sigma_y}{2}\right) + \left(\frac{\sigma_x - \sigma_y}{2}\right) \cos 2\theta$$

Also multiplying (iii) by $\cos\theta$ and 4 by $\sin\theta$ and then adding.
$$-\sigma_x \sin\theta \cos\theta + \tau \cos^2\theta + \sigma_y \cos\theta \sin\theta + \tau \sin^2\theta = 0$$
$$\tau = (\sigma_x + \sigma_y) \sin\theta \cos\theta$$

or
$$\tau = \left(\frac{\sigma_x + \sigma_y}{2}\right) \sin 2\theta$$

Q. 9: Obtain the expression for normal and tangential stresses on a plane *BE* inclined at an angle θ with *BC* subjected to compound stresses as shown in fig 15.9.

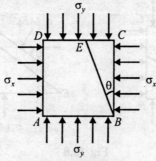

Fig 15.9

Sol.: Consider the unit thickness of the element. Applying equations of equilibrium to the free body diagram of wedge *BCE*.

Let σ_n and τ be the normal and shear stresses on the plane BE,
$$\Sigma Fx = 0$$
$$\sigma_x (BC) - \tau_{xy} (EC) - \tau (EB) \cos(90 - \theta) - \sigma_n (EB) \cos \theta = 0$$
$$\Sigma Fy = 0$$
$$\sigma_x (EC) + \tau_{xy} (BC) + \tau (EB) \sin(90 - \theta) - \sigma_n (EB) \sin \theta = 0$$
Dividing both equations by BE replacing

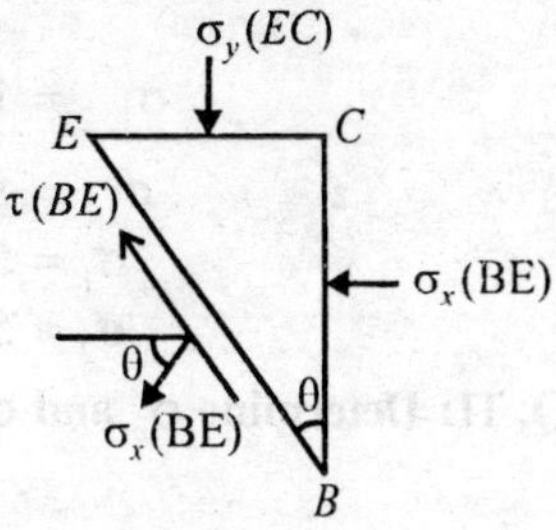

$$\sin \theta = \frac{EC}{BE} \text{ and } \cos \theta = \frac{BC}{BE}$$

Fig. 15.10

$$\sigma_x \cos \theta + \tau_{xy} \sin \theta - \tau \sin \theta - \sigma_n \cos \theta = 0 \quad ...(i)$$
$$\sigma_y \sin \theta + \tau_{xy} \cos \theta + \tau \cos \theta - \sigma_n \sin \theta = 0 \quad ...(ii)$$

Multiplying equation (i) and $\cos \theta$, (ii) by $\sin \theta$ and then adding,
$$\sigma_x \cos^2 \theta + \tau_{xy} \sin \theta \cos \theta - \sigma_n \cos^2 \theta + \sigma_y \sin^2 \theta + \tau_{xy} \cos \theta \sin \theta - \sigma_n \sin^2 \theta = 0$$
$$\sigma_x \cos^2 \theta + \sigma_y \sin^2 \theta + 2\tau_{xy} \sin \theta \cos \theta = \sigma_n$$
or
$$\sigma_n = \sigma_x \cos^2 \theta + \sigma_y \sin^2 \theta + \tau_{xy} \sin 2\theta$$

Putting
$$\cos^2 \theta = \frac{1 + \cos 2\theta}{2} \text{ and } \sin^2 \theta = \frac{1 - \cos 2\theta}{2}$$

$$\sigma_n = \sigma_x \left[\frac{1 + \cos 2\theta}{2}\right] + \sigma_y \left[\frac{1 - \cos 2\theta}{2}\right] + \tau_{xy} \sin 2\theta$$

i.e.,
$$\sigma_n = \left(\frac{\sigma_x + \sigma_y}{2}\right) + \sigma_y \left[\frac{\sigma_x - \sigma_y}{2}\right] \cos 2\theta + \tau_{xy} \sin 2\theta$$

Multiplying equation (i) by $\sin \theta$, (ii) by $\cos \theta$ and then subtracting (ii) from (i)
$$(\sigma_x \cos \theta \sin \theta + \tau_{xy} \sin^2 \theta - \tau \sin^2 \theta)(\sigma_y \sin \theta \cos \theta + \tau_{xy} \cos^2 \theta + \tau \cos^2 \theta = 0$$
$$(\sigma_x - \sigma_y) \cos \theta \sin \theta + \tau_{xy} (\sin^2 \theta - \cos^2 \theta) - \tau = 0$$

$$\tau = \frac{\sigma_x + \sigma_y}{2} \sin 2\theta - \tau_{xy} \cos \theta \quad (\cos^2 \theta - \sin^2 \theta = \cos 2\theta$$

Q. 10: Find the principal stresses for the state of stress given below *(May–02 (C.O.))*

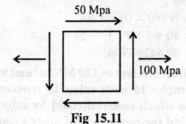

Fig 15.11

Sol.: Given that
$$\sigma_x = 100 \text{MPa}$$
$$\sigma_y = 0$$
$$\tau_{xy} = 50 \text{MPa}$$
Since we known that; the principal stresses are given by
$$\sigma_{1 \text{ or } 2} = \frac{1}{2}\left[(\sigma_x + \sigma_y) \pm \sqrt{(\sigma_x - \sigma_y)^2 + (4\tau_{xy}^2)}\right]$$

$$\sigma_{1,2} = 1/2\ [(100 + 0) \pm \{(100 - 0)^2 + 4 \times (50)^2\}^{1/2}]$$
$$\sigma_{1,2} = 50 \pm 70.71$$
$$\sigma_1 = 50 + 70.71 = 120.71 MN/m^2 \quad\quad\quad ....ANS$$
$$\sigma_2 = 50 - 70.71 = -20.71 MN/m^2 \quad\quad\quad ....ANS$$

Q. 11: Determine σ_n and σ_t for a plane at $\theta = 25°$, for the element shown in figure.   *(Dec–03 (C.O.))*

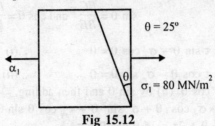

Fig 15.12

Sol.: Given that:
$$\theta = 25°$$
$$\sigma_1 = \sigma_x = 80 MN/m^2$$
$$\sigma_y = 0$$

Since we know that
(i) Normal stress

$$\sigma_y = \left(\frac{\sigma_x + \sigma_y}{2}\right) + \left(\frac{\sigma_x - \sigma_y}{2}\right)\cos 2\theta$$
$$= (80 + 0)/2 + \frac{1}{2}(80 - 0)\cos 50°$$
$$= 40 + 40 \cos 50°$$
$$= 65.711 MN/m^2 \quad\quad\quad ....ANS$$

(ii) Shear stress

$$\tau = \frac{1}{2}(\sigma_x - \sigma_y)\sin 2\theta$$
$$= \frac{1}{2}(80 - 0) \sin 50°$$
$$= 40 \sin 50°$$
$$= 30.642 MN/m^2 \quad\quad\quad ....ANS$$

Q. 12: In an elastic material, the direct stresses of 120 MN/m² and 90 MN/m² are applied at a certain point on planes at right angles to each other in tension and compressive respectively. Estimate the shear stress to which material could be subjected, if the maximum principal stress is 150 MN/m². Also find the magnitude of other principal stress and its inclination to 120 MN/m².   *(May–01 (C.O.))*

Sol.: Given that
$$\sigma_x = 120 MN/m^2 \text{ (tensile i.e + ive)}$$
$$\sigma_y = 90 MN/m^2 \text{ (Compressive i.e. - ive)}$$
$$\tau_{xy} = ?$$
$$\sigma_1 = 150 MN/m^2$$

Since we have

$$\sigma_1 = \frac{1}{2}\left[(\sigma_x - \sigma_y) + \sqrt{(\sigma_x - \sigma_y)^2 + (4\tau_{xy}^2)}\right]$$

$150 = \frac{1}{2}[(120 - 90) + \{(120 - (-90))^2 + 4(\tau_{xy})^2\}^{1/2}]$

$\tau_{xy} = 84.85 \text{MN/m}^2$ ....ANS

Now the magnitude of other principal stress σ_2

$$\sigma_2 = \frac{1}{2}\left[(\sigma_x - \sigma_y) - \sqrt{(\sigma_x - \sigma_y)^2 + (4\tau_{xy}^2)}\right]$$

$\sigma_2 = \frac{1}{2}[(120 - 90) - \{(120 - (-90))^2 + 4(84.85)^2\}^{1/2}]$

$\sigma_2 = -120 \text{MN/m}^2$ ....ANS

The direction of principal planes is:

$$\tan 2\theta = \frac{2\tau_{xy}}{\sigma_x - \sigma_y}$$

$\tan 2\theta = (2 \times 84.85)/(120 - (-90))$

$2\theta = 38.94°$ or $2218.94°$

$\theta = 19.47°$ or $109.47°$ ....ANS

Q. 13: **A load carrying member is subjected to the following stress condition;**
Tensile stress σ_x = 400MPa;
Tensile stress σ_y = – 300MPa;
Shear stress τ_{xy} = 200MPa (Clock wise);
Obtain
(1) Principal stresses and their plane
(2) Maximum shearing stress and its plane. *(Dec–00 (C.O.))*

Sol.: Since Principal stresses are given as:

Major Principle stress = $\sigma_1 = \frac{1}{2}\left[(\sigma_x + \sigma_y) + \sqrt{(\sigma_x - \sigma_y)^2 + (4\tau_{xy}^2)}\right]$

Minor Principle stress = $\sigma_2 = \frac{1}{2}\left[(\sigma_x + \sigma_y) - \sqrt{(\sigma_x - \sigma_y)^2 + (4\tau_{xy}^2)}\right]$

or;

$$\sigma_{1 \text{ or } 2} = \frac{1}{2}\left[(\sigma_x + \sigma_y) \pm \sqrt{(\sigma_x - \sigma_y)^2 + (4\tau_{xy}^2)}\right]$$

Given that;

$\sigma_x = 400\text{MPa};$
$\sigma_y = -300\text{MPa};$
$\tau_{xy} = 200\text{MPa (Clock wise)};$
$\sigma_{1,2} = \frac{1}{2}[(400 - 300) \pm \{(400 + 300)^2 + 4 \times (200)^2\}^{1/2}]$
$\sigma_1 = 453.11\text{MPa}$ ....ANS
$\sigma_2 = -353.11\text{MPa}$ ....ANS

The direction of principal planes is:

$$\tan 2\theta = \frac{2\tau_{xy}}{\sigma_x - \sigma_y}$$

$\tan 2\theta = (2 \times 200)/(400 - (-300))$

$2\theta = 29.04°$ or $209.04°$
$\theta = 14.52°$ or $104.52°$ ....ANS

Since Maximum Shear stress is at $\theta = 45°$

$$\tau = \frac{1}{2}(\sigma_1 - \sigma_2)\sin 2\theta$$
$= \frac{1}{2}(453.11 + 353.11)\sin 90°$
$= 403.11 \text{MPa}$ ....ANS

Now plane of maximum shear
$\theta_s = \theta_p + 45°$
$\theta_s = 14.52° + 45°$ or $104.52° + 45°$
$\theta_s = 59.52°$ or $149.52°$ ....ANS

Q. 14: A piece of steel plate is subjected to perpendicular stresses of 50 N/mm² tensile and 50N/mm² compressive as in the fig 5.13. Calculate the normal and shear/stresses at a plane making 45°. *May–03 (C.O.))*

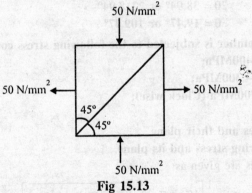

Fig 15.13

Sol.: Given that:
$\sigma_x = 50 \text{ N/mm}^2$
$\sigma_y = -50 \text{ N/mm}^2$
$\theta = 45°$
$\tau = 0$

Normal stress is given as:

$$\sigma_n = \left(\frac{\sigma_x + \sigma_y}{2}\right) + \left(\frac{\sigma_x - \sigma_y}{2}\right)\cos 2\theta + \tau_{xy}\sin 2\theta$$

$\sigma_n = \frac{1}{2}(50 - 50) + \frac{1}{2}(50 - (-50))\cos 90° + 0 \times \sin 90°$
$\sigma_n = 0$ ....ANS

Shear stress at a plane is given by the following equation:

$$\tau = \left(\frac{\sigma_x - \sigma_y}{2}\right)\sin 2\theta - \tau_{xy}\cos 2\theta$$

$\tau = \frac{1}{2}(50 - (-50))\sin 90° - 0 \times \cos 90°$
$\tau = 50 \text{N/m}^2$ ....ANS

Q. 15: The state of stress at a point in a loaded component principal stresses is found to be as given below : $\sigma_x = 50$ GN/m²; $\sigma_y = 150$ GN/m²; $\tau_{xy} = 100$ GN/m²; Determine the principal stresses and maximum shearing stress. Find the orientations of the planes on which they act. *(Dec–03 (C.O.))*

Sol.: Principal stresses is given by the equation:

$$\sigma_{1,2} = \frac{1}{2}\left[(\sigma_x + \sigma_y) \pm \sqrt{(\sigma_x - \sigma_y)^2 + (4\tau_{xy}^2)}\right]$$

$\sigma_{1,2} = \frac{1}{2}[(50 + 150) \pm \{(50 - 150)^2 + 4 \times (100)^2\}^{1/2}]$

$\sigma_{1,2} = 100 \pm 111.8$

$\sigma_1 = 211.8$ GN/m² ....ANS

$\sigma_2 = -11.8$ GN/m² ....ANS

Now Maximum shear stress is given by the equation:

$$t = \frac{1}{2}(\sigma_1 + \sigma_2)\sin 2\theta$$

(Since Maximum Shear stress is at $\theta = 45°$)

$= \frac{1}{2}(211.8 + 11.8)\sin 90°$

$= 111.8$ GN/m² ....ANS

The orientation of the planes on which they act is given by the equation:

$$2\theta = \frac{2\tau_{xy}}{\sigma_x - \sigma_y}$$

$\tan 2\theta = (2 \times 100)/(50 - 150))$

$2\theta = -63.43°$

$\theta = -31.72°$

Major principal plane = $\theta = -31.72°$ ....ANS
Miner principal plane = $\theta + 90° = 58.28°$ ....ANS

Q. 16: A plane element is subjected to following stresses $\sigma_x = 120$ KN/m² (tensile), $\sigma_y = 40$ KN/m² (Compressive) and $\tau_{xy} = 50$ KN/m² (counter clockwise on the plane perpendicular to x-axis) find

(1) Principle stress and their direction
(2) Maximum shearing stress and its directions.
(3) Also, find the resultant stress on a plane inclined 40° with the x-axis. *(May–05 (C.O.))*

Sol.: Given that:

$\sigma_x = 120$ KN/m²
$\sigma_y = -40$ KN/m²
$\tau_{xy} = -50$ KN/m²

(*i*) Calculation for Principle stress and their direction
Principal stresses is given by the equation:

$$\sigma_{1,2} = \frac{1}{2}\left[(\sigma_x + \sigma_y) \pm \sqrt{(\sigma_x - \sigma_y)^2 + (4\tau_{xy}^2)}\right]$$

$\sigma_{1,2} = \frac{1}{2}[(120 - 40) \pm \{(120 + 40)^2 + 4 \times (-50)^2\}^{1/2}]$

$\sigma_{1,2} = 40 \pm 94.34$

$\sigma_1 = 134.34$ KN/m² ....ANS

$\sigma_2 = -54.34$ KN/m² ....ANS

The Direction of the plane is given by the equation:

$$\tan 2\theta = \frac{2\tau_{xy}}{\sigma_x - \sigma_y}$$

$$\tan 2\theta = (2 \times (-50))/(120 + 40))$$
$$2\theta = -32°$$
$$\theta = -16°$$

Major principal plane = $\theta = -16°$ ....ANS
Miner principal plane = $\theta + 90° = 74°$ ....ANS

(ii) Maximum shear stress is given by the equation:

$$\tau = \frac{1}{2}(\sigma_1 - \sigma_2)\sin 2\theta$$

(Since Maximum Shear stress is at $\theta = 45°$)
$$= \frac{1}{2}(134.34 + 54.34)\sin 90°$$
$$= 94.34 \text{ KN/m}^2 \quad\text{........ANS}$$

(iii) Resultant stress on a plane inclined at 40° with x-axis
Since

$$\sigma_n = \left(\frac{\sigma_x + \sigma_y}{2}\right) + \left(\frac{\sigma_x - \sigma_y}{2}\right)\cos 2\theta + \tau_{xy}\sin 2\theta$$

$$\sigma_n = \frac{1}{2}(120 - 40) + \frac{1}{2}(120 + 40)\cos 80° - 50\sin 80°$$
$$= 40 + 13.89 - 49.24$$
$$= 4.65 \text{ KN/m}^2$$

$$\tau = \frac{\sigma_x - \sigma_y}{2}\sin 2\theta - \tau_{xy}\cos 2\theta$$

$$\tau_t = \frac{1}{2}(120 + 40)\sin 80° + 50\cos 80°$$
$$\tau_t = 87.47 \text{ KN/m}^2$$

Resultant stress is given by the equation

$$\sigma_r = \sqrt{\sigma_n^2 + \tau_t^2}$$
$$\sigma_r = (4.65^2 + 87.47^2)^{1/2}$$
$$\sigma_r = 87.6 \text{ KN/m}^2 \quad\text{........ANS}$$

Q. 17: Using Mohr's circle, derive expression for normal and tangential stresses on a diagonal plane of a material subjected to pure shear. Also state and explain mohr's theorem for slope and deflection. *(Dec–00, May–01 (C.O.))*

Sol.: Mohr circle is a graphical method to find the stress system on any inclined plane through the body. It is a circle drawn for the compound stress system. The centre of the circle has the coordinate $\left(\frac{\sigma_x - \sigma_y}{2}, 0\right)$

and radius of circle is $\sqrt{\left(\frac{\sigma_x - \sigma_y}{2}\right)^2 + \tau^2}$ by drawing the mohr's circle of stress the following three systems can be determined.

(a) The normal stress, shear stress, and resultant stress on any plane.
(b) Principal stresses and principal planes.
(c) Maximum shearing stresses and their planes along with the associated normal stress.
Mohr's circle can also be drawn for compound strain system.

1. *Mohr's circle of stress with reference to two mutually perpendicular principal stresses acting on a body consider both dike stresses.*

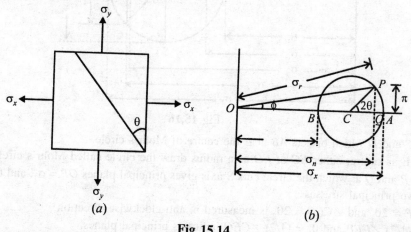

Fig 15.14

Both principal stresses may be considered as (a) Tensile and (b) Compressive.
Let us for tensile
i.e.; $\sigma_x > \sigma_y$
Steps:
1. Mark $OA = S_x$ and $OB = S_y$, along x-axis (on +ve side if tensile and -ve side if compressive).
2. Draw the circle with BA as diameter, called mohr's circle of stress.
3. To obtain stress on any plane, as shown in fig. 15.14 (a) measure angle $ACP = 2_s$ in counter-clock wise direction.
4. The normal stress on the plane is $S_n = OQ$
   Shear stress is $\tau = PQ$
   Resultant stress $= \sigma_r = OP$
   And Angle $POQ = \varphi$ is known as angle of obliquity.

2. Mohr's stress circle for a two dimensional compound stress condition shown in fig.
Let for a tensile
$\sigma_x > \sigma_y$
Steps:
1. Mark $OA = \sigma x$ and $OB = \sigma y$ along x axis
   [on +ve side if tensile and –ve side if compessive.]
2. Mark $AC = \tau_{xy}$ and $BD = \tau_{xy}$

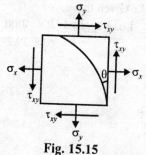

Fig. 15.15

406 / *Problems and Solutions in Mechanical Engineering with Concept*

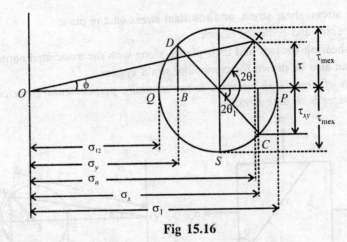

Fig 15.16

3. Join C and D which bisects AB at E the centre of Mohr's circle.
4. With E as centre, either EC or ED as a radius draw the circle called Mohr's circle of stress.
5. Point P and Q at which the circle cuts S axis gives principal planes $OP = \sigma_1$, and $OQ = \sigma_2$ gives the two principal stresses.
6. $\angle CEP = 2\theta_1$ and $\angle CEQ = 2\theta_2$ is measured in anti-clockwise direction.
   $\theta_1 = (1/2) \angle CEP$ and $\theta_1 = (1/2) \angle CEQ$ indicates principal planes.
7. $ER = EC = \tau_{max}$ is the maximum shearing stress.
   $\theta'_1 = (1/2) \angle CER$ and $\theta'_2 = (1/2) \angle CES$
   is measured anti-clockwise direction gives maximum shearing planes.
8. To obtain stress on any plane 'θ' measure $\angle CEX = 2\theta$ in anticlockwise direction.
   Normal stress on the plane is,
   $\quad\quad\quad\quad\quad\quad\sigma_n = OY$
   Shear Stress, $\quad\quad\tau = XY$
   Resultant Stress is $\quad\sigma_r = OX$
   And
   $\angle XOY = \varphi$ is known as angle of obliquity.

Q. 18: A uniform steel bar of 2 cm × 2 cm area of cross-section is subjected to an axial pull of 40000 kg. Calculate the intensity of "normal stress, shear stress and resultant stress on a plane normal to which is inclined at 30° to the axis of the bar. Solve the problem graphically by drawing Mohr Circle. *(Dec–01 (C.O.))*

Sol.: Given that :
Load applied 'P' = 4000 × 9.8 = 39.2 kN
$\sigma_x = 39.2/(2 \times 10^{-2})^2 = 98 MN/m^2$

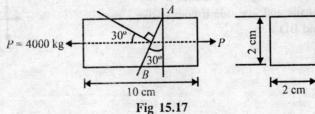

Fig 15.17

Steps to draw Mohr's circle.

Step1: Take origin 'O' and draw a horizontal line OX.

Step2: Cut off OA equal to σ_x by taking scale
1 mm = 1MN/m²

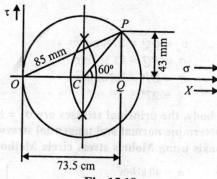

Fig 15.18

Step 3: Bisect OA at C

Step 4: With C as center and radius CA draw a circle.

Step 5: At C draw a line CP at an angle 2θ with OX meeting the circle. At P (θ is angle made by oblique plane with minor principle stress, here zero).

Step 6: Through P draw perpendicular to OX, it intersect OX at Q, join OP. Measure OQ, PQ and OQ as σ, τ and σ_r respectively. Therefore;

Normal stress on the plane σ = OQ = 73.5 × 1 = 73.5MN/m²

Tangential or shear stress on the plane τ = PQ = 43 × 1 = 43 MN/m²

And Resultant stress σ_r = OP = 85 × 1 = 85 MN/m².

Q. 19: The stresses on two mutually perpendicular planes are 40N/mm² (Tensile) and 20N/mm² (Tensile). The shear stress across these planes is 10N/mm². Determine by Mohr's circle method the magnitude and direction of resultant stress on a plane making an angle 30° with the plane of first stress.

Sol.: For the given stress system, the mohr's circle has been drawn and this depicts as shown in fig.

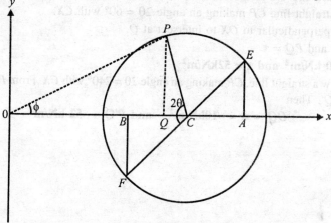

Fig 15.19

$OA = \sigma x = 40\text{N/mm}^2$
$OB = \sigma y = 20\text{N/mm}^2$
$AE = BF = \tau = 10\text{N/mm}^2$
$2\theta = 120°$

Scale 1cm = 5N/mm²
From Measurement:

$\sigma_n = OQ = 27\text{N/mm}^2$
$\sigma_t = PQ = 13.5\text{N/mm}^2$
$\sigma_r = OP = 30\text{N/mm}^2$

And $\varphi = 27°$

Q. 20: At a point in a stressed body, the principal stresses are σ_x = 80 kN/m² (tensile) and σ_y = 40 kN/m² (Compressive). Determine normal and tangential stresses on planes whose normal are at 30° arid 120° with x-axis using Mohr's stress circle Method.

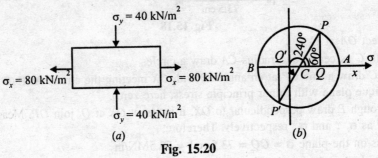

Fig. 15.20

Sol.:

Steps:
1. Taking origin at O draw the axes.
2. Cut off $OA = \sigma_x$ = 80 kN/m² and $OB = -\sigma_y$ = –40 kN/m². Let us choose a suitable scale, say 1 mm = 2 kN/m²
3. Bisect AB at C.
4. With C as centre and CA as radius, draw a circle, which is Mohr's circle.
5. At C draw a straight line CP making an angle 2θ = 60° with CX.
6. From P draw perpendicular to OX to intersect at Q.
7. Then $OQ = \sigma$ and $PQ = \tau$
It is found σ = 50 kN/m² and τ = 52kN/m² ....ANS

Similarly, at C draw a straight line, CP making an angle 2θ = 240° with CX. From P' draw perpendicular to OX to intersect at Q'. Then

$OQ' = \sigma$ = –10kN/m² and $P'Q'$ = –52 kN/m² ....ANS

CHAPTER 16

PURE BENDING OF BEAM

Q. 1: Explain the concept of centre of gravity and centroid.

Sol.: A point may be found out in a body through which the resultant of all such parallel forces acts. This point through which the whole weight of the body acts irrespective of the position of the body is known as centre of gravity. Every body has one and only one centre of gravity.

The plane fig like rectangle, circle, triangle etc., have only areas, but no mass. The centre of area of such figure is known as centroid.

The method of finding out the centroid of a fig is same as that of finding out the C.G. of a body.

$$X = \Sigma A_i \cdot x_i / A$$
$$Y = \Sigma A_i \cdot y_i / A$$
$$C.G. = (X, Y)$$

Q. 2: Explain the following terms:
(i) Area moment of inertia
(ii) Theorem of perpendicular axis,
(iii) Theorem of parallel axis.
(iv) Radius of Gyration
(v) Axis of symmetry.

(May–02, Dec–01 (C.O.))

(i) Area Moment of Inertia

Moment of a force about a point is the product of the force (F) and the perpendicular distance (d) between the point and the line of action of the force i.e. F.d. This moment is also called first moment of force.

If this moment is again multiply by perpendicular distance (d) between the point and the line of action of the force i.e.; $F.d^2$. This quantity is called moment of moment of a force or second moment of force or force moment of inertia. If we take area instead of force it is called Area Moment of inertia.

Unit of area moment of inertia ($A.d^2$) = m^4

For rectangular body: M.I. about X-X axis; $I_{GXX} = bd^3/12$
For rectangular body: M.I. about Y-Y axis; $I_{Gyy} = db^3/12$
For circular body: $I_{GXX} = I_{Gyy} = \pi D^4/64$
For hollow circular body: $I_{GXX} = I_{Gyy} = \pi(D^4 - d^4)/64$

(ii) Theorem of perpendicular axis

It states "If I_{XX} and I_{YY} be the M.I. of a plane section about two mutually perpendicular axes meeting at a point. The M.I. I_{ZZ} perpendicular to the plane and passing through the intersection of X-X and Y-Y is given by the relation"

$$I_{ZZ} = I_{XX} + I_{YY}$$

I_{ZZ} also called as Polar moment of inertia.

(iii) Theorem of parallel axis
It states "If the M.I. of a plane area about an axis passing through its C.G. be denoted by I_G. The M.I. of the area about any other axis AB, parallel to first and a distance 'h' from the C.G. is given by"

$$I_{AB} = I_G + a.h^2$$

Where;
I_{AB} = M.I. of the area about an axis AB
I_G = M.I. of the area about its C.G.
a = Area of the section
h = Distance between C.G. of the section and the axis AB
This formula is reduced to;
$I_{XX} = I_G + a.h^2$; h = distance from x – axis i.e.; Y – y
$I_{YY} = I_G + a.h^2$; h = distance from y – axis i.e.; X – x

(iv) Radius of Gyration (K)
The radius of gyration of a given lamina about a given axis is that distance from the given axis at which all elemental parts of the lamina would have to be placed so as not to alter the moment of inertia about the given axis.

$$K = (I/A)^{1/2}$$
$$K_{xx} = (I_{xx}/A)^{1/2}$$
$$K_{yy} = (I_{yy}/A)^{1/2}$$

Where;
K_{xx} = Radius of gyration of the area from x-x axis
K_{yy} = Radius of gyration of the area from y-y axis

(v) Axis of symmetry
If in a diagram half part of the diagram is mirror image of next half part, then there is a symmetry in the diagram. The axis at which the symmetry is create, called axis of symmetry.

C.G. of the body will lies on axis of symmetry.
If symmetrical about Y axis, then X = 0
If symmetrical about X axis, then Y = 0

Q. 3: What is bending stress ?
The bending moment at a section tends to bend or deflect the beam and the internal stresses resist its bending. The process of bending stops when every cross section sets up full resistance to the bending moment. The resistance offered by the internal stresses to the bending is called the bending stress.

Q. 4: Write down the different assumptions in simple theory of bending. *(Dec–05 (C.O.))*
The following assumptions are made in the theory of simple bending:
1. The material of the beam is homogeneous (i.e.; uniform in density, strength etc.) and isotropic (i.e.; possesses same elastic property in all directions.)
2. The cross section of the beam remains plane even after bending.

3. The beam in initially straight and unstressed.
4. The stresses in the beam are within the elastic limit of its material.
5. The value of Young's modulus of the material of the beam in tension is the same as that in compression.
6. Every layer of the beam material is free to expand or contract longitudinally and laterally.
7. The radius of curvature of the beam is very large compared to the cross section dimensions of the beam.
8. The resultant force perpendicular to any cross section of the beam is zero.

Q. 5: What is simple bending or pure bending of beam?   *(Dec–01, 04, 05 (C.O.))*

If portion of a beam is subjected to constant bending moment only and no shear force acts on that portion as shown in the Fig. 16.1, that portion of the beam is said to be under simple bending or pure bending.

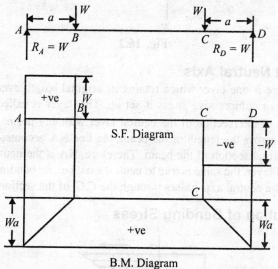

Fig 16.1

A simply supported beam loaded symmetrically as shown in the figure, will be subjected to a constant bending moment over the length BC and on this length shear force is nil. So the portion BC is said to be under simple bending.

Q. 6: Define the following:
  (1) Nature of bending stress
  (2) Neutral layer and Neutral axis
  (3) Nature of Distribution of bending stress.

(1) Nature of Bending Stress

If a beam is not loaded, it will not bend as shown in Fig. 16.2 (a). But if a beam is loaded, it will bend as shown in Fig. 16.2 (b), whatever may be the nature of load and number of loads. Due to bending of the beam, its upper layers are compressed and the lower layers are stretched. Therefore, longitudinal compressive stresses are induced in the upper layers and longitudinal tensile stresses are induced in the lower layers. These stresses are bending stress. In case of cantilevers, reverse will happen, i.e., tensile stresses will be induced in the upper layers and compressive stresses will be induced in the lower layers.

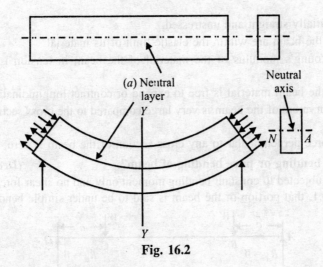

Fig. 16.2

(2) Neutral Layer and Neutral Axis

In a beam or cantilever, there is one layer which retains its original length even after bending. So in this layer neither tensile stress nor compressive stress is set up. This layer is called neutral layer.

Neutral axis is the line of intersection of the neutral layer with any normal section of the beam. If a cutting plane YY is passed across the length of the beam, the line NA becomes the line of intersection of the neutral layer and the normal section of the beam. Therefore, NA is the neutral axis. Since no bending stress is set up in the neutral layer, the same is true to neutral axis, i.e., no bending stress is set tip in neutral axis. It will be proved that the neutral axis passes through the *C.G.* of the section of the beam or cantilever.

(3) Nature of Distribution of Bending Stress

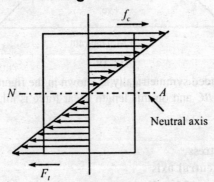

Fig 16.3 *Nature of stress distribution in the section of a beam*

It will be proved that the bending stress at any layer of the section of a beam varies directly as the distance of the layer from the neutral axis. The bending stress is maximum at a layer whose distance from the neutral axis is maximum. The bending stress is gradually reduced as the neutral axis more and more becomes nearer, it becomes zero at the neutral axis, and then again the bending stress increases in the reverse direction (see Fig. 16.3) as the distance of the laver from the neutral axis $N - A$ is increased. The arrows indicating the magnitudes of the bending stress at different layers of a section above and below the neutral axis have been given in opposite directions just to show the difference in nature of stresses in these areas.

Q. 7: Differentiate between direct stress and bending stress.

Direct tensile and compressive stress is set up due to load applied parallel to the length of the object, and direct shear stress is set up in the section which is parallel to the line of action of the shear load. But bending stress is set up due to load at right angles to the length of the object subjected to bending.

In case of direct stress, nature and intensity of stress is the same at any layer in the section of the object subjected to direct stress, but in case of bending stress nature of stress is opposite on opposite sides of the neutral axis, and intensity of stress is different at different layers of the section of the object subjected to bending.

In case of direct stress, intensity of stress is the same in a section taken through any point of the object. But in case of bending stress, intensity of stress is different at the same layer of the section taken through different points of the object.

Q. 8: Explain Moment of resistance?

Sol.: Two equal and unlike parallel forces whose lines of action are not the same, form a couple. The resultant compressive force (P_c) due to compressive stresses on one side of the neutral layer (or neutral axis) is equal to the resultant tensile force (P_t) due to tensile stresses on the other side of it. So these two resultant forces form a couple, and moment of this couple is equal and opposite to the bending moment at the section where the couple acts. This moment is called moment of resistance (M.R.).

Fig 16.4

So, moment of resistance at any section of a beam is defined as the moment of the couple, formed by the longitudinal internal forces of opposite nature and of equal magnitude, set up at that section on either side of the neutral axis due to bending.

In magnitude moment of resistance of any section of a beam is equal to the bending moment at that section of it.

Q. 9: Derive the bending equation i.e.; $M/I = \sigma/y = E/R$. *(Dec–04)*

Sol.: With reference to Fig. 16.5 (a), let us consider any two normal sections AB and CD of a beam at a small distance δL apart (i.e., $AC = BD = \delta L$). Let AB and CD intersect the neutral layer at M and N respectively.

Let;

M = bending moment acting on the beam

θ = Angle subtended at the centre by the arc.

R = Radius of curvature of the neutral layer $M'N'$.

At any distance 'y' from the neutral layer MN, let us consider a layer EF.

Fig. 16.5 (b) shows the beam due to sagging bending moment. After bending, $A'B'$, $C'D'$, $M'N'$ and $E'F'$ represent the final positions of AB, CD, MN and EF respectively.

When produced, $A'B'$ and $C'D'$ intersect each other at O subtending an angle θ radian at O, which is the centre of curvature.

Since δL is very small, arcs $A'C'$, $M'N'$, $E'F'$ and $B'D'$ may be taken as circular.
Now, strain in the layer EF due to bending is given by $e = (E'F' - EF)/EF = (E'F' - MN)/MN$
Since MN is the neutral layer, $MN = M'N'$

$$e = \frac{E'F' - M'N'}{M'N'} = \frac{(R+y)\theta - R\theta}{R\theta} = \frac{y\theta}{R\theta} = \frac{y}{R} \qquad \ldots(i)$$

Let; σ = stress set up in the layer EF due to bending
E = Young's modulus of the material of the beam.

Then $\quad E = \dfrac{\sigma}{e} \quad$ or, $\quad e = \dfrac{\sigma}{e} \qquad \ldots(ii)$

Equate equation (i) and (ii); we get

$$\frac{y}{R} = \frac{\sigma}{E}$$

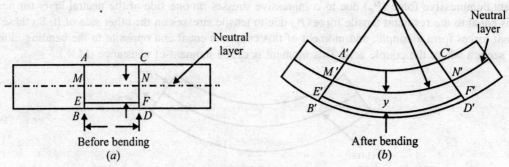

Fig. 16.5

or, $\quad\quad\quad\quad\quad\quad\quad\quad\quad \sigma/y = E/R \qquad \ldots(iii)$

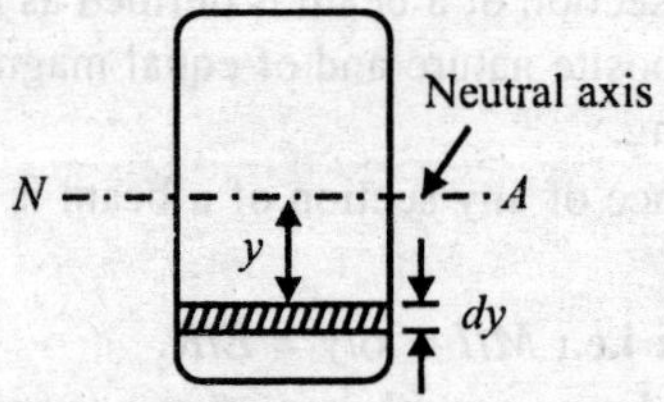

Fig 16.6

With reference to Fig. 16.6.
At the distance 'y', let us consider an elementary strip of very small thickness dy. We have already assumed that 'σ' is the bending stress in this strip.
Let dA = area of this elementary strip.
Then, force developed in this strip = $\sigma.dA$.
Then, elementary moment of resistance due to this elementary force is given by $dM = f.dA.y$
Total moment of resistance due to all such elementary forces is given by

$$\int dM = \int \sigma \times dA \times y$$

or, $\quad\quad\quad\quad\quad M = \int \sigma \times dA \times y \qquad \ldots(iv)$

From Eq. (*iii*), we get

$$\sigma = y \times \frac{E}{R}.$$

Putting this value of f in Eq. (*iv*), we get

$$M = \int y \times \frac{E}{R} \times dA \times y = \frac{E}{R} \int dA \times y^2$$

But $\int dA \cdot y^2 = I$

where I = Moment of inertia of the whole area about the neutral axis N-A.

$$M = (E/R) \cdot I$$
$$M/I = E/R$$

Thus; $M/I = \sigma/y = E/R$

Where;

M = Bending moment
I = Moment of Inertia about the axis of bending i.e; I_{xx}
y = Distance of the layer at which bending stress is consider
   (We take always the maximum value of y, i.e., distance of extreme fibre from N.A.)
E = Modulus of elasticity of the beam material.
R = Radius of curvature

Q. 10: What is section modulus (Z)? What is the value of Bending moment in terms of section modulus?

(Dec–01, May–02)

Sol.: Section modulus is the ratio of M.I. about the neutral axis divided by the outer most point from the neutral axis.

$$Z = I/y_{max}.$$

For Circular Shaft $(Z) = I/y = (\pi D^4/64)/D/2 = (\pi D^3/32)$
For Hollow Shaft $(Z) = I/y = \{\pi(D^4 - d^4)/64\}/D/2 = (\pi/32)(D^4 - d^4)/D$
For Rectangular section $(Z) = I/y = (bd^3/12)/d/2 = bd^2/6$

Section modulus represent the strength of the section of the beam.

Since; $M = \sigma.I/y = \sigma.Z$

Stress at the outer fiber will be maximum. i.e.;

$$M = \sigma_{max}.(I/y_{max}) = \sigma_{max}.Z$$

Q. 11: What is the relation between maximum tensile stress and maximum compressive stress in any section of a beam?

Sol.: For generalization, let us assume inverted angle section of a beam as shown below:

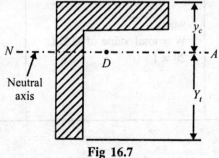

Fig 16.7

In case of a beam, maximum compressive stress will be set up in the topmost layer, and maximum tensile stress will be set up in the bottom most layer due to bending.

Let

σ_c = maximum compressive stress

σ_t = maximum tensile stress

y_C = distance of the topmost layer from the neutral axis N-A

v_t = distance of the bottommost layer from the neutral axis N-A.

Then, according to bending equation, we get

$M/I = \sigma/y$

where

σ = maximum bending stress,

y = distance of the layer at which maximum bending stress occurs, the distance being measured from the neutral axis,

M = maximum moment of resistance
  = maximum $B.M.$

I = Moment of inertia ($M.I.$) of the section of the beam about the neutral axis.

$$M/I = \sigma_c/y_c$$

Also; $\qquad M/I = \sigma_t/y_t$

Equate both we get;

$$\sigma_c/y_c = \sigma_t/y_t$$

or; $\qquad \sigma_c/\sigma_t = y_c/y_t$

Q. 12: Write down the basic formula for maximum bending moment in some ideal cases.

Sol.:

	Berm together with nature of load	Maximum B.M.	Section when maximum B.M. occurs
1.	Cantilever loaded with one point load wat the free end.	WL	It occurs at the fixed end.
2.	Cantilever loaded with U.D.L. over the enitre length.	$\dfrac{WL^2}{2}$ where W = total value of U.D.L. 20 × l	It occurs at the fixed end. w/unit length

Contd...

	Berm together with nature of load	Maximum B.M.	Section when maximum B.M. occurs
3.	Boam loaded with one point load at the mid-span.	$\frac{WL}{4}$ where W = point load placed at the mid-span.	If occurs at the mid-span.
4.		$\frac{WL^2}{8}$ where W = total value of U.D.L. = 20 × l.	If occurs at the mid-span. w/unit length

Q. 13: Find out the M.I. of T section as shown in fig 16.8 about X-X and Y-Y axis through the C.G. of the section.

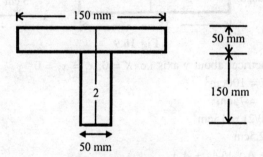

Fig 16.8

Sol.: Since diagram is symmetrical about y axis i.e. $X = 0$

$A_1 = 150 \times 50 = 7500 \text{ mm}^2$

$A_2 = 50 \times 150 = 7500 \text{ mm}^2$

$y_1 = (150 + 50/2) = 175 \text{ mm}$

$y_2 = 150/2 = 75 \text{ mm}$

$Y = (A_1 y_1 + A_2 y_2)/(A_1 + A_2)$

$= (7500 \times 175 + 7500 \times 75)/(7500 + 7500) = 125 \text{ mm}$

C.G. = (0, 125)

Moment of inertia (M.I.) about x-x axis = $I_{XX} = I_{XX1} + I_{XX2}$

$I_{XX1} = I_{GXX1} + A_1 h_1^2 = (bd^3/12)_1 + A_1(Y - y_1)^2 = 150 \times 50^3/12 + 150 \times 50 (125 - 175)^2$
$= 20.31 \times 10^6 \text{ mm}^4$ ...(i)

$I_{XX2} = I_{GXX2} + A_2 h_2^2 = (bd^3/12)_2 + A_2(Y - y_2)^2 = 50 \times 150^3/12 + 50 \times 150 (125 - 75)^2$
$= 32.8125 \times 10^6 \text{ mm}^4$ ...(ii)

$I_{XX} = I_{XX1} + I_{XX2}$
$= 20.31 \times 10^6 + 32.8125 \times 10^6 = 53.125 \times 10^6 \text{ mm}^4$ ...(iii)

Moment of inertia (M.I.) about y-y axis = $I_{yy} = I_{yy1} + I_{yy2}$

Since $X = 0$ i.e.; $X_1 = X_2 = 0$

$I_{yy1} = I_{Gyy1} + A_1 h_1^2 = (db^3/12)_1 + A_1(X - X_1)^2 = (db^3/12)_1 = 50 \times 150^3/12$
$= 14 \times 10^6$ mm^4 ...(iv)

$I_{yy2} = I_{Gyy2} + A_2 h_2^2 = (db^3/12)_2 = 150 \times 50^3/12$
$= 1.5 \times 10^6$ mm^4 ...(v)

$I_{yy} = I_{yy1} + I_{yy2}$
$= 14 \times 10^6 + 1.5 \times 10^6 = 15.5 \times 10^6$ mm^4 ...(vi)

$I_{XX} = 53.125 \times 10^6$ mm^4; $I_{yy} = 15.5 \times 10^6$ mm^4 ....ANS

Q. 14: Find the greatest and least moment of inertia of an inverted T–section shown in fig 16.9.

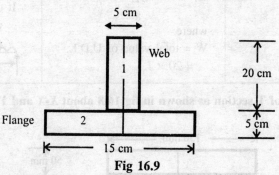

Fig 16.9

Sol.: Since diagram is symmetrical about y axis i.e. $X = 0$, $x_1 = x_2 = 0$

$A_1 = 5 \times 20 = 100$ cm^2
$A_2 = 15 \times 5 = 75$ cm^2
$y_1 = (5 + 20/2) = 15$ cm
$y_2 = 5/2 = 2.5$ cm
$Y = (A_1 y_1 + A_2 y_2)/(A_1 + A_2)$
$= (100 \times 15 + 75 \times 2.5)/(100 + 75) = 9.64$ cm
C.G. $= (0, 9.64)$

Moment of inertia (M.I.) about x-x axis $= I_{xx} = I_{XX1} + I_{XX2}$

$I_{XX1} = I_{GXX1} + A_1 h_1^2 = (bd^3/12)_1 + A_1(Y - y_1)^2 = 5 \times 20^3/12 + 5 \times 20 (9.64 - 15)^2$
$= 6206.09$ cm^4 ...(i)

$I_{XX2} = I_{GXX2} + A_2 h_2^2 = (bd^3/12)_2 + A_2(Y - y_2)^2 = 15 \times 5^3/12 + 15 \times 5 (9.64 - 2.5)^2$
$= 3979.72$ cm^4 ...(ii)

$I_{xx} = I_{XX1} + I_{XX2}$
$= 6206.09 + 3979.72 = 10186.01$ cm^4 ...(iii)

Moment of inertia (M.I.) about y-y axis $= I_{yy} = I_{yy1} + I_{yy2}$; $h = 0$
Since $X = 0$ i.e.; $X_1 = X_2 = 0$

$I_{yy1} = I_{Gyy1} + A_1 h_1^2 = (db^3/12)_1 + A_1(X - X_1)^2 = (db^3/12)_1$
$I_{yy2} = I_{Gyy2} + A_2 h_2^2 = (db^3/12)_2$
$I_{yy} = I_{yy1} + I_{yy2}$
$= 20 \times 5^3/12 + 5 \times 15^3/12 = 1614.25$ cm^4 ...(iv)

Greatest Moment of inertia $= I_{XX} = $ **10186.01 cm^4** ....ANS

Q. 15: An I section has the following dimensions Top flange = 8 cm × 2 cm; Bottom flange = 12 cm × 2 cm; Web = 12 cm × 2 cm; Over all depth of the section = 16 cm. Determine the *MI* of the *I* section about two centroidal axis.

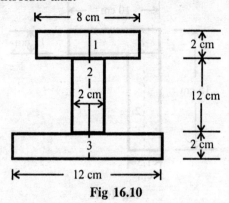

Fig 16.10

Sol.: Since diagram is symmetrical about y axis i.e. $X = 0$, $x_1 = x_2 = 0$

$A_1 = 8 \times 2 = 16$ cm^2

$A_2 = 12 \times 2 = 24$ cm^2

$A_3 = 12 \times 2 = 24$ cm^2

$y_1 = (2 + 12 + 2/2) = 15$ cm

$y_2 = (2 + 12/2) = 8$ cm

$y_2 = 2/2 = 1$ cm

$Y = (A_1 y_1 + A_2 y_2 + A_3 y_3)/(A_1 + A_2 + A_3)$

$= (16 \times 15 + 24 \times 8 + 24 \times 1)/(16 + 24 + 24) = 7.125$ cm

C.G. = (0, 7.125)

Moment of inertia (M.I.) about x-x axis = $I_{XX} = I_{XX1} + I_{XX2}$

$I_{XX1} = I_{GXX1} + A_1 h_1^2 = (bd^3/12)_1 + A_1(Y - y_1)^2 = 8 \times 2^3/12 + 16 (7.125 - 15)^2$

$= 997.58$ cm^4 ...(i)

$I_{XX2} = I_{GXX2} + A_2 h_2^2 = (bd^3/12)_2 + A_2(Y - y_2)^2 = 2 \times 12^3/12 + 24 (7.125 - 8)^2$

$= 306.375$ cm^4 ...(ii)

$I_{XX3} = I_{GXX3} + A_3 h_3^2 = (bd^3/12)_3 + A_2(Y - y_3)^2 = 12 \times 2^3/12 + 24 (7.125 - 1)^2$

$= 908.375$ cm^4 ...(iii)

$I_{XX} = I_{XX1} + I_{XX2} + I_{XX3}$

$= 997.58 + 306.375 + 908.375 = 2212.33$ cm^4 ...(iv)

Moment of inertia (M.I.) about y-y axis = $I_{yy} = I_{yy1} + I_{yy2}$; $h = 0$

Since $X = 0$ i.e.; $X_1 = X_2 = 0$

$I_{yy} = I_{yy1} + I_{yy2} + I_{yy3}$

$= (db^3/12)_1 + (db^3/12)_2 + (db^3/12)_3$

$= 2 \times 8^3/12 + 12 \times 2^3/12 + 2 \times 12^3/12 = 381.33$ cm^4 ...(iv)

$I_{XX} = 2212.33$ cm^4; $I_{yy} = 381.33$ cm^4 ....ANS

Q. 16: Determine the *MI* of an unequal angle section 15cm × 10cm × 1.5cm with longer leg vertical and flange upwards.

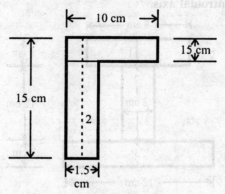

Fig 16.11

Sol.:
$A_1 = 10 \times 1.5 = 15$ cm^2
$A_2 = 15 \times 1.5 = 22.5$ cm^2
$y_1 = (13.5 + 1.5/2) = 14.25$ cm
$y_2 = 13.5/2 = 6.75$ cm
$x_1 = (10/2) = 5$ cm
$x_2 = 1.5/2 = 0.75$ cm
$X = (A_1x_1 + A_2x_2)/(A_1 + A_2)$
$\quad = (15 \times 5 + 22.5 \times 0.75)/(15 + 22.5) = 2.56$ cm
$Y = (A_1y_1 + A_2y_2)/(A_1 + A_2)$
$\quad = (15 \times 14.25 + 22.5 \times 6.75)/(15 + 22.5) = 9.94$ cm
C.G. = (2.56, 9.94)

Moment of inertia (M.I.) about x-x axis = $I_{xx} = I_{xx1} + I_{xx2}$
$I_{xx} = (bd^3/12)_1 + A_1(Y - y_1)^2 + (bd^3/12)_2 + A_2(Y - y_2)^2$
$\quad = 10 \times 1.5^3/12 + 15 (9.94 - 14.25)^2 + 1.5 \times 13.5^3/12 + 22.5 (9.94 - 6.75)^2$
$\quad = 795.07$ cm^4 ....(i)

Moment of inertia (M.I.) about y-y axis = $I_{yy} = I_{yy1} + I_{yy2}$
$I_{yy} = I_{yy1} + I_{yy2}$
$\quad = [(db^3/12)_1 + A_1(X - X_1)^2] + [(db^3/12)_2 + A_1(X - X_2)^2]$
$\quad = 1.5 \times 10^3/12 + 15 (2.56 - 5)^2 + 13.5 \times 1.5^3/12 + 22.5 (2.56 - 0.75)^2$
$\quad = 284.44$ cm^4 ...(ii)
$I_{xx} = 795.07$ cm^4; $I_{yy} = 284.44$ cm^4 ....ANS

Q. 17: A beam made of *C.I.* having a section of 50mm external diameter and 25 mm internal diameter is supported at two points 4 m apart. The beam carries a concentrated load of 100N at its centre. Find the maximum bending stress induced in the beam. *(Dec–01 (C.O.))*

Sol.:

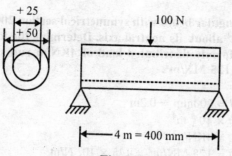

Fig 16.12

This problem is the case of simply supported beam with load at its mid point, and in this case maximum bending moment = $M = WL/4$

$$= (100 \times 4)/4 = 100 \text{ Nm} = 100 \times 10^3 \text{ N.mm} \qquad ...(i)$$

Let I = Moment of inertia = $\frac{\pi}{64}(d_0^4 - d_i^4) = \pi/64 \,(50^4 - 25^4) = 287621.4 \text{ mm}^4$ ...(ii)

$$y = d/2 = 50/2 = 25 \text{mm} \qquad ...(iii)$$

We know that
Maximum stress during Bending moment (at $y = 25$mm) = $M.y/I$
$$= (100 \times 10^3 \times 25)/287621.4 = 8.69 \text{ N/mm}^2$$

Maximum bending stress = 8.69 N/mm² ....ANS

Q. 18: A steel bar 10 cm wide and 8 mm thick is subjected to bending moment. The radius of neutral surface is 100 cm. Determine maximum and minimum bending stress in the beam. *(May–02)*

Sol.:

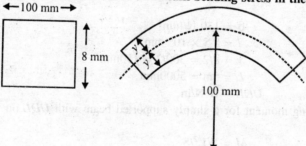

Fig 16.13

Assume for steel bar $E = 2 \times 10^5$ N/mm²

$y_{max} = 4$mm
$R = 1000$mm
$\sigma_{max} = E.y_{max}/R = (2 \times 10^5 \times 4)/1000$

We get maximum bending moment at lower most fiber, Because for a simply supported beam tensile stress (+ive value) is at lower most fiber, while compressive stress is at top most fiber (–ive value).

$\sigma_{max} = 800$ N/mm² ....ANS
$y_{min} = -4$mm
$R = 1000$mm
$\sigma_{min} = E.y_{min}/R = (2 \times 10^5 \times - 4)/1000$
$\sigma_{max} = -800$ N/mm² ....ANS

Q. 19: A simply supported rectangular beam with symmetrical section 200mm in depth has moment of inertia of 2.26 x 10⁻⁵ m⁴ about its neutral axis. Determine the longest span over which the beam would carry a uniformly distributed load of 4KN/m run such that the stress due to bending does not exceed 125 MN/m². *(May–03)*

Sol.: Given data:
Depth $\quad d = 200\text{mm} = 0.2\text{m}$
I = Moment of inertia = 2.26×10^{-5} m⁴
UDL = 4KN/m
Bending stress $\quad \sigma = 125$ MN/m² = 125×10^6 N/m²
Span = ?

Since we know that Maximum bending moment for a simply supported beam with UDL on its entire span is given by = $WL^2/8$
i.e; $\quad M = WL^2/8$ ...(i)

Now from bending equation $M/I = \sigma/y_{max}$
$\quad y_{max} = d/2 = 0.2/2 = 0.1$m
$\quad M = \sigma.I/y_{max} = [(125 \times 10^6) \times (2.26 \times 10^{-5})]/0.1 = 28250$ Nm ...(ii)

Substituting this value in equation (i); we get
$\quad 28250 = (4 \times 10^3)L^2/8$
$\quad L = 7.52$m $\quad$ ....ANS

Q. 20: A rectangular beam 300 mm deep is simply supported over a span of 5 m. What uniformly distributed load per meter the beam may carry? If the bending stress is not to exceed 130N/mm². Take $I = 8.5 \times 10^6$ mm⁴. *(Sep–(C.O.)03)*

Sol.: Given data:
$\quad \sigma = 130$ N/mm²
$\quad I = 8.5 \times 10^6$ mm⁴
$\quad y = d/2 = 300/2 = 150$mm
$\quad L = 5$m = 5000mm
Let $\quad$ UDL = W N/m

Maximum bending moment for a simply supported beam with UDL on its entire span is given by = $WL^2/8$
i.e; $\quad M = WL^2/8$ ...(i)

Now from bending equation $M/I = \sigma/y_{max}$
$\quad M = \sigma.I/y_{max} = [(130) \times (8.5 \times 10^6)]/150 = 7366666.67$ Nmm ...(ii)

Substituting this value in equation (i); we get
$\quad 7366666.67 = W(5000)^2/8$
$\quad W = 2.357$ N/mm = 2357.3 N/m $\quad$ ....ANS

Q. 21: A rectangular beam of 200 mm in width and 400 mm in depth is simply supported over a span of 4m and carries a distributed load of 10 KN/m. Determine maximum bending stress in the beam. *(Dec –03)*

Sol.:

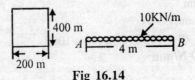

Fig 16.14

Given data:
$$b = 200 \text{ mm} = 0.2 \text{ m}$$
$$d = 400 \text{ mm} = 0.4 \text{ m}$$
$$L = 4 \text{ m}$$
$$W = 10 \text{KN/m}$$
$$\sigma_{max} = ?$$

We know that $\sigma_{max}/y = M/I$; $\sigma_{max} = y.M/I$
Here;
$$y = d/2 = 0.2 \text{ m}$$
$$M = WL^2/8 = (10 \times 10^3 \times 4^2)/8 = 20000 \text{ Nm}$$
$$I = bd^3/12 = (0.2 \times 0.4^3)/12 = 0.001066 \text{ m}^4$$

Putting all the value we get;
$$\sigma_{max} = (0.2 \times 20000)/0.001066 = 3750000 \text{ N/m}^2$$
$$= 3.75 \text{MPa} \quad ....\text{ANS}$$

Q. 22: A wooden beam of rectangular cross section is subjected to a bending moment of 5KNm. If the depth of the section is to be twice the breadth and stress in wood is not to exceed 60N/cm². Find the dimension of the cross section of the beam. *(Dec–05)*

Sol.: Since We know that $\sigma/y = M/I$;
Where
$$\sigma = 60 \text{ N/cm}^2 = 60 \times 10^4 \text{ N/m}^2$$
$$d = 2b$$
$$y = d/2$$
$$M = 5 \text{KNm} = 5 \times 10^3 \text{ Nm}$$
$$I = bd^3/12; \text{ for rectangular cross section}$$

Substituting all the values we get
$$60 \times 10^4 / (d/2) = 5 \times 10^3/(bd^3/12)$$
$$d = 2b;$$
$$60 \times 10^4 / (2b/2) = 5 \times 10^3/(b.(2b)^3/12)$$
$$b = 0.232 \text{ m} = 23.2 \text{ cm}$$
$$d = 2b = 46.4 \text{ cm}$$
$$\mathbf{b = 23.2 \text{ cm}, d = 46.4 \text{ cm}} \quad ....\text{ANS}$$

Q. 23: Find the dimension of the strongest rectangular beam that can be cut out of a log of 25 mm diameter.

Sol.:
$$b^2 + d^2 = 25^2$$
$$d^2 = 25^2 - b^2$$
Since $M/I = \sigma/y$; $M = \sigma(I/y) = \sigma.Z$
M will be maximum when Z will be maximum
$$Z = I/y = (bd^3/12)/(d/2) = bd^2/6 = b.(25^2 - b^2)/6$$
The value of Z maximum at $dZ/db = 0$;
i.e.; $d/db[25^2 b/6 - b^3/6] = 0$
$$25^2/6 - 3b^2/6 = 0 \quad b^2 = 625/3;$$
$$\mathbf{b = 14.43 \text{ mm}} \quad ....\text{ANS}$$
$$\mathbf{d = 20.41 \text{ mm}} \quad ....\text{ANS}$$

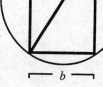

Fig. 16.15

Q. 24: In previous question specify the safe maximum Spain for the simply supported beam of rectangular section when it is to carry a *UDL* of 2.5KN/m and bending stress are limited to 10MN/m².

Sol.: Since $M/I = \sigma/y; M = \sigma(I/y) = \sigma(bd^3/12)/(d/2) = \sigma(bd^2)/6$
$$M = \{(10 \times 10^6) \times 14.43 \times 20.41^2 \times 10^{-6}\}/6 = 10KN\text{–}m \qquad ...(i)$$

Since Maximum bending moment for a simply supported beam with *UDL* on its entire span is given by
$$M = WL^2/8$$
$$10 = 2.5.L^2/8;$$
$$L = 5.66 \text{ m} \qquad \text{........ANS}$$

Q. 25: A beam having *I* – section is shown in fig is subjected to a bending moment of 500 Nm at its Neutral axis. Find maximum stress induced in the beam.

Sol.: Since diagram is symmetrical about *y*–axis.
$$Y = (A_1y_1 + A_2y_2 + A_3y_3)/(A_1 + A_2 + A_3)$$

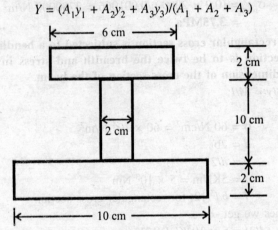

Fig 16.16

$A_1 = 6 \times 2 = 12\text{cm}^2$
$A_2 = 10 \times 2 = 20\text{cm}^2$
$A_3 = 10 \times 2 = 20\text{cm}^2$
$y_1 = 2 + 10 + 1 = 13\text{cm}$
$y_2 = 2 + 5 = 7 \text{ cm}$
$y_1 = 1 = 1 \text{ cm}$

putting all the values; we get
$$Y = \{12 \times 13 + 20 \times 7 + 20 \times 1\}/(12 + 20 + 20)$$
$$Y = 6.08 \qquad ...(i)$$

Moment of inertia about an axis passing through its *C.G.* and parallel to *X – X* axis.
$$I = I_{XX1} + I_{XX2} + I_{XX3}$$
$$I_{XX1} = I_{G1} + A_1h_1^2$$
$$= bd^3/12 + A_1(Y - y_1)^2$$
$$I_{XX2} = I_{G2} + A_2h_2^2$$
$$= bd^3/12 + A_2(Y - y_2)^2$$
$$I_{XX3} = I_{G3} + A_3h_3^2$$
$$= bd^3/12 + A_3(Y - y_3)^2$$

$$I = [bd^3/12 + A_1(Y-y_1)^2] + [bd^3/12 + A_2(Y-y_2)^2] +$$
$$[bd^3/12 + A_3(Y-y_3)^2]$$
$$= [(6 \times 2^3)/12 + 12 \times (6.08-13)^2] + [(2 \times 10^3)/12 + 20 \times$$
$$(6.08-7)^2] + [(10 \times 2^3)/12 + 20 \times (6.08-1)^2]$$
$$I = 1285 \text{ cm}^4$$

Distance of C.G. from upper extreme fiber $y_c = 14 - 6.08 = 7.92$ cm
Distance of C.G. from lower extreme fiber $y_t = 6.08$ cm
Therefore we will take higher value of y i.e.; $y = 7.92$,
which gives the maximum value of stress but compressive in nature.

At $y_t = 6.08$ cm, we get the tensile stress, but at this value of y, we don't get the maximum value of stress, Since Our aim is to find out the maximum value of stress in the beam which we get either on top most fiber or on bottom most fiber, depending upon the distance of fiber from centre of gravity. We always take the maximum value of y, because;

$$M/I = \sigma_{max}/y; \quad \sigma_{max} = M.y/I,$$
$$\sigma_{max} = 500 \times 10^2 \times 7.92/1285 = 308.2 \text{ N/cm}^2$$
$$\sigma_{max} = 308.2 \text{ N/cm}^2 \quad \text{........ANS}$$

Q. 26: A cast iron bracket subjected to bending has cross section of I – form with unequal flanges. If maximum Bending moment on the section is 40 MN-mm, determine Maximum bending stress. What should be the nature of stress?

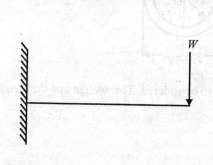
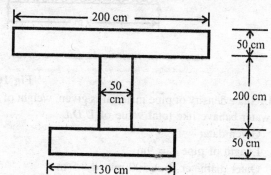

Fig 16.17

Sol.:

Since diagram is symmetrical about y–axis.
$$Y = (A_1y_1 + A_2y_2 + A_3y_3)/(A_1 + A_2 + A_3)$$
$A_1 = 200 \times 50 = 10000 \text{ mm}^2$
$A_2 = 50 \times 200 = 10000 \text{ mm}^2$
$A_3 = 130 \times 50 = 6500 \text{ mm}^2$
$y_1 = 50 + 200 + 25 = 275$ mm
$y_2 = 50 + 100 = 150$ mm
$y_1 = 25 = 25$ mm

putting all the values; we get
$$Y = \{10000 \times 275 + 10000 \times 150 + 6500 \times 25\}/(10000 + 10000 + 6500)$$
$$Y_C = 166.51 \text{ mm (N.A. from bottom face)} \quad ...(i)$$

Moment of inertia about an axis passing through its C.G. and parallel to X–X axis.

$I = I_{XX1} + I_{XX2} + I_{XX3}$

$I = [bd^3/12 + A_1(Y - y_1)^2] + [bd^3/12 + A_2(Y - y_2)^2] + [bd^3/12 + A_3(Y - y_3)^2]$

$= [(200 \times 50^3)/12 + 10000 \times (166.51 - 275)^2] + [(50 \times 200^3)/12$
$\qquad + 10000 \times (166.51 - 150)^2] + [(130 \times 50^3)/12 + 6500 \times (166.51 - 25)^2]$

$I = 287360458.3$ mm^4

Since the beam is cantilever, so Tensile stress is at top most fiber and compressive stress is at bottom.

Distance of C.G. from upper extreme fiber $y_t = 300 - 166.51 = 133.49$ mm

Distance of C.G. from lower extreme fiber $y_c = 166.51$ mm

Therefore we will take higher value of y i.e.; $y_c = 166.51$ mm, and we get compressive stress.

$M/I = \sigma_{max}/y; \sigma_{max} = M.y_c/I,$

$\sigma_{max} = 40 \times 10^6 \times 166.51/284907234.9 = 23.37$ N/mm^2

$\sigma_{max} = 23.37$ N/mm^2(compressive) ....ANS

Q. 27: A *C.I.* water pipe 450 mm bore and 500 mm outer dia is supported at two points 9 m apart. Find maximum stress when pipe is running full. $\sigma_{CI} = 7.2$ gm/c.c., and that of water is 1000 Kg/m^3.

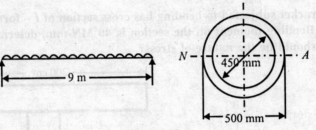

Fig 16.18

Sol.: Since density of pipe material is given, weight of the pipe is to considered. This weight and the weight of water behave like total value of *U.D.L.*

Given data:

Length of pipe $L = 9$m

Outer diameter $d_o = 500$ mm $= 0.5$ m

Inner diameter $d_i = 450$ mm $= 0.45$ m

Density of C.I. = 7.2 gm/c.c. = 7.2 $\times 10^6/1000$ =7200 Kg/m^3

Density of water = 1000 kg/m^3

Volume of C.I. pipe $V_{CI} = \pi/4(d_0^2 - d_i^2) \times L = \pi/4(0.5^2 - 0.45^2) \times 9 = 0.3357$ m^3

Weight of C.I. pipe $= \pi_{CI}.V_{CI}.g = 7200 \times 0.3357 \times 9.81 = 23715.29$ N

Volume of water contained in the pipe $V_W = \pi/4(d^2) \times L = \pi/4(0.45^2) \times 9 = 1.43$ m^3

Weight of water pipe $= \pi_W.V_W.g = 1000 \times 1.43 \times 9.81 = 14041.95$ N

Total value of *UDL* on the pipe $(W \times L)$ = self weight of the pipe + weight of water
$\qquad\qquad\qquad\qquad\qquad = 23715.29 + 14041.95 = 37757.24$ N

Now a beam loaded with *UDL* over the entire length, maximum M.M. is given by

$M = WL^2/8 = WL.L/8 = (37757.24 \times 9000)/8 = 42476895$ Nmm

Let σ = required maximum bending stress up in the pipe material.

Then; $\qquad M/I = \sigma/y$

Where;
$$I = \pi/64(d_0^4 - d_i^4) = \pi/64(500^4 - 450^4) = 1055072000 \text{ mm}^4$$
y = distance of farthest layer from the neutral axis = 500/2 = 250 mm
putting all the value we get
$\qquad 42476895/1055072000 = \sigma/250;$
$$\sigma = 10.06 \text{ N/mm}^2 \qquad \text{........ANS}$$

Q. 28: A beam is freely supported on supports as shown in fig 16.19 carries a *UDL* of 12 KN/m and a concentrated load. If the stress in beam is not to exceed 8N/mm². Design a suitable section making the depth twice the width.

Sol.: For finding out Maximum bending moment which is at point 'C'
$\qquad \Sigma M_A = 0;$
$-R_B \times 6 + 9000 \times 2.5 + 12000 \times 6 \times 3 = 0$
Since $\qquad R_A + R_B = 9000 + 12000 \times 6 = 81000$ N
$\qquad\qquad R_B = 39750$ N;
$\qquad\qquad R_A = 41250$ N

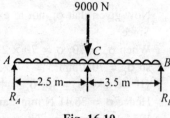

Fig. 16.19

For shear force equation;
Consider right hand side of the section; we get
$$SF_{1-1} = 39750 - 12000 \cdot X = 0; \ X = 3.31 \text{ m}$$
Maximum *B.M.* at X = 3.31m
$$= R_B \cdot X - 12000 \cdot X^2/2$$
Maximum *B.M.* = 65836 × 10³ N–mm
$$\sigma = y.M/I \text{ or}; \ M = \sigma.I/y = \sigma[(bd^3/12)/d/2] = \sigma.bd^2/6$$
$$65836 \times 10^3 = [8 \times d/2 \times d^2]/6$$
$$d = 462 \text{ mm}, \ b = 231 \text{ mm} \qquad \text{........ANS}$$

Q. 29: A *C.I.* beam 2.75m long is shown in the fig 16.20 with support reaction $R_A = 0.375W$ and $R_B = 0.625W$, with loads. The beam has *T*–section. If tensile and compressive stress are not to exceed 40N/mm², and 70 N/mm² respectively. Find the safe concentrated load '*W*' that can be applied at the rigid end of the beam.

Sol.: Maximum bending moment for beam
$\qquad M = W \times 0.75 = 0.75W$ N–m $\qquad\qquad$ ...(*i*)
$\qquad Y = [A_1y_1 + A_2y_2]/[A_1 + A_2]$
$\qquad\quad = [(150 \times 20)(90) + (20 \times 80)(40)]/[(150 \times 20) + (20 \times 80)]$
$\qquad\quad = 72.61$ mm
Y from top most fiber = 100 – 72.61 = 27.39 mm
For moment of inertia *I*;
$\qquad I = I_{xx1} + I_{XX2}$
$\qquad\quad = [150 \times 20^3/12 + 150 \times 20 (72.61 - 90)^2] + [20 \times 80^3/12 + 20 \times 80 (72.61 - 40)^2]$
$\qquad\quad = 1007236.3 + 2554792.69 = 3562028.99 \text{ mm}^4 \qquad$ ...(*ii*)

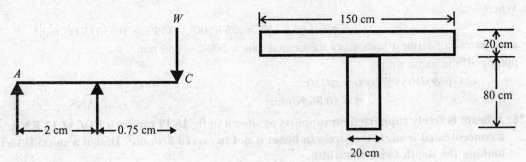

Fig 16.20

Now given that σ_C not to exceed 70 N/mm² & σ_t not to exceed 40 N/mm²
Since $\sigma_t / \sigma_C = y_t/y_c$
When $\sigma_C = 70$; $\sigma_t = 70 \times 27.39/72.61 = 26.41$ N/mm², which is less than 40N/mm²;
so $\sigma_C = 70$N/mm²;
{If we take $\sigma_t = 40$; $\sigma_C = 40 \times 72.71/27.39 = 106.03 > 70$; Not satisfied.}
Hence $\sigma_t = 26.41$ N/mm²; and $\sigma_C = 70$ N/mm²;
We take always maximum stress i.e; $\sigma = 70$ N/mm²;
Maximum moment of resistance $M = \sigma.I/y = 70 \times 3562028.99/72.61$
$$= 3434037 \text{ N-mm} = 343.037 \text{ N-m} \quad ...(iii)$$
This value is equal to 0.75W;
$$0.75W = 343.037$$
$$W = 4.579 \text{KN} \quad ....\text{ANS}$$

Q. 29: Cross section of beam is shown in fig 16.21, permissible stress in compression and tension are 1000 Kg/cm² and 1400 Kg/cm². It is subjected to a moment causing compression at top and tension at bottom. Calculate compression in top flange and tension in bottom flange.

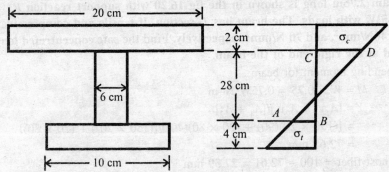

Fig 16.21

Sol.:
$$Y = [A_1y_1 + A_2y_2 + A_3y_3]/[A_1 + A_2 + A_3]$$
$$[(20 \times 8)(36) + (6 \times 28)(18) + (10 \times 4)(2)]/[(20 \times 8) + (6 \times 28) + (10 \times 4)]$$
$$= 24.1 \text{cm} \quad ...(i)$$
$$I = [20 \times 8^3/12 + 20 \times 8(24.1 - 36)^2] + [6 \times 28^3/12 + 6 \times 28(24.1 - 18)^2] +$$
$$[10 \times 4^3/12 + 10 \times 4 (24.1-2)^2]$$
$$= 60470 \text{ cm4} \quad ...(ii)$$

Now; $y_t = 24.1$ cm
$y_C = 40 - 24.1 = 15.9$ cm
Let design for $\sigma_c = 1000$ Kg/cm²
$$\sigma_t/\sigma_c = y_t/y_C$$
$$\sigma_t = y_t/y_C \cdot \sigma_c$$
$$= 24.1 \times 1000/ 15.9 = 1514.9 \text{ kg/cm}^2$$
but 1514.9 > 1400; hence design not safe
Now design for $\sigma_t = 1400$ kg/cm²
$$\sigma_C/\sigma_t = y_C/y_t$$
$$\sigma_C = y_C/y_t \cdot \sigma_t$$
$$= 1400 \times 15.9/ 24.1 = 923.65 \text{ kg/cm}^2$$
Since 923.65 > 1000; hence design safe.
So $\sigma_t = 1400$ kg/cm²; $\sigma_C = 923.65$ kg/cm²
Now ask that compression in top flange and tension in bottom flange

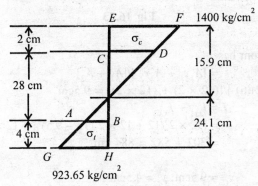

Fig 16.22

For compressive stress σ_C at CD;
At EF $\sigma_C = 923.65$ at $y_C = 15.9$ cm
Now σ_C at $y_C = 15.9 - 8 = 7.9$ cm
$$\sigma_{CD} = 923.65 \times 7.9 /15.9 = 458.9 \text{ Kg/cm}^2$$
Now total compression in top flange = P_C = Area × average compressive stress
$$= (8 \times 20) \times \{(923.65 + 458.9)/2\}$$
$$P_C = 110640 \text{ Kg} \qquad \text{.......ANS}$$
For tensile stress σ_t at AB;
At GH $\sigma_t = 1400$ at $y_t = 24.1$ cm
Now σ_t at $y_t = 24.1 - 4 = 20.1$ cm
$$\sigma_{AB} = 1400 \times 20.1 /24.1 = 1167.6 \text{ Kg/cm}^2$$
Now total Tension in bottom flange = P_t = Area × average tensile stress
$$= (10 \times 4) \times \{(1400 + 1167.6)/2\}$$
$$P_t = 51350 \text{ Kg} \qquad \text{.......ANS}$$

Q. 30: A *T* section with 14 cm overall depth having flange width of 12 cm as shown in fig 16.23. The beam rests on two supports. If allowable tensile and compressive stresses are 10KN/cm² & 6KN/cm². Calculate *UDL* covering the entire length of 5 m which can be safely carried by the beam.

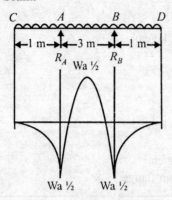
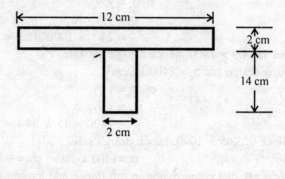

Fig 16.23

Sol.:

Y = Distance of *C.G.* from base

$$= [A_1 y_1 + A_2 y_2]/[A_1 + A_2]$$

$[(12 \times 2)(13) + (12 \times 2)(6)]/[(12 \times 2) + (12 \times 2)] = 9.5 \text{cm}$ ...(i)

$$I = I_{xx1} + I_{XX2}$$
$$= [12 \times 2^3/12 + 12 \times 2 (13 - 9.5)^2] + [2 \times 12^3/12 + 2 \times 12 (9.5 - 6)^2]$$
$$= 302 + 582 = 884 \text{ cm}^4 \qquad ...(ii)$$

For *CA* & *BD*

$y_C = 9.5 \text{cm}; \ y_t = 4.5 \text{cm}$

Let design for $\sigma_t = 10 \text{ kN/cm}^2$

$$\sigma_C/\sigma_t = y_C/y_t$$
$$\sigma_C = y_C/y_t \cdot \sigma_t$$
$$= 10 \times 9.59/ 4.5 = 21.11 \text{ kN/cm}^2$$

Since 21.11 > 6; So tensile stress can not be allowed to be 10KN/cm², hence design safe.

Now design for $\sigma_c = 6 \text{ KN/cm}^2$

$$\sigma_t/\sigma_c = y_t/y_C$$
$$\sigma_t = y_t/y_C \cdot \sigma_c$$
$$= 6 \times 4.5/ 9.5 = 2.84 \text{ kN/cm}^2$$

This value 2.84 < 10; hence design safe for $\sigma_t = 2.84 \text{ kN/cm}^2$

Hence $\sigma_t = 2.84 \text{ kN/cm}^2$ and $\sigma_C = 6 \text{ kN/cm}^2$

Now $M/I = \sigma/y$;

Where; $\sigma = 6 \text{ kN/cm}^2$, because we take always maximum value, hence design for compression

$I = 884 \text{ cm}^4$

$y = 9.5$ cm, always take maximum distance of fiber from *C.G.*

$$M = 6 \times 884/9.5 = 558.32 \text{ KN/cm}^2 \qquad ...(iii)$$

Since *M* = Maximum bending moment

$$= Wa^2/2 = W(100)^2/2 = 5000W \text{ N/cm}^2 = 50W \text{ KN/cm}^2 \qquad ...(iv)$$

Equate equation (iii) and (iv) we get
$$558.32 = 50W$$
$$W = 11.16 \text{ KN/cm} \quad \text{.......ANS}$$

Portion AB react like a simply supported beam

Now $\quad y_t = 9.5$ cm; $y_c = 4.5$ cm

Let design for $\quad \sigma_t = 10$ kN/cm^2

$$\sigma_c/\sigma_t = y_c/y_t$$
$$\sigma_c = y_c/y_t \cdot \sigma_t$$
$$= 10 \times 4.5/9.5 = 4.73 \text{ kN/cm}^2$$

Since 4.73 < 6; hence design safe.

Hence $\quad \sigma_t = 10$ kN/cm^2 and $\sigma_c = 4.73$ kN/cm^2

Now $\quad M/I = \sigma/y$;
$$M = 884 \times 10/9.5 = 930.53 \text{ KN/cm}^2$$

Maximum Bending moment = $Wa^2/2 = W(100)^2/2 = 50W$
$$50 W = 930.53$$
$$W = 18.6 \text{KN/cm} \quad \text{.......ANS}$$

Q. 31: A steel wire of 5 mm diameter is bend into a circular shape of 6 m radius. Determine the maximum stress in the wire. Take $E = 2.0 \times 10^6$ kg/cm^2. (UPTUQB)

Sol.: The steel wire after bending, forms a curved beam. Hence we apply the flexural formula

$$\sigma/y = E/R, \text{ in which}$$
$$y = d/2 = 5/2 \text{ mm} = 0.25 \text{cm}$$
$$E = 2 \times 10^6 \text{ Kg/cm}^2$$
$$R = 6\text{m} = 600\text{cm}$$

Using the relation $\sigma/y = E/R$
$$\sigma/0.25 = 2 \times 10^6/600,$$
$$\sigma = 833.33 \text{ Kg/cm}^2 \quad \text{.......ANS}$$

CHAPTER 17

TORSION

Q. 1: What is a shaft? What duty is performed by a shaft? What is its usual cross section and of what material it is usually made.?

Sol.: The shafts are usually cylindrical in section, solid or hollow. They are made of mild steel, alloy steel and copper alloys. Shaft may be subjected to the following loads:
1. Torsional load
2. Bending load
3. Axial load
4. Combination of above three loads.

The shaft are designed on the basis of strength and rigidity.

Shafts are also used to transmit power from a motor to a pump or compressor, from an engine or turbine to a generator, and from an engine to axle in automobiles. During power transmission, the shaft is subjected to torque which causes twist of the shaft.

A shaft needs to be designed that the excessive twist is avoided and the induced shear stress is within prescribed limits.

Q. 2: Differentiate between torque and torsion. List few examples of torsion in engineering practice.

Sol.: When a structural or machine member is subjected to a moment about its longitudinal axis, the member twists and shear stress is induced in every cross section of the member.

Such a mode of loading is called TORSION. And the twisting moment is referred to as TORQUE.

Example: Door knob, Screw driver, drill bit and shaft.

Q. 3: What is meant by pure Torsion? *(Dec–01)*

Generally two types of stresses are induced in a shaft.
1. Torsional (Shear) stresses due to transmission of torque.
2. Bending stresses due to weight of pulley, gear etc mounted on shaft.

A circular shaft is said to be in a state of pure torsion when it is subjected to torque only, without being acted upon by any bending moment or axial force.

OR; if the shaft is subjected to two opposite turning moment it is said to be in pure torsion. and it will exhibit the tendency of shearing off at every cross-section which is perpendicular to longitudinal axis.

Q. 4: Define Section modulus. What is torsional section modulus or polar modulus? *(Dec–01)*

Polar Moment of Inertiaj(J)

The M.I. of a plane about an axis perpendicular to the plane of the area is called polar moment of inertia of the area with respect to the point at which the axis intersects the plane.

Polar moment of inertia of solid body $(J) = I_{xx} + I_{yy} = \pi/32 \cdot D^4$
Polar moment of inertia of hollow body $(J) = I_{xx} + I_{yy} = \pi/32 \cdot (D_o^4 - D_i^4)$

Section Modulus(Z)
Section modulus is the ratio of *MI* about the neutral axis divided by the most distant point from the neutral axis.

$$Z = I/y_{max}$$

Section modulus for circular solid shaft $(Z) = (\pi/64 \cdot D^4)/D/2 = \pi/32 \cdot D^3$
Section modulus for circular hollow shaft $(Z) = [\pi/64 \cdot (D_o^4 - D_i^4)]/D_o/2 = \pi/32 \cdot (D_o^4 - D_i^4)/D_o$
Section modulus for rectangular section $(Z) = (bd^3/12)/d/2 = bd^2/6$

Polar Modulus(Z_p) *(Dec–04, 05)*
It is the ratio of polar moment of inertia to outer radius.

$$Z_p = J/R$$

Polar Modulus of solid body $(Z_p) = [\pi/32 \cdot D^4]/D/2 = \pi \cdot D^3/16$
Polar modulus of hollow body $(Z_p) = [\pi/32 \cdot (D_o^4 - D_i^4)]/D/2 = [\pi \cdot (D_o^4 - D_i^4)/D_o]/16$

Q. 5: What are the assumption made in deriving the torsional formulas? *(Dec–02 (C.O.))*
Sol.: The torsion equation is based on the following assumptions:
1. The material of the shaft is uniform throughout.
2. The shaft circular in section remains circular after loading.
3. A plane section of shaft normal to its axis before loading remains plane after the torques have been applied.
4. The twist along the length of shaft is uniform throughout.
5. The distance between any two normal cross-sections remains the same after the application of torque.
6. Maximum shear stress induced in the shaft due to application of torque does not exceed its elastic limit value.

Q. 6: Derive the Torsional equation $T/J = \tau/R = G\theta/L$

Or

Derive an expression for shear stress in a shaft subjected to a torque. *(Dec–02, May–02)*
Sol.: Let,
T = Maximum twisting torque or twisting moment
D = Diameter of the shaft
R = Radius of the shaft
J = Polar moment of Inertia
τ = Max. Permissible Shear stress (Fixed for a given material)
G = Modulus of rigidity
θ = Angle of twist (Radians) = angle $D'OD$
L = Length of the shaft.
Φ = Angle $D'CD$ = Angle of Shear strain

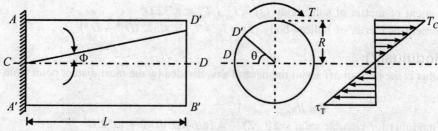

Fig 17.1

Than **Torsion equation is:** $T/J = \tau/R = G.\theta/L$

Let the shaft is subjected to a torque or twisting moment 'T'. And hence every C.S. of this shaft will be subjected to shear stress.

Now distortion at the outer surface = DD'

Shear strain at outer surface = Distortion/Unit length

$$\tan \Phi = DD'/CD$$

i.e. shear stress at the outer surface $(\tan \Phi) = DD'/L$

or $\quad\quad\quad\quad\quad\quad \Phi = DD'/L$ ...(i)

Now $\quad\quad\quad\quad\quad DD' = R.\theta \quad$ or $\quad \Phi = R \cdot \theta/L$ ...(ii)

Now G = Shar stress induced/shear strain produced

$$G = \tau/(R.\theta/L);$$

or; $\quad\quad\quad\quad\quad \tau/R = G.\theta/L \quad\quad\quad$ ...(A);

This equation is called Stiffness equation.

Hear G, θ, L are constant for a given torque 'T'.

i.e., **τ is proportional to R**

If τ_r be the intensity of shear stress at any layer at a distance 'r' from canter of the shaft, then;

$$\tau_r/r = \tau/R = G.\theta/L$$

Now Torque in terms of Polar Moment of Inertia

From the fig 17.2

Area of the ring $(dA) = 2\pi r \cdot dr$

Since, $\quad\quad\quad\quad\quad \tau_r = (\tau/R) \cdot r$

Turning force on Elementary Ring; $= (\tau/R) \cdot r \cdot 2\pi r dr$.

$\quad\quad\quad\quad\quad\quad\quad\quad = (\tau/R).2\,\pi r^2.dr$ ...(i)

Turning moment $dT = (\tau/R).2\,\pi r^2.dr.r$

$$dT = (\tau/R) \cdot r^2 \cdot 2\pi.r \cdot dr = (\tau/R) \cdot r^2 \cdot dA$$

$$T = (\tau/R)\int_0^R r^2 \cdot dA \quad\quad\quad ...(ii)$$

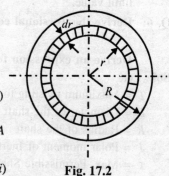

Fig. 17.2

$\int_0^R r^2 \cdot dA$ = M.I. of elementary ring about an axis perpendicular to the plane passing through center of circle.

$\int_0^R r^2 \cdot dA = J$ Polar Moment of Inertia

Now from equation (ii) $T = (\tau/R) \cdot J$
or $\qquad \tau/R = T/J;$ ...(B)
This equation is called strength equation
Combined equation A and B; we get
$$T/J = \tau/R = G.\theta/L$$
This equation is called Torsion equation.
From the relation $\quad T/J = \tau/R$;; We have $T = \tau.J/R = \tau.Z_P$

For a given shaft I_P and R are constants and I_P/R is thus a constant and is known as **POLAR MODULUS(Z_P)**. of the shaft section.

Polar modulus of the section is thus measure of strength of shaft in torsion.

TORSIONAL RIGIDITY or Torsional Stiffness (K): $= G.J/L = T/\theta$

Q. 7 Derive an expression for strain energy due to torsion. *(Dec–02, May–02)*
Sol.: The work done in straining the shaft with in the elastic limit is called strain energy.

consider a shaft of diameter D, and Length L, subjected to a gradually applied torque T. Let θ be the angle of twist. Energy is stored in the shaft due to this angular distortion. This is called torsional energy or the Torsional resilience.

Torsional energy or strain energy = W.D. by the torque = average torque X angular twist.
$$= (T/2).\theta$$
$$= \tfrac{1}{2}.T.\theta$$
$$= \tfrac{1}{2}. (\tau.J/R)(\tau.L/R.G)$$
$$= \tfrac{1}{2}. (\tau^2/G).(J/R^2).L$$
$$= \tfrac{1}{2}. (\tau^2/G).[(\pi D^4/32)/(D/2)^2].L$$
$$= \tfrac{1}{2}. (\tau^2/G).(\pi D^2/8).L = \tfrac{1}{2}. (\tau^2/G).(\pi.4R^2/8).L$$
$$= \tfrac{1}{2}. (\tau^2/G).(\pi. R^2.L/2) = \tfrac{1}{4}. (\tau^2/G).\text{Volume}$$
$$= \tfrac{1}{4}. (\tau^2/G).\text{Vor};$$
$$U/V = \tfrac{1}{4}. (\tau^2/G)$$

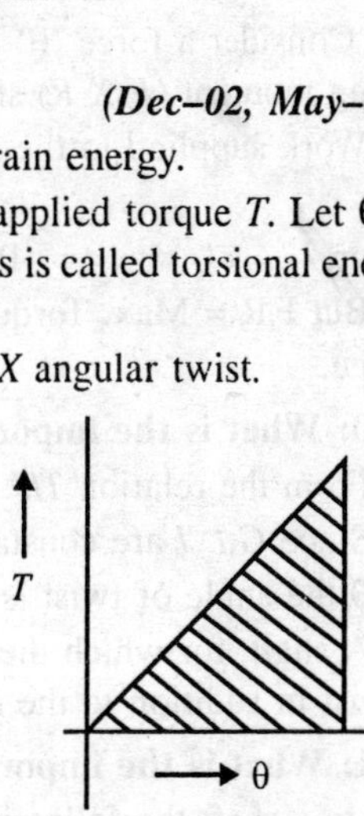

Fig. 17.3

So; Strain energy per unit volume is $1/4^{th}$ ratio of square of shear stress to modulus of rigidity.
For a hollow shaft : $U = [\tau^2.(D^2 + d^2)/4.G.D^2].\text{Volume of shaft}$
For a Solid shaft ($d =$: $U = [\tau^2/4.G].\text{Volume of shaft}$
very thin hollow shaft ($d = 0$): $U = [\tau^2/2.G].\text{Volume of shaft}$

Q. 8: How you evaluate the strength of solid and hollow circular shaft (T_{max})?
Sol.: Strength of a shaft may be defined as the maximum torque which can be applied to the shaft without exceeding allowable shear stress and angle of twist.
FOR SOLID SHAFT
From the torsion equation: $T/J = \tau/R = G.\theta/L$
Since $\qquad T = \tau.J/R$
$\qquad\qquad = \tau. [(\pi D^4/32)/(D/2)]$
$\qquad\qquad = \tau.(\pi/16)D^3$
$\qquad T_{max} = (\pi/16)\, \tau_{max}.D^3$ ...(i)
Again since $T/J = G.\theta/L;\ T = J.G.\theta/L = (\dot{A}D^4/32)\,.G.\theta/L$
Or; $\qquad T_{max} = (\pi D^4/32)\,.G.\theta_{max}/L$ ...(ii)

Where;
T_{max} = Maximum torque
θ_{max} = Maximum angle of twist
τ_{max} = Maximum shear stress

Note: The strength of the shaft is the minimum value of T_{max} from equation (i) and (ii). And for finding out diameter of shaft we take maximum value of dia obtained from equation (i) and (ii)
As the same for hollow shaft

$$T_{max} = (\pi/16)\tau_{max}.(D^4 - d^4)/D \quad \ldots(iii)$$
$$= (\pi/32)(D^4 - d^4).G.\theta_{max}/L \quad \ldots(iv)$$

Q. 9: How you evaluate the power transmitted by a shaft (P)?

Sol.: Consider a force 'F' Newton's acting tangentially on the shaft of radius 'R'. If the shaft due to this turning moment (F X R) starts rotating at N rpm then.

Work supplied to the shaft/sec = F. Distance moved/sec.
$$= F.2\pi.R.N/60 \text{ Nm/sec}$$
or, Power $(P) = F.R.2\pi.N/60$ Nm/sec or watt.
But F.R = Max. Torque (T) in N-m
i.e. $P = 2\pi.N.T_{max}/60$ watts

Q. 10: What is the importance of angle of twist?

From the relation $T/J = G.\theta/L$;; we have $\theta = T.L/G.J$
Since G, L, J are constants for a given shaft,
θ the angle of twist is directly proportional to the twisting moment.

A shaft, for which the angle of twist is significant, should always be designed or checked for angle of twist in addition to the design stresses in shafts.

Q. 11: What is the importance of stresses in shaft?

Sol.: In a shaft the following significant stresses occur.

1. A maximum shear stress occurs on the cross-section of the shaft at its outermost surface.
2. The maximum longitudinal shear stress occurs at the surface of the shaft on the longitudinal planes passing through the longitudinal axis of the shaft.
3. The maximum tensile stress (i.e. major principal stress) occurs at planes 45° to the maximum shearing stress planes at the surface of the shaft. This stress is equal to the maximum shear stress on the cross section of the shaft.
4. The maximum compressive stress (i.e. minor principal stress) occurs on the planes at 45° to the longitudinal and the cross-sectional planes at the surface of the shaft. This stress is equal to the maximum shear stress on the cross section.
5. These stresses are important/ significant because they govern the failure of the shaft. These stresses develop simultaneously and therefore they should be considered simultaneously for design purposes.
6. For most engineering materials, fortunately the shear strength is the smallest as compared to the tensile and compressive stresses and in such cases only the maximum shear stress on the cross-section of the shaft is the significant stress for design.
7. For materials for which tensile and compressive strengths are lower than the shear strength, the shaft design should be carried for the lowest strength.

Q. 12: What is Modulus of Rupture?

Sol.: The maximum fictitious shear stress calculated by the torsion formula by using the experimentally found maximum torque (ultimate torque) required to rupture a shaft.

If,
τ_r = Modulus of rupture in torsion (Also called computed ultimate twisting strength)
T_u = Ultimate torque at failure
R = Outer radius of the shaft
Then, $\qquad T_u/J = \tau_r/R$

The above expression for $\ddot{A}_r$ gives fictitious value of shear stress at the ultimate torque because the torsion formula $T/J = \tau/R$ is not applicable beyond the limit of proportionality.

The actual shear stress at the ultimate torque is quite different from the shearing modulus of rupture because the shear stress does not vary linearly from zero to maximum but it is uniformly distributed at the ultimate torque.

Q. 13: Compare a solid shaft with a hollow shaft, by strength and by their weight.

Sol.: (*a*) **Comparison by strength:** Assume that both the shaft have same length, material, same weight and hence the same maximum shear stress.

Let,
D_S = Diameter of the solid shaft.
D_H = External Diameter of the Hollow shaft
d_H = Internal diameter of the hollow shaft
A_S = Cross sectional Area of solid shaft
A_H = Cross-sectional area of hollow shaft
T_S = Torque transmitted by the solid shaft
T_H = Torque transmitted by the hollow shaft

$$T_H/T_S = 1.44$$

This show that the torque transmitted by the hollow shaft is greater than the solid shaft, thereby proving that the hollow shaft is stronger than the solid shaft.

(*b*) **Comparison by weight:** Assume that both the shaft have same length, material, Now if the torque applied to both shafts is same, then the maximum shear stress will also be same in both the cases.

Let W_H = Weight of hollow shaft
W_S = Weight of Solid shaft
Then, $\qquad W_H/W_S = A_H/A_S$
$\qquad\qquad W_H/W_S = 0.7829$

Q. 14: A solid shaft transmits power at the rate of 2000KW at the speed of 60RPM. If the safe allowable stress is 80MN/m², find the minimum diameter of the shaft.

Sol.:

$$P = 2000KW = 2000 \times 10^3 \text{ W}$$
$$N = 60RPM$$
$$\tau = 80MN/m^2 = 80 \times 10^6 \text{ N/m}^2$$
$$d = ?$$

Using the relation: $P = 2\pi.N.T_{max}/60$ watts
$$2000 \times 10^3 = 2\pi.60.T_{max}/60$$
$$T_{max} = 318309.88 \text{ N–m} \qquad\qquad \text{.......ANS}$$

Now using the relation $T_{max} = (\pi/16) \tau_{max}.D^3$
$$318309.88 = (\pi/16) \times 80 \times 10^6.D^3$$
$$D = 0.2726 \text{ m} \quad \text{or} \quad D = 272.63 \text{mm} \quad \text{......ANS}$$

Q. 15: A solid circular shaft transmits 75kW power at 200 rpm. Calculate the shaft diameter, if the twist in the shaft is not to exceed 1° in 2m length and the shear strength is limited to 50 MN/m². Take G = 100GN/m². Dec – 2003

Sol.:
$$P = 75\text{KW} = 75 \times 10^3 \text{ W}$$
$$N = 200 \text{RPM}$$
$$\theta = 1° = \pi/180 \text{ radian}$$
$$L = 2\text{m}$$
$$\tau = 50\text{MN/m}^2 = 50 \times 10^6 \text{ N/m}^2$$
$$G = 100\text{GN/m}^2 = 100 \times 10^9 \text{ N/m}^2$$

Using the relation
$$P = 2\pi.N.T_{max}/60 \text{ watts}$$
$$75 \times 10^3 = 2\pi.200.T_{max}/60$$
$$T_{max} = 3581 \text{ N-m} \qquad ...(i)$$

CASE – 1: Shaft diameter when allowable shear stress is considering
$$T_{max} = (\pi/16) \tau_{max}.D^3$$
$$3581 = (\pi/16).50 \times 10^6.D^3$$
$$D = 0.0714\text{m or } 71.4\text{mm} \qquad ...(ii)$$

CASE – 2: Shaft diameter when twist angle is considering
$$T = (\pi D^4/32).G.\theta/L$$
$$3581 = (\pi D^4/32).[100 \times 10^9(\pi/180)]/2$$
$$D = 80.4\text{mm} \qquad ...(iii)$$

For suitable value always take larger value of diameter
$$D = 80.4\text{mm} \quad \text{......ANS}$$

Q. 16: A torque of 1 KN-m is applied to a 40 mm diameter rod of 3 m length. Determine the maximum shearing stress induced and the twist produced. Take G = 80GPa. *(May–03)*

Sol.:
$$T = 1 \text{ KN-m} = 1000 \text{ N-m}$$
$$d = 40 \text{ mm} = 0.04 \text{ m}$$
$$L = 3 \text{ m}$$
$$\tau_{max} = ?$$
$$\theta = ?$$
$$G = 80\text{GPa} = 80 \times 10^9 \text{ N/m}^2$$

Using the relation $T_{max} = (\pi/16) \tau_{max}.D^3$
$$1000 = (\pi/16) \tau_{max}.(0.04)^3$$
$$\tau_{max} = 7.96 \times 10^7 \text{ N/m}^2 \quad \text{......ANS}$$

Now using the relation $T = (\pi D^4/32).G.\theta/L$
$$\theta = T.L/[(\pi D^4/32).G]$$
$$= (1000 \times 3)/[\{\pi(0.04)^4/32\} \times 80 \times 10^9]$$
$$\theta = 0.1494 \text{ rad}$$
$$= 0.1494 (180/\pi)$$
$$\theta = 8.56° \quad \text{......ANS}$$

Q. 17: A circular shaft of 10 cm diameter is subjected to a torque of 8×10^3 Nm. Determine the maximum shear stress and the consequent principal stresses induced in the shaft.

(Dec-02 (C.O.))

Sol.:
$$d = 10 \text{ cm} = 0.1 \text{ m}$$
$$T = 8 \times 10^3 \text{ Nm}$$
$$\tau_{max} = ?$$

Using the relation $T_{max} = (\sigma/16)\,\tau_{max}.D^3$
$$\tau_{max} = 16T/\pi.D^3$$
$$\tau_{max} = (16 \times 8 \times 10^3)/(\pi.(0.1)^3)$$
$$\tau_{max} = 40.74 \times 10^6 \text{ N/m}^2 \quad \text{......ANS}$$

Principal stress = $2.\,\tau_{max}$
Principal stress = $2 \times 40.74 \times 10^6$
Principal stress = 81.48×10^6 N/m² ......ANS

Q. 18: A solid shaft of mild steel 200 mm in diameter is to be replaced by hollow shaft of alloy steel for which the allowable shear stress is 22% greater. If the power to be transmitted is to be increased by 20% and the speed of rotation increased by 6%, determine the maximum internal diameter of the hollow shaft. The external diameter of the hollow shaft is to be 200 mm.

(Dec-00 (C.O.))

Sol.: Solid shaft dia $d = 200$mm
Outer dia of Hollow shaft = d_o = 200mm
Inner dia of Hollow shaft = d_i = ?
Allowable shear stress $\tau_H = 1.22\,\tau_S$
Power transmitted $P_H = 1.20 P_S$
Shaft speed $N_H = 1.06 N_S$
Now use $P_H = 1.20 P_S$
i.e.; $\quad 2\pi.N_H T_H/60 = 1.2\,[2\pi.N_S T_S/60]$
$$N_H T_H = 1.2\,N_S T_S$$
Since $N_H = 1.06 N_S$, putting this values
$$1.06\,N_S T_H = 1.2\,N_S T_S$$
$$T_S = (1.06/1.2) T_H \qquad ...(i)$$
Since $\quad T = (\sigma/16)D^3.\tau_{max}$
$$(\sigma/16)d^3.\tau_S = (1.06/1.2)\,[(\sigma/16)(d_O^4 - d_i^4)/d_O].\tau_H$$
putting $\quad \tau_H = 1.22\,\tau_S$
$$(\sigma/16)d^3.\tau_S = (1.06/1.2)\,[(\sigma/16)(d_O^4 - d_i^4)/d_O].\,1.22\,\tau_S$$
$$200^3 = (1.06/1.2)(1.22)\,[(200^4 - d_i^4)/200]$$
$$d_i = 104 \text{ mm} \quad \text{......ANS}$$

Q. 19: A solid shaft is replaced by a hollow one. The external diameter of which is 5/4 times the internal diameter. Allowing the same intensity of torsional stress in each, compare the weight and the stiffness of the solid with that of the hollow shaft. *(Dec-01 (C.O.))*

Sol.: Diameter of solid shaft = d
Outer Diameter of hollow shaft = d_O
Inner Diameter of hollow shaft = d_i
Given that $d_O = 1.25 d_i$

Same intensity of torsional stress i.e: $\tau_S = \tau_H$
$$W_S/W_H = ?$$
$$K_S/K_H = ?$$
Since: $\quad \tau_S = 16T/\pi d^3$ ...(i)
and $\quad \tau_H = 16T/[\pi(d_o^4 - d_i^4)/d_o]$ ...(ii)

Equate equation (i) and (ii)
$16T/\pi d^3 = 16T/[\pi(d_o^4 - d_i^4)/d_o]$, putting the value $d_o = 1.25 d_i$
we get $d = 1.048\, d_i$ or $0.838\, d_o$

Now for ratio of weight
Since $\quad \rho_S = \rho_H,\ L_S = L_H$
$$W_S = \rho_S \cdot g(\text{volume})_S = \rho_S \cdot g \cdot A_S \cdot L_S$$
$$W_H = \rho_H \cdot g(\text{volume})_H = \rho_H \cdot g \cdot A_H \cdot L_H$$
Now; $\quad W_S/W_H = \rho_S \cdot g \cdot A_S \cdot L_S / \rho_H \cdot g \cdot A_H \cdot L_H = A_S/A_H = (\rho/4)d^2 /[(\pi/4)(d_o^2 - d_i^2)]$
Now putting the value $d_i = d/1.048$; $d_o = d/0.838$; we get
$$W_S/W_H = 1.9525 \qquad \text{......ANS}$$

Now for ratio of torsional stiffness
$$K = T/\theta = G.J/L$$
$$K_S = G_S.J_S/L_S$$
$$K_H = G_H.J_H/L_H$$
$$K_S/K_H = G_S/G_H[\{(\pi/4)(d^4)\}/\{(\pi/4)(d_o^4 - d_i^4)\}]$$
If $\quad G_H = G_S$ and $d_i = d/1.048$; $d_o = d/0.838$, than
$$K_S/K_H = 0.837 \qquad \text{......ANS}$$

Q. 20: For one propeller drive shaft, compute the torsional shear stress when it is transmitting a torque of 1.76 kN-m. The shaft is a hollow tube having an outside diameter of 60 mm and an inside diameter of 40 mm. Find the stress at both the outer and inner surfaces.
(Dec–01 (C.O.))

Sol.:
$$T = 1.76\ \text{KN-m} = 1.76 \times 10^6\ \text{N-mm}$$
$$d_o = 60\text{mm}$$
$$d_i = 40\text{mm}$$
$$\tau_{outer} = ?$$
$$\tau_{inner} = ?$$
For hollow $\quad T = [(\pi/16)(d_o^4 - d_i^4)/d_o]\,\tau$
$$1.76 \times 10^6 = [(\pi/16)(60^4 - 40^4)/60]\,\tau$$
$$\tau = 51.7\ \text{N/mm}^2$$
This is shear stress at outer surface i.e.; $\tau_{outer} = 51.7\ \text{N/mm}^2$ ......ANS

Now the stress is varies linearly along the diameter of the shaft, the shear stress at inner diameter of the shaft i.e.: $\tau_{inner} = \tau_{outer} \times d_i/d_o = 51.7 \times 40/60 = 34.48\ \text{N/mm}^2$
$$\tau_{inner} = 34.48\ \text{N/mm}^2 \qquad \text{......ANS}$$

Q. 21: The diameter of a shaft is 20cm. Find the safe maximum torque which can be transmitted by the shaft if the permissible shear stress in the shaft material be 4000 N/cm². and permissible angle of twist is 0.2 degree per meter length. Take $G = 8 \times 10^6$ N/cm². If the shaft rotates at 320 r.p.m. what maximum power can be transmitted by the shaft. *(Dec–05 (C.O.))*

Sol.: Given that:
$$d = 20\text{cm}$$

$T_{max} = ?$
$\iota_{max} = 4000$ N/cm^2
$\theta = 0.2°$/meter length $= 0.2 \times (\pi/180) = 0.0035$ rad
$G = 8 \times 10^6$ N/cm^2
$N = 320$ r.p.m.
$P = ?$

CASE – 1: When $\iota_{max} = 4000$ N/cm^2
$T = (\pi/16)(\iota_{max} \cdot d^3)$
$T = (\pi/16) \times 4000 \times (20)^3$
$T = 6283185.3$ N – cm ...(i)

CASE – 2: When $\theta = 0.2°$/meter length
$T = G.J. \theta/L = (\pi/32)d^4 \cdot G.\theta/L$
Let $L = 100$cm
$T = (\pi/32)(20)^4 \cdot 8 \times 10^6 \cdot (0.0035)/100$
$T = 4386490.85$ N – cm ...(ii)

The permissible torque is the minimum of (i) and (ii)
i.e.; $T = 4386490.85$ N – cm $= 43.86$ KN – m ....ANS
Power $= 2\pi NT/60$ Watt, T in N
$= \{2\pi(320)(43.86 \times 10^3)\}/60$
$= 1469762.7$ Watt
$P = 1469.76$ KW ....ANS

Q. 22: A propeller shaft 100 mm in diameter, is 45 m long, transmits 10 MW at 80 rpm. Determine the maximum shearing stress in shaft. Also calculate the stress at 20 mm, 40 mm, 60 mm and 80 mm diameters. Show the stress variation. *(May–05 (C.O.))*

Sol.:
$d = 100$mm
$L = 45$m
$P = 10$MW $= 10 \times 10^6$ W
$N = 80$RPM
$\iota_{max} = ?$
$\iota_{20} = ?, \iota_{40} = ?, \iota_{60} = ?, \iota_{80} = ?$

Using the relation
$P = 2\pi.N.T_{max}/60$ watts
$10 \times 10^6 = 2\pi.80.T_{max}/60$
$T_{max} = 1193662.073$ N ...(i)

For solid shaft $\iota = 16T_{max}/\pi d^3$
$\iota = 16(1193662.073)/[\pi(100)^3]$
$\iota = 6.079$ N/mm^2

This is shear stress at outer surface i.e.; $\iota_{100} = 6.079$ N/mm^2 ....ANS

Now the stress is varies linearly along the diameter of the shaft, the shear stress at inner diameter of the shaft i.e.;

$\iota_{20} = \iota_{100} \times d_{20}/d_{100} = 6.079 \times 20/100 = 1.22$ N/mm^2
$\iota_{20} = 1.22$ N/mm^2 ....ANS

$\iota_{40} = \iota_{100} \times d_{40}/d_{100} = 6.079 \times 40/100 = 2.44$ N/mm²
$\iota_{40} = 2.44$ N/mm² ....ANS
$\iota_{60} = \iota_{100} \times d_{60}/d_{100} = 6.079 \times 60/100 = 3.66$ N/mm²
$\iota_{60} = 3.66$ N/mm² ....ANS
$\iota_{80} = \iota_{100} \times d_{80}/d_{100} = 6.079 \times 80/100 = 4.88$ N/mm²
$\iota_{80} = 4.88$ N/mm² ....ANS

Q. 23: 20 kNm Torque is applied to a shaft of 7 cm diameter. Calculate:

(i) Maximum shear stress in the shaft.

(ii) Angle of twist per unit length of shaft. Take $G = 10^5$ N/mm². *(Dec–04 (C.O.))*

Sol.: $T_{max} = 20$ KN-m $= 20 \times 10^3$ N-m
$d = 7$cm $= 0.07$m
$\iota_{max} = ?$
$\theta = ?$

L = unit length say 1m
$G = 10^5$ N/mm² $= 10^{11}$ N/m²
Using torsion equation; $T/J = \iota/R = G.\theta/L$
Or;
$$T_{max} = (\pi/16) \iota_{max}.D^3 \quad ...(i)$$
$$= (\iota D^4/32).G.\theta_{max}/L \quad ...(ii)$$

Where;
T_{max} = Maximum torque
θ_{max} = Maximum angle of twist
ι_{max} = Maximum shear stress

Using equation (i)
$20 \times 10^3 = (\pi/16) \iota_{max}.(0.07)^3$
$\iota_{max} = $ **296965491.5 N/m²** ....ANS

Using equation (ii)
$T_{max} = (\pi D^4/32).G.\theta_{max}/L$
$20 \times 10^3 = (\pi (0.07)^4/32).10^{11}.\theta_{max}/1$
$\theta_{max} = 0.08484$ radian $= 0.08484 \times (180/\pi)$
$\theta_{max} = $ **4.86°** ....ANS

Q. 24: What do you meant by compound shaft? How compound shaft be classified? *(Dec–01 (C.O.))*

Sol.: Shaft made up of two or more materials are called compound shaft. These shaft may be connected in series or in parallel.

SHAFT IN SERIES: In order to form a composite shaft sometimes two shafts are connected in series. In such cases each shaft transmits the same torque. The angle of twist is the sum of the angle of twist of the two shaft connected in series.

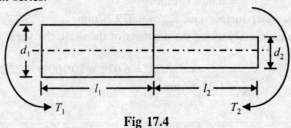

Fig 17.4

Total angle of twist $(\theta) = \theta_1 + \theta_2 = TL_1/C_1I_{P1} + TL_2/C_2I_{P2}$
Where, T = torque transmitted by each shaft
L_1, L_2 = Respective lengths of the two shafts.
G_1, G_2 = Respective moduli of rigidity
I_{P1}, I_{P2} = Respective polar moment of inertia
→ When shaft are made of same material: than $G_1 = G_2 = G$
than,
$$\theta = \theta_1 + \theta_2 = T/C[L_1/I_{P1} + L_2/I_{P2}]$$

SHAFT IN PARALLEL: The shaft are said to be in parallel when the driving torque is applied at the junction of the shafts and the resisting torque is at the other ends of the shafts. Here, the angle of twist is same for each shaft, but the applied torque is divided between the two shaft.

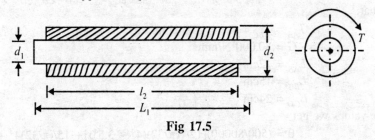

Fig 17.5

i.e. $\theta_1 = \theta_2$ and $T = T_1 + T_2$
if the shaft are made of same material than $G_1 = G_2$

Such a situation $(\theta_1 = \theta_2$ and $T = T_1 + T_2)$ also arises when the shaft ends are fixed and are subjected to a torque at the common junction. As shown in fig 17.6

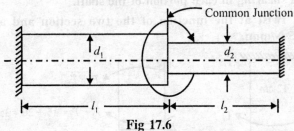

Fig 17.6

Q. 25: A steel shaft as shown below is subjected to equal and opposite torque at the ends. Find the maximum permissible value of d for the maximum shearing stress in AB not to exceed that in CD. Calculate the total angle of twist, if the torque applied is 500N-m. $G = 80,000$ N/mm^2.

Sol.: We know that:
$$T/J = \tau/r = \tau/D/2 = 2\tau/D$$

For shaft AB
$$\tau_{AB} = 16.T.d_o/\pi(d_o^4 - d_i^4) = 16.T.4/\pi(4^4 - d^4)$$
$$\tau_{AB} = 64.T/\pi(256 - d^4) \qquad \ldots(i)$$

For shaft CD
$$\tau_{CD} = 16.T/\pi(3.5^3) = 16.T/\pi(42.875) \quad \tau_{CD} = 16.T/\pi.42.875 \qquad \ldots(ii)$$

Equating equation (i) and (ii); we get

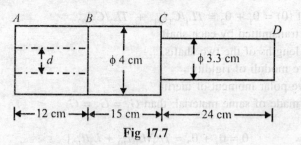

Fig 17.7

$d = 3.03$ cm   ......ANS

Now all shafts are in series i.e.: $\theta = \theta_1 + \theta_2 + \theta_3$

$$\theta = [T_{AB}.L_{AB}/J_{AB}.G + T_{BC}.L_{BC}/J_{BC}.G + T_{CD}.L_{CD}/J_{CD}.G]$$

Since given that;

$T_{AB} = T_{BC} = T_{CD} = 500$ N.m
$G = 80,000$ N/mm^2
$L_{AB} = 12$cm
$L_{BC} = 15$cm
$L_{CD} = 24$cm

Putting all the values we get

$$\theta = (500/80000)[12/\{(\pi/32)(4^4 - 3.5^4)\} + 15/(\pi/32)4^4 + 24/(\pi/32)(3.5)^4]$$
$\theta = 1.05°$ ......ANS

Q. 26: A compound shaft 1.5m long fixed at one end is subjected to a torque of 15KN-m at the free end and of 20KNm at the Junction point as shown in fig 17.11. Determine

(a) The maximum shearing in each portion of the shaft.

(b) The angle of twist at the junction of the two section and at the free ends. Take $G = 0.82 \times 10^5$ N/mm^2.

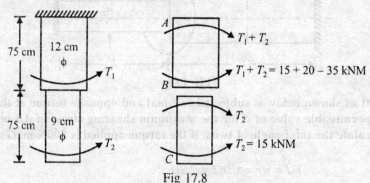

Fig 17.8

Sol.: Torque in $BC = 15$ KN-m

Torque in $AB = 35$ KN-m

Shaft are in series i.e.: $\theta = \theta_1 + \theta_2$

$\theta = (T.L/G.J)_{AB} + (T.L/G.J)_{BC}$
$= (35 \times 10^6 \times 750)/\{0.82 \times 10^5 \times (\pi/32)(120)^4\} + (15 \times 10^6 \times 750)/$
$\{0.82 \times 10^5 \times (\pi/32)(90)^4\}$

$\theta = 0.037$ rad $= 2.12°$ ......ANS

Now from torque equation $T/J = \iota/R$; $\iota = T.R/J = 16T/\pi d^3$

$\iota_{BC} = (16 \times 15 \times 10^6)/\pi(90)^3$

$\iota_{BC} = 104.85$ N/mm² ....ANS

$\iota_{AB} = (16 \times 35 \times 10^6)/\pi(120)^3$

$\iota_{AB} = 103.21$ N/mm² ....ANS

Q. 27: A compound shaft is made up of a steel rod of 50 mm diameter surrounded by a closely fitted brass tube. When a torque of 9 KN-m is applied on this shaft, its 60% is shared by the steel rod and the rest by brass tube. If shear modulus for steel is 85 GPa and for brass it is 45GPa. Calculate

(a) The outside diameter of brass tube.

(b) Maximum shear stress induced in steel and brass. *(May–05 (C.O.))*

Sol.:

Diameter of steel d_s = 50mm

Total torque applied = 9KNm

Torque sheared by steel rod $T_s = 9 \times 60/100 = 5.4$KN-m

Torque sheared by brass rod $T_b = 9 \times 40/100 = 3.6$KN-m

$G_S = 85$GPa, $G_b = 45$GPa

Let; Outer diameter of brass = d_{ob} = ? Inner diameter of brass = d_{ib} = 50mm Maximum shear stress induced in steel = ι_s Maximum shear stress induced in brass = ι_b Since shaft are parallel i.e.; $\theta_s = \theta_b$ & $T = T_s + T_b$ Now applied $\theta_s = \theta_b$

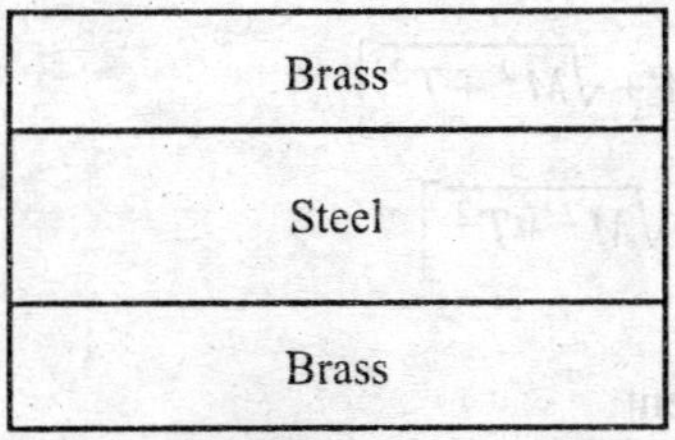

Brass
Steel
Brass

Fig 17.9

$T_S.L_S/G_S.J_S = T_b.L_b/G_b.J_b$; $L_S = L_b$

We get; $T_S/G_S.J_S = T_b/G_b.J_b$

Or; $J_b/J_S = T_b.G_S/T_S.G_b$

Putting all the values

$\{(\pi/32)(d_{ob}^4 - 50^4)\} / \{(\pi/32).50^4\} = (3.6 \times 85)/(5.4 \times 45)$

$d_{ob} = 61.29$mm ....ANS

Again $T/J = \iota/R = T_S/J_S = \iota_S/R_S$

$\iota_S = T_S.R_S/J_S = \{(5.4 \times 10^3)(25 \times 10^{-3})\}/\{(\pi/32)(50 \times 10^{-3})^4\}$

$\iota_S = 2.2 \times 10^8$ N/m²

$\iota_S = 220$ *GPa* ....ANS

Q. 28: Derive the equation subjected to combined bending and torsion for finding maximum shear stress and equivalent twisting moment. *(Dec–03, May–03 (C.O.))*

Sol.: Generally we assume that shaft is subjected to torsion only but in actual practice due to weight of the pulley, couplings, pull in belts or ropes etc, the shaft is subjected to bending too. Thus actually in the shaft, both the shear stress due to torsion and direct stress due to bending are induced.

From bending equation,
$$\frac{M}{I} = \frac{\sigma_b}{y}; \quad \sigma_b = \frac{My}{I} = \frac{M \times d/2}{\pi d^4/32} = \frac{16T}{\pi d^3}$$

From torsion equation,
$$\frac{T}{J} = \frac{\tau}{R}; \quad \tau = \frac{TR}{J} = \frac{T \times d/2}{\pi d^4/32} = \frac{16T}{\pi d^3}$$

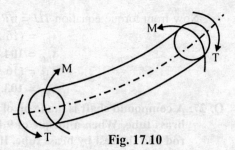

Fig. 17.10

(i) The maximum principal stress at the section
$$= \frac{\sigma_1 + \sigma_2}{2} + \sqrt{\left(\frac{\sigma_1 - \sigma_2}{2}\right)^2 + \tau^2}$$

$$= \frac{\sigma_b}{2} + \sqrt{\left(\frac{\sigma_b}{2}\right)^2 + \tau^2} \qquad (\because \sigma_1 = \sigma_b \text{ and } \sigma_2 = 0)$$

$$= \frac{16M}{\pi d^3} + \sqrt{\left(\frac{16M}{\pi d^3}\right)^2 + \left(\frac{16T}{\pi d^3}\right)^2}$$

$$= \frac{16}{\pi d^3} + \left[M + \sqrt{M^2 + T^2}\right]$$

The bending moment corresponding to the maximum principal stress is termed as equivalent bending moment, M_e. Thus;

$$\frac{32 M_e}{\pi d^3} = \frac{16}{\pi d^3}\left[M + \sqrt{M^2 + T^2}\right]$$

or
$$M_e = \frac{1}{2}\left[M + \sqrt{M^2 + T^2}\right]$$

Where;
M_e = Equivalent bending Moment
M = Maximum Bending Moment from the bending moment diagram
T = Torque transmitted

(ii) The maximum shear stress developed at the section

$$\sigma_{max} = \sqrt{\left(\frac{\sigma_1 - \sigma_2}{2}\right)^2 + \tau^2} = \sqrt{\left(\frac{\sigma_b}{2}\right)^2 + \tau^2} = \sqrt{\left(\frac{32M}{2\pi d^3}\right)^2 + \left(\frac{16T}{\pi d^3}\right)^2}$$

$$= \frac{16}{\pi d^3}\left[\sqrt{M^2 + T^2}\right]$$

The twisting moment corresponding to the maximum shear stress on the surface of the shaft is called equivalent twisting moment, T_e. Thus;

$$\frac{16 T_e}{\pi d^3} = \frac{16}{\pi d^3}\left[\sqrt{M^2 + T^2}\right]$$

or
$$T_e = \sqrt{M^2 + T^2}$$

Where T_e = Equivalent Twisting moment

Finally;
Equivalent Bending moment

$$M_e = \frac{1}{2}\left[M + \sqrt{M^2 + T^2}\right]$$

Equivalent Twisting moment

$$T_e = \sqrt{M^2 + T^2}$$

Maximum principal stress σ_{max}

$$= \frac{16}{\pi d^3}\left[M + \sqrt{M^2 + T^2}\right]$$

Minimum Principal stress σ_{min}

$$= \frac{16}{\pi d^3}\left[M - \sqrt{M^2 + T^2}\right]$$

Maximum shear stress τ_{max}

$$= \frac{16}{\pi d^3}\sqrt{M^2 + T^2}$$

Q. 29: A shaft is to be designed for transmitting 100 kW power at 150 rpm. Shaft is supported in bearings 3 m apart and at 1 m from one bearing a pulley exerting a transverse load of 30 KN on shaft is mounted. Obtain the diameter of shaft if the maximum direct stress is not to exceed 100 N/mm². *May–01 (C.O.))*

Sol.:
$R_a + R_b = 30$ KN
$R_a = 20$ KN
$R_b = 10$ KN

Maximum bending moment occurs at Point 'C'.
$M_C = 20 \times 1 = 20$ KN–m   ...(i)

Power transmitted by shaft $P = 2\pi NT/60$
$100 \times 10^3 = 2\pi.150.T/60$
$T = 6366.19$ N–m   ...(ii)

Equivalent bending moment $Me = \frac{1}{2}[M + (M^2 + T^2)^{1/2}]$
$= \frac{1}{2}[20{,}000 + (20{,}000^2 + 6369^2)^{1/2}]$
$M_e = 20494.38$ N–m   ...(iii)

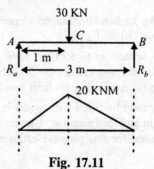

Fig. 17.11

From bending equation $M_e/I = \sigma/y$
$2044.38 \times 10^3/[(\pi/64)d^4] = 100/d/2$
$d = \mathbf{127.8}$ **mm**   ....**ANS**

APPENDIX 1

TUTORIAL SHEET – 1

BASIC CONCEPT, DEFINITION AND ZEROTH LAW

Q.1. Calculate the work done in a piston cylinder arrangement during an expansion process $1 m^3$ to $4 m^3$ where process is given by $p = (V^2 + 6V)$ bar. [6.6 ms]

Q.2. Determine pressure of steam flowing through a steam pipe when the U – tube manometer connected to it indicated as shown in fig1. During pressure measurement some steam gets condensed in manometer tube and occupies a column of height 2cm (AB) while mercury gets raised by 10cm (CD) in open limb. Consider barometer reading as 76cm of Hg, density of mercury & water as 13.6×10^3 kg/m³ and 1000 kg/m³ respectively. [114.54 kPa]

Q.3. A mass of 2 kg fall on a paddle and turns it, consequent upon which the paddle stirs 2 kg water kept in a container. Determine the height from which the mass should fall so as to rise the temperature of water by 2^0 C. Take water equivalent of the container as 10 gm. Assume any data if required. [89.62 m]

Q.4. In a thermoelectric thermometer for t^0C temperature, the emf is given as $E = 0.003t - 5 \times 10^{-7} t^2 + 0.5 \times 10^{-3}$ volts. Thermometer is having reference junction at ice point and is calibrated at ice point and steam points. What temperature shall be shown by the thermometer for a substance at 30^0C. [33.23^0C]

Q.5. A spherical balloon of 5 m diameter is filled with hydrogen at 27^0C and atmospheric pressure of 1.013 bar. It is supposed to lift some load if the surrounding air is at 17^0C. Estimate the max load that can be lifted. [74.344kg]

Q.6. Determine the pressure of 5 kg carbon dioxide contained in a vessel of 2 m³ capacity at 27^0C considering it as, (a) perfect gas (b) real gas. [1.417×10^5 N/m² , 1.408×10^5 N/m²]

Q.7. A metal block of 1 kg mass is heated up to 80^0C in open atmosphere & subsequently submerged in 10 kg of water so as to raise its temperature by 5^0C the initial temperature of water is 25^0C. Determine specific heat of metal considering no heat loss to surroundings. [4.18 kJ/kg K]

Q.8. A boiler of 5000 litre capacity is to be filled with water in 45 min, using a feed water pump. The boiler is installed at 10 cm height and feed pump at, 1-meter height from the ground level. Determine the power of feed pumps required for this purpose efficiency of the pump is 85%

Q.9. The pressure in a chamber was recorded as 50 cm of water above atmospheric pressure when barometer reading was 765 mm of Hg determine the absolute pressure in the vessel in bar.

[1.07 bar]

Q.10. The temperature of human body is 98.6 F. determine the temperature in degree Celsius (^{0}C), Rankine (R) and Kelvin (K) scales. [37^{0}C, 558.27 R, 310.15 K]

TUTORIAL SHEET – 2

FIRST LAW & SECOND LAW OF THERMODYNAMICS

(A) First Law

Q.1. How much work is done when $0.566 \, m^3$ of air initially at a pressure of 1.0335 bar and temperature of 7°C undergoes an increase in pressure upto 4.13 bar in a closed vessel.

Q.2. Hydrogen from cylinder is used for inflating a balloon to a volume of $35 \, m^3$ slowly. Determine the work done by hydrogen if the atmospheric pressure is 101.325 kPa. [3.55 MJ]

Q.3. Water flowing through a pipe as shown in fig 3 develops power at 2.5 kW. Between the section 2 and 3, $d_1 = 15$ cm, $d_2 = d_3 = 7.5$ cm $p_1 = 1.9$ bar, $v_1 = 3$ m/s. Assuming Du = 0 and neglecting the change in KE between section 2 & 3, determine the velocity v_2, pressure p_2 & p_3

(B) Second Law

Q.4. A refrigerator has COP one half as that of a carnot refrigerator operating between reservoirs at temperature of 200 K & 400 K, and absorbs 633 kJ from low temperature reservoirs. How much heat is rejected to the high temperature reservoirs. [1899 kJ]

Q.5. A reversible heat engine operates between two reservoirs at 827°C and 27°C engine drives a carnot refrigerator maintaining – 13°C and rejecting heat to reservoirs at 27 °C. Heat input to the engine is 2000 kJ and the net work available is 300 kJ. Determine the heat transferred to refrigerant and total heat rejected to reservoirs at 27°C

Q.6. A cold storage plant of 40 tones of refrigeration capacity runs with it's performance just 1/4 of its carnot COP. Inside temperature is –15°C atmospheric temperature is 35°C. Determine the power required to run the plant. [Take one ton of refrigeration as 3.52 kW]

Q.7. A domestic refrigerator maintain temperature of – 8°C when the atmospheric air temperature is 27°C. Assuming the leakage of 7.5 kJ/hr from outside to refrigerator. Determine power required to run this refrigerator. Consider refrigerator as carnot refrigerator.

TUTORIAL SHEET –3

PROPERTIES OF STEAM AND THERMODYNAMICS CYCLES

Q.1. Water vapour mixture at 100^0C is contained in a rigid vessel of .5 m³ capacity. Water is now heated till it reaches critical state. What was the mass & volume of water initially.

(mass = 158.48 kg, volume = $0.1652 m^3$)

Q.2. A spherical vessel of 0.4 m³ capacity contains 2 kg of wet steam at a pressure of 600 kPa. Calculate (a) the volume and mass of liquid (b) Volume and mass of vapour.

(Ans m_l = 0.736 kg, m_v = 1.264 kg, v_{1-} = 0.0008 m³, v_2 = 0.3992 m³)

Q.3. Steam flow rate through a steam turbine is 1.5 kg/s. the steam pressure & temperature of the entry of the turbine are 20 bar and 350 ^0C and at the exit 1 bar and dry saturated condition. The velocity of the steam at the exit is 200 m/s. Determine the power output if the heat transfer from the turbine is 8.5 kW. [655.7kW]

Q.4. Steam at 0.8 MPa, 250^0C and flowing at the rate of 1 kg/s passes into a pipe carrying wet steam at 0.8 MPa, 0.95 dry after adiabatic mining the flow rate is 2.3 kg/s. Determine the condition of steam after mining. (509.9 m/s)

Q.5. A steam boiler initially contains 5 m³ of steam and 5 m³ of water at 1 MPa steam is taken out at constant pressure until 4 m³ of water is left. What is the heat transferred during the process.

(1752.676 MJ)

Q.6. Steam at 10 bar and 200^0C is cooled till it becomes dry saturated and then throttled to 1 bar pressure. Find

(a) Change in enthalpy and heat transferred in each process.

(b) The quality of steam at the end of throttling process. (Ans Dh = 180 kJ/kg, DT = 44.9 ^0C)

Q.7. Water at 60^0C is supplied to a boiler to generated steam at 10 bar and 300^0C this steam is supplied to a turbine after throttling up to 8 bar the enthalpy of steam in the adiabatic turbine is reduced by 400 kJ/kg. The steam then passes through a converging – diverging nozzle where the enthalpy is further reduced by 400 kJ/kg. If the steam flow rate is 10,000 kg/hr, find

(a) Heat supplied in the boiler

(b) Condition of steam at the inlet to the steam turbine.

(c) Power generated by the turbine.

(d) Velocity of steam at the exit of the nozzle.

(26910 x 10^3 kJ, 78^0C, 1111 kW, 694.4 m/s)

TUTORIAL SHEET – 4

INTRODUCTION TO I.C. ENGINE

Q.1. An engine is 75 mm bore and stroke and has a compression ratio of 5.6, what amount of metal should be machined off from the cylinder head face if the compression ratio is to be raised to 6.
(1.3 mm)

Q.2. An Otto cycle takes in air at 1 bar and 15°C the compression ratio is 8 to 1 and 2000 kJ/kg of energy is released to air in each cycle. To what value must the compression ratio be increased to increase the network per cycle by 70%.
(16.95)

Q.3. A diesel engine receives air at 0.1 MPa & 300 K in the beginning of compression stroke. The compression ratio is 16; heat added per kg of air is 1500 kJ/kg. Determine the fuel cut-off ratio & cycle thermal efficiency. Assume C_p = 1 kJ/Kg K, R = .286 KJ/Kg K
(2.65, 58.39%)

Q.4. In a diesel engine during the compression process, pressure is soon to be 138 kPa at 1/8th of stroke and 1.38 MPa at 7/8th of stroke. The cut off occurs at 1/15th of stroke. Calculate air standard efficiency and compression ratio assuming indicated thermal efficiency to be half of ideal efficiency, mechanical efficiency as .8, calorific value of fuel = 4.800 KJ/kg and g = 1.4. Also find fuel consumption BHP/hr.
(63.25%, 19.37, .255 kg)

Q.5. In a I.C. engine using air as working fluid, total 1700 KJ/kg K of heat is added during combustion and maximum pressure in cylinder does not exceed 5 MPa. Compare the efficiency two cycles used by engine.
(a) Cycle in which combustion takes place isochorically
(b) Cycle in which half of heat is added at constant volume and half at constant pressure temperature Pressure at the beginning of compression are 100°C and 103 kPa. Compression and expansion processes are adiabatic. Specific heat at constant pressure and volume are 1.003 KJ/Kg ° K and 0.71 KJ/kg ° K.
(50.83%, 56.47%)

Q.6. An SI Engine having a clearance volume of 250 cc has a compression ratio of 8. Initial pressure in its cycle is 1 bar and the ratio of pressure rise at constant volume is 4. Taking g = 1.4, determine
(a) The work done per cycle
(b) Indicated mean effective pressure.
(1.946 kj/cycle, 11.2 bar)

Q.7. An Otto cycle operate between maximum and minimum pressure of 600 Kpa and 100 Kpa respectively the minimum and maximum temperature in the cycle are 27°C and 1600 ° K respectively. Determine thermal efficiency of cycle and also shown it on T-S & P-V diagram
(48%)

Q.8. The stroke and diameter of 2 stroke petrol engine are 14 cm & 10 cm respectively. The clearance volume is 157 cm^3. If the exhaust port opens after 140° crank rotation from TDC, find the actual air – standard efficiency of the cycle.
(54.7%)

TUTORIALS SHEET – 5

CONCURRENT FORCE SYSTEM

Q.1. Two smooth sphere each of radius 150mm and weight 250N rest in a horizontal channel having vertical walls, the distance between which is 560mm. Find the reaction at the points of contact A, B, C and D as shown in Fig (1). R_A=434.3N, R_B=501.1N, R_C=500N

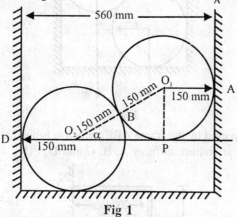

Fig 1

Q.2. The weight and radii of the three cylinders piled in a rectangular ditch as shown in fig (2). W_A=80N, W_B=160N, W_C=80N, R_A=100mm, R_B= 200mm, R_C=100mm. Assuming all contact surfaces to be smooth. Determine the reactions acting on cylinder C.   138.6N, 320N.

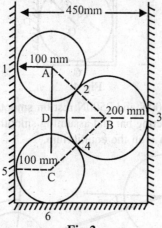

Fig 2

Q.3. Two smooth spheres each of weight W and each of radius r are in equilibrium in a horizontal channel of width b = 4r and vertical sides as shown in fig 3. Find the three reactions from the sides of the channel, which are all smooth. Find also the force exerted by each sphere on the other. Calculate these values if r =25cm, b=90cm and W=1000N. *(Dec–05)*

(ANS: $P_o = 2W_r/(4br-b^2)^{1/2}$, $R_B = (b-2r)/(4br-b^2)^{1/2}$, $R_A = W(b-2r)/(4br-b^2)^{1/2}$, $R_C = 2W$, $P_o = 1.67KN$, $R_B = 1.33KN$, $R_A = 1.3KN$, $R_C = 2KN$)

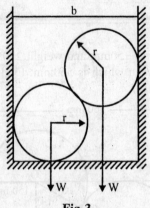

Fig 3

Q.4. A 500N Cylinder is supported by the frame ABC, which is hinged at A, and rests against wall AD. Determine the reactions at contact surfaces A, B, C and D. ($R_A = 527N$, $R_B = R_C = 166.67N$)

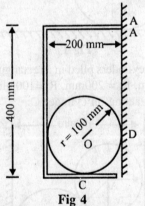

Fig 4

Q.5. Cylinder A and B weighing 5000N and 2500N rest on smooth incline planes as shown in fig (5). Neglecting the weight of connecting bar and assuming smooth pin connections, find the force P to be applied such that the system is in the equilibrium. (P=2427.25N)

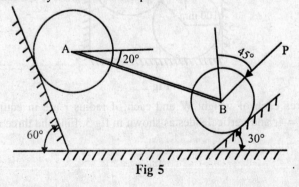

Fig 5

Q.6. Two steel cylinders are supported in a right angled wedge support as shown in Fig. 6. The side OL makes an angle of 30° with the horizontal. The diameters of the cylinders are 250 mm and 500 mm; their weights being 100 and 400 N respectively. Deter-mine the reaction R between the smaller cylinder and the side OL. [Ans. 157 N]

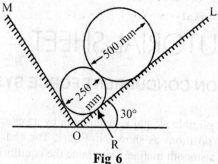

Fig 6

Q. 7. The body shown in fig-7, is acted upon by four forces. Determine the resultant.

(ANS: 5.724KN, 53.56°)

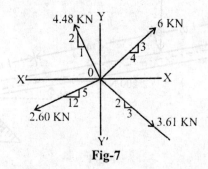

Fig-7

Q. 8. Concurrent forces 3P, 7P and 5P act respectively along three directions, which are parallel to the side of an equilateral triangle taken in order. Determine the magnitude and direction of the resultant.

(ANS : $12^{1/2}$P, 210^0 with the 3P force)

Q.9. ABCDEF is a regular hexagon. Forces 2, P, Q, 4 and 3KN act along AB, CA, DA, AE and FA respectively and are in equilibrium. Determine the value of P and Q.

(ANS: P=4.65KN, Q =1.044KN)

Q.10. For the cylinder shown in fig (8). Determine reaction at C.

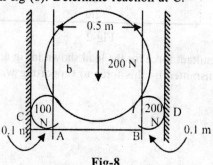

Fig-8

TUTORIAL SHEET – 6

(A) NON CONCURRENT FORCE SYSTEM

Q.1. A bar 4 m long and of negligible weight is acted upon by a vertical load of 400 N and a horizontal load of 200 N applied at positions as shown, in Fig. 1. The ends of the bar are in contact with, a smooth vertical wall and a smooth incline. Determine the equilibrium position of the bar as defined by the angle θ, it makes with the horizontal. (ANS : θ = 28.6°)

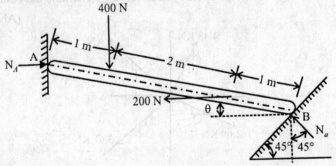

Fig 1

Q.2. A rod 2m long & of negligible wt. Rest in horizontal position on two smooth inclined planes. Determine distance 'x' at which the load Q = 100N should be placed from point B to keep the bar horizontal. As shown in fig 2. (ANS-X = 0.81m)

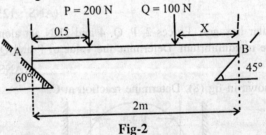

Fig-2

Q.3. Compute the simplest resultant force for the load shown in fig acting on the cantilever beam. What force and moment is transmitted by this force to supporting wall at A? (May - 2005)

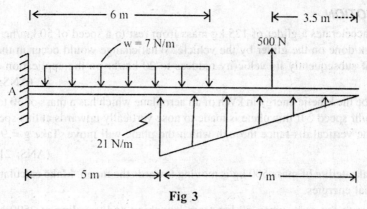

Fig 3

Q.4. Find the support reactions in the beam as shown in fig 4. (Dec-00-01)

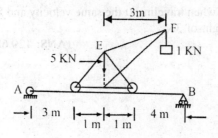

Fig 4

Q.5. One end A of a split horizontal beam ACB is fixed into a wall and the other end B rest on a roller support. A hinge is at a point C. A crane weighing W = 50,000N is mounted on the beam and is lifting a load of P = 10,000N at the end L. The C.G. of the crane acts along the vertical line CD and KL = 4m as shown in fig 5, Neglecting the weight of the beam, find the reaction/moments at A and B. (May–00-01)

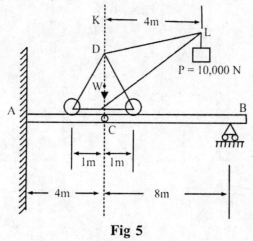

Fig 5

(B) LAW OF MOTION

Q.6. A vehicle accelerates a glider of 125 kg mass from rest to a speed of 50 km/hr. Make calculations for the work done on the glider by the vehicle. What change would occur in the kinetic energy of the glider if subsequently its velocity reduces to 20 km/hr on the application of brakes?

(ANS: 12058J, 10126J)

Q.7. What will be the kinetic energy in kWh of an aeroplane which has a mass of 30 tons and is traveling at 1000 km/hr speed ? If this plane is made to nose vertically upwards at this speed with power off, calculate the vertical distance through which the plane will move. Take g = 9.80 m/s^2.

(ANS: 21.43KWH, 7.87m)

Q.8. An artificial satellite of mass 500 kg is moving towards the moon. Make calculations for the kinetic and potential energies.

(i) Relative to the earth when 50 km from launching and traveling at 2500 km/hr. Take earth's gravitational field equal to 7.9 m/s^2.

(ii) Relative to the moon when traveling at the same velocity and 50 km from its destination where 1 kg mass has a weight of 3N.

(ANS: 120.6MJ, 197.5MJ, 75MJ, 120.6MJ)

TUTORIAL SHEET – 7

FRICTION

(A) Friction on Horizontal Plane

Q.1. A body weighing 500N is just moved along a horizontal plane by a pull of 141.41N making 45° with horizontal. Find the coefficient of friction. (Ans: 0.25)

Q.2. Force required to pull a body of 50N on a rough horizontal plane is 20N. Determine the coefficient of friction if force is applied at an angle of 25° with X- axis. (Ans: 0.35)

Q.3. For the arrangement shown in fig (1), find the force F needed to cause impending motion to 3KN weight, coefficient of friction for all the contact surfaces being 0.3. What is the tension in the cable. (Ans: 1.92KN, 1.7KN)

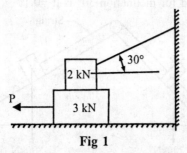

Fig 1

Q.4. A man wishing to slide a stone block of weight 1000N over a horizontal concrete floor ties a rope to the block and pulls it in a direction inclined upward at an angle of 20^0 to the horizontal. Calculate the minimum pull necessary to slide the block if the co-efficient of friction = 0.6. Calculate also the pull required if the inclination of the rope with the horizontal is equal to the angle of friction and prove that tie is the least force required to slide the block. (Ans: P=524N, 514.5N)

(B) Friction on Inclined Plane

Q.5. As Shown in fig (2), what should be the minimum weight of W so that the block of 1000N will not slide down the plane? Assume the pulley to be smooth and µ = 0.3. (W = 272.73N)

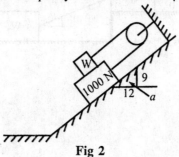

Fig 2

Q.6. Two blocks connected by a horizontal link AB are supported on two rough planes as shown in fig-3. The coefficient of friction for the block on the horizontal plane is 0.4. The limiting angle of friction for block B on the inclined plane is 20°. What is the smallest weight W of the block A for which equilibrium of the system can exist if weight of block B is 5KN?

(Ans: W = R_2 = 10.49KN)

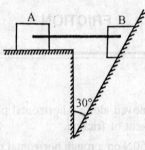

Fig 3

Q.7. Two inclined planes have a common vertex and a string passing over a smooth pulley at the vertex, supporting two bodies of weight 200KN and 800KN as shown in fig-4. A cord fixes the weight of 200KN to the inclined plane of inclination 45°. Determine the tension in this cord. Assume that $\mu = 0.2$ for plane at 45° and for inclination 30° is $\mu = 0.1$.

(ANS: T_2 = 161.02KN)

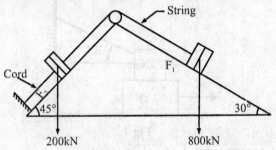

Fig 4

Q.8. Two blocks of weight W_1 and W_2 are connected by a string and rest on a horizontal plane as shown in fig(5). Find the magnitude and direction of the least force 'P' that should be applied to the upper block to induce sliding. The coefficient of friction for each block is to be taken as μ

(ANS: P = (W_1 + W_2) sin θ, act at an angle of θ = Φ)

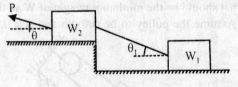

Fig 5

Q.9. Two identical blocks of weight W are supported by a rod inclined at 450 with the horizontal as shown in fig (5.43). If both the blocks are in limiting equilibrium, find the coefficient of friction, assuming it to be the same at floor as well as wall. (ANS: 0.414)

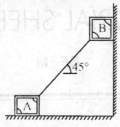

Fig 6

(C) Ladder Friction

Q.10. For a ladder of length 4m, determine the minimum horizontal force to be applied at A to prevent slipping. $\mu = 0.2$ between the wall and the ladder, and 0.3 for the floor and the ladder. The ladder weight 200N and a man weighing 600N is at 3m from A. (Point A is on floor.)   (ANS: 61.77N)

Q.11. A uniform ladder of weight 800N and length 7m rests on a horizontal ground and leans against a smooth vertical wall. The angle made by the ladder with the horizontal is 60^0. When a man of weight 600N stands on the ladder at a distance 4m from the top of the ladder, the ladder is at the point of sliding. Determine the coefficient of friction between the ladder and the floor.

(ANS: $\mu = 0.27$)

Q.12. A 7.0m ladder rests against a vertical wall making an angle of 45°. If a man, whose weight is one half of that of the ladder, climbs it, at what distance along the ladder is he about to slip. For both the surfaces coefficient of friction is μ.

ANS: $X = 14 \left[1 - 3/2 \left\{ (1 - \mu)/(1 + \mu^2) \right\} \right]$

Q.13. A uniform ladder rests with one end against a smooth vertical wall and the other on the ground, the coefficient of friction being 0.75 and the inclination of the ladder on the ground being 45°. Show that a man whose weight is equal to that of the ladder can just ascent to the top of the ladder without its slipping.

Q.14. A uniform ladder 3m long weights 20N. It is placed against a wall making an angle of 60^0 with the floor. The coefficient of friction between the wall and the ladder is 0.25 and that between the floor and ladder is 0.35. The ladder, in addition to its own weight, has to support a man of 100N at its top at B calculate:

(1) The horizontal force P to be applied to ladder at the floor level to prevent slipping.

(2) If the force P is not applied, what should be the minimum inclination of the ladder with the horizontal so that there is no slipping of it, with the man at its top?

(ANS: P = 18.35N, 68°57)

TUTORIAL SHEET – 9

BEAM

(A) Simply Supported Beam

Q.1. A horizontal beam 6m long is loaded as shown in fig 1. Construct the shear force and bending moment diagrams.

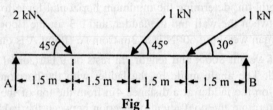

Fig 1

Q.2. Draw SF and BM diagram for the beam with a central moment M as shown in fig 2.

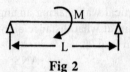

Fig 2

Q.3. Determine the SF and BM diagrams for the simply supported beam shown in fig 3. Determine the maximum bending moment.

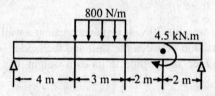

Fig 3

Q.4. Draw the SF and BM diagram for the simply supported beam loaded as shown in fig 4.

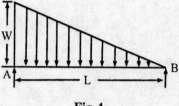

Fig 4

Q.5. A log of wood is floating in water with a weight W placed at its middle as shown in Fig.5 Neglecting the weight of log, draw shear force (SF) and bending moment (BM) diagrams of the log.

(Dec–02)

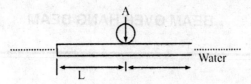

Fig 5

Q.6. Draw the shear force diagrams for the beam AB (Fig.6) loaded through the attached strut.

(May–03)

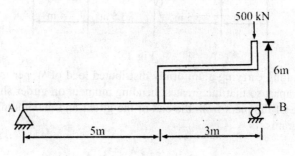

Fig 6

Q.7. Draw SF diagram for simply supported beam shown in fig 7.   *Dec-03(C.O.)*

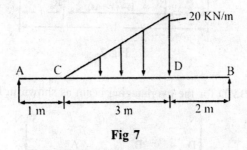

Fig 7

(B) Cantilever Beam

Q.8. A cantilever is shown in fig 8. Draw the BMD and SFD.

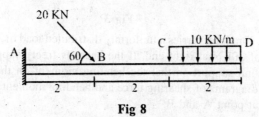

Fig 8

TUTORIAL SHEET – 10

BEAM OVER HANG BEAM

Q.1. Draw the SF and BM diagram for the simply supported beam loaded as shown in fig 1.

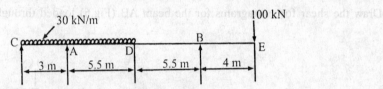

Fig 1

Q.2. A girder of 10m long carrying a uniformly distributed load of W per meter run is to be supported on two piers 6m apart so that the greatest bending moment on girder shall be as small as possible. Find the distance of the piers from the end of the girder and maximum bending moment. Plot the BM and SF diagrams.

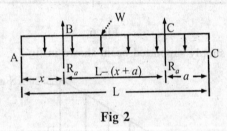

Fig 2

Q.3. Draw the SFD and BMD for the overhanging beam as shown in fig 3.

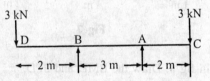

Fig 3

Q.4. A horizontal beam, 30m long carries a uniformly distributed load of 10KN/m over whole length and concentrated load of 30KN at right end. If the beam is freely supported at the left end, find the position of the second support so that the bending moment on the beam should be as small as possible. Draw the diagrams of shearing force and bending moment and insert the principal values. R_A, R_B = reaction at point A and B.

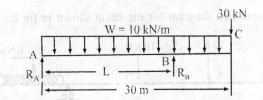

Fig 4

Q.5. A beam ABCD 20m long is loaded as shown in fig 5. The beam is supported at B and C and has a overhang of 2m to the left of support B and an overhang of K meters to the right of support C which in the right hand half of the beam. Determine the value of K if the mid point of beam is a point of inflexion and for this arrangement, plot BM and SF diagrams indicating the principle numerical values.

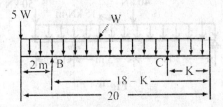

Fig 5

Q.6. Figure 6 shows a beam pivoted at A and simply supported at B and carrying a load varying from 0 at A to 12 kN/m at B. Determine the reactions at A and B; and draw the bending moment (BM) diagram. *(Dec–02)*

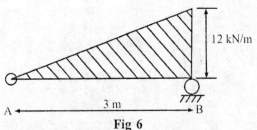

Fig 6

Q.7. A uniformly loaded beam with equal overhang on both sides of the supports is shown in the fig.7. Draw the bending moment diagram, when a = l/4. *(May–03)*

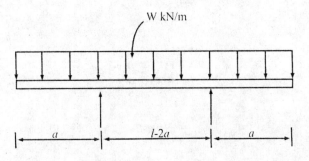

Fig 7

Q.8. Draw the shear force and BM diagram for the beam shown in fig 8. *(Dec–03)*

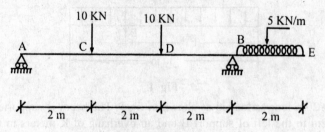

Fig 8

Q.9. Draw the shear force and bending moment diagram for the beam shown in fig 9. *(May–05)*

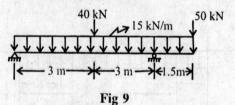

Fig 9

TUTORIAL SHEET – 11

TRUSS

Q.1. Using the method of joints, find the axial forces in all the members of truss with the loading as shown in fig (1).

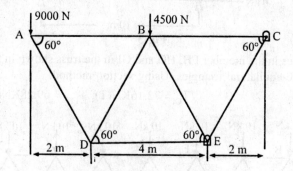

Fig 1

Q.2. A framed structure of 6m span is carrying a central load of 10KN as shown in fig 2. Find by any method, the magnitude and nature of forces in all members of the structure.

(T_{AD}, T_{DB} = 7.08(T); T_{AC}, T_{CB} = 11.19(C); T_{CD} = 10(T))

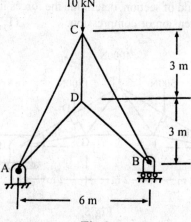

Fig-2

Q.3. Determine the force in member ED, EG, and BD of the truss shown in fig-3. Using section method.

$R_{DE} = 60KN(T)$, $R_{EG} = 35.6KN(C)$, $R_{BD} = 35.6KN(T)$

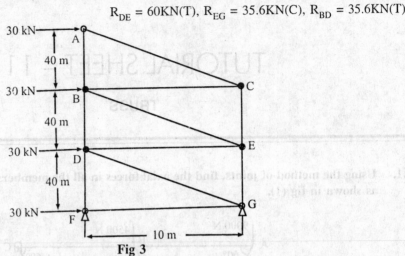

Fig 3

Q.4. Determine the forces in the member FH, HG and GI in the truss shown in fig 4. Each load is 10KN and all triangles are equilateral triangles. Using section method.

$(T_{GI} = 72.16KN(T)$, $T_{FH} = 69.28KN(C)$, $T_{GH} = 5.77KN(C))$

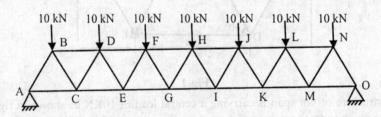

Fig 4

Q.5. The roof truss shown in fig-5, is supported at A and B carries vertical loads at each of the upper chord points. Using the method of section, determine the forces in the members CE and FG of truss, stating whether they are in tension or compression. $(T_{FG} = 2422N(T)$, $T_{CE} = 3852N(C))$

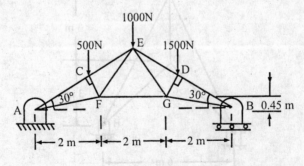

Fig 5

Q.6. Compute the forces in member CD, KD and KJ of the truss shown in fig-6.

$T_{CD} = 250KN(c)$, $T_{KJ} = 300KN(T)$, $T_{KD} = 70.71KN(c)$

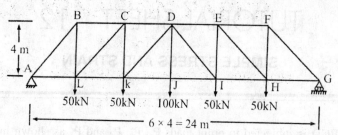

Fig 6

Q.7. In the truss shown in fig 7, compute the forces in the member CD, DH and HG.

($T_{DH} = 70.78KN(T)$, $T_{GH} = 89.49(T)$, $T_{CD} = 100KN(C)$)

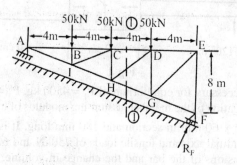

Fig 7

Q.8. For the overhanging truss shown in fig 8. Compute the member force in BC, CE and DE.

($T_{DE} = 6KN(C)$, $T_{CE} = 20KN(C)$, $T_{BC} = 24.04KN(T)$)

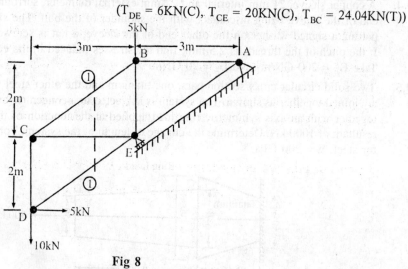

Fig 8

TUTORIAL SHEET – 12

SIMPLE STRESS AND STRAIN

Q.1. A member ABCD is subjected to point loads P_1, P_2, P_3 and P_4 as shown in fig 1.

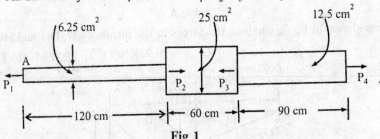

Fig 1

Calculate the force P_2 necessary for equilibrium, if P_1 = 4500 kg, P_3 = 45000 kg and P_4 = 13000 kg. Determine the total elongation of the member, assum-ing modulus of elasticity to be 2.1×10^6 kg/cm².

Q.2. A bar of steel is 60 mm x 60 mm in section and 180 mm long. It is subjected to a tensile load of 300kN along the longitudinal axis and tensile loads of 750kN and 60 kN on the lateral faces. Find the change in the dimensions of the bar and the change in volume.

Q.3. The piston of a steam engine is 300 mm in diameter and the piston rod is of 50 mm diameter. The steam pressure is 1N/mm². Find the stress in the piston rod and elongation in a length of 800 mm E = 200GPa.

Q.4. A copper sleeve, 21 mm internal and 27 mm external diameter, surrounds a 20 mm steel bolt, one end of the sleeve being in contact with the shoulder of the bolt. The sleeve is 60 mm long. After putting a signed washer on the other end of the sleeve, a nut is screwed on the bolt through 10^0. If the pitch of the threads is 2.5 mm, find the stresses induced in the copper sleeve and steel bolt: Take E_s = 200 GN/m² and E_c = 90 GN/m².

Q.5. Two solid circular cross-section bars, one titanium and the other steel each in the form of a cone, are joined together as shown. The system is subjected to a concentric axial tensile force of 500 kN together with an axis symmetric ring load applied at the junction of the bars having a horizontal resultant of 1000 kN. Determine the change of length of the system. For titanium, E = 110 GPa and for steel, E = 200 GPa.

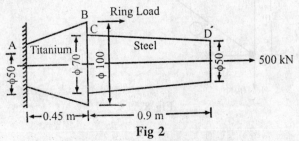

Fig 2

Q.6. Calculate the strain energy stored in a bar 250 cm long, 5 cm wide and 4 cm thick. When it is subjected to a tensile load of 6 tons. Take $E = 2.0 \times 10^6$ kg/cm²

Q.7. An unknown weight falls through a height of 10 mm on a collar rigidly attached to the lower end of a vertical bar 5 m long and 600 mm² in section. If the maximum extension of the rod is to be 2mm, what is the corresponding stress and magnitude of the unknown weight ? Take: $E = 200$ GN/m².

Q.8. A vertical compound member fixed rigidly at its upper end consists of a steel rod 2.5 m long and 20 mm diameter placed within an equally long brass tube 21 mm internal diameter and 30 mm external diameter. The rod and the tube are fixed together at the ends. The compound member is then suddenly loaded in the tension by a weight of 10 kN falling through a height of 3 mm on to a flange fixed to its lower end. Calculate the maximum stress in steel and brass.

Assume $E_s = 200$ GN/m² and $E_b = 100$ GN/m².

Q.9. A C.I. Flat, 300 mm long and of 30 mm x 50 mm uniform section, is acted upon by the following forces uniformly distributed over the respective cross-section ; 25kN in the direction of length (tensile); 350kN in the direction of the width (compressive) ; and 200kN in the direction of thickness (tensile). Determine the change in volume of the flat. Take $E = 140$ GN/m² and $m = 4$.

Q.10. For a given material, Young's modulus is 1.2×10^6 kg/cm². Find the bulk modulus and lateral contraction of a bound bar of 50 mm diameter and 2.5m long, when stretched 2.5mm. Take Poison's ratio as 0.25.

Q.11. A bronze specimen has $E = 1.2 \times 10^6$ kg/cm² and modulus of rigidity 0.45×10^6 kg/cm². Determine the Poisson's ratio of the material.

TUTORIAL SHEET – 13

COMPOUND STRESS

Q.1. A piece of steel plate is subjected to perpendicular stresses of 50 N/mm² both tensile. Calculate the normal and tangential stresses and the interface whose normal makes an angle of 30°C with the axis of the second stress.

Q.2. Draw Mohr's circle for principal stresses of 80 N/mm² tensile and 50 N/mm² compressive, and find the resultant stresses on planes making 22^0 and 64^0 with the major principal plane. Find also the normal and tangential stresses on these planes.

Q.3. A propeller shaft of 30 cm external diameter and 15 cm internal diameter transmits 1800 kW power at 1200 rpm There is at the same time a bending moment of 12 KN-m and an end thrust of 300 IN. Find:

(*i*) The principal stresses and their planes

(*ii*) The maximum shear stress

(*iii*) The stress which acting along will produce the same maximum strait Take Poison's ratio = 0.3.

Q.4. The principle stresses at a point in a bar are 200 N/mm²(tensile) and 100N/mm² (compressive). Determine the resultant stress in magnitude and direction on a plane inclined at 60° to the axis of the major principle stress.

Q.5. A tension member is formed by connecting with glue two wooden scantling each 7.5 x 10⁶ kg/cm² at their ends, which are cur at an angle of 60° as shown in Fig.

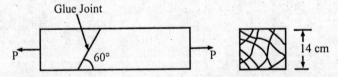

The member is subjected to a force P. Calculate the safe value of P, if the permissible normal shear stress in glue are 14 kg/cm² and 7 kg/cm² respec-tively.

Q.6. A point in a strained material is subjected to two mutually perpendicular tensile stresses of 3000 kg/cm² and 1000 kg/cm². Determine the intensities of normal and resultant stresses on a plane included at 30 to the axis of the minor stress.

Q.7. A piece of material is subjected to tensile stresses of 60 N/mm² and 30 N/mm² at right angles to each other. Find resultant stress on a plane which makes an angle of 40 with 60N/mm² stress.

Q.8. The stressed on two mutually perpendicular planes are 40 N/mm²(tensile) and 20N/mm² (tensile). The shear stress across planes is 10 N/mm². Find using Mohr's circle, the magnitude and direction of the resultant stress on plane making an angle of 30° with the plane of tire fire stress.

Q.9. Two wooden blocks 50 min x 1000 mm are joined together along the joint AB as shown in Fig. Determine the normal stress and shearing stress in the joint if P = 200.

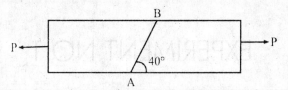

Q.10. A piece of material is subjected to two compressive stresses at right angles, their values being 40 MN/m^2 and 60 MN/m^2. Find the position of the plane across which the resultant stress is most inclined to the normal, and the determine the value of this resultant stress.

Q.11. Direct stress of 120 MN/m^2 in tension and 90 MN/m^2 in compression are applied to an elastic material at a certain point on planes at right angles to another. If the Maximum principal stress is not to exceed 150 MN/m^2 in tension, to what shearing stress can the material be subjected?

What is their the maximum resulting shearing stress in the material? Also find the magnitude of the other principal stress and its inclination to 120MN/m^2 stress.

Lab Manual

EXPERIMENT NO.1

OBJECT: Study of fire and water Boilers.
APPARATUS : Models of Babcock Wilcox, Lancashire and Locomotive Boilers.

INTORDUCTION :

(a) BOILERS: Boilers may be defined as closed pressure vessels, which are used to generate steam at the pressure much higher than the atmospheric pressure by transfer of heat produced by burning of fuel to water.

The construction and appearance of steam generators depends on the arrangement made for burning the fuel and for transfer of heat to water. The steam produced by steam generator is used in following applications :

(1) For operating steam turbine of power plants for power generation.
(2) Locomotive Steam engines for traction.
(3) As process steam in industries.
(4) For heating applications.
(5) For running large steam propelled ships.

(b) SPECIFICATIONS : The boilers are specified as per following parameters :
- Steam Generation Rate : for example 50 000 kg /hr.
- Max. Pressure : for example 150 kg/ sq.cm
- Dimensions of steam drum
- Horizontal / Vertical
- Water / Fire Tube
- Natural / Forced Circulation
- Type of Fuel Used : Coal / Oil / Gas etc.

(c) ESSENTIALS OF A GOOD BOILER :
- Must be capable of producing the required Steam at the required pressure for minimum fuel required.
- Should be capable of withstanding load variations.
- Must not take a long time for starting
- Easy maintenance and must be accessible.
- Tubes must be strong enough
- Must comply all safety regulations.

(d) FACTORS AFFECTING BOILER SELECTION
- Working Pressure and amount of steam required.
- Floor area required.
- Operating and Maintenance Required & Inspection facilities.

- Fuel required.
- Water requirements.

(e) BOILER CLASSIFICATION:

(1) **According to Tube Contents :** Boilers may be water tube or fire tube type. In water tube boilers the water flows in the tubes and in Fire Tube Boilers hot flue gasses circulate through the tube. Babcox & Wilcox Boiler is water tube boiler where as Locomotive and Lancashire Boiler are examples of Fire tube Boiler.

(2) **According to the Axis of Boiler :** The Boiler may be Horizontal axis or Vertical Axis type. Babcox & Wilcox, Locomotive and Lancashire Boilers are all examples of Horizontal axis boilers.

(3) **According to Number of Boiler tubes** : May be single or multiple tubes boilers. Cornish Boiler is a single tube boiler whereas other boilers are multi-tube boilers.

(4) **According to Position of Furnace :** Externally / internally fired boilers. In externally fired boiler the furnace is external to the boiler shell.

(5) **On the Basis of Mobility** : Stationary / Moving Boilers.

(6) **On the Basis of Water Circulation** : Natural / Forced Circulation.

(7) **On the Basis of Draught** : Forced / Induced draught.

(8) **On the Basis of utility of steam** : Example : Power Plant, Marine and Locomotive boilers etc.

(f) IMPORTANT TERMS :

Fire Box : Place where fuel is burnt.
Grate : The fuel is burnt on this.
Baffles: Plates used for directing the flow of flue gasses.
Chimney: For ensuring that spent flue gasses exit the boiler at reasonable height. It helps in creation of natural draught.
Damper: Regulates the amount of air through chimney.
Headers: Pipes connected to tubes and drums.
Tubes: For enabling heat transfer from flue gasses to water.
Shell: Cylindrical vessel which contains water to be converted to steam.
Mountings: For ensuring safe and satisfactory performance
Safety Valve: It releases steam if pressure inside boiler exceeds certain design limits. May be of dead weight, spring loaded and lever types.
Pressure Gauge : For indicating steam pressure
Water Level Indicator: for indicating water level.
Steam Stop Valve : Permits flow of steam from boiler as and when desired.
Feed Check Valve : Permits flow of water to boiler.
Blow off Cock : For removing settlements collected at the bottom of boiler.
Manhole: Permits entering of operator / inspection staff.
Fusible Plug: For avoiding any explosion due to overheating.
Boiler Accessories : Auxiliary equipment for efficient boiler operation.
Preheater: A heat exchanger which is used to heat air entering the boiler with waste flue gases .
Economiser: A heat exchanger to heat feed water by means of outgoing flue gasses.
Superheater: To super heat the steam coming from boiler.
BABCOCK & WILCOX BOILER: It is a horizontal, externally fired, natural draught, stationary, water tube boiler. It is generally a high capacity boiler and can produce steam at a pressure of 4 Mpa at a rate of 40000 kg/hr.

CONSTRUCTION FEATURES: Consists of a water drum which is connected to two headers at the front and back ends. The headers are connected by a large number of water tubes which are inclined upwards from downtake header to uptake header. Further, a mud drum, in which heavier sediments of water settle down and are blown off from the blow off pipe, is also provided. The combustion chamber is located below the boiler drum. A chain grate with stoker is provided fro burning coal. Hot flue gasses produced after combustion of fuel are made to pass over the water tubes in several passes with the help of baffles. The draught is regulated using dampers provided. Finally, chimney is provided to permit exit of waste flue gasses.

A superheater is also provided to superheat dry and saturated steam. It consists of U tubes and is placed in the path of flue gasses.

Several mountings which ensure safe and efficient operation of boiler are also provided. These include water level indicator, Pressure Gauge, Dead weight safety valve etc.

WORKING :Water is filled in boiler shell through the feed valve upto 2/3 rd of shell. Water flows down through the downtake header via descending water tubes and rises upwards through the uptake header and ascending tubes. Inclined tubes help in setting up water circulation currents. The hot flue gasses coming from combustion chamber are made to pass over the tubes, under side of the drum and superheater tubes. Baffle plates ensure longer contact time of flue gasses with the tubes. When water in tubes is heated, it moves upwards through the ascending tubes and cold water takes its place through the descending tubes. In this way convection currents are set up.

The hottest gases being raised from the grate come in with that portion of water tube which are located on the highest side near the upper header. Water begins to evaporate and mixture of water and steam thus formed reaches into the boiler drum. Steam is collected in space above water. This steam is then lead to superheater tubes for superheating. Superheated steam thus produced flows out through the stop valve.

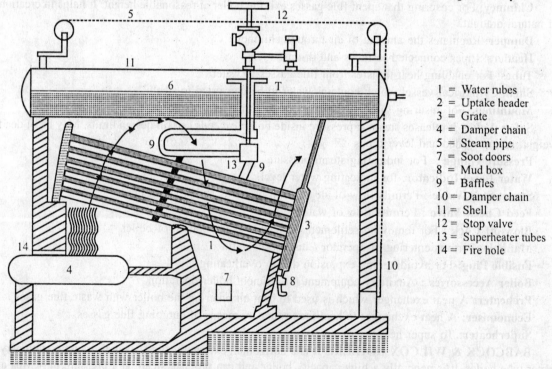

Fig A-1: BABCOCK & WILCOX BOILER:

LACASHIRE BOILER : It is a fire tube, internally fired, stationary, horizontal, natural circulating type boiler. Evaporative capacity may be upto 8500 kg/hr. and can operate with working pressures upto 1.5 Mpa. Normally it is used in sugar mills and chemical plants ie in moderate conditions.

CONSTRUCTION: Consists of a boiler shell which contains water and steam. Diameter may vary from 1.75 to 2.75 m and length may be 7.25 – 9 m. It has two side channels connected to rear end of the boiler shell and then finally to the chimney. To provide larger heating surface area it has two large diameter flue gas tubes. The tubes are tapered with larger dia in front and smaller dia at back. The taper is provided to accommodate the grate. Two grates are provided at the front end of the flue gas tubes. For controlling the flow of the flue gases two dampers are also provided at the rear end. Further for cleaning and inspection of the drum, manhole is also provided. Other boiler mountings like feed check valve, pressure gauge, water level gauge, steam stop valve, blow off cock are also provided.

WORKING : Water is filled through the feed check valve. On burning the fuel over the grate, the hot gasses are produced which move in the flue gas tubes. The flue gasses reach the rear end of the boiler and then made to deflect and pass through the bottom central chamber. On reaching the front end the flue gasses again get deflected and then pass through the side chambers. Finally the flue gasses are discharged to atmosphere through chimney. The flow of flue gasses are controlled by dampers which are operated by boiler operator using chain pulley arrangement.

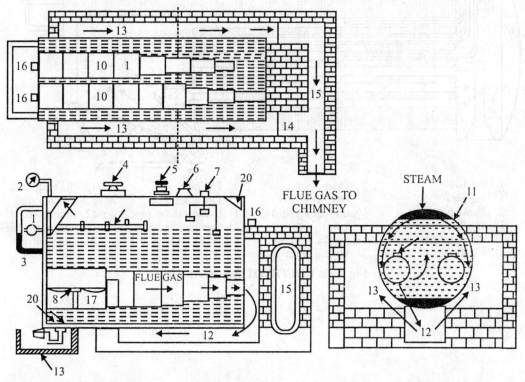

1. Feed check vlave, 2. Pressure gauge, 3. Water level gauge, 4. Dead weight safety valve, 5. Steam stop valve, 6. Main hole, 7. Low water high steam safety valve, 8. Fire grate, 9. Fire bridge, 10. Flue tubes, 11. Boiler shell, 12. Bottom flue, 13. Side flue, 14. Dampers, 15. Main flue, 16. Doors, 17. Ashpit, 18. Blow off cock, 19. Blow off pit, 20. Gossel stays, 21. Perforated leed pipe, 22. Anti priming pipe, 23. Fusible plug.

Fig A-2: LACASHIRE BOILER

LOCOMOTIVE BOILER : It is a horizontal, natural circulation, fire tube boiler. It can produce steam at the rate of 7000 kg/hr. at a pressure of upto 2.5 Mpa. Earlier it was used mostly in railways for traction.

CONSTRUCTION & WORKING : The boiler barrel is a cylindrical shell and consists of a large number of flue tubes. The barrel consists of a rectangular fire box at one end and a smoke box at the other end. The coal is introduced in the fire box through the fire hole and is made to burn on the grate. Water is filled into cylindrical boiler shell upto about ¾ level. Hot gases generated as a result of coal burning rise and get deflected by a fixed fire brick lining. The hot flue gases heat the water and then reach the smoke box at tbe other end. Finally the flue gasses pass to the atmosphere through a short chimney. The steam produced is stored in the steam space and the steam dome. A throttle valve is provided in the steam dome. The throttle valve is controlled by a regulating rod from outside. For superheating, the steam passes through the throttle valve to the super heater.

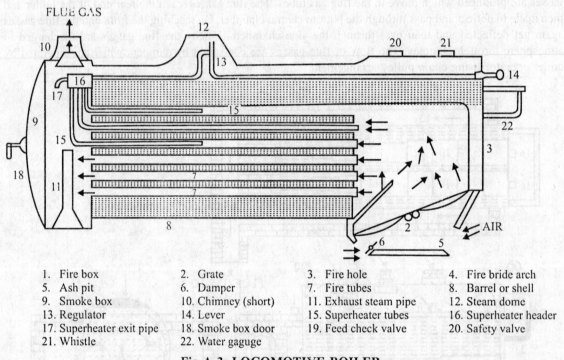

1. Fire box	2. Grate	3. Fire hole	4. Fire bride arch
5. Ash pit	6. Damper	7. Fire tubes	8. Barrel or shell
9. Smoke box	10. Chimney (short)	11. Exhaust steam pipe	12. Steam dome
13. Regulator	14. Lever	15. Superheater tubes	16. Superheater header
17. Superheater exit pipe	18. Smoke box door	19. Feed check valve	20. Safety valve
21. Whistle	22. Water gaguge		

Fig A-3: LOCOMOTIVE BOILER

EXPERIMENT No.2

OBJECT : To study steam engine and steam turbine models.
APPARATUS USED : Models of steam engines and steam turbines.
INTRODUCTION : Steam engine is a heat engine that converts heat energy to mechanical energy. The pressure energy of steam acts on the piston and piston is moved to & fro in a cylinder. The to & fro motion is then converted to rotary motion using a suitable mechanism. Steam engines find application in locomotives, drives for process equipment, steam hoists, pump drives etc.

CLASSIFICATION :

(1) **According to Cylinder Axis** : Engine may be vertical /horizontal.
(2) **According to action of steam** : May be single acting / double acting.
(3) **According to number of cylinders:** May be single /multi cylinder (Compounded).
(4) **According to the method of steam exhaust** : Condensing / Non condensing. In condensing type engines the engine discharges into a condesnser where exhaust is converted to water.
(5) **According to speed of crankshaft** : Low (< 100 rpm), Medium (100 – 200 rpm) and High (> 200 rpm).
(6) **According to Valve gear used** : Slide Valve (D type) / Poppet valve.
(7) **According to type of service** : Stationary / Mobile
(8) **According to cylinder Arrangement** : Tandem Type / Cross Compounded.
(9) **According to Method of Governing** : Throttling Steam Engine / Automatic Cut off engine.

STEAM ENGINE MAIN PARTS :

1. **Frame** : It supports the various moving parts.
2. **Cylinder** : It forms a chamber in which the piston moves to and fro.
3. **Steam Chest** : Integrally casted with cylinder and is closed with a cover.
4. **Piston** : It is a cylindrical part which reciprocates to and fro in the cylinder under the action of steam pressure.
5. **Piston Rod** : It connects the piston with cross head and is made of mild steel.
6. **Cross Head :** Fixed between connecting rod and piston rod. The cross head slides between guide bars.
7. **Connecting Rod** : It connects cross head with the crank on the other side.
8. **Crank Shaft** : It is the output shaft on which the mechanical power is available for doing work. It gets rotated due to twisting of crank. The crank is rigidly attached to the crank shaft.
9. **Stuffing Box** : It is placed at the point where the piston rod comes out of the cylinder cover. It prevents the leakage of steam from the cylinder to the atmosphere.
10. **Fly Wheel** : It is provided to minimize speed fluctuations or crank shaft. It is keyed to the crank shaft.
11. **Eccentric** : It is mounted on the crank shaft and converts rotary motion of the crank shaft to reciprocating motion of eccentric rod.
12. **Eccentric Rod and valve rod :** The eccentric rod connects the valve rod guide and the eccentric. The valve connects the guide with the D slide valve. The eccentric and eccentric rod converts

the rotary motion of the eccentric in to reciprocating motion which is transmitted to the valve through the valve rod.

13. **D slide valve :** It controls the admission of steam in to cylinder through the ports alternatively to act on both sides of the piston in a double acting steam engine. It is installed inside the steam chests and slides over the machined surface of cylinder by the action of valve rod and eccentric rod.

14. **Port :** These are the rectangular openings provided in the cylinder to allow steam inlet and exhaust.

STEAM ENGINE TERMINOLOGY

1. **Cylinder Bore :** Inside diameter of cylinder.
2. **Cover End and Crank End of cylinder :** In horizontal cylinders the end which is farthest from the crank is called cover end and the end which is nearest to the crank is called crank end. In case of vertical engines top end is called front end and lower end is called bottom end.
3. **Piston Stroke :** The distance traveled or moved by the piston from one end to the other end is called stroke. In one stroke of piston the crank shaft makes one half revolution.
4. **Crank Through :** It is the distance between center of the crank shaft and the center of the crank pin.
5. **Dead Centre :** It is the position of the piston at the end of the stroke. At dead center the center line of the piston rod, connecting rod and crank are in the same straight line. A horizontal engine has inner and outer dead centers where as the vertical engine has top and bottom dead centers.
6. **Valve Travels :** It is the total distance that the valve travels in one direction.

WORKING : The high pressure steam from the boiler is supplied to the steam chest. The high pressure steam enters the cylinder through inlet port and exerts pressure on the piston and drives it towards the other end. At the same time the steam on the other side of the piston is exhausted through exhaust port. The D slide valve also moves gradually in opposite direction. The D slide valve then closes the inlet port and steam supply is cut off. The entrapped steam expands and does work to push the piston further. With further movement of the D slide valve, the inlet port of other side is connected to the steam inlet and thus high pressure steam enters other side of the piston which provides the return stroke of the piston.

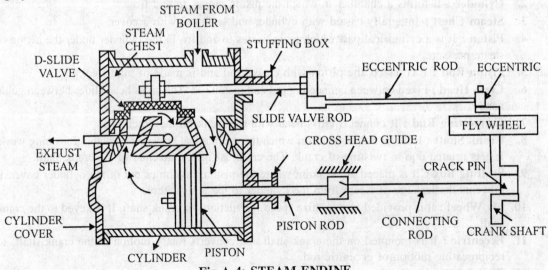

Fig A-4: STEAM ENDINE

Steam Turbine

INTRODUCTION : Steam turbine is prime mover in which rotary motion is obtained by a gradual change of momentum of steam. The force exerted on the blade is due to rate of change of momentum of steam. The curved blades change the direction of steam. The pressure of steam rotates the vanes. The turbine blades are curved in such a way that the steam directed upon them enters without shock.

CLASSIFICATION: The steam turbines are classified as follows :
1. **According to method of steam expansion:** (a) Impulse (b) Reaction.
2. **According to direction of steam flow :** (a) Axial (b) Radial (c) Tangential.
3. **According to Number of stages :** (a) Single Stage (b) Multi Stage.
4. **According to Steam Exhaust conditions :** (a) Condensing Type (b) Non Condensing
5. **According to pressure of steam :** (a) High Pressure (b) Medium Pressure (c) Low Pressure.

CONSTRUCTION DETAILS :
1. **Casing :** It is made of cast steel. The casing consists of rotor inside it.
2. **Rotor:** It carries the blades or buckets.
3. **Nozzle :** It provides flow passage for steam where the expansion of steam takes place.
4. **Frame:** It provides support to rotors, stator and all other mountings. It may be integral part of stator in case of small tubines.

IMPULSE TURBINE : The steam turbine in which steam expands while passing through the nozzle and remains at constant pressure over the blades is called impulse turbine. In figure single stage impulse turbine is shown. In this type of turbine there is one set of fixed nozzles which is followed by the one ring of moving blades. The blades are attached over the rim of wheel, which is keyed to the shaft. The steam expands from its initial pressure to the final pressure in only one set of nozzle. The jet of steam with a very high velocity enters the moving blades. The jet of steam is deflected when passing over these blades, exerts force on them and in this way the rotor starts rotating.

De-laval, Curtis, Zoelly, Rateau are the examples of this type of turbine. Delaval turbine is the impulse turbine which is suitable for low pressure steam suply. But the only disadvantage of this type of turbine is its very high speed (generally 30000rpm) so its use is limited. It is as shown in figure. The steam is expanded from the boiler pressure to the condenser pressure in a single stage only, its velocity will be extremely high. But this speed is too high for practical applications.

The method in which multiple system of rotors are keyed to a common shaft, in series and the steam pressure of jet velocity is absorbed in stages as it flows over the rotor blades, is known as compounding. The velocity compounding is as shown in figure. Curtiz turbine is an example of such type of turbine. Three rings of moving blades (keyed to shaft) are separated by rings of fixed or guide blades. The ring of fixed blades are attached to the turbine casing which is stationary. The steam is expanded from the boiler pressure to the condenser pressure in the nozzle. The high velocity jet of steam first enters the first row of moving blades, where some portion of this high velocity is absorbed by this blade ring. The remaining being exhausted on the next ring of fixed blades. These fixed blades change the direction of jet. The jet is in turn passed to the next ring of moving blades. This process is repeated as the steam flows over the remaining pairs of blades until practically whole of the velocity of the jet is absorbed.

REACTION TURBINE: The turbine in which the steam expands while passing over the moving blades as well as while passing over the fixed blades and the pressure of steam decreases gradually throughout the flow is called reaction turbine. Parsons turbine is an example of reaction turbine. In these turbines the pressure drop during the expansion of steam occurs within the moving and fixed blades. Thus the rotation of shaft is due to both impulsive and reactive forces in the steam.

One stage of a reaction turbine consists of one row of fixed blades followed by a row of moving blades. The fixed blades acts as nozzle. Fixed blades are attached to the inside of the cylinder whereas the moving blades are fixed with the rotor. The rotor is further mounted on the shaft.

In the impulse turbine the steam is expanded, causing pressure and heat drop in nozzle only and the moving blades only direct the steam through an angle. The impeller blades are symmetrical so pressure of steam remains constant while passing over blades. While in reaction turbines the steam is expanded both in the fixed and moving blades continuously as the steam passes over them. So the pressure drops gradually and continuously over both moving as well as fixed blades. The blades of reaction turbines are symmetrical and thicker at one end which provides suitable shape for steam to expand.

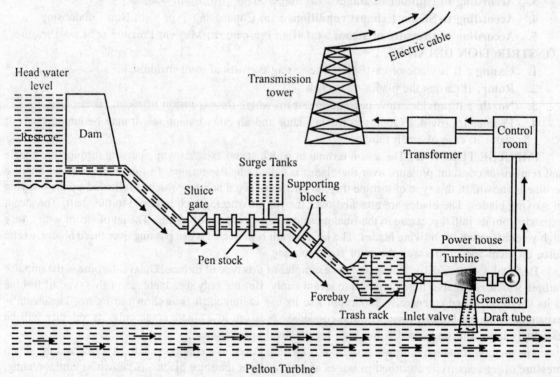

Pelton Turbine

Fig A-5

EXPERIMENT No.3

OBJECT : Study of two stroke and four stroke internal combustion engine models.

APPARATUS USED : Two stroke and four stroke engine models.

INTRODUCTION : The engines which develop power by combustion of fuel within the engine itself are called internal combustion engines. The examples of internal combustion engines are petrol and diesel engines used in cars, trucks etc. The engines in which power is developed by combustion of fuel outside the engine are called external combustion engines eg. steam engine. In an internal combustion engine power is developed from the combustion of fuel which is a chemical reaction. Due to combustion of fuel hot gases are produced at sufficiently high pressure. This pressure is used to move the piston linearly. This linear motion of piston is then converted into rotary motion.

CLASSIFICATION :

1. According to number of strokes per cycle :
   (a) Two stroke   (b) Four stroke.
2. According to the fuel being used :
   (a) Petrol Engine   (b) Diesel Engine   (c) Gas engine   (d) Dual fuel engines
   (e) Liquefied Petroleum Gas Engines.
3. According to working cycles :
   (a) Otto Cycle (Constant volume)
   (b) Diesel Cycle (Constant pressure)   (c) Dual Cycle
4. According to number of cylinder :
   (a) Single Cylinder engine
   (b) Multi cylinder
5. According to the type of Cooling :
   (a) Air cooled engine
   (b) Water cooled engine
6. According to the engine RPM :
   (a) slow speed (< 1000 rpm)
   (b) Medium Speed (1000 – 3000 rpm)
   (c) High Speed (> 3000 rpm)
7. Arrangement of cylinder :
   (a) Radial Engine   (b) Inline engine   (c) V engine
8. According to the type of ignition system:
   (a) Spark ignition   (b) Compression ignition

MAIN PARTS:

1. **Cylinder :** It provides a cylindrical closed space to allow movement of piston and to admit the charge. It is made of grey cast iron or iron alloyed with other elements as nickel, chromium etc. The fuel is burnt inside the cylinder. The internal diameter of cylinder is called bore.

2. **Piston**: The piston reciprocates within the cylinder and transmits the force exerted by expanding gases to crank via connecting rod. The piston is accurately machined to running fit in the cylinder bore and is provided with several grooves in which piston rings are fitted.
3. **Piston Rings**: Piston is equipped with piston rings to provide a good sealing between the cylinder valves and piston. The rings are installed in the grooves in the piston. The rings are of two types a) Compression rings b) Oil control ring.
4. **Connected Rod**: It is attached to the piston at its small end by means of a gudgeon pin. The big end bearing is connected to crank pin. It is made of forged steel.
5. **Crank Pin**: Crank pin is the region on crank shaft on which the big end of connecting rod is attached. These pins are eccentrically located with respect to the axis of the crank shaft. The eccentricity is called throw of the crank.
6. **Crank Shaft**: It is a rotating member which receives the power transmitted by piston connecting rod assembly via crank. It is made of forged alloy steel or carbon steel.
7. **Crank Web or counter weights**: It is provided in the crank shaft to counter act the tendency of bending of the crank shaft due to centrifugal action.

TERMINOLOGY :

1. **Top dead and Bottom dead Centre**: These are two extreme positions between which the piston reciprocates in side the cylinder. TDC & BDC have relevance to opening and closing of valves and the crank shaft rotation.
2. **Bore**: The inner diameter of cylinder is bore. In automobile engines it varies from 40 to 120 mm.
3. **Stroke**: Displacement of piston with in a cylinder between TDC and BDC is called stroke.
4. **Swept Volume (V_s)**: Volume of charge sucked into cylinder when piston travels from TDC to BDC during suction stroke.
5. **Clearance Volume (V_c)**: It is the volume occupied by charge in the space provided between TDC and end of the cylinder.
6. **Engine Capacity (V_E)**: Capacity of the engine is defined as the sum of swept volume of all cylinders.
7. **Compression Ratio**: Ratio of initial volume to the final compressed volume is called compression ration. CR for petrol engine varies from 6.5 – 12 where as for diesel engine it varies from 16 – 23.

WORKING PRINCIPLE :

(a) **Working Principle of Petrol Engine : OTTO CYCLE**

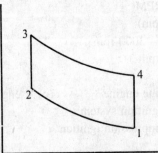

5-1 Suction Stroke
1-2 Adiabatic Compression Stroke
2-3 Heat Addition at constant Volume.

3-4 Adiabatic Expansion
4-1 Heat rejection at constant volume.
1-5 Exhaust stroke at constant pressure.

(b) Diesel (Constant Pressure Cycle):
5-1 Suction Stroke at constant pressure
1-2 Adiabatic Compression
2-3 Heat Addition at constant pressure
3-4 Adiabatic Expansion
4-1 Heat rejection at constant volume
1-5 Exhaust at constant pressure.

Two Stroke and Four Stroke Engines:

In two stroke engines the thermodynamic cycle is completed in One revolution of the crank shaft whereas in four stroke engines the cycle is completed in two revolutions of the crank shaft. For same capacity and speed nearly two times power is developed in two stroke engine as compared to four stroke engine. But as high compression ratios can not be achieved in two stroke engines, the efficiency of two stroke engine is less as compared to four stroke engines. Further, the engines can also be classified as Spark Ignition engines which operate with petrol as fuel and Compression ignition engines which operate with diesel as fuel.

Constructional Details of Two Stroke Engine:

Main components are cylinder, piston, piston rings, piston liners, connecting rod, crank pin, crank shaft, counter weight etc.

Spark Plug : They are mounted on cylinder head for ignition of the charge. The electrode gap in spark plug is maintained between 0.5 to 0.8 mm A voltage of 15000 to 24000 volts is required for creation of spark for ignition.

Ports : They are openings in the cylinder block for ensuring of flow of charge to and from the cylinder. There are three such ports in a two stroke engine viz. inlet port, exhaust port and transfer port.

Deflector Type Piston : In the two stroke engines the piston used is of deflector type. Deflector ensures that fresh incoming charge is not exhausted from the exhaust port with out combustion. The piston is provided with two types of rings also : Compression ring and oil scrapper rings. In two stroke engines the opening and closing of ports is also carried out by pistons and there are no valves or valve operating mechanisms.

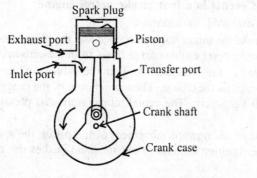
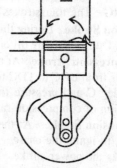

Fig A-6

WORKING : The various operating stages are given below :
1. **Suction Stroke :** The inlet port is opened and a mixture of petrol and air enters the crankcase.
2. **Compression Stroke:** When transfer and exhaust ports are closed the compression takes place.
3. **Power Stroke :** When spark is produced by the spark plug ignition of charge takes place and the charge expands pushing the piston towards the bottom dead center.
4. **Exhaust Stroke :** It takes place when the exhaust port is opened. At this juncture, the transfer port is also opened and a fresh charge is being supplied to the cylinder. Due to deflector shape of the piston the charge goes upwards and not to the exhaust port. Further, to avoid escaping of fresh charge without burning, exhaust port is made a little (1 – 2 mm) above the transfer port.

Construction of Four Stroke Engines:

The single cylinder four stroke engine consists of cylinder, piston, piston rings, crank shaft etc. Further following parts are also provided in a four stroke engine :

Flywheel : It is mounted on the crank shaft to reduce fluctuations in speed during operation as there is only one power stroke for every two revolutions of the crank shaft.

Inlet Valve : For letting in the fresh charge as and when desired on the basis of cylinder – piston relative position.

Exhaust Valve : It allows scavenging of burnt charge.

Valve Operating Mechanism : Consists of tappet, push rod, rocker arm and valve spring. The valve operating mechanism is operated with the help of engine cam shaft which in turn is operated by the crank shaft. The cam shaft operates the tappet which in turn operates the push rod. Push rod pushes the rocker arm which then presses the poppet valve against the spring. Thus the valve is opened.

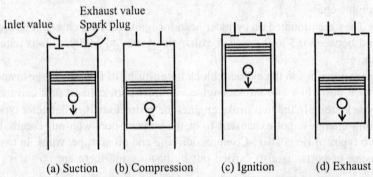

(a) Suction   (b) Compression   (c) Ignition   (d) Exhaust

Fig A7. Cycle of events in a four stroke petrol engine

WORKING : The four strokes are executed as follows :
1. **Suction Stroke :** During this stroke the intake valve is opened and the piston moves from TDC to BDC. Due to pressure difference the combustible charge flows from the carburetor to the cylinder.
2. **Compression Stroke :** At the end of the suction stroke both the valves are closed and the piston moves from BDC to TDC to compress the charge. The temperature of the charge rises to approx. 300 deg C and pressure to 6-9 kg/ sq.cm. The actual temperature and pressure achieved is a function of compression ratio.
3. **Expansion Stroke :** At the end of the upward movement of the piston the spark pug creates a spark to ignite the charge. After ignition the charge expands and pushes the piston downward. This is the power stroke.
4. **Exhaust Stroke :** During this stroke the piston moves up again and pushes out the burnt gasses through the exhaust valve which is kept open during this stroke.

EXPERIMENT No.4

OBJECT: Study of Diesel Engine.

APPARATUS USED: Prototypes of diesel engines.

FIAT ENGINE: The fiat engine is a four cylinder, four stroke engine. Main components of the same are as follows:

Cylinder Block: It is the basic structure of the engine made of grey cast iron. Now a days cylinder blocks of aluminum alloys are also available. It is generally a single piece casting consisting of cylinders, water jacket, passageways, openings for inlet and outlet valves etc. The cylinders are generally provided with liners which may be of wet and dry type depending on whether water comes into direct contact with liner or not.

Cylinder Head: It is mounted on the cylinder block and valves, spark plug etc. are fitted in it.

Manifolds: These are passages which permit entry of fresh charge to cylinders. Further, exhaust manifold is also provided to facilitate removal of burnt gasses.

Piston: These are cylindrical members which reciprocate in the cylinder. Pistons are made of either cast iron or aluminum alloys. Ring grooves are provided to accommodate piston rings. For effective lubrication, drilled holes are provided at periphery.

Piston rings: Piston rings are provided to prevent escape of hot gases from combustion chamber to crank case. Thus piston rings acts as seals.

Connecting Rod: It converts the reciprocating motion of the piston to rotary motion of the crank shaft. It is normally manufactured by forging.

Gudgeon Pin: For connecting piston and the connecting rod.

Crank shaft: Crank shaft receives power from the piston through connecting rod for on ward transmission to the gear box through the clutch. It is made of carbon alloy steel by forging.

Cam Shaft: It has a number of integral cams on it. It is driven by the crank shaft through timing gears. Cam shaft in turn operates the valve operating mechanism.

Timing Gears: For providing drive from crank shaft to cam shaft a pair of timing gears is used. The timing gears reduce the speed of cam shaft to half that of crank shaft.

Gasket: A gasket is provided between cylinder block and cylinder head. These are generally made of copper and asbestos.

Flywheel: Mounted on engine crank shaft for storing and releasing energy as and when required.

Specifications:

Bore :	70 mm
Stroke Length	64.9 mm
Power	45 BHP at 5500 rpm
No. of Cylinder	4

The Diesel Engine:

The diesel engine components are generally heavier than the components of petrol engines. Some of the components like cylinder, piston, cylinder head, connecting rod, crank shaft, cam shaft etc. are some components which are common to both the engines. However, the fuel injection system and combustion chamber of the diesel engine is entirely different.

Fuel Injection system : Mostly these days the diesel engines use air less injection system. In this system only liquid fuel is injected. The pump used is required to develop pressure varying from 140 kg/sq.cm to 400 kg/sq.cm. The layout of fuel injection system is as shown in figure. The fuel is stored in fuel tank from where it is lifted by means of a fuel feed pump through a filter. The feed pump supplies the fuel to the injection pump. Injection pump boosts the pressure of the fuel and then send metered quantity of the fuel to the injectors through high pressure pipelines. The fuel is then injected in the combustion chamber at the end of the compression stroke. Since after compression the temperature of charge is increased considerably, the fuel sprayed in the combustion chamber is ignited automatically. Such type of ignition is therefore called as compression ignition.

Combustion Chamber : Direct type injection system is used. No separate combustion space is provided. A depression is provided in the piston which forms a combustion space.

As in case of petrol engine the thermodynamic cycle is completed in four strokes of the piston ie two revolutions of the crank shaft. These strokes are

Suction stroke: Only filtered air is sucked in the cylinder.

Compression Stroke : The air sucked is compressed to high pressure. The compression ratio is of the order of 16 – 22.

Power stroke : Just before the end of the compression stroke high pressure fuel coming from injection pump is injected in the combustion chamber by the injector. The fuel so injected is ignited and thus a power stroke is obtained.

Exhaust Stroke : The burnt gases are exhausted through the exhaust port.

Comparision Between Diesel and Petrol Engines:

- In the petrol engine a mixture of petrol and air is drawn in the cylinder and then compressed. In case of diesel engine only fresh air is drawn in the combustion chamber.
- Compression ratio for petrol is 7 : 1 to 10 : 1 whereas the same for diesel engines is much higher ie 12:1 to 22: 1.
- The thermal efficiency of diesel engines is higher because of higher combustion ratios.
- The fuel for diesel engines ie diesel is much cheaper than petrol used in petrol engines.
- Components like carburetor, spark plugs, ignition system etc used in petrol engines are not required in diesel engines. However, the diesel engines require a precision and complicated injection system.
- In view of higher combustion efficiencies of diesel engines, there is less emission of carbon mono-oxide and un-burnt hydrocarbons.
- The initial cost of diesel engine is higher.
- The diesel engines are more noisy and create heavy vibrations during operations.
- A governor is required with diesel engines for speed control. No such governor is needed for petrol engines.
- During cold starting conditions it is difficult to start the diesel engines.

Lab Manual / 489

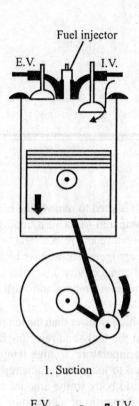
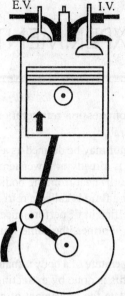

1. Suction   2. Compression

I.V. — Inlet valve
E.V. — Exhaust value

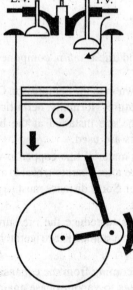
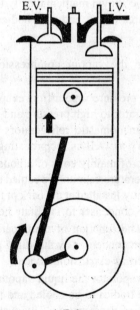

3. Power   4. Exhaust

I.V. — Inlet valve
E.V. — Exhaust value

Fig A-8

EXPERIMENT No.5

OBJECT : Study of vapour compression refrigeration unit.

APPARATUS USED: Refrigerator

INTRODUCTION : Refrigerator may be defined as a machine which is used to remove the heat from a particular space or system which is already at lower temperature as compared to its surroundings. The working fluid in the refrigerator absorbs heat from the bodies at low temperature and rejects heat to bodies at higher temperature. Such a working fluid is called a refrigerant. A good refrigerant must have high latent heat, low specific volume, high coefficient of performance, good thermal conductivity, low viscosity and ease of detection if the same leaks. Further, the same must not harm the environment and must not be harmful to humans.

As shown in figure suppose temperature of a body is maintained at T_2 which is lower than the temperature of thermal sink (atmosphere) T_1. This is done by extracting Q_2 heat from the body and rejecting heat Q_1 to the atmosphere. In order to facilitate the movement of heat from low temperature to high temperature a compressor does additional work W_r. Coefficient of Performance is used to judge the efficiency of the refrigerators. COP is defined as the amount of heat extracted from the cold body to the amount of work done by the compressor. It can be shown mathematically that COP is equal to ratio of temperature of cold body to difference in temperatures of hot and cold bodies.

Contructional Features:

The refrigerators work on the vapour compression cycles and their main components are as follows :

- **Compressor :** The low pressure vapour from evaporator is drawn in to the compressor. Compressor compresses the refrigerator to high pressure and high temperature. Normally, hermitically sealed compressor are used in domestic refrigerators and the same are installed at the base of the refrigerator. Both rotary as well as reciprocating compressors are used.
- **Condenser:** It consists of mainly coils of a good conducting material like copper in which high pressure and high temperature vapours are cooled and condensed after liberating heat to atmosphere. The condenser is normally installed at the back of the refrigerator. Some distance must be maintained between walls and the condenser to facilitate better cooling.
- **Expansion Devices:** The function of the expansion device is to reduce the pressure of liquid refrigerant coming from condenser by throttling process. In most domestic refrigerators very fine capillary tubes are used as expansion devices.
- **Evaporator :** In the evaporator the liquid vapour refrigerant coming from the expansion devices picks up heat from the bodies to be cooled and thus evaporates to vapour phase again. The heat absorbed is latent heat of vaporization. The evaporator is generally placed inside the cabinet of the refrigerator at top. Often to facilitate better cooling circulating fans are also provided.

Vapour Compression Refrigration Cycle

The vapour compression refrigeration cycle is completed in four steps as given below :

- Step 1 – 2   Isentropic Compression
- Step 2 – 3   Condensation at constant temperature as refrigerant gives up heat
- Step 3 – 4   Expansion Process
- Step 4 – 1   Evaporation at constant temperature.

The above processes can be shown on Temperature – Entropy and pressure – enthalpy diagrams.

SPECIFICATIONS The domestic refrigerators are specified as follows :
- Cooling Capacity :   0.5 Ton, 1.0 Ton etc.
- Cooling Space Volume: 165 litre, 215 lt. etc.
- Refrigerant Used :   Freon 12, Freon 22, R- 50, R 717 etc.
- Voltage   160 to 260 volts
- Power Supply   AC 230 Volts, 50 Hz

WORKING : The low pressure refrigerant vapours (R12, R22) are drawn through the suction line to the compressor . The accumulator which is placed between evaporator and the compressor collects any liquid refrigerant coming out from the evaporator due to incomplete evaporation. The compressor then compresses the vapour refrigerant to high pressure and high temperature. The compressed vapour through the discharge line to condenser where vapour refrigerant condenses again to liquid phase and liberates heat to the atmosphere. Thus liquid refrigerant at high pressure is obtained. The high pressure liquid then enters the capillary tube which is an expansion device and the liquid refrigerant is throttled. Due to throttling pressure of the liquid is decreased. Now low pressure liquid refrigerant enters evaporator where it picks up heat from low temperature bodies and evaporates to vapour stage. The heat is picked up by latent heat of vapourization at constant temperature. The vapours so formed are again sucked by the compressor and cycle goes on.

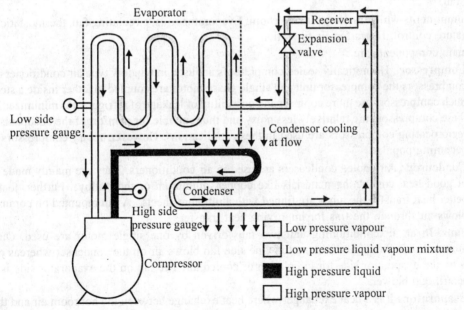

Vapour Compression Refrigeration System

Fig-A-9

EXPERIMENT No.6

OBJECT : Study of window type air conditioner
APPARATUS : Window type air conditioner
INTRODUCTION: The air conditioner is a machine that artificially maintains conditions of temperature and humidity in a closed room or building. Window type airconditioners are commonly used in offices and residential buildings for regulating temperature and humidity conditions. Since such air conditioners are fitted in windows they are called window type air conditioners. The comfort of occupants of a room or a building depends on certain factors like temperature, humidity, air motion and air purity. Comfort air conditioning is maintenance of the above parameters with in certain specified limits which are as furnished below:

Temperature :   22 to 26 deg C
Relative Humidity: 40 to 60 %
Air Velocity :   5 to 8 m/min.
Pure, dirt and Dust free air

CONSTRUCTION: The air conditioner mainly consists of following :

(a) Components Kept outside the room : Hermitically sealed motor compressor unit, condenser and condenser fan.

(b) Components which are kept inside room: Evaporator coils, circulating fan, thermostatic controls for temperature control, filters for air filteration

The main components are :

- **Compressor** : Hermitically sealed compressors are used in window type air conditioners. In such compressors the compressor unit and single phase motor are coupled together inside a steel dome. Such compressors are more compact and possibility of leakage of refrigerant is minimized. Further, these compressors are relatively less noisy and thus suitable for comfort of the occupants. Mostly reciprocating compressors are used in these applications. However, rotary compressors are also becoming popular.
- **Condenser** : Air cooled condensers are used in air conditioners. They are mainly made of tubes of good heat conducting materials like copper, aluminum or their alloys. Further, to facilitate better heat transfer the tubes are finned with aluminum sheets. A fan mounted on common shaft blows air through the fins for improving heat transfer.
- **Fans**: In an air conditioning unit two fans driven by one single motor are used. One fan is mounted on either side of the motor. One side fan blows air on the condenser whereas other fan is on the evaporator side for circulation of cooled air. The fan on the evaporator side is infact a centrifugal blower.
- **Evaporator** : It is a heat exchanger where heat exchange between the hot room air and the liquid refrigerant takes place. The refrigerant absorbs latent heat and vapourises thus cooling the room air which is then re-circulated. The evaporator is also made of copper or aluminum tubes and fins of aluminum sheets are provided for better heat transfer.
- **Capillary Tubes** : It is throttling device and reduces the pressure of liquid refrigerant coming from condenser.

- **Filters**: Filter pads of different materials like jute, polymer fibers etc are installed just before air is sucked in through the evaporator to remove any dust / dirt particles suspended in air.
- **Control System** : Two types of controls are provided :

    (a) **Temperature Controls :** For temperature control a thermostat switch is provided. The bulb of the thermostat switch is mounted in the passage of return air. The bulb senses the return air temperature and controls it within the desired limits by cutting in or cutting out the compressor.

    (b) **Air Movement controls :** The front panel is provided with supply and return air grill. The supply side of the grill has adjustable louvers for controlling the air direction flow. Motorized air louvers are also provided in some models.

Specifications:
Capacity 0.75 , 1.0, 1.5 , 2.0 , 2.5 Tons
Power Supply AC, 220 Volts, 50 Hz

Working
The room air is sucked through the air filter element and any suspended particle is removed from the air. The air then comes in to contact with evaporator coils of the air conditioner where the liquid refrigerant (R22) picks up the heat and vapourises. The low pressure refrigerant vapours then enter the compressor where their pressure and temperature is increased. The compressor discharge line takes high pressure and high temperature vapours to the condenser which is outside the room to be cooled. In condenser the refrigerant looses heat to the atmosphere and thus it is converted to liquid phase. The liquid refrigerant than enters the capillary tube where it is throttled to low pressure cool liquid phase refrigerant.

Further, as shown in figure on the board the same unit may be designed to work as room heater with the help of a reversing valve. The reversing valve is two position valve with four ports. The discharge and suction lines of the compressor are connected to two ports. The other two ports are connected to the inlet side of the condenser (outdoor coil) and the suction outlet of the evaporator coil (indoor coil). With the change in position of the reversing valve the role of condenser and evaporator gets inter-changed and accordingly the indoor coil which works as evaporator for normal cooling use start working as condensing coil and heat is transferred from outdoor to indoor for heating applications.

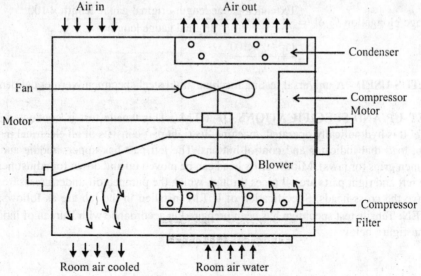

Window type air-conditioner
Fig A-10

EXPERIMENT NO –7

OBJECT-To conduct the tensile strength test of a given mild steel specimen on U.T.M.

THEORY-Various machine and structural component are subjected to tensile loading in numerous applications. For safe design of these components, their ultimate tensile strength and ductility are to be determined before actual use. For that the above test is conducted. Tensile test can be conducted on U.T.M.

A material when subjected to a tensile load, resists the applied load by developing internal resisting force. This resistance comes due to atomic bonding between atoms of the material. The resisting force per unit normal cross-sectional area is known as stress. The value of stress in material goes on increasing with an increase in applied tensile load, but it has a certain maximum (finite) limit too. The maximum stress at which a material fails, is called **ultimate tensile strength**.

All known materials are elastic in nature and so is the steel specimen also. Its initial length increases with increase in applied load followed with corresponding decrease in its lateral dimensions. Increase in length is called elongation which is a measure of **ductility**. The 'change in length' over the 'original length' is called **strain**. The ratio of stress to strain within elastic limit is termed as **Modulus of Elasticity**. The end of elastic limit is indicated by the yield point (load). With increase in loading beyond the elastic limit (in inelastic or plastic region), original area of cross-sectional A_0 goes on decreasing and finally reduces to its minimum value when the specimen breaks. The ultimate tensile strength s_{ult} and elongation dl are computed with the help of follow

1-Ultimate tensile strength $\sigma_{ult} = \dfrac{\text{Ultimate load}}{\text{original cross-sectional area}} = P_{max}/A_0$

2-percentage elongation % $\delta l = \dfrac{\text{Extended gauge length-original gauge length} \times 100}{\text{original gauge length}}$

$= \dfrac{(l_1 - l_0) \times 100}{l_0}$

APPARATUS USED - A universal testing machine, mild steel specimen, vernier caliper/micrometer, dial gauge.

TEST SET UP AND SPECIFICATIONS OF UTM : This tensile test is conducted on U.T.M as shown in figure. It is hydraulically operated machine. Its right part consists of an electrical motor, a pump, oil in oil sump, load dial indicator and control buttons. The left part has upper, middle and lower cross heads i.e specimen grips (or jaws). Middle cross head can be moved up and down for adjustment. The pipes connecting the left and right parts are oil pipes through which the pumped oil under presssure flows on the left part to move the cross-heads. Specification of U.T.M installed in the lab are as follows:

SPECIMEN: Tensile test specimen has been prepaired in accordance with Bureau of indian standards as shown in the figure below

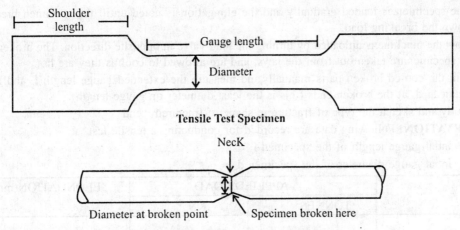

Tensile Test Specimen

Broken pieces of specimen joined togehter

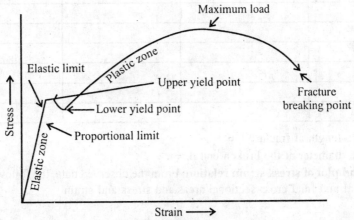

Fig A-11

PROCEDURE: First of all the gauge length is marked on the specimen .It is diameter and gauge length is also measured.Now the following sequential operation are performed.

1. The load pointer is set at zero by adjusting the initial setting knob.
2. The load range of machine's operation is selected .Say it is upto 40 Tonnes.The load range selection is base on approximate calculations (P = A). It should be sufficiently higher than the expected value.
3. The dial gauge is fixed on the specimen for measuring elongation of small amounts (of the order of micron or hundred part of a millimeter).
4. Now the specimen is gripped between upper and middle cross – head jaws of the machine.
5. The machine is switched by pressing the approximate button, and applying the load by turning the load valve, gradually.
6. The elongation of specimen is recorded for a certain specified load .reading may be taken at an interval of 1Tonnes upto yield point.
7. When the load reaches the yield point (asindicated by fluctuation of the live point), The upper and lower yield loads are recorded. Dial gauge can be removed now because the elongation is too much, and hence the change in length may be recorded/noted by a linear scale.

8. The specimen is loaded gradually and the elongation is noted untill the specimen break. Note down the breaking load.
9. Now the machine is unloaded by turning the load valve in opposite direction. The broken pieces of specimenare taken out from the jaws, and are allowed to cool as they are hot.
10. Join the cooled broken parts manually,and measure the extended gauge length l_f and reduced diameter d_f at the broken ends (this is the least diameter on gauge length).
11. Study and sketch the type of fracture ie shape of fractured end

OBSERVATION: Following data are recorded for conducting a tensile test.

Intial gauge length of the specimen l_o =

Intial gauge diameter of the specimen d_o =

S.NO	APPLIED LOAD		ELONGATION(mm)
	TONNES	KN	

Extended gauge length at fracture l_f =

Reduced gauge diameter at the broken end d_f =

Calculation and plot of stress-strain relation: From the observed data, the following calculations are done to obtain initial and final cross-sectional areas, and stress and strain

$A_0 = \pi/4 \, d_0^2$

$A_f = \pi/4 \, d_f^2$

S.No	Stress $\sigma = P/A_0 \, (kN/mm^2)$	Strain $\varepsilon = \delta l / l_o$

Result :

Precautions :

EXPERIMENT No. 8

Object : To conduct the compression test and determine the ultimate compressive strength for a given specimen.

Apparatus : Universal Testing Machine, cylindrical shaped specimen of Brittle Material Like cast iron, vernier caliper, linear scale.

Theory: Several machines and structural components such as columns and studs are subjected to compressive loads in applications. These components are made of high compressive strength materials. All the materials are not strong in compression. Several materials which are good in tension are poor in compression. Contrary to this many materials are poor in tension but very strong in compression. Cast iron is one such example. Hence determination of ultimate compressive strength is essential before using a material. This strength is determined by conducting a compression test. Compression test is just opposite in nature to a tensile test. Nature of deformation an fracture is quite different from that in the tensile tests. Compressive loads tends to squeeze the specimen. Brittle materials are generally weak in tension but strong in compression. Hence this test is normally performed on cast iron, cement concrete etc which are brittle materials. But ductile materials like aluminum and mild steel which are strong in tension, are also tested in compression. From compression test we can

- Draw Stress – Strain curve in compression
- Determine Young's Modulus in compression.
- Determine ultimate compressive strength.
- Determine percentage reduction in length.

However, during this experiment only ultimate compressive strength needs to be determined.

Test Setup, Specification of Machine and Sepciman Details:

A Compression test can be performed on UTM by keeping the test piece on base block and moving down the central grip to apply load. The UTM is hydraulically operated and runs on 420 volts, 3 phase, 50 HZ AC supply and has four load measuring ranges. For compressive test the machine is operated in 0-40 tonnes range. Test is to be performed on a cylindrical test piece of cast iron given. In cylindrical specimen, it is essential to keep height/diameter ratio must not be more than 2 to avoid lateral instability.

Procedure:

- Measure dimensions of the test piece i.e. its diameter at three locations and find the average diameter.
- Find the cross sectional area of the specimen using average diameter.
- Ensure that ends of the specimen are plane.
- Keep the specimen on the base plate (lower cross head) of UTM.
- Bring down the middle cross head mechanically so that it is about to touch the specimen.
- Note initial load reading of the machine dial if any.

- Apply compressive load hydraulically by opening the control valve slowly.
- Ensure relief valve is closed.
- Load is applied until the specimen fractures. Note the final applied load from the red colour pointer.

Observation & Calculation:
- Diameter of the specimen (do) mm
- Area of cross section $3.14 do^2/4$ sq.mm.
- Load at the time of fracture : kgf.
- Ultimate compressive Strength = Load / Area Kgf/mm^2

S.N	Load	Area	Compressive Strength

EXPERIMENT NO 9

OBJECT- To conduct the impact test (Izod / Charpy) on the impact testing machine.

Theory- In manufacturing locomotive wheel, coin, connecting rods etc., the component are subjected to impact (shock) load. These load are applied suddenly. The stress induced in these component are many times more than the stress produced by gradual loading. Therefore impact tests are performed to assess shock absorbing capacity of materials subjected to suddenly applied loads. These capabilities are expressed as (i) rupture energy (ii) modulus of rupture, and (iii) notch impact strength.

Two types of notch impact test are commonly conducted these are

1-Charpy test

2-Izod test

In both tests, standard specimen is in the form of a notched beam. In charpy test, the specimen is placed as 'simply supported beam' while in izod test it is kept as a 'cantilever beam'. The specimens have V shape notch of 45Degree. U shaped notch is also common. The notch is located on tension side of specimen during Impact loading. Depth of notch is generally taken as t/5 to t/3 where t is the thickness of the specimen.

APPARATUS USED - Impact testing machine, Izod and charpy test specimen of mild steel and/or aluminum, vernier caliper.

Specifications of Machine Used:
- Impact Capacity   :   200 Joules
- Weight of hammer   18.75 kg
- Angle of Hammer   160 degrees
- Striking Velocity   5.6 m / sec

Sketch of Specimen Given:

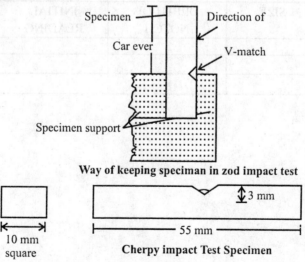

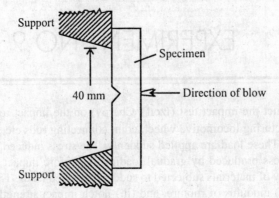

Way of keeping specimen in Cherpy impact test

Fig A-12

Procedure: Following procedure should be adopted to conduct the test.
1. First measure the length, width and thickness of the specimen.
2. Set the machine at 30kg-m dial reading and lock the striking hammer in its top position.
3. Now press down the 'pendulum release lever so that the hammer falls and swing past the bottommost position. Note down the reading on dial. Let this is 'x'. This is initial reading. Remember that this reading is without any specimen and indicates frictional and wind age (air) loss of energy of the hammer.
4. Now put the test-piece on support in proper manner. Release the lever so that the hammer strikes the test piece and breaks it. Note down this reading. This is final reading .let this be 'Y'.
5. Make use of brake handle to stop the motion of hammer after its swing.
6. Repeat the experiment on other specimen.
7. Study the type of fracture mode of the broken pieces.

Observation Table:

SPECIMEN NO.	SIZE	DEPTH OF NOTCH	INITIAL READING	FINAL READING

Calculations:

Results:

EXPERIMENT NO –10

OBJECT- To determine the hardness of a given specimen using Brinell / Rockwell / Vicker testing machine

THEORY- Hardness is a surface property. It is defined as the resistance of material against permanent deformation of the surface in the form of scratch, cutting, indentation, or mechanical wear. The need of hardness test arises from the fact that in numerous engineering application, two components in contact are made to slide or roll over each other. In due course, their surfaces are scratched and they may fail due to mechanical wear. This result in not only a quick replacement of both parts but also incurs a big loss in terms of money.

For example, piston ring of an I.C ingine remain in sliding contact with the cylinder body when the piston reciprocates with in the cylinder. If proper care is not taken in selection of materials for then, The piston rings and cylinder will wear soon.

In this case the replacement or repairing of cylinder block will involve much time, trouble and money. Therefore, the material of piston rings and cylinder block should be taken such that the wear is least on the cylinder. Thus in case of repairing, comparatively cheaper piston rings can be easily replaced. This envisages that material of cylinder block should be harder than the material of piston rings so that the cylinder wears. The least. This can be ascertained by conduct of a hardness test. That is why it is essential to known as to how this test can be conducted.

APPARATUS USED – Brinell / Rockwell / Vicker testing machine, specimen of mild steel/cast iron/ non-ferrous metal, optical microscope

Specifications of Hardness Testing Machines and Indentors:

The Brinel cum Rockwell Tester has following specifications :
- Ability to determine hardness upto 500 BHN
- Diameter of Ball D 2.5, 5 & 10 mm
- Maximum Load 30 D^2

Brinell Hardness Test:
- Insert Ball of diameter D in ball holder of the machine.
- Make the specimen surface free from dust and dirt.
- Make contact between specimen surface and ball by rotating the wheel
- Turn the lever to apply load.
- Wait for 30 seconds.
- Remove the specimen from the support table and check the indentation through optical microscope.

Rockwell Hardness Test:

The specimen is subjected to a major load for about 15 seconds after the initial load. ASTM says 13 scales for testing of wide range of materials. These scales are A,B, C... etc. B-scale is preferred for soft steel and Aluminum alloys while C scale is used for hard steel. B scale uses a ball of 1/16" while cone indenter is used for c scale.

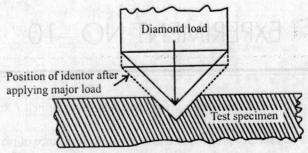

Principle of Rockwell Testing

Fig A-13

Vicker's Hardness Test:

This test is similar to the Brinell test but uses a different indenter. A square based pyramid of cone angle 136 degrees is used. The applied load may be 5, 10, 30, 50 ... etc. kgf. The same test procedure is adapted.

Obervation Table

Sp.No.	BALL DIA	LOAD APPLIED	INDENTATION DIA (BY MICROSCOPE)	DIAMETER OF INDENTATION IN MM
1				
2				
3				

Calculation:

Result:

Previous Year Question Papers

B. Tech
Second Semester Examination, 2004-2005
Mechanical Engineering

1. **Attempt any two parts of the following:** (10 × 2 = 20)
   - (a) (i) What do you understand by thermodynamic equilibrium?
     - (ii) What do you understand by flow work? Is it different from displacement work? How?
     - (iii) A pump dischanges a liquid into a drum at the rate of 0.032 m³/s. The drum, 1.50 m in diameter and 4.20 m in length, can hold 3500 kg of the liquid. Find the denisty of the liquid and the mass flow rate of the liquid handled by the pump.
   - (b) Derive steady flow energy equation:
     The steam supply to an engine comprises two streams which mix before entering the engine. One stream is supplied at the rate of 0.01 kg/s with an enthalpy of 2950 kj/kg and a velocity of 20 m/s. The other stream is supplied at the rate of 0.1 kg/s with an enthalpy of 2565 kj/kg and a velocity of 120 m/s. At the exit from the engine the fluid leaves as two streams, one of water at the rate of 0.001 kg/sec. with an enthalpy of 421 kj/kg and the other of steam; the fluid velocities at the exit are negligible. The engine develops a shaft power of 25 KW. The heat transfer is negligible. Evaluate the enthalpy of the second exit stream.
   - (c) Two identical bodies of constant heat capacity are at the same initial temperature T_1. A refrigerator operates between these two bodies until one body is cooled to temperature T_2. If the bodies remain at constant pressure and undergo no change of phase, find the minimum amount of work deeded to do this, in terms of T_4, T_2 and heat capacity.

2. **Attempt any two parts of the following:** (10 × 2 = 20)
   - (a) Explain Rankine cycle with the help of $p - v$, $T - s$ and $h - s$ diagram.
   - (b) Steam at 10 bar, 250°C flowing with negligible velocity at the rate of 3 kg/min mixes adiabatically with steam at 10 bar, 0.7 quality, flowing also with negligible velocity at the rate of 5 kg/min. The combined stream of steam is throttled to 5 bar and then expanded isentropically in a nozzle to 2 bar. Determine (a) the state of steam after mixing (b) the state of steam after throttling (c) the increase in entropy due to throttling (d) the exit area of the nozzle. Neglect the kinetic energy of the steam at the inlet to the nozzle.
   - (c) (i) What is C. I. engine? Why it has more compression ratio compared to S.I. engines?
     - (ii) A diesel engine operating on Air Standard Diesel cycle operates on 1 kg of air with an initial pressure of 98 kPa and a temperature of 36°C. The pressure at the end of compression is 35 bar and cut off is 6% of the stroke. Determine (i) Thermal efficiency (ii) Mean effective pressure.

3. Attempt any two of the following (10 × 2 = 20)

(a) Compute the simplest resultant force for the loads shown acting on the cantilever beam in Fig. 1. What force and moment is transmitted by this force to supporting wall at A?

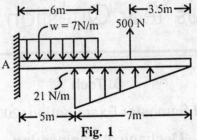

Fig. 1

(b) The pulley, in Fig. 2, at D has a mass of 200 kg. Neglecting the weights of the bars ACE and BCD, find the force transmitted from one bas to the other at C.

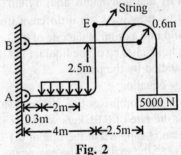

Fig. 2

(c) Block C, shown in Fig. 3 has a mass of 100 kg. and identical blcoks A and B have masses of 75 kg. each are placed on the floor as shwon. If coefficient of friction is $\mu = 0.2$ for all mating surfaces, can the arrangement shown in the diagram remain in equilibrium?

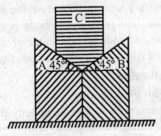

Fig. 3

4. Attempt any two of the following: (10 × 2 = 20)

(a) Draw the shear force and bending moment diagram for the beam shown in Fig 4.

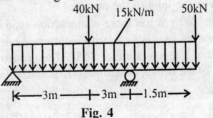

Fig. 4

(b) Find the force in members HF, FH, FE and FC of the truss shown in Fig. 5.

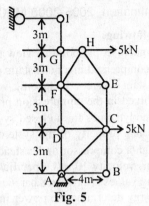

Fig. 5

(c) (i) Derive the relation E = 2G (1 + v) where
E = Young's modulus, G = Modulus of rigidity
v = Poisson's ratio.

(ii) A 1 m long steel rod of rectangular section 80 mm × 40 mm is subjected to an axial tensile load of 200 kN. Find the strain energy and maximum stress produced in it for the following cases when load is applied gradually and when load falls through a height of 100 mm. Take E = 2 × 10^5 nm^2.

5. Attempt any two of the following: (10 × 2 = 20)

(a) A plane element is subjected to following stresses σ_x = 120 kN/m^2 (tensile), σ_y = 40 kN/m^2 (compressive) any T_{xy} = 50 kN/m^2 (counter clockwise on the plane perpendicular to x-axis). Find:

(i) Principal stresses and their directions.

(ii) Maximum shearing stress and its direction.

(iii) Also, find the resultant stress on a plane inclined 40° with the x-axis.

(b) (i) If cross-sectional area of the beam shown in Fig. 4 is as shown in Fig. 6, find the maximum bending stress.

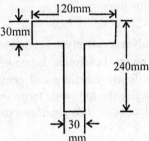

(ii) What are the assumptions taken in the theory of pure bending.

(c) (i) Draw stress-strain diagram for Aluminium and Cast iron.

(ii) A compound shaft is made up of a steel rod of 50 mm diameter surrounded by a closely fitted brass tube. When a torque of 9 kN-m is applied on this shaft, its 60% is shared by the steel rod for steel is 85 GPa and for brass it is 45 GPa. Calculate (a) the outside diameter of brass tube (b) Maximum shear stress induced in steel and brass.

B.Tech
Special Carry Over Examination, 2005-2006 MECHANICAL ENGINEERING

1. **Attempt any four parts of the following:** (4 × 5 = 20)
   (a) Explain microscopic and macroscopic point of view to study the subject of thermodynamics.
   (b) A mass of 1.5 kg of air is compressed in a quasistatic process from 1.1 bar to 10 bar according to the law $P_v\, 1.25$ = constant where v is specific volume. The initial density of air is. 1.2 kg/m^3. Find the work involved in the compression process.
   (c) State the first law of thermodynamics for a closed system undergoing a change of state. Also show that the total energy is a property of the system.
   (d) 0.8 kg/s of air flows through a compressor under steady state conditions. The properties of air at entry are: pressure 1 bar, velocity 10 m/s. specific volume 0.95 m^3/kg and internal energy 30 kj/kg. The corresponding values at exit are 8 bar, 6 m/s. 0.2 m^3/kg. and 124 kj/kg. Neglecting the change in potential energy determine the power input and pipe diameter at entry and exit.
   (e) Determine the sp. work of compression when air flows steadily through a compressor from 1 bar and 30° to 0.9 MPa according to
      (i) isothermal process
      (ii) adiabatic process
   (f) Show that entropy is a point function. Calculate the entropy change, during the complete evaporation of water at 1 bar to dry saturated steam at the same pressure and temperature.

2. **Attempt any two parts of the following:** (2 × 10 = 20)
   (a) Calculate the change in output and efficiency of a theoretical Rankine cycle when its condenser pressure is changed from 0.2 bar to 0.1 bar. Inlet condition is 40 bar & 400°C.
   (b) Air enters at a condition of 1 bar and 30° C to an air standard Diesel cycle and compressed to 20 bar. Cut-off takes place at 6% of stroke. Calculate
      (i) Power output
      (ii) Heat input
      (iii) Air standard efficienc.
      (iv) P-V and T-S diagram for cycle.
   (c) With the help of neat sketches explain the working of 2-stroke CI engine.

3. **Attempt any two parts of the following:** (2 × 10 = 20)
   (a) State the transmissibility of forces.
   Forces 2, $\sqrt{3}$, 5 $\sqrt{3}$ and 2 kN respectively act at one of the angular points of a regular-hexagon towards five other angular points. Determine the magnitude and direction of the resultant force.
   (b) (i) State the necessary and sufficient conditions of equilibrium of a system of complanar non-concurrent force system. Define the following terms in connection with friction: (i) Coefficient of friction, (ii) Angle of friction, .(iii) Angle of respose, (iv) Cone of friction and (v) Limiting friction.
   (c) A uniform ladder of length 15 m rests 'against a vertical wall making an angle of 60° with the horizontal. Coefficient of friction between wall and ladder and the ground and ladder are 0.3 and 0.25 respectively. A man weighing 65 kg ascends the ladder. How high will he be able to go before the ladder slips? Find the magnitude of weight to be put at the bottom of the ladder so as to make it just sufficient to permit the man to go to the top. Take ladder's weight = 900 N.

4. **Attempt any two parts of the following:** (2 × 10 = 20)
   (a) Define a beam. What is a cantilever, a simply supported and a overhung beam? What is the point of contraflexure? Draw the shear force and bending moment diagram for the beam as shown in Fig. 4.1.

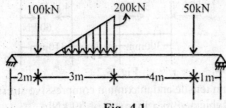

Fig. 4.1

   (b) (i) Derive the relationship between shear force, bending moment and intensity of loading at any section of a beam.
   (ii) State the assumptions made while making an analysis of a framed structure.
   (c) Determine the magnitude and nature of forces in the various members of the truss shown in Fig. 4.2.

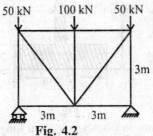

Fig. 4.2

5. **Attempt any four parts of the following:** (4 × 5 = 20)
   (a) Explain the stress-strain diagram for a ductile and brittle material under tension on common axes single diagram.
   (b) A rectangular element is subjected to a plane stress system as shown in Fig. 5.1. Determine the principal planes, principal stresses and maximum shear stress by Mohr's Circle method only.

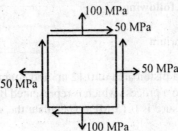

   (c) Explain:
   (i) Strain energy
   (ii) complementary shear stress

(d) A steel bar is subjected to loads as shown in Fig. 5.2. Determine the change in length of the bar ABCD of 18 cm diameter E = 180 kN/mm².

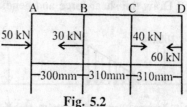

Fig. 5.2

(e) Calculate the maximum tensile and maximum compressive stress developed in the cross-section of beam in Fig. 5.3 subjected to a moment of 30 kNm.

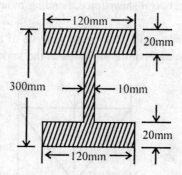

Fig. 5.3

(f) A propeller shaft 100 mm in diameter is 45 m long, transmits 10 MW at 80 rpm. Determine the maximum shearing stress in shaft. Also calculate the stress at 20 mm, 40 mm, 60 mm and 80 mm diameters. Show the stress variation.

B. Tech.

First Semester Examination, 2005-2006
TMT–101: Mechanical Engineering

1. Attempt any four parts of the following: (5 × 4 = 20)
   (a) Explain the following: .
      (i) Thermodynamic Equilibrium
      (ii) Quasi-Static Process
   (b) An engine cylinder has a piston area of 0.12 m² and contains gas at a pressure of 1.5 MPa. The gas expands according to a process which is represented by a straight line on a pressure-volume diagram. The final pressure is 0.15 MPa. Calculate the workdone by the gas on the piston if the stroke is 0.3 m.
   (c) A system undergoes a cyclic process through four states 1-2, 2-3, 3-4 and 4-1. Find the values of x_1, x_2, y_1, y_2 and y_3, in the following table:

Process	Heat Transfer KJ/Min	Work Transfer KW	Change of Internal Energy
1-2	800	5.0	y_1
2-3	400	x_1	600
3-4	−400	x_2	y_2
4-1	0	3.0	y_3

(d) A reversible heat engine operates between two reservoirs at temperature of 600°C and 40°C. The engine drives a reversible refrigeration which operates between reserves at temperature of 40°C and − 20°C. The heat transfer to the heat engine is 2000KJ and net work output of combined engine − refrigerator plant is 360KJ. Evaluate the heat transfer to the refrigerator and the net heat transfer to the reservoir at 40°C.

(e) 5 kg of ice at −10°C is kept in atmosphere which is at 30°C. Calculate the change of entropy of universe when it melts and comes into thermal equilibrium with the atmosphere. Take latent heat of fusion as 335 KJ/Kg and specific heat of ice is half of that of water.

(f) Explain:
  (i) Zeroth Law of thermodynamics and its application in temperature measurement
  (ii) Clausius inequality

2. Attempt any two of the following questions: (10 × 2 = 20)

(a) (i) With the help of neat sketches explain working of two-stroke CI engine.
  (ii) Define the following terms with reference to phase change for water: Saturation state, triple point, critical point, dryness fraction, compressed or subcooled liquid.

(b) Explain the Rankine cycle with the help of flow diagram or water/steam in various components. Also draw the cycle on pay and T-s diagram. Obtain the net output and thermal efficiency of a theoretical Rankine cycle in which boiler pressure is 40 bar and it is generating steam at 300°C. Condenser pressure is 0.1 bar.

(c) Air enters at 1 tiar and 230°C in an engine running on Diesel cycle whose compression ratio is 18, Maximum temperature of the cycle is limited to 1500°C. Calculate
  (i) cut-off ratio,
  (ii) heat supplied per kg of air,
  (iii) cycle efficiency.

3. Attempt any two of the following questions: (10 × 2 = 20)

(a) Explain the following:
  (i) Principle of transmissibility of a force.
  (ii) Necessary and sufficient conditions of equilibrium of a system of coplanar force system.
  (iii) Laws of static friction.
  (iv) Useful uses of friction.

(b) Two smooth sphare seach of weight W and each of radius 'r' are in equilibrium in a horizontal channel of width 'b' (b<4r) and vertical sides as shown in figure:

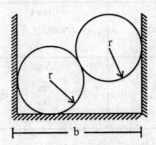

Fig. 1

Find the three reactions from the sides of channel which are all smooth. Also find the force exerted by each sphere on the other.

(c) A ladder of length 'I' rests against a wall, the angle of inclination being 45°. If the coefficient of friction between the ladder and the ground that between the ladder and the wall be 0. 5 each what will be the maximum distance on ladder to which a man whose weight is 1.5 times the weight of ladder may ascend before the ladder begins to slip?

4. Attempt any two of the following questions: $(10 \times 2 = 20)$

(a) (i) Define a beam. What are the different types of beams and different type of loading? What do you understand by the term 'point of contraflexure'?

(ii) How the trusses are classified? What assumptions are made while determining stresses in a truss?

(b) Each member of following truss given in Fig. 2 is 2m logn. The truss is simply supported at the ends. Determine forces in all members clearly showing whether they are the tension or compression.

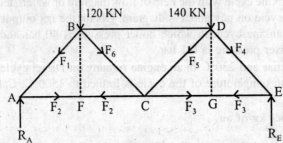

Fig. 2

(c) A simply supported beam is subjected to various loadings as shown in Fig 3. Sketch the shear force and bending moment diagrams showing their values at significant locations.

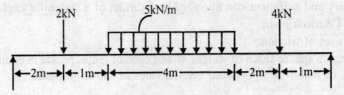

Fig. 3

5. Attempt any four of the following questions: (4 × 5 = 20)
   (a) Explain the following:
       (i) Poisson 's' ratio and its significance.
       (ii) Complementary shear stress.
   (b) A steel bar is subjected to loads as shown in Fig 4. If young's modulus for the bar material is 200kN/mm² determine the change in length of bar. The bar is 200 mm in diameter.

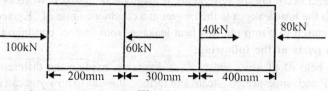

Fig. 4

   (c) In an elastic material the direct stresses of 100 MN/m² and 80MN/m² are applied at a certain point on plans at right angle to each other in tension and compression respectively. Estimate the shear stress to which material can be subjected. If the maximum principle stress is 130 MN/m². Also find the magnitude of other principle stress and its inclination to 100 MN/m² stress.
       (i) What do you mean by simple bending? What assumption are made in simple bending stress analysis?
       (ii) What do you mean by polar modules and tensional rigidity?
   (e) A wooden beam of rectangular cross-section is subjected to a bending moment of 5 KNm. If the depth of the section is to be twice the breadth and stress in wood is not to exceed 60 N/cm². Find the dimension of the cross-section of the beam.
   (f) The diameter of a shaft is 20 cm. Find the safe maximum torque which can be transmitted by the shaft if the permissible shear stress in the shaft material be 4000 N/cm² and permissible angle of twist is 0.2 degree per meter length. Take $G = 8 \times 10^6$ N/cm². If the shaft rotates at 320 rpm, what maximum power can be transmitted by the shaft?

B. Tech.
Second Semister Examination, 2005-2006
TME : 201 Mechanical Engineering

1. Attempt any four of the following: (5 × 4 = 20)
   (a) Define the terms, 'system', 'surroundings', 'boundary' and 'universe', as related to thermodynamics and distinguish between 'open' 'closed', and 'isolated' systems.
   (b) State zeroth low of thermodynamics giving its practical importance, and explain, how this low can be used to establish equality of temperature of two bodies without bringing them in direct contact.
   (c) State the first low of thermodynamics as applied to closed system and prove that for a non flow process, it leads to the energy equation $Q = \Delta u + W$. Also explain the difference between a non flow and a steady flow process, in brief.
   (d) The internal energy of a certain substance is expressed by the equation; $u = 3.62pv + 86$, where u is expressed in KJ/Kg, p is kPa and v is in m^3/Kg. A system composed of 5Kg of this substance expands from an initial pressure final processor of 125 kPa. In a process in which

512 / *Problems and Solutions in Mechanical Engineering with Concept*

pressure and volume are related by $pv^{1.2}$ = constant. If the expression process is quasistatic, determine Q, Δu and W, for this process.

(e) State the clausius and Kelvin Plank statements being used for second low of thermodynamics. Further, define: efficiency of a heat engine, COP of a refrigerator and COP of a heat pump, and show that: $(COP)_{Heat\ pump} = 1 + (COP)_{Refrigerator}$.

(f) Describe a cannot cycle with the help of (P – v) and (T – s) diagram, in brief. The temperature of the freezer of a domestic refrigerator is maintained' at –16°C whereas the ambient temperature is 35°C. If the heat leaks in to the freezer at a continuous rate of 2Kg/sec., what is the minimum power required to pump out this heat leakage from freezer continuously?

2. Attempt any two parts of the following: (10 × 2 = 20)

(a) With the help of (T – v) and (T – s) diagrams, explain the difference between 'wet' 'dry' 'saturated' and 'superheated' steam and further show, as to how you can calculate their properties with the help of steam tables and motlier ? 5Kg of steam is generated at a pressure of 10 bar from feed water at a temperature of 25°C. Starting from the basic principles and taking the help of steam table only, calculate the enthalpy and entropy of steam, if :
   (i) Steam is dry and saturated
   (ii) Steam is superheated up to a temperature of 300°C. Take Cp for steam as 2.1 KJ/Kg°K and Cp for water as" 4.187 KJ/Kg°K.

(b) Which are the' four basic components of a steam power plant? Draw a basic layout of a steam power plant and explain the working of a simple Rankine cycle with the help of (T – S) diagram.

A steam power plant working on Ranking cycle, pressure of 0.5 bar. If the initial condition of supply steam is dry and saturated, calculate the carnot and Rankine efficiencies of the cycle, neglecting pump work.

(c) How can you define IC engine and how they are classified? What is the basic difference between SI and CI engines? Further explain the working of a 4 stroke SI engine with the help of neat sketch.

3. Attempt any two parts of the following: (10 × 2 = 20)

(a) Enumerate different laws of motion, discussing the significance of each them. What do you understand by transfer of force to parallel position? Also explain varignon's theorem of moments, in brief.

(b) What do you understand by resultant of a force system and which are the methods used for determining the resulting of coplanar concurrent force system? Four forces having magnitudes of 20N, 40N, 60N and 80N respectively, are acting along the four sides (1m each), of a square ABCD taken in order, as shown in figure. Determine the magnitude and direction of the resultant force.

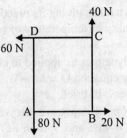

(c) What is the characteristics of frictional force? Describe the laws of coulomb friction, explaining the concept of equilibrium of bodies involvring dry friction.

A body of weight 500N is pulled up along an inclined plane having an inclination of 30° with the body and the plane is 0.3 and the force is applied parallel to the inclined plane, determine the force required.

4. Attempt any two parts of the following: (10 × 2 = 20)

(a) Define a bean and classify the different types of beams on the basis of support conditions and loading. What do you understand by 'shear force' and 'Bending-moment' and what is their importance in beam design? What do you understand by statically determinate beam?

(b) Explain how shearing force and Bending moment diagrams are drawn for a beam. Also, draw the shear force and bending moment diagrams for the cantilever beam shown in figure.

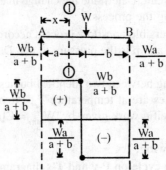

(c) Define a truss and differentiate between perfect, deficient and Redundant trusses. A truss having a span of 6m, carries a load of 30KH and is shown in figure. Find the forces in members AB, AC, BC and AD.

5. Attempt any two parts of the following: (10 × 2 = 20)

(a) (i) Define stress, stain and elasticity and differentiate between normal stress and share stress. Draw the stress - strain diagram for mild steel showing salient points on it.

(ii) A bar of 25mm diameter is subjected to a pull of 60KN. The measured extension over a gauge length of 250mm is 0.15mm and change in diagram is 0.004mm. Calculate the modulus of elasticity, modules of rigidity and Poisson's Ratio.

(b) What do you understand by principle planes and principal stresses? What is Mohr's circle? Explain the construction of Mohr's uncle and clearly indicate, how will you find out major principle stress minor principal stress and max shear stress with the help of Mohr's circle.

(c) (i) What do you understand by Pure Bending of beams and how it differs from simple binding? Plot the variation of bending stress across the section; of a solid circular beam, a T section beam and a rectangular beam, indicating the salient features on it.

(ii) What do you understand by Pure-torsion and Tensional rigidity?

A solid circular shaft to transmit 160 KW at 180rpm. What will be the suitable diameter of this shaft of permissible stress in the shaft material should not exceed 2×10^6 Pa and turists per unit length should not exceed 2°. Take G = 200GPa.

B. Tech.
First Semester Examination, 2006-2007
Mechancial Engineering

1. Attempt any two of the following; (10 × 2)
   (a) Define the following
   - Thermodynamic State, Path and Process
   - Continuum
   - Thermodynamic equilibrium
   - Kelvin Planck statement of IInd law of thermodynamics

   (b) A fluid undergoes a reversible adiabatic compression from 0.5 MPa, 0.02 m³ to 0.05 m³ according to the law $pv^{1.3}$ = Constant. Determine the change in enthalpy, internal energy, entropy and heat of work transfer during the process.

   (c) (i) A gas undergoes a reversible non-flow process according to the relation p = (–3V + 15) where V is the volume in m³ and p is the pressure in bar. Determine the work done when the volume changes trom 3 to 6m³.

   (ii) A refrigerator operating between two identical bodies cools one of the bodies to a temperature T_2. Initially both bodies are at temperature T_c. Prove that for this operation, the refrigerator needs a minimum specific work given by, $W_{min} = C[(T_c^2/T_2) + T_2 - 2T_c]$, where C = Sp. heat capacity

2. Attempt any two of the following; [10 × 2]
   (a) (i) Draw Otto and Diesel cycle on P-v and T-s diagram.
   (ii) Estimate the enthalpy, entropy and specific volume for following states of steam,
   Steam at 4 MPa and 80 % wet
   Steam at 10 MPa and 550D°C

   (b) Steam at 0.8 MPa, 250D°C and flowing at the rate of lkg/s passes into a pipe carrying wet steam at 0.8 MPa, 0.95 dry. After adiabatic mixing the flow rate is 2.3 kg/s. Determine the condition of steam after mixing.

   (c) (i) Differentiate between the 2 stroke and 4 stroke SI engines.
   (ii) Write short notes on the following, Sensible heating, Latent heating, Critical point, Triple point, Compressed liquid.

3. Attempt any two of the following, [10 × 2]
   (a) Find the reaction at A and B for the beam shbwn below

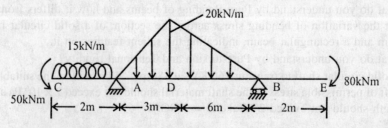

   (b) What is the least value of 'P' required to cause the motion impend in the arrangement shown below ? Assume the coefficient of friction on all contact surfaces as 0.2. Weight of blocks A and B are 840 N and 560 N respectively.

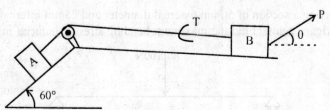

(c) Prove that for a flat belt passing over a pulley as shown below, $(T_1/T_2) = e^{\mu\beta}$

4. **Attempt any two of the following.** [10 × 2]

   (a) A truss is loaded and supported as shown in figure below. Find the value of loads P which would produce a force of magnitude 3KN in the member AC.

   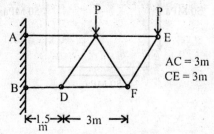

   (b) A beam 5m long, hinged at both the ends is subjected to a moment M = 60 kNm at a point 3m from end A as shown below.
   Draw the shear force and bending moment diagram.

   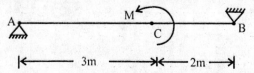

   (c) Draw bending moment diagram of the beam shown below.

   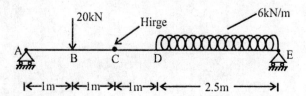

5. **Attempt any two of the following** (10 × 2)

   (a) (i) Define Hooks law and Poisson's ratio.
   (ii) Principal Plane and Principal Stresses.
   (iii) Prove that the deflection of free end of a uniform bar of X-section area A, caused under own weight when suspended vertically is WL/(2AE), where W and 2L are the weight and length of the bar.

(b) A beam having a section of 50mm external diameter and 25mm infernal diameter is loaded as shown in figure below. Find the maximum bending stresses induced in the beam.

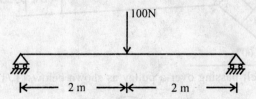

(c) (i) Draw the shear stress distribution for a hollow shaft under pure torsion.
(ii) Draw the Mohr's circle for the state of stress shown below.

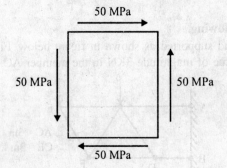